Important Formulas

1 The **distance** $d(P_1, P_2)$ between $P_1(x_1, y_1)$ an

$$d(P_1, P_2) = \sqrt{(x_2 - x_1)^2 \cdot}$$

2 The **equation of a circle** with center $C(h, k)$

$$(x - h)^2 + (y - k)^2 = r^2$$

3 The **slope** m of the line through $P_1(x_1, y_1)$ and $P_2(x_2, y_2)$:

$$m = \frac{y_2 - y_1}{x_2 - x_1}$$

4 The **point-slope form** for the equation of a line through $P_1(x_1, y_1)$ with slope m:

$$y - y_1 = m(x - x_1)$$

5 The **slope-intercept form** for the equation of a line with slope m and y-intercept b:

$$y = mx + b$$

6 **Linear function** f: $f(x) = ax + b$

7 **Quadratic function** f: $f(x) = ax^2 + bx + c$

8 **Polynomial function** f: $f(x) = a_n x^n + a_{n-1} x^{n-1} + \cdots + a_1 x + a_0$

9 **Rational function** f: $f(x) = \dfrac{h(x)}{g(x)}$, where h and g are polynomial functions

10 **Exponential function** f with base $a > 0$: $f(x) = a^x$

11 **Logarithmic function** f with base $a > 0$: $f(x) = \log_a x$, where $y = \log_a x$ if and only if $a^y = x$

12 **Summation notation:** $\displaystyle\sum_{i=1}^{m} a_i = a_1 + a_2 + a_3 + \cdots + a_m$

13 **Binomial Theorem:** $(a + b)^n = \displaystyle\sum_{r=0}^{n} \binom{n}{r} a^{n-r} b^r$, where $\dbinom{n}{r} = \dfrac{n!}{r!(n-r)!}$

**EARL W.
SWOKOWSKI**

MARQUETTE UNIVERSITY

**FUNCTIONS
AND
GRAPHS**

**THIRD
EDITION**

**PRINDLE, WEBER & SCHMIDT
BOSTON, MASSACHUSETTS**

Prindle, Weber & Schmidt is a division of Wadsworth, Inc.

Library of Congress Cataloging in Publication Data

Swokowski, Earl William
    Functions and graphs.

    Includes index.
    1. Functions.   2. Algebra--Graphic methods.
    3. Trigonometry.   I. Title.
QA331.3.S95   1980   512'.1   79-25088
ISBN 0-87150-283-6

Third printing: October 1980

Cover design by Novum, Inc. Text design and production by Nancy Blodget. Text composition in Monophoto Times Roman and Univers by Technical Filmsetters Europe Limited. Cover printed by New England Book Components, Inc. Text printed and bound by Halliday Lithograph.

PREFACE

The aim of *Functions and Graphs* is to meet the wide-ranging mathematical needs of students who enroll in courses at the precalculus level. Numerous examples and figures, together with a large number of graded exercises, should help students of average ability gain a firm grasp of material that is required in both advanced mathematics and applied areas.

The development in this text differs from that found in more traditional precalculus books. Specifically, topics such as factoring, solving equations, simplifying algebraic expressions, and laws of exponents are integrated into the text whenever the need arises, instead of appearing in separate drill sections in the first chapter. This procedure should make it easier for instructors to generate and maintain student interest early in the course.

Chapter 1 provides a review of real numbers, inequalities, absolute value, and graphs. Basic properties of functions, linear functions, and inverse functions are discussed next, in Chapter 2. Chapter 3 contains a rather detailed examination of polynomial and rational functions. Several theorems concerning the location of zeros of polynomials and a section on partial fractions were added to this edition.

Chapter 4 begins with a thorough review of exponents and radicals. Succeeding sections contain many practical illustrations and exercises involving exponential and logarithmic functions. Optional exercises, to be solved using hand-held calculators, appear in this and the next two chapters.

In Chapter 5, an accessible introduction to trigonometric functions using a unit circle is managed by emphasizing examples and keeping the discussion relatively brief. In addition, the angular approach to trigonometry appears much earlier than in the previous edition. Later in the chapter examples are again emphasized to help students learn to sketch graphs rapidly.

Most of Chapter 6 consists of work with trigonometric identities and equations. The last section of this chapter contains an introduction to geometric and algebraic properties of vectors in two dimensions. Complex numbers are defined in Chapter 7, and the theory of zeros of polynomials with real or complex coefficients is then completed.

Matrices are introduced in Chapter 8 as a means for solving systems of linear equations. The chapter ends with a study of determinants and a brief introduction to linear programming. Chapter 9 has sections on mathematical induction, the Binomial Theorem, summation notation, and infinite sequences. The standard topics in analytic geometry are included in Chapter 10.

In answer to the requests of some instructors, an appendix was added containing guidelines for solving applied (word) problems and a long list of exercises.

There is a review section at the end of each chapter consisting of a list of important topics and pertinent exercises. The review exercises are similar in scope to those which appear throughout the chapter and may be used to prepare for examinations. Answers to odd-numbered exercises are given at the end of the text. An answer booklet for the even-numbered exercises may be obtained from the publisher.

There are sufficient topics so that instructors may vary the emphasis of the course. The diagram below indicates possible sequences of chapters which may be followed.

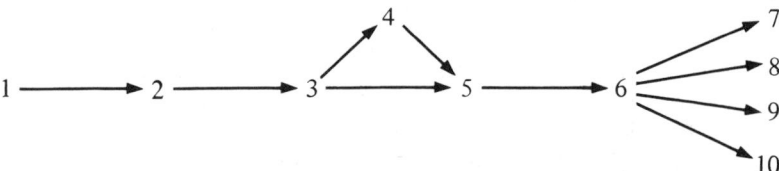

It should be noted that Chapters 7-10 are, for the most part, independent, and may be taken in any order, provided certain exercises are not assigned. For example, several exercises in Chapters 8 and 9 require the use of complex numbers or logarithms, which are discussed in Chapters 7 and 4, respectively. If desired, many sections of Chapters 3 and 4 may be omitted without disrupting the continuity of the course.

This revision has benefited from comments and suggestions of users of the second edition. I wish to thank the following individuals who reviewed all or parts of the manuscript and offered helpful suggestions: Alberto Beron (Moorpark College), Daniel D. Bonar (Denison University), Joseph Cleary (Massasoit Community College), Gregory M. Dotseth (University of Northern Iowa), Joe W. Fitzpatrick (University of Texas at El Paso), Walter M. Jensen, Jr. (Central New England College), Vladik A. Miculka (University of Texas at El Paso), John W. Riner, Jr. (Ohio State University), Junior Stein (University of Toledo), Andrew L. Walker (North Seattle Community College), and Susan S. Wood (J. Sargeant Reynolds Community College). In addition, many instructors were kind enough to respond to a survey conducted by my publisher. E. Ray Bobo (Georgetown University), Elvy L. Fredrickson (Lewis and Clark College), Robert L. Hamilton (Spokane Falls Community College), Frederick M. Liss (University of Wisconsin, Marathon County Center), Cecil W. McDermott (Hendrix College), Janice Meyer (Western Illinois University), Thomas S. Meyer (Saddleback College), Charles J. Miller (Foothill College), Janet P. Ray (Seattle Community College), Jerry Sue Townsend (Augusta College), Visutdhi Upatisringa (Humboldt State University), Peter R. Weidner (Edinboro State College), and Eric L. Wilson (Wittenberg University) were among those who sent in valuable information and ideas.

I am also grateful to the staff of Prindle, Weber & Schmidt, for their painstaking work in the production of this book. In particular, Nancy Blodget, John Martindale, and Mary Lu Walsh gave valuable assistance with the revision. Finally, I wish to thank my family, whose patience was called upon so often during the time this text was written.

Earl W. Swokowski

TABLE OF CONTENTS

5 THE TRIGONOMETRIC FUNCTIONS

6 ANALYTIC TRIGONOMETRY

7 COMPLEX NUMBERS

8 SYSTEMS OF EQUATIONS AND INEQUALITIES

9 SEQUENCES AND SERIES

10 TOPICS IN ANALYTIC GEOMETRY

1 FUNDAMENTAL CONCEPTS

In this chapter we introduce some basic concepts which are used throughout the remainder of the text. Included in our discussion are properties of real numbers, inequalities, absolute values, *and* graphs *in two dimensions.*

1.1 REAL NUMBERS

Real numbers are used considerably in all phases of mathematics, and you are undoubtedly well acquainted with symbols which are used to represent them, such as

$$1, \quad 73, \quad -5, \quad \frac{49}{12}, \quad \sqrt{2}, \quad 0, \quad \sqrt[3]{-85}, \quad 0.33333\ldots, \quad 596.25.$$

The real numbers are said to be **closed** relative to operations of addition (denoted by $+$) and multiplication (denoted by $\cdot$). This means that to every pair a, b of real numbers there corresponds a unique real number $a + b$ called the **sum** of a and b and a unique real number $a \cdot b$ (also written ab) called the **product** of a and b. These operations have the following properties, where all lower-case letters denote arbitrary real numbers, and where 0 and 1 are special real numbers referred to as **zero** and **one**, respectively.

Commutative Properties

$$a + b = b + a, \qquad ab = ba$$

Associative Properties

$$a + (b + c) = (a + b) + c, \qquad a(bc) = (ab)c$$

Identities

$$a + 0 = a = 0 + a, \qquad a \cdot 1 = a = 1 \cdot a$$

1

Inverses

For every real number a, there is a real number denoted by $-a$ such that

$$a + (-a) = 0 = (-a) + a.$$

For every real number $a \neq 0$, there is a real number denoted by $1/a$ such that

$$a\left(\frac{1}{a}\right) = 1 = \left(\frac{1}{a}\right)a.$$

Distributive Properties

$$a(b + c) = ab + ac, \qquad (a + b)c = ac + bc$$

The equals sign, $=$, means, of course, that the expressions immediately to the right and left of the sign represent the same real number. The real numbers 0 and 1 are sometimes referred to as the **additive identity** and **multiplicative identity**, respectively. We call $-a$ the **additive inverse** of a (or the **negative** of a). If $a \neq 0$, then $1/a$ is called the **multiplicative inverse** of a (or the **reciprocal** of a). The symbol a^{-1} is often used in place of $1/a$. Thus, by definition,

$$a^{-1} = \frac{1}{a}.$$

If $a = b$ and $c = d$, then, since a and b are merely different names for the same real number, and likewise for c and d, it follows that $a + c = b + d$ and $ac = bd$. This is often called the **substitution principle**, since we may think of replacing a by b and c by d in the expressions $a + c$ and ac. As a special case, using the fact that $c = c$ gives us the following.

If $a = b$, then $a + c = b + c$.
If $a = b$, then $ac = bc$.

We sometimes refer to these properties by the statements "any number c may be added to both sides of an equality" and "both sides of an equality may be multiplied by the same number c."

The following results can be proved (see Exercises 54 and 55).

$a \cdot 0 = 0$ for every real number a.
If $ab = 0$, then either $a = 0$ or $b = 0$.

The last two statements imply that $ab = 0$ *if and only if* either $a = 0$ or $b = 0$. The phrase "if and only if," which is used throughout mathematics, always has a two-fold character. Here it means that if $ab = 0$, then $a = 0$ or $b = 0$ and, *conversely*, if $a = 0$ or $b = 0$, then $ab = 0$. Consequently, if both $a \neq 0$ and $b \neq 0$, then $ab \neq 0$; that is, *the product of two nonzero real numbers is always nonzero*.

The following rules for negatives can also be established.

$$-(-a) = a$$
$$(-a)b = -(ab) = a(-b)$$
$$(-a)(-b) = ab$$
$$(-1)a = -a$$

The operation of **subtraction** (denoted by $-$) is defined by

$$a - b = a + (-b).$$

If $b \neq 0$, then **division** (denoted by $\div$) is defined by

$$a \div b = a\left(\frac{1}{b}\right) = ab^{-1}.$$

The symbol a/b is often used in place of $a \div b$, and we refer to it as the **quotient of a by b** or the **fraction a over b**. The numbers a and b are called the **numerator** and **denominator**, respectively, of the fraction. It is important to note that since 0 has no multiplicative inverse, a/b is not defined if $b = 0$; that is, *division by zero is not permissible*. Also note that

$$1 \div b = \frac{1}{b} = b^{-1}.$$

The following properties of quotients may be established, where all denominators are nonzero real numbers.

$$\frac{a}{b} = \frac{c}{d} \quad \text{if and only if } ad = bc$$
$$\frac{a}{b} = \frac{ad}{bd}, \qquad \frac{a}{-b} = \frac{-a}{b} = -\frac{a}{b}$$
$$\frac{a}{b} + \frac{c}{b} = \frac{a+c}{b}, \qquad \frac{a}{b} + \frac{c}{d} = \frac{ad+bc}{bd}$$
$$\frac{a}{b} \cdot \frac{c}{d} = \frac{ac}{bd}, \qquad \frac{a}{b} \div \frac{c}{d} = \frac{a}{b} \cdot \frac{d}{c} = \frac{ad}{bc}$$

The **positive integers** 1, 2, 3, 4,... may be obtained by adding the real number 1 successively to itself. The negatives, $-1, -2, -3, -4,...$, of the positive integers are referred to as **negative integers**. The **integers** consist of the totality of positive and negative integers together with the real number 0.

If $a, b,$ and c are integers and $c = ab$, then a and b are called **factors**, or **divisors**, of c. For example the integer 6 may be written as

$$6 = 2 \cdot 3 = (-2)(-3) = 1 \cdot 6 = (-1)(-6).$$

Hence, 1, -1, 2, -2, 3, -3, 6, and -6 are factors of 6.

A positive integer p different from 1 is **prime** if its only positive factors are 1 and p. The first few primes are 2, 3, 5, 7, 11, 13, 17, and 19. One of the reasons for the importance of prime numbers is that every positive integer a different from 1 can be expressed in one and only one way (except for order of factors) as a product of primes. The proof of this result, which is called **The Fundamental Theorem of Arithmetic**, will not be given in this book. As examples, we have

$$12 = 2 \cdot 2 \cdot 3, \quad 126 = 2 \cdot 3 \cdot 3 \cdot 7, \quad 540 = 2 \cdot 2 \cdot 3 \cdot 3 \cdot 3 \cdot 5.$$

A real number is called a **rational number** if it can be written in the form a/b, where a and b are integers and $b \neq 0$. Real numbers that are not rational are called **irrational**. The ratio of the circumference of a circle to its diameter is an irrational real number and is denoted by π. It is often approximated by the decimal 3.1416 or by the rational number 22/7. We use the notation $\pi \approx 3.1416$ to indicate that π is *approximately equal* to 3.1416. To cite another example, a real number b such that $b^2 = 2$, where b^2 denotes $b \cdot b$, is not rational. There are two such irrational numbers denoted by $\sqrt{2}$ and $-\sqrt{2}$. In general, if a is *any* nonnegative real number, then by definition, the **principal square root of a**, denoted by $\sqrt{a}$, is the *nonnegative* real number b such that $b^2 = a$. For simplicity $\sqrt{a}$ is referred to as the *square root of a*. Some square roots are rational. For example,

$$\sqrt{25} = 5, \qquad \sqrt{\frac{9}{4}} = \frac{3}{2}, \qquad \sqrt{16} = 4.$$

Others, such as $\sqrt{3}, \sqrt{5}, \sqrt{7/2}$, are irrational. Note that *square roots of positive real numbers are positive*; that is,

$$\sqrt{25} \neq \pm 5, \qquad \sqrt{\frac{9}{4}} \neq \pm \frac{3}{2}, \qquad \sqrt{16} \neq \pm 4.$$

This convention is not always followed in elementary arithmetic where, for example, $\sqrt{25}$ denotes either 5 or -5. More will be said about roots in Chapter Four.

Real numbers may be represented by decimal expressions. Decimal representations for rational numbers either terminate or are nonterminating and repeating. For example, it can be shown by long division that a decimal representation for 7434/2310 is 3.2181818..., where the digits 1 and 8 repeat indefinitely. The rational number 5/4 has the terminating decimal representation 1.25. Decimal representations for irrational numbers may also be obtained; however, they are always nonterminating and nonrepeating. The process of finding decimal representations for irrational numbers is usually difficult. Often some method of successive approximation is employed. For example, the device learned in arithmetic for extracting square roots can be used to find a decimal representation for $\sqrt{2}$. Using this technique we successively obtain the approximations 1, 1.4, 1.41, 1.414, 1.4142,... .

It is possible to associate the set of real numbers with the set of all points on a line l in such a way that for each real number a there corresponds one and only one point, and conversely, to each point P on l there corresponds precisely one real number. Such an association between two sets is referred to as a **one-to-one correspondence**. We first choose an arbitrary point O, called the **origin**, and associate with it the real number 0. Points associated with the integers are then determined by laying off successive line segments of equal length on either side of O as illustrated in Figure 1.1. The points corresponding to rational numbers such as 23/5 and $-\frac{1}{2}$ are obtained by subdividing the equal line segments. Points associated with certain irrational numbers, such as $\sqrt{2}$, can be found by geometric construction. For other irrational numbers such as π, no construction is possible. However, the point corresponding to π can be approximated to any degree of accuracy by locating successively the points corresponding to 3, 3.1, 3.14, 3.141, 3.1415, 3.14159,... .

It can be shown that to every irrational number there corresponds a unique point on l and conversely, every point that is not associated with a rational number corresponds to an irrational number.

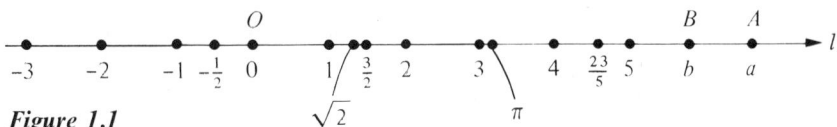

Figure 1.1

The number a that is associated with a point A on l is called the **coordinate** of A. An assignment of coordinates to points on l is called a **coordinate system** for l, and l is called a **coordinate line**, or a **real line**. A direction can be assigned to l by taking the **positive direction** along l to the right and the **negative direction** to the left. The positive direction is noted by placing an arrowhead on l as shown in Figure 1.1.

The numbers which correspond to points to the right of 0 in Figure 1.1 are called **positive real numbers**, whereas those which correspond to points to the left of 0 are **negative real numbers**. The real number 0 is neither positive nor negative. The set of positive real numbers is **closed** relative to addition and multiplication; that is, if a and b are positive, then so are the sum $a + b$ and the product ab.

Note that if a is positive, then $-a$ is negative. Similarly, if $-a$ is positive, then $-(-a) = a$ is negative. A common error is to think that $-a$ is always a negative number; however, this is not necessarily the case. For example, if $a = -3$, then $-a = -(-3) = 3$, which is positive. Indeed, whenever a is negative the number $-a$ is positive.

If a and b are real numbers, and $a - b$ is positive, we say that **a is greater than b** and write $a > b$. An equivalent statement is **b is less than a**, written $b < a$. The symbols $>$ or $<$ are called **inequality signs** and expressions such as $a > b$ or $b < a$ are called **inequalities**. From the manner in which we constructed the coordinate line l in Figure 1.1, we see that if A and B are points with coordinates a and b, respectively, then $a > b$ (or $b < a$) *if and only if A lies to the right of B.* The following definition is stated for reference, where a and b denote real numbers.

> $a > b$ means $a - b$ is positive.
>
> $b < a$ means $a - b$ is positive.

Observe that $a > b$ and $b < a$ have exactly the same meaning. As illustrations we may write

$$5 > 3 \text{ since } 5 - 3 = 2 \text{ is positive;}$$
$$-6 < -2 \text{ since } -2 - (-6) = -2 + 6 = 4 \text{ is positive;}$$
$$-\sqrt{2} < 1 \text{ since } 1 - (-\sqrt{2}) = 1 + \sqrt{2} \text{ is positive;}$$
$$2 > 0 \text{ since } 2 - 0 = 2 \text{ is positive;}$$
$$-5 < 0 \text{ since } 0 - (-5) = 5 \text{ is positive}$$

The last two illustrations are special cases of the following general properties.

$a > 0$ if and only if a is positive.
$a < 0$ if and only if a is negative.

We sometimes refer to the **sign** of a real number as being positive or negative according as the number is positive or negative.

There are several other useful symbols which involve inequality signs. In particular, $a \geq b$, which is read **a is greater than or equal to b**, means that either $a > b$ or $a = b$ (but not both). As an illustration we have $a^2 \geq 0$ for every real number a. The symbol $a \leq b$ is read **a is less than or equal to b** and means that either $a < b$ or $a = b$. The expression $a < b < c$ means that both $a < b$ and $b < c$, in which case we say that **b is between a and c**. We may also write $c > b > a$. For instance,

$$1 < 5 < \frac{11}{2}, \quad -4 < \frac{2}{3} < \sqrt{2}, \quad 3 > -6 > -10.$$

Other variations of the inequality notation are used. For example, $a < b \leq c$ means both $a < b$ and $b \leq c$. Similarly, $a \leq b < c$ means both $a \leq b$ and $b < c$. Finally, $a \leq b \leq c$ means both $a \leq b$ and $b \leq c$.

If a is a real number, then it is the coordinate of some point A on a coordinate line l, and the symbol $|a|$ is used to denote the number of units (or distance) between A and the origin, without regard to direction. Referring to Figure 1.2 we see that for the point with coordinate -4 we have $|-4| = 4$. Similarly, $|4| = 4$. In general, if a is negative we change its sign to find $|a|$, whereas if a is nonnegative, then $|a| = a$; that is,

$$|a| = \begin{cases} a & \text{if } a \geq 0. \\ -a & \text{if } a < 0. \end{cases}$$

The nonnegative number $|a|$ is called the **absolute value** of a.

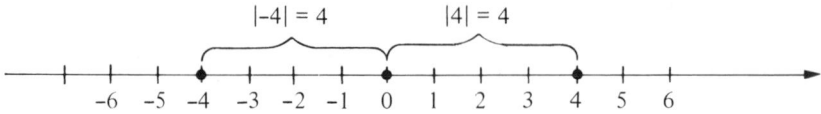

Figure 1.2

Example 1   Find $|3|$, $|-3|$, $|0|$, $|\sqrt{2} - 2|$, and $|2 - \sqrt{2}|$.

Solution   Since 3, $2 - \sqrt{2}$, and 0 are nonnegative, we have

$$|3| = 3, \quad |2 - \sqrt{2}| = 2 - \sqrt{2}, \quad \text{and} \quad |0| = 0.$$

Since -3 and $\sqrt{2} - 2$ are negative, we use the formula $|a| = -a$ to obtain

$$|-3| = -(-3) = 3 \quad \text{and} \quad |\sqrt{2} - 2| = -(\sqrt{2} - 2) = 2 - \sqrt{2}. \qquad \blacksquare$$

Note that in Example 1, $|-3| = |3|$ and $|2 - \sqrt{2}| = |\sqrt{2} - 2|$. It can be shown in general that

$$\boxed{|a| = |-a| \text{ for every real number } a.}$$

We shall use the concept of absolute value to define the distance between any two points on a coordinate line. Let us begin by noting that the distance between the points with coordinates 2 and 7 shown in Figure 1.3 equals 5 units on l. This distance is the difference, $7 - 2$, obtained by subtracting the smaller coordinate from the larger. If we employ absolute values, then since $|7 - 2| = |2 - 7|$, it is unnecessary to be concerned about the order of subtraction. We shall use this as our motivation for the following definition.

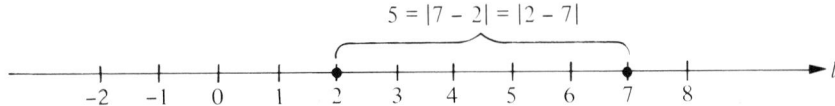

Figure 1.3

Definition

Let a and b be the coordinates of two points A and B, respectively, on a coordinate line l. The **distance between A and B**, denoted by $d(A, B)$, is defined by

$$d(A, B) = |b - a|.$$

The number $d(A, B)$ is also called the **length of the line segment AB**.
Observe that, since $d(B, A) = |a - b|$ and $|b - a| = |a - b|$, we may write

$$d(A, B) = d(B, A).$$

Also note that the distance between the origin O and the point A is

$$d(O, A) = |a - 0| = |a|$$

which agrees with the geometric interpretation of absolute value illustrated in Figure 1.2. The formula $d(A, B) = |b - a|$ is true regardless of the signs of a and b, as illustrated in the next example.

Example 2   Let A, B, C, and D have coordinates -5, -3, 1, and 6, respectively, on a coordinate line l (see Figure 1.4). Find $d(A, B)$, $d(C, B)$, $d(O, A)$, and $d(C, D)$.

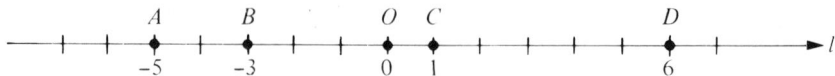

Figure 1.4

Solution   Using the definition of the distance between points,

$$d(A, B) = |-3 - (-5)| = |-3 + 5| = |2| = 2$$
$$d(C, B) = |-3 - 1| = |-4| = 4$$
$$d(O, A) = |-5 - 0| = |-5| = 5$$
$$d(C, D) = |6 - 1| = |5| = 5.$$

These answers can be checked geometrically by referring to Figure 1.4.   ∎

The concept of absolute value has uses other than that of finding distances between points. Generally, it is employed whenever we are interested in the "magnitude" or "numerical value" of a real number without regard to its sign.

EXERCISES 1.1

In each of Exercises 1–10, justify the equality by stating only one of the following properties: (a) commutative, (b) associative, (c) identity, (d) inverse, (e) distributive.

1   $(4 \cdot 5) \cdot 4 = 4 \cdot (4 \cdot 5)$

2   $3 \cdot (4 + 5) = (4 + 5) \cdot 3$

3   $(4 \cdot 5) \cdot 4 = 4 \cdot (5 \cdot 4)$

4   $(4 + 5) \cdot 3 = 4 \cdot 3 + 5 \cdot 3$

5   $3 \cdot (5 + 0) = 3 \cdot 5$

6   $3 + (-3) = 0$

7   $1 \cdot (2 + 3) = 2 + 3$

8   $(1 + 2) + 1 = 1 + (1 + 2)$

9   $(1/4)4 = 1$

10   $0 \cdot 1 = 0$

In each of Exercises 11–20, write the expression as a rational number with least positive numerator.

11   $\dfrac{3}{4} + \dfrac{5}{3}$

12   $\dfrac{1}{5} + \dfrac{3}{2}$

13   $\dfrac{5}{6} \cdot \dfrac{2}{3}$

14   $\dfrac{4}{9}\left(\dfrac{1}{2} + \dfrac{3}{4}\right)$

15   $\dfrac{1}{2} + \dfrac{1}{4} + \dfrac{1}{6}$

16   $\dfrac{3}{-2} \cdot \dfrac{-5}{6}$

17   $\dfrac{13}{4} \div \left(\dfrac{3}{2} + 1\right)$

18   $\dfrac{2}{5} + \dfrac{7}{4} + \dfrac{3}{2}$

19   $\dfrac{5}{7}\left(\dfrac{3}{2} - \dfrac{5}{6}\right)$

20   $\dfrac{10}{11} \cdot \dfrac{11}{10}$

21   Show, by means of examples, that the operation of subtraction on $\mathbb{R}$ is neither commutative nor associative.

22 Show, by means of examples, that the operation of division, as applied to nonzero real numbers, is neither commutative nor associative.

In Exercises 23–26, replace the $\square$ with either $<$, $>$, or $=$.

23 (a) $-7 \square -4$ (b) $3 \square -1$
(c) $1 + 3 \square 6 - 2$

24 (a) $-3 \square -5$ (b) $-6 \square 2$
(c) $1/4 \square 0.25$

25 (a) $1/3 \square 0.33$ (b) $125/57 \square 2.193$
(c) $22/7 \square \pi$

26 (a) $1/7 \square 0.143$ (b) $(3/4) + (2/3) \square 19/12$
(c) $\sqrt{2} \square 1.4$

Express the statements in Exercises 27–38 in terms of inequalities.

27 -8 is less than -5.

28 2 is greater than 1.9.

29 0 is greater than -1.

30 $\sqrt{2}$ is less than π.

31 x is negative.

32 y is positive.

33 a is between 5 and 3.

34 b is between $1/10$ and $1/3$.

35 b is greater than or equal to 2.

36 x is less than or equal to -5.

37 c is not greater than 1.

38 d is nonnegative.

Rewrite the numbers in Exercises 39–42 without using symbols for absolute value.

39 (a) $|4 - 9|$ (b) $|-4| - |-9|$
(c) $|4| + |-9|$

40 (a) $|3 - 6|$ (b) $|0.2 - (1/5)|$
(c) $|-3| - |-4|$

41 (a) $3 - |-3|$ (b) $|\pi - 4|$
(c) $(-3)/|-3|$

42 (a) $|8 - 5|$ (b) $-5 + |-7|$
(c) $(-2)|-2|$

In each of Exercises 43–46, the given numbers are coordinates of three points A, B, and C (in that order) on a coordinate line l. For each, find: (a) $d(A,B)$ (b) $d(B,C)$ (c) $d(C,B)$ (d) $d(A,C)$.

43 -6, -2, 4   **44** 3, 7, -5

45 8, -4, -1 **46** -9, 1, 10

Rewrite the expressions in Exercises 47–50 without using symbols for absolute value.

47 $|5 - x|$ if $x > 5$ **48** $|x^2 + 1|$

49 $|-4 - x^2|$ **50** $|a - b|$ if $a < b$

In Exercises 51–60, all letters denote real numbers and no denominators are zero.

51 Prove or disprove: $\dfrac{1}{a} + \dfrac{1}{b} = \dfrac{1}{a + b}$

52 Prove or disprove: $\dfrac{a}{b + c} = \dfrac{a}{b} + \dfrac{a}{c}$

Prove the rules in Exercises 53–56.

53 $a = -a$ if and only if $a = 0$.

54 $a \cdot 0 = 0$ (*Hint:* Write $a \cdot 0 = a \cdot (0 + 0)$ $= a \cdot 0 + a \cdot 0$ and then add $-(a \cdot 0)$ to both sides.)

55 If $ab = 0$, then either $a = 0$ or $b = 0$.

56 If $a + b = 0$, then $b = -a$.

57 $a(b - c) = ab - ac$

58 $\left(\dfrac{a}{b}\right)^{-1} = \dfrac{b}{a}$ **59** $\dfrac{-a}{-b} = \dfrac{a}{b}$

60 $\dfrac{a}{b} + \dfrac{c}{b} = \dfrac{a + c}{b}$ (*Hint:* Write

$\dfrac{a}{b} + \dfrac{c}{b} = ab^{-1} + cb^{-1}$ and use the Distributive Properties.)

1.2 INEQUALITIES

In order to shorten explanations it is often convenient to use the notation and terminology of sets. A **set** may be thought of as a collection of objects of some type. The objects are called **elements** of the set. Capital letters $A, B, C, R, S, \ldots$ will often be used to denote sets. Lowercase letters $a, b, x, y, \ldots$ will represent elements of sets. Throughout our work $\mathbb{R}$ will denote the set of real numbers, and $\mathbb{Z}$ the set of integers. If S is a set, then $a \in S$ means that a is an element of S, whereas $a \notin S$ signifies that a is not an element of S. If every element of a set S is also an element of a set T, then S is called a **subset** of T. For example, $\mathbb{Z}$ is a subset of $\mathbb{R}$. Two sets S and T are said to be **equal**, written $S = T$, if S and T contain precisely the same elements. The notation $S \neq T$ means that S and T are not equal.

If the elements of a set S have a certain property, then we write $S = \{x : \_\_\_\}$ where the property describing the arbitrary element x is stated in the space after the colon. For example, $\{x : x > 3\}$ may be used to represent the set of all real numbers greater than 3.

Finite sets are sometimes denoted by listing all the elements within braces. For example, if S consists of the first five letters of the alphabet, we write $S = \{a, b, c, d, e\}$. When sets are given in this way, the order used in listing the elements is considered irrelevant, and we could also write $S = \{a, c, b, e, d\}$, $S = \{d, c, b, e, a\}$, etc.

We frequently make use of symbols to denote arbitrary elements of a set. Thus we may use x to denote a real number, although no *particular* real number is specified. A letter which is used to represent any element of a given set is sometimes called a **variable**. Throughout this text, unless otherwise specified, variables will represent real numbers. The **domain of a variable** is the set of real numbers represented by the variable. The domain of a variable x is often referred to as the set of "permissible" or "allowable" values for x. To illustrate, given the expression $\sqrt{x}$, we note that in order to obtain a real number we must have $x \geq 0$, and hence in this case the domain of x is assumed to be the set of nonnegative real numbers. Similarly, when working with the expression $1/(x - 2)$ we must exclude $x = 2$ (Why?), and consequently, we take the domain of x as the set of all real numbers different from 2. If x is a variable, then expressions such as

$$x + 3 = 0, \quad x^2 - 5 = 4x, \quad \text{or} \quad (x^2 - 9)\sqrt[3]{x + 1} = 0$$

are called **equations** in x. If certain numbers are substituted for x in these equations, true statements are obtained, whereas other numbers produce false statements. For example, the equation $x + 3 = 0$ leads to a false statement for every value of x except -3. If 2 is substituted for x in the equation $x^2 - 5 = 4x$ we obtain $4 - 5 = 8$, or $-1 = 8$, a false statement. However, if we let $x = 5$, then we obtain $(5)^2 - 5 = 4 \cdot 5$, or $20 = 20$, which is true. In general, given any equation in x, if a true statement is obtained when x is replaced by some real number a from the domain of x, then a is called a **solution** or a **root** of the equation. We also say that a **satisfies** the equation. To **solve** an equation means to find all the solutions.

Sometimes every number in the domain of the variable is a solution of an equation. In this case the equation is called an **identity**. For example,

$$x^2 - 4 = (x + 2)(x - 2)$$

is an identity since it is true for every number in the domain of x. If there are numbers in the domain of x which are not solutions, then the equation is called a **conditional equation**. In the course of our work we shall discuss techniques for solving various types of equations.

It is often necessary to consider inequalities which involve variables. As a first illustration, let us consider

$$x^2 - 3 < 2x + 4.$$

If certain numbers such as 4 or 5 are substituted for x, we obtain the false statements $13 < 12$ or $22 < 14$, respectively. Other numbers, such as 1 or 2, produce the true statements $-2 < 6$ or $1 < 8$, respectively. In general, if we are given an inequality involving one variable x, and if a true statement is obtained when x is replaced by a real number a, then a is called a **solution** of the inequality. Thus 1 and 2 are solutions of the inequality $x^2 - 3 < 2x + 4$, whereas 4 and 5 are not solutions. To **solve** an inequality means to find all solutions. We say that two inequalities are **equivalent** if they have exactly the same solutions.

A standard method for solving an inequality is to replace it with a chain of equivalent inequalities, terminating in one for which the solutions are obvious. Among the main tools used in applying this method are the following properties, where a, b and c denote real numbers.

(i)   If $a > b$ and $b > c$, then $a > c$.

(ii) If $a > b$, then $a + c > b + c$.

(iii) If $a > b$ and $c > 0$, then $ac > bc$.

(iv) If $a > b$, and $c < 0$, then $ac < bc$.

We can supply proofs for all four properties by making use of the fact that both the sum and product of any two positive real numbers are positive. Thus to prove (i) we first note that if $a > b$ and $b > c$, then $a - b$ and $b - c$ are both positive. Consequently, the sum $(a - b) + (b - c)$ is positive. Since the sum reduces to $a - c$, we see that $a - c$ is positive, which means that $a > c$.

To establish (ii) we again note that if $a > b$, then $a - b$ is positive. Since $(a + c) - (b + c) = a - b$, it follows that $(a + c) - (b + c)$ is positive; that is, $a + c > b + c$.

To prove (iii) observe that if $a > b$ and $c > 0$, then $a - b$ and c are both positive and hence so is the product $(a - b)c$. Consequently, $ac - bc$ is positive; that is, $ac > bc$.

Finally, to prove (iv) we first note that if $c < 0$, then $0 - c$, or $-c$, is positive. In addition, if $a > b$, then $a - b$ is positive and hence the product $(a - b)(-c)$ is positive. However, $(a - b)(-c) = -ac + bc$ and hence $bc - ac$ is positive. This means that $bc > ac$, or $ac < bc$.

As an illustration of property (ii), note that if we add 2 to both sides of $7 > 3$ we obtain

$$7 + 2 > 3 + 2 \quad \text{or} \quad 9 > 5.$$

Similarly, adding -2 to both sides or, equivalently, *subtracting* 2 from both sides, gives us

$$7 - 2 > 3 - 2 \quad \text{or} \quad 5 > 1.$$

As indicated by property (iii), we can also multiply both sides of an inequality by the same positive real number. For example, multiplying both sides of $7 > 3$ by 2 gives us

$$7 \cdot 2 > 3 \cdot 2 \quad \text{or} \quad 14 > 6.$$

If we multiply by a negative real number, then, as in (iv), the inequality sign is reversed. To illustrate, multiplying both sides of $7 > 3$ by -2 gives us

$$7(-2) < 3(-2) \quad \text{or} \quad -14 < -6.$$

Similar results hold for the symbol $<$. Specifically, the following can be proved.

(i)   If $a < b$ and $b < c$, then $a < c$.
(ii)   If $a < b$, then $a + c < b + c$.
(iii) If $a < b$ and $c > 0$, then $ac < bc$.
(iv) If $a < b$ and $c < 0$, then $ac > bc$.

If x represents a real number, then adding the same expression in x to both sides of an inequality leads to an equivalent inequality. We may multiply both sides of an inequality by an expression containing x if we are certain that the expression is positive for all values of x under consideration. To illustrate, multiplication by $x^4 + 3x^2 + 5$ would be permissible since this expression is always positive. If we multiply both sides of an inequality by an expression that is always negative, such as $-7 - x^2$, then the inequality sign is reversed.

The reader should supply reasons for the solutions of the following inequalities.

Example 1   Solve the inequality $-3x + 4 < 11$.

Solution   The following inequalities are all equivalent.

$$-3x + 4 < 11$$
$$(-3x + 4) + (-4) < 11 + (-4)$$
$$-3x < 7$$
$$(-\tfrac{1}{3})(-3x) > (-\tfrac{1}{3})(7)$$
$$x > -\tfrac{7}{3}$$

Since the last inequality is equivalent to the first, the solutions of $-3x + 4 < 11$ consist of all real numbers x such that $x > -7/3$. ∎

Example 2 Solve the inequality $4x - 3 < 2x + 5$.

Solution The following is a list of equivalent inequalities:

$$4x - 3 < 2x + 5$$
$$(4x - 3) + 3 < (2x + 5) + 3$$
$$4x < 2x + 8$$
$$4x + (-2x) < (2x + 8) + (-2x)$$
$$2x < 8$$
$$\tfrac{1}{2}(2x) < \tfrac{1}{2}(8)$$
$$x < 4.$$

Hence the solutions of the given inequality consist of all real numbers x such that $x < 4$. ■

It is convenient to give graphical interpretations for solutions of inequalities. By the **graph** of a set of real numbers we mean the collection of all points on a coordinate line l which correspond to the numbers. To **sketch a graph** we darken (or color) an appropriate portion of l. The graph corresponding to the solutions of Example 2 consists of all points to the left of the point with coordinate 4 and is sketched in Figure 1.5. The parenthesis in the figure indicates that the point corresponding to 4 is not part of the graph. If we wish to include such an **endpoint** of a graph we use a bracket instead of a parenthesis. This will be illustrated in Example 4.

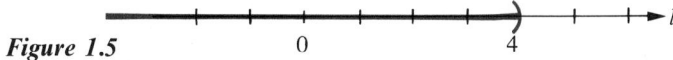

Figure 1.5

If $a < b$, the symbol (a, b) is sometimes used for all real numbers between a and b. This may be expressed in set notation as follows:

$$(a, b) = \{x : a < x < b\}$$

The set (a, b) is called an **open interval**, and a and b are called **endpoints** of the interval. The graph of (a, b) consists of all points on a coordinate line which lie between the points corresponding to a and b. In Figure 1.6 we have sketched the graph of a general open interval (a, b) and also the special open intervals $(-1, 3)$ and $(2, 4)$. The parentheses in the figure indicate that the endpoints of the intervals are not to be included. For convenience we shall use the terms *open interval* and *graph of an open interval* interchangeably.

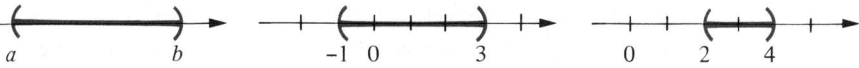

Figure 1.6. *Open intervals* (a, b), $(-1, 3)$ *and* $(2, 4)$

The solutions of an inequality may often be described in terms of intervals, as illustrated in the next example.

Example 3 Solve the inequality $-6 < 2x - 4 < 2$ and represent the solutions graphically.

Solution   A real number x is a solution of the given inequality if and only if it is a solution of *both* of the inequalities

$$-6 < 2x - 4 \quad \text{and} \quad 2x - 4 < 2.$$

The first inequality is equivalent to each of the following:

$$-6 + 4 < (2x - 4) + 4$$
$$-2 < 2x$$
$$\tfrac{1}{2}(-2) < \tfrac{1}{2}(2x)$$
$$-1 < x.$$

The second inequality is equivalent to each of the following:

$$2x - 4 < 2$$
$$2x < 6$$
$$x < 3.$$

Thus x is a solution of the given inequality if and only if *both*

$$-1 < x \quad \text{and} \quad x < 3,$$

that is,

$$-1 < x < 3.$$

Hence the solutions are all numbers in the open interval $(-1, 3)$. The graph is the middle graph sketched in Figure 1.6.

An alternate (and shorter) solution may be given by working with both inequalities simultaneously as follows:

$$-6 < 2x - 4 < 2$$
$$-6 + 4 < 2x < 2 + 4$$
$$-2 < 2x < 6$$
$$-1 < x < 3.$$

The last inequality agrees with the first solution.   ■

For convenience we refer to the interval $(-1, 3)$ as the **interval solution** of the inequality in Example 3, meaning that the solutions of the inequality consist of all numbers in that interval. We shall use this terminology in other problems, whenever applicable.

If we wish to include an endpoint of an interval, a bracket is used instead of a parenthesis. **Closed intervals**, denoted by $[a, b]$, and **half-open intervals**, denoted by $[a, b)$ or $(a, b]$, are defined as follows:

$$[a, b] = \{x : a \le x \le b\}$$
$$[a, b) = \{x : a \le x < b\}$$
$$(a, b] = \{x : a < x \le b\}$$

Typical graphs are sketched in Figure 1.7. A bracket in a figure indicates that the corresponding endpoint is part of the graph.

Figure 1.7

Example 4 Solve the inequality $-5 \le \dfrac{4 - 3x}{2} < 1$ and illustrate the solutions graphically.

Solution A number x is a solution if and only if it satisfies both of the inequalities

$$-5 \le \frac{4 - 3x}{2} \quad \text{and} \quad \frac{4 - 3x}{2} < 1.$$

We could work with each of these separately or proceed as in the alternate solution of Example 3. Let us employ the latter technique as follows:

$$-5 \le \frac{4 - 3x}{2} < 1$$
$$-10 \le 4 - 3x < 2$$
$$-10 - 4 \le -3x < 2 - 4$$
$$-14 \le -3x < -2$$
$$\left(-\frac{1}{3}\right)(-14) \ge \left(-\frac{1}{3}\right)(-3x) > \left(-\frac{1}{3}\right)(-2)$$
$$\frac{14}{3} \ge x > \frac{2}{3}$$
$$\frac{2}{3} < x \le \frac{14}{3}.$$

Hence the interval solution of the inequality is $(2/3, 14/3]$. The graph is sketched in Figure 1.8.

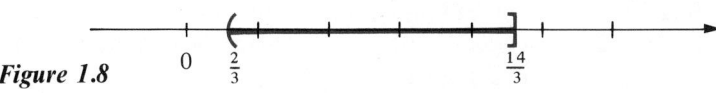

Figure 1.8   ■

The next example is an elementary illustration of an applied problem involving inequalities.

Example 5 The temperature readings on the Fahrenheit and Celsius scales are related by the equation $C = \frac{5}{9}(F - 32)$. What range of F corresponds to $30 \leq C \leq 40$?

Solution   The following inequalities are equivalent:

$$30 \leq C \leq 40$$
$$30 \leq \tfrac{5}{9}(F - 32) \leq 40$$
$$\tfrac{9}{5}(30) \leq F - 32 \leq \tfrac{9}{5}(40)$$
$$54 \leq F - 32 \leq 72$$
$$86 \leq F \leq 104.$$

Thus a Celsius temperature range from $30°\,C$ to $40°\,C$ is the same as a Fahrenheit range from $86°\,F$ to $104°\,F$. ∎

In order to describe solutions of inequalities such as $x < 4$, $x > -2$, $x \leq 7$, or $x \geq 3$, it is convenient to employ **infinite intervals**. In particular, if a is a real number we define

$$(-\infty, a) = \{x : x < a\}$$

that is, $(-\infty, a)$ denotes the set of all real numbers less than a. The symbol ∞ (**infinity**) is merely a notational device and is not to be interpreted as representing a real number. To illustrate, the interval solution in Example 2 is the infinite interval $(-\infty, 4)$. The graph of $(-\infty, 4)$ is sketched in Figure 1.5. If we wish to include the point corresponding to a we write

$$(-\infty, a] = \{x : x \leq a\}.$$

Other types of infinite intervals are defined by

$$(a, \infty) = \{x : x > a\}.$$
$$[a, \infty) = \{x : x \geq a\}.$$

Example 6 Solve $\dfrac{1}{x - 2} > 0$ and illustrate the solutions graphically.

Solution   Since the numerator is positive, the given fraction is positive if and only if $x - 2 > 0$ or equivalently, $x > 2$. Hence the interval solution is $(2, \infty)$. The graph is sketched in Figure 1.9.

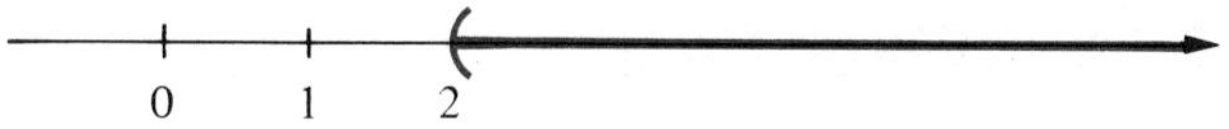

Figure 1.9

EXERCISES 1.2

1 Given $-6 < -4$, what inequality is obtained if
   (a) 5 is added to both sides?
   (b) 5 is subtracted from both sides?
   (c) both sides are multiplied by 1/2?
   (d) both sides are multiplied by $-1/2$?

2 Given $3 > -3$, what inequality is obtained if
   (a) 3 is added to both sides?
   (b) -3 is added to both sides?
   (c) both sides are divided by 3?
   (d) both sides are divided by -3?

In each of Exercises 3–12, express the inequality in interval notation and sketch a graph of the interval.

3 $2 < x < 5$ **4** $-1 < x < 2$

5 $-1 < x \leq 3$ **6** $-2 \geq x \geq -3$

7 $4 \geq x \geq 1$ **8** $0 < x \leq 4$

9 $x > -1$ **10** $x < 1$

11 $x \leq 2$ **12** $x \geq -3$

Express each of the intervals in Exercises 13–18 as an inequality in the variable x.

13 $(-1, 7)$ **14** $[5, 12]$

15 $(8, 9]$ **16** $[-5, 0)$

17 $(5, \infty)$ **18** $(-\infty, 7]$

Solve the inequalities in Exercises 19–42 and express the solutions in terms of intervals.

19 $5x - 6 > 11$ **20** $3x - 5 < 10$

21 $2 - 7x \leqslant 16$ **22** $7 - 2x \geq -3$

23 $3x + 1 < 5x - 4$ **24** $6x - 5 > 9x + 1$

25 $4 - \frac{1}{2}x \geqslant -7 + \frac{1}{4}x$ **26** $\frac{1}{3}x - 4 \leq \frac{1}{4}x - 6$

27 $-4 < 3x + 5 < 8$ **28** $6 > 2x - 6 > 4$

29 $3 \geq \dfrac{7 - x}{2} \geq 1$ **30** $-2 \leq \dfrac{5 - 3x}{4} \leq \dfrac{1}{2}$

31 $0 < 2 - \frac{3}{4}x \leq \frac{1}{2}$ **32** $-3 < \frac{1}{2}x - 4 \leqslant 0$

33 $(3x + 1)(2x - 4) < (6x - 2)(x + 5)$

34 $(x + 2)(x - 5) < (x - 1)^2$

35 $(x + 2)^2 > x(x - 2)$

36 $2x(8x - 1) \geq (4x + 1)(4x - 3)$

37 $\dfrac{5}{3x + 7} > 0$ **38** $\dfrac{1}{6 - 2x} < 0$

39 $(1 - 4x)^{-1} < 0$ **40** $(x + 4)^{-1} > 0$

41 $\dfrac{1}{(x - 1)^2} > 0$ **42** $\dfrac{1}{x^2 + 1} < 0$

43 The relationship between the Fahrenheit and Celsius temperature scales is given by $C = (5/9)(F - 32)$. If $60 \leq F \leq 80$, express the corresponding range for C in terms of an inequality.

44 In the study of electricity, Ohm's Law states that $R = E/I$ where E is measured in volts, I in amperes, and R in ohms. If $E = 110$, what values of R correspond to $I \leq 10$?

45 According to Hooke's Law, the force F (in pounds) required to stretch a certain spring x inches beyond its natural length is given by $F = (4.5)x$. If $10 \leq F \leq 18$, what is the corresponding range for x?

46 Boyle's Law for a certain gas states that $pv = 200$, where p denotes the pressure (lbs/in^2) and v denotes the volume (in^3). If $25 \leq v \leq 50$, what is the corresponding range for p?

47 If $0 < a < b$, prove that $(1/a) > (1/b)$. Why is the restriction $0 < a$ necessary?

48 If $0 < a < b$, prove that $a^2 < b^2$. Why is the restriction $0 < a$ necessary?

49 If $a < b$ and $c < d$, prove that $a + c < b + d$.

50 If $a < b$ and $c < d$, is it always true that $ac < bd$? Explain.

1.3 MORE ON INEQUALITIES

Among the most important inequalities which occur in calculus are those involving absolute values. From the discussion in Section 1.1, if a represents a real number, then $|a|$ is the distance on a coordinate line between the origin and the point corresponding to a. It follows that if we are given an inequality of the form $|a| < 5$, then the solutions consist of all real numbers a such that $-5 < a < 5$, or equivalently, the open interval $(-5, 5)$. In general, if b is any positive real number, then it can be shown that

(i)	$|a| < b$	if and only if $-b < a < b$
(ii)	$|a| > b$	if and only if either $a > b$ or $a < -b$
(iii)	$|a| = b$	if and only if $a = b$ or $a = -b$.

Properties (ii) and (iii) are also true if $b = 0$. The next three examples illustrate uses of these important properties.

Example 1   Solve each of the following:

(a) $|x| < 4$ (b) $|x| > 4$ (c) $|x| = 4.$

Solution

(a)   Using property (i) with $a = x$ and $b = 4$, we see that

$$|x| < 4 \quad \text{if and only if} \quad -4 < x < 4.$$

Hence the interval solution is the open interval $(-4, 4)$.

(b)   By property (ii), $|x| > 4$ means that either $x > 4$ or $x < -4$. Thus the solutions consist of all real numbers in the two infinite intervals $(-\infty, -4)$ and $(4, \infty)$.

(c)   By (iii),

$$|x| = 4 \quad \text{if and only if} \quad x = 4 \text{ or } x = -4. \qquad \blacksquare$$

For solutions of the type obtained in (b) of Example 1 it is sometimes convenient to use the union symbol, $\cup$, and write

$$(-\infty, -4) \cup (4, \infty)$$

to denote all real numbers x such that either x is in $(-\infty, -4)$ or x is in $(4, \infty)$.

Example 2   Solve the inequality $|x - 3| < 0.1.$

Solution   By property (i), with $a = x - 3$ and $b = 0.1$, the given inequality is equivalent to

$$-0.1 < x - 3 < 0.1$$

and hence to $$-0.1 + 3 < (x - 3) + 3 < 0.1 + 3$$

or $$2.9 < x < 3.1.$$

Consequently, the interval solution of the given inequality is $(2.9, 3.1)$. ∎

Example 3 Solve $|2x + 3| > 9$ and illustrate the solutions graphically.

Solution By property (ii), with $a = 2x + 3$ and $b = 9$, the solutions of the given inequality are the solutions of the *two* inequalities

$$2x + 3 > 9 \quad \text{and} \quad 2x + 3 < -9.$$

The first inequality is equivalent to $2x > 6$, or $x > 3$. This gives us the infinite interval $(3, \infty)$. The second inequality is equivalent to $2x < -12$, or $x < -6$, which leads to the interval $(-\infty, -6)$. Consequently, the solutions of $|2x + 3| > 9$ consist of the numbers in the union $(-\infty, -6) \cup (3, \infty)$. The graph is sketched in Figure 1.10.

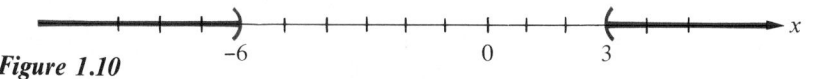

Figure 1.10   −6   0   3   ∎

Before proceeding, let us make an important remark about notation. The solutions in Example 3 consist of all real numbers x such that either $x > 3$ or $x < -6$. Some amateurs employ the *incorrect* form $3 < x < -6$ to denote these inequalities. The notation $a < b < c$ should be used if and only if b satisfies *both* of the conditions $a < b$ and $b < c$. In particular, there is no real number x such that $3 < x < -6$, since it is impossible for x to satisfy $3 < x$ and $x < -6$ simultaneously.

In Chapter Three we shall make a detailed study of **polynomials** in a variable x. In particular, a **polynomial of degree 1** (or a **linear polynomial**) is an expression of the form $ax + b$, where a and b are real numbers with $a \neq 0$. A **polynomial in x of degree 2** (or a **quadratic polynomial**) has the form $ax^2 + bx + c$, where a, b and c are real numbers with $a \neq 0$. It is often possible to express a quadratic polynomial as a product of two linear polynomials with real coefficients. For example,

$$6x^2 + x - 2 = (2x - 1)(3x + 2).$$

We call $2x - 1$ and $3x + 2$ **factors** of $6x^2 + x - 2$, and the process of finding them is called **factoring**. The next two examples illustrate how factoring may be employed to find solutions of inequalities which involve quadratic polynomials.

Example 4 Solve $2x^2 < x + 3$.

Solution The following inequalities are equivalent:

$$2x^2 < x + 3$$
$$2x^2 - x - 3 < 0$$
$$(2x - 3)(x + 1) < 0.$$

Consequently, x is a solution of the given inequality if and only if the product $(2x - 3)(x + 1)$ is negative. In order for the latter to occur, the factors $2x - 3$ and $x + 1$ must have opposite signs. Let us therefore examine the signs of each factor.

The factor $2x - 3$ is positive whenever x is a solution of any of the following inequalities:

$$2x - 3 > 0, \quad 2x > 3, \quad x > \frac{3}{2}.$$

Hence $2x - 3$ is positive if x is in the infinite interval $(3/2, \infty)$. In like manner, $2x - 3$ is negative if x is in the interval $(-\infty, 3/2)$.

The factor $x + 1$ is positive if

$$x + 1 > 0 \quad \text{or} \quad x > -1;$$

that is, if x is in the interval $(-1, \infty)$. Similarly, $x + 1$ is negative if x is in the interval $(-\infty, -1)$.

It is convenient to use the graphical technique shown in Figure 1.11 to display the intervals in which the factors $2x - 3$ and $x + 1$ are positive or negative. Referring to the figure we see that the factors have opposite signs, and hence the product is negative, whenever $-1 < x < 3/2$. Hence the interval solution of the given inequality is $(-1, 3/2)$.

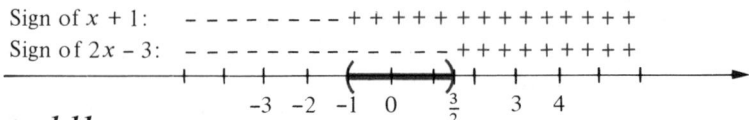

Figure 1.11

Example 5   Solve the inequality $x^2 - 7x + 10 > 0$ and represent the solutions graphically.

Solution   Since the given inequality may be written

$$(x - 5)(x - 2) > 0$$

it follows that x is a solution if and only if both factors $x - 5$ and $x - 2$ are positive, or both are negative. The diagram in Figure 1.12 indicates the signs of these factors for various real numbers. Evidently, both factors are positive if x is in the interval $(5, \infty)$, and both are negative if x is in $(-\infty, 2)$. Hence the solutions are given by $(-\infty, 2) \cup (5, \infty)$. The graph is sketched in Figure 1.12.

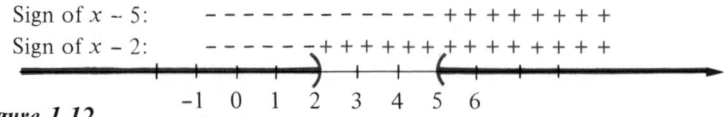

Figure 1.12

Example 6 Solve $\dfrac{x+1}{x+2} \leq 3$ and represent the solutions graphically.

Solution The following inequalities are equivalent:

$$\frac{x+1}{x+2} \leq 3$$

$$\frac{x+1}{x+2} - 3 \leq 0$$

$$\frac{x+1-3x-6}{x+2} \leq 0$$

$$\frac{-2x-5}{x+2} \leq 0$$

$$\frac{2x+5}{x+2} \geq 0.$$

The fraction $\dfrac{2x+5}{x+2}$ equals zero if $2x+5 = 0$, or $x = -5/2$. Hence $-5/2$ is a solution. It remains to determine the solutions of the inequality

$$\frac{2x+5}{x+2} > 0.$$

Since a quotient is positive if and only if the numerator and denominator have the same signs, we must determine where both $2x + 5$ and $x + 2$ are positive or where both are negative. Referring to the diagram in Figure 1.13, we see that both factors are positive if x is in the interval $(-2, \infty)$ and both are negative if x is in the interval $(-\infty, -5/2)$. Since $-5/2$ is also a solution, it follows that the solutions of the given inequality are given by $(-\infty, -5/2] \cup (-2, \infty)$. The graph is sketched in Figure 1.14.

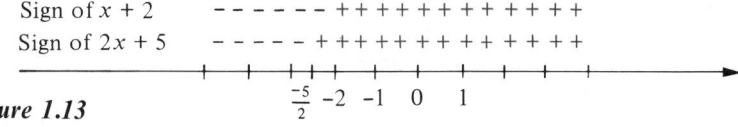

Figure 1.13

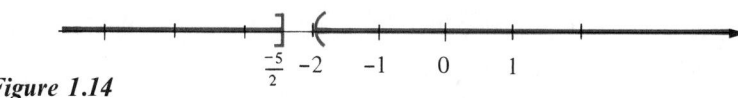

Figure 1.14

It would have been incorrect to solve Example 6 merely by multiplying both sides of $(x+1)/(x+2) \leq 3$ by $(x+2)$, obtaining $(x+1) \leq 3(x+2)$, since $x + 2$ is not always positive for every x under consideration. If one wishes to begin a solution in this way, it is necessary to consider the following two cases: (i) $x + 2 > 0$ and (ii) $x + 2 < 0$.

Example 7 Solve the inequality $(x + 2)(x - 1)(x - 5) > 0$ and represent the solutions graphically.

Solution   If a number is substituted for x in $(x + 2)(x - 1)(x - 5)$, the result will be positive provided all three factors are positive or provided two of the factors are negative and one is positive. As in the previous sections we find that $x + 2$ is negative in $(-\infty, -2)$ and positive in $(-2, \infty)$. Similarly, the factor $x - 1$ is negative in $(-\infty, 1)$ and positive in $(1, \infty)$. Finally, $x - 5$ is negative in $(-\infty, 5)$ and positive in $(5, \infty)$. These facts are displayed in Figure 1.15. We see from this diagram that all factors are positive if x is in $(5, \infty)$ and that two factors are negative and one is positive if x is in $(-2, 1)$. Hence the solutions are given by $(-2, 1) \cup (5, \infty)$. The graph is sketched in Figure 1.16

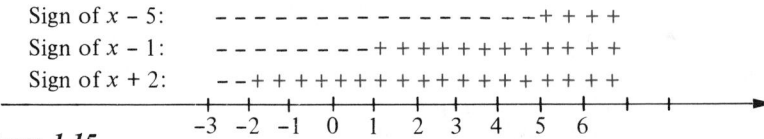

Figure 1.15

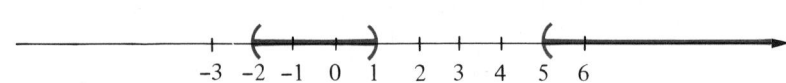

Figure 1.16   ∎

EXERCISES 1.3

Solve the inequalities in Exercises 1–38 and express the solutions in terms of intervals.

1 $|x| < 2$

2 $|x| < 25$

3 $|x| > 6$

4 $|x| > 1/2$

5 $|-x| \leq 10$

6 $|x| \geq 7$

7 $|x - 10| < 0.05$

8 $|x - 5| < 0.001$

9 $\left|\dfrac{x + 4}{3}\right| \leq 2$

10 $\left|\dfrac{x + 1}{5}\right| \leq 1$

11 $|5 - 3x| < 7$

12 $|7x + 4| < 10$

13 $|2x + 4| > 8$

14 $|6x - 7| \geq 4$

15 $\dfrac{3}{|5 - 2x|} \leq 1$

16 $\left|\dfrac{4 - x}{2}\right| > 9$

17 $x^2 - x - 6 < 0$

18 $x^2 + 6x + 5 < 0$

19 $x(2x + 3) > 5$

20 $x(3x - 1) > 4$

21 $x^2 \leq 10x$

22 $4x^2 \geq x$

23 $\dfrac{2x + 1}{10 - 3x} < 0$

24 $\dfrac{9 - 4x}{x + 1} \geq 0$

25 $\dfrac{6}{x^2 - 9} < 0$

26 $\dfrac{10}{25x^2 - 16} > 0$

27 $\dfrac{7x}{x^2 - 16} \geq 0$

28 $\dfrac{x - 2}{x^2 - 9} < 0$

29 $3x^2 + 7x - 6 > 0$

30 $10x^2 + 3x - 4 < 0$

31 $\dfrac{x + 4}{2x - 1} < 3$

32 $\dfrac{1}{x - 2} > \dfrac{3}{x + 1}$

33 $x(x^2 - 1) < 0$

34 $(x^2 + 4)(x^2 - 1) > 0$

35 $x^3 - x^2 - 4x + 4 \geq 0$

36 $(x^2 - 4x + 4)(3x - 7) < 0$

37 $\dfrac{x^2 - x - 2}{x^2 - 4x + 3} \geq 0$ **38** $\dfrac{x^2 + 2x - 3}{x^2 - 4x} \leq 0$

39 If a projectile is fired straight upward from level ground with an initial velocity of 72 ft/sec, its altitude s (in feet) after t seconds is given by $s = -16t^2 + 72t$. During what time interval will the projectile be at least 32 feet above the ground?

40 The period T(sec) of a simple pendulum of length l(cm) is given by $T = 2\pi\sqrt{l/g}$, where g is a physical constant. If, under certain conditions, $g = 980$ and $98 \leq l \leq 100$, what is the corresponding range for T?

1.4 COORDINATE SYSTEMS IN TWO DIMENSIONS

In Section 1.1 we indicated how coordinates may be assigned to points on a straight line. Coordinate systems can also be introduced in planes by means of ordered pairs. The term **ordered pair** refers to two real numbers, of which one is designated as the "first" number and the other as the "second." The symbol (a, b) is used to denote the ordered pair consisting of the real numbers a and b where a is first and b is second. There are many uses for ordered pairs. They were used previously to denote open intervals. In this section we shall use ordered pairs to represent points in a plane. Although ordered pairs are employed in different situations, there is little chance for confusion, since it should always be clear from the discussion whether the symbol (a, b) represents an interval, a point, or some other mathematical object. We consider two ordered pairs (a, b) and (c, d) equal, and write

$$(a, b) = (c, d) \quad \text{if and only if} \quad a = c \text{ and } b = d.$$

This implies, in particular, that $(a, b) \neq (b, a)$ if $a \neq b$.

A **rectangular**, or **Cartesian,*** **coordinate system** may be introduced in a plane by considering two perpendicular coordinate lines in the plane which intersect in the origin O on each line. Unless otherwise specified, the same unit of length is chosen on each line. Usually one of the lines is horizontal with positive direction to the right, and the other line is vertical with positive direction upward, as indicated by the arrowheads in Figure 1.17. The two lines are called **coordinate axes** and the point O is called the **origin**. The horizontal line is often referred to as the **x-axis** and the vertical line as the **y-axis**, and they are labeled x and y, respectively. The plane is then called a **coordinate plane** or, with the preceding notation for coordinate axes, an **xy-plane**. Although the symbols x and y are used to denote lines as well as numbers, there should be no misunderstanding as to what these letters represent when they appear alongside of coordinate lines as in Figure 1.17. In certain applications different labels such as d, t, etc., are used for the coordinate lines. The coordinate axes divide the plane into four parts called the **first**, **second**, **third**, and **fourth quadrants** and labeled I, II, III, and IV, respectively, as shown in Figure 1.17.

* The term "Cartesian" is used in honor of the French mathematician and philosopher René Descartes (1596–1650), who was one of the first to employ such coordinate systems.

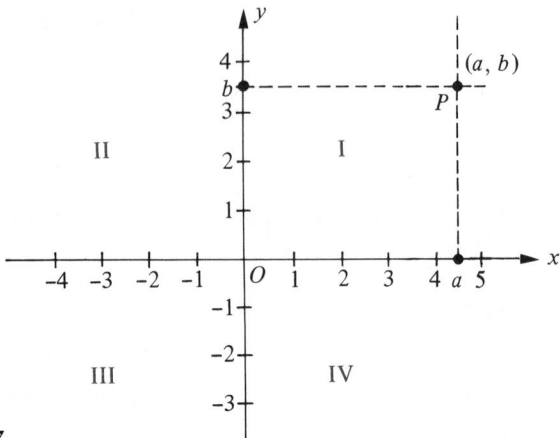

Figure 1.17

Each point P in an xy-plane may be assigned a unique ordered pair. If vertical and horizontal lines through P intersect the x- and y-axes at points with coordinates a and b, respectively (see Figure 1.17), then P is assigned the ordered pair (a, b). The number a is called the **x-coordinate** (or **abscissa**), of P, and b is called the **y-coordinate** (or **ordinate**), of P. We sometimes say that P *has coordinates* (a, b). Conversely, every ordered pair (a, b) determines a point P in the xy-plane with coordinates a and b. Specifically, P is the point of intersection of lines perpendicular to the x-axis and y-axis at the points having coordinates a and b, respectively. This establishes a one-to-one correspondence between the set of all points in the xy-plane and the set of all ordered pairs. It is sometimes convenient to refer to the *point* (a, b) meaning the point with abscissa a and ordinate b. The symbol $P(a, b)$ will denote the point P with coordinates

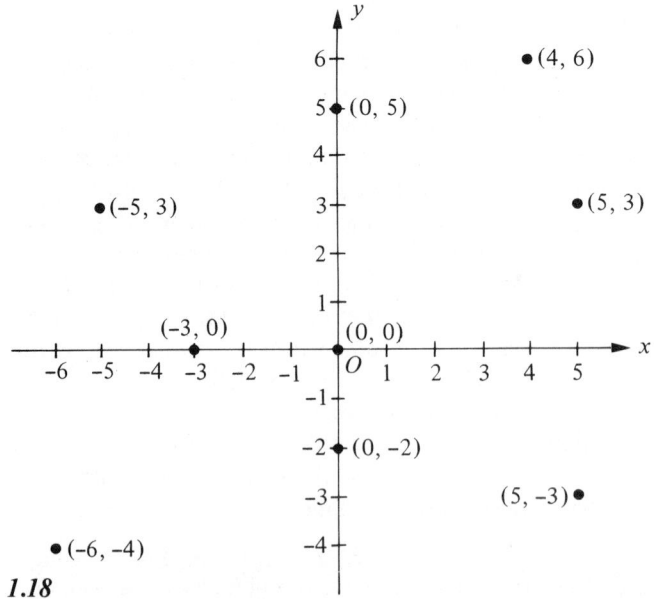

Figure 1.18

(a, b). To **plot a point** $P(a, b)$ means to locate, in a coordinate plane, the point P with coordinates (a, b). This point is represented by a dot in the appropriate position.

Note that abscissas are positive for points in quadrants I or IV and negative for points in quadrants II or III. Ordinates are positive for points in quadrants I or II and negative for points in quadrants III or IV. Some typical points in a coordinate plane are plotted in Figure 1.18.

We shall next derive a formula for finding the distance between any two points in a coordinate plane. The distance between two points P and Q will be denoted by $d(P, Q)$. If $P = Q$, then we agree that $d(P, Q) = 0$, whereas, if $P \neq Q$, the distance is positive. Let us consider any two points $P_1(x_1, y_1)$ and $P_2(x_2, y_2)$ in the plane. If the points lie on the same horizontal line then $y_1 = y_2$, and we may denote the points by $P_1(x_1, y_1)$ and $P_2(x_2, y_1)$. If lines through P_1 and P_2 parallel to the y-axis intersect the x-axis at $A_1(x_1, 0)$ and $A_2(x_2, 0)$, as shown in (i) of Figure 1.19, then we see that $d(P_1, P_2) = d(A_1, A_2)$. However, from Section 1.1, $d(A_1, A_2) = |x_2 - x_1|$ and hence

$$d(P_1, P_2) = |x_2 - x_1|.$$

Since $|x_2 - x_1| = |x_1 - x_2|$, the last formula is valid whether P_1 lies to the left of P_2 or to the right of P_2. Moreover, the formula is independent of the quadrants in which the points lie.

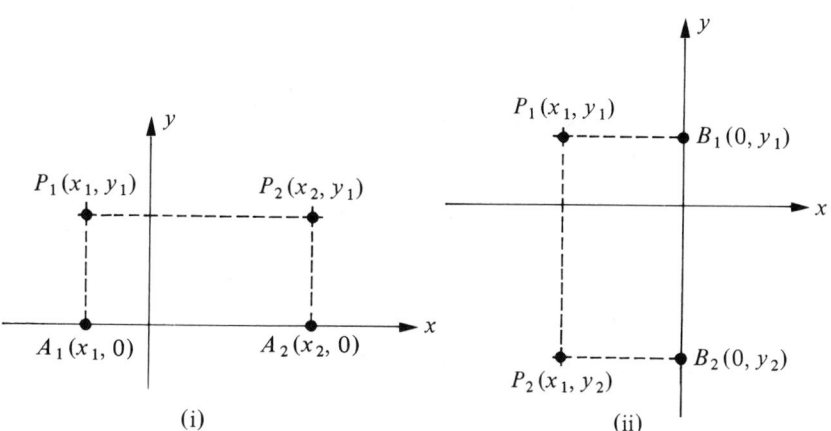

(i)   (ii)

Figure 1.19

In similar fashion, if P_1 and P_2 are on the same vertical line, then $x_1 = x_2$, and we may denote the points by $P_1(x_1, y_1)$ and $P_2(x_1, y_2)$. If we consider the points $B_1(0, y_1)$ and $B_2(0, y_2)$ on the y-axis as shown in (ii) of Figure 1.19, then

$$d(P_1, P_2) = d(B_1, B_2) = |y_2 - y_1|.$$

Finally, let us consider the general case, in which the points $P_1(x_1, y_1)$ and $P_2(x_2, y_2)$ do not lie on the same horizontal or vertical line. The line through $P_1(x_1, y_1)$ parallel to the x-axis and the line through $P_2(x_2, y_2)$ parallel to the y-axis intersect at some point P_3. Since P_3 has the same y-coordinate as P_1 and the same x-coordinate as P_2, we can denote it by $P_3(x_2, y_1)$ as in Figure 1.20. From the previous discussion,

$d(P_1, P_3) = |x_2 - x_1|$ and $d(P_3, P_2) = |y_2 - y_1|$. Since P_1, P_2, and P_3 form a right triangle with hypotenuse from P_1 to P_2, we have, by the Pythagorean Theorem,

$$[d(P_1, P_2)]^2 = [d(P_1, P_3)]^2 + [d(P_3, P_2)]^2$$

and hence

$$[d(P_1, P_2)]^2 = |x_2 - x_1|^2 + |y_2 - y_1|^2.$$

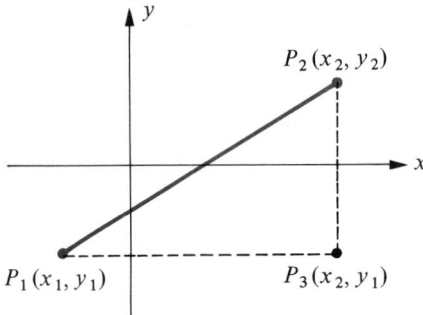

Figure 1.20

Using the fact that $d(P_1, P_2)$ is nonnegative and that $|a|^2 = a^2$ for every real number a, we obtain the following important formula.

Distance Formula

If $P_1(x_1, y_1)$ and $P_2(x_2, y_2)$ are points in a coordinate plane, then the **distance between P_1 and P_2** is given by

$$d(P_1, P_2) = \sqrt{(x_2 - x_1)^2 + (y_2 - y_1)^2}.$$

Although we referred to the special case indicated in Figure 1.20, the argument used in the proof of the distance formula is independent of the positions of P_1 and P_2.

Example 1   Plot the points $A(-1, -3)$, $B(6, 1)$, $C(2, -5)$, and prove that the triangle with vertices A, B, C is a right triangle.

Solution   The points and the triangle are shown in Figure 1.21. From plane geometry, a triangle is a right triangle if and only if the sum of the squares of two of its sides is equal to the square of the remaining side. Using the Distance Formula, we obtain

$$d(A, B) = \sqrt{(-1 - 6)^2 + (-3 - 1)^2}$$
$$= \sqrt{49 + 16} = \sqrt{65}$$
$$d(B, C) = \sqrt{(6 - 2)^2 + (1 + 5)^2}$$
$$= \sqrt{16 + 36} = \sqrt{52}$$
$$d(A, C) = \sqrt{(-1 - 2)^2 + (-3 + 5)^2}$$
$$= \sqrt{9 + 4} = \sqrt{13}.$$

Since $[d(A, B)]^2 = [d(B, C)]^2 + [d(A, C)]^2$, the triangle is a right triangle with hypotenuse joining A to B.

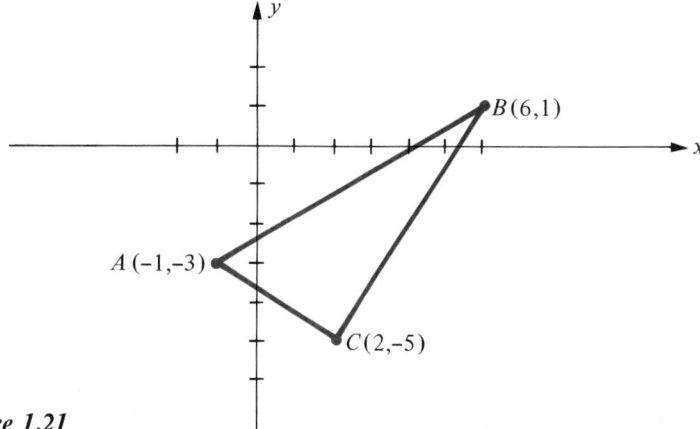

Figure 1.21 ■

It is easy to obtain a formula for the midpoint of a line segment. Let $P_1(x_1, y_1)$ and $P_2(x_2, y_2)$ be two points in a coordinate plane, and let M be the midpoint of the segment $P_1 P_2$. The lines through P_1 and P_2 parallel to the y-axis intersect the x-axis at $A_1(x_1, 0)$ and $A_2(x_2, 0)$ and, from plane geometry, the line through M parallel to the y-axis bisects the segment $A_1 A_2$ (see Figure 1.22). If $x_1 < x_2$, then $x_2 - x_1 > 0$, and hence

$$d(A_1, A_2) = x_2 - x_1.$$

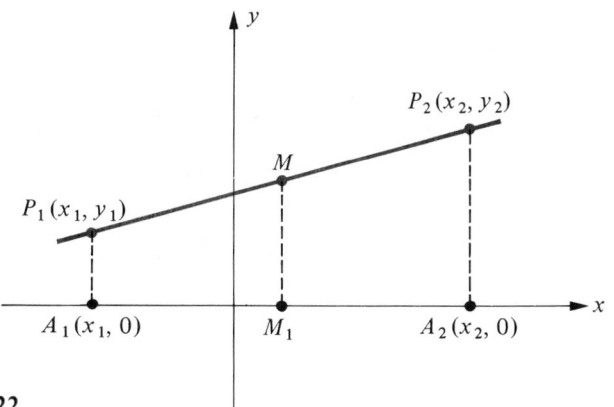

Figure 1.22

Since M_1 is halfway from A_1 to A_2, the abscissa of M_1 is

$$x_1 + \tfrac{1}{2}(x_2 - x_1) = x_1 + \tfrac{1}{2}x_2 - \tfrac{1}{2}x_1$$
$$= \tfrac{1}{2}x_1 + \tfrac{1}{2}x_2$$
$$= \frac{x_1 + x_2}{2}.$$

It follows that the abscissa of M is also $(x_1 + x_2)/2$. It can be shown in similar fashion that the ordinate of M is $(y_1 + y_2)/2$. Moreover, these formulas hold for all positions of P_1 and P_2. This gives us the following result.

Midpoint Formula

> The midpoint of the line segment from $P_1(x_1, y_1)$ to $P_2(x_2, y_2)$ is
>
> $$\left(\frac{x_1 + x_2}{2}, \frac{y_1 + y_2}{2}\right).$$

Example 2   Find the midpoint M of the line segment from $P_1(-2, 3)$ to $P_2(4, -2)$. Plot the points P_1, P_2, and M, and verify that $d(P_1, M) = d(P_2, M)$.

Solution   Applying the Midpoint Formula, the coordinates of M are

$$\left(\frac{-2 + 4}{2}, \frac{3 + (-2)}{2}\right) \quad \text{or} \quad \left(1, \frac{1}{2}\right).$$

The three points P_1, P_2, and M are plotted in Figure 1.23. Using the Distance Formula, we obtain

$$d(P_1, M) = \sqrt{(-2 - 1)^2 + (3 - \tfrac{1}{2})^2} = \sqrt{9 + (\tfrac{25}{4})}$$
$$d(P_2, M) = \sqrt{(4 - 1)^2 + (-2 - \tfrac{1}{2})^2} = \sqrt{9 + (\tfrac{25}{4})}.$$

Hence $d(P_1, M) = d(P_2, M)$.

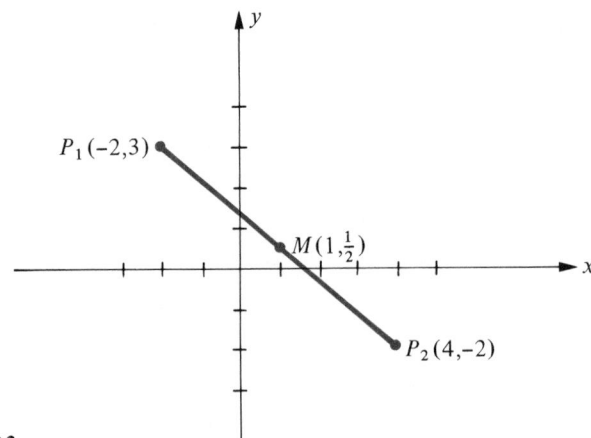

Figure 1.23

EXERCISES 1.4

1 Plot the following points on a rectangular coordinate system: $A(5, -2)$, $B(-5, -2)$, $C(5, 2)$, $D(-5, 2)$, $E(3, 0)$, $F(0, 3)$.

2 Plot the points $A(-3, 1)$, $B(3, 1)$, $C(-2, -3)$, $D(0, 3)$, and $E(2, -3)$ on a rectangular coordinate system and then draw the line segments AB, BC, CD, DE, and EA.

3 Plot $A(0,0)$, $B(1,1)$, $C(3,3)$, $D(-1,-1)$, and $E(-2,-2)$. Describe the set of all points of the form (x,x) where x is a real number.

4 Plot $A(0,0)$, $B(1,-1)$, $C(2,-2)$, $D(-1,1)$, and $E(-3,3)$. Describe the set of all points of the form $(a,-a)$ where a is a real number.

5 Describe the set of all points $P(x,y)$ in a coordinate plane such that
(a) $x=3$ (b) $y=-1$ (c) $x \geq 0$
(d) $xy > 0$ (e) $y < 0$

6 Describe the set of all points $P(x,y)$ in a coordinate plane such that
(a) $y=0$ (b) $x=-5$ (c) $x/y < 0$
(d) $xy = 0$ (e) $y > 1$

In Exercises 7–12, find (a) the distance $d(A,B)$ between the given points A and B; (b) the midpoint of the segment AB.

7 $A(4,-3), B(6,2)$ **8** $A(-2,-5), B(4,6)$

9 $A(-5,0), B(-2,-2)$

10 $A(6,2), B(6,-2)$

11 $A(7,-3), B(3,-3)$ **12** $A(-4,7), B(0,-8)$

In Exercises 13 and 14 prove that the triangle with the indicated vertices is a right triangle and find its area.

13 $A(8,5), B(1,-2), C(-3,2)$

14 $A(-6,3), B(3,-5), C(-1,5)$

15 Prove that the following points are vertices of a square: $A(-4,2), B(1,4), C(3,-1), D(-2,-3)$.

16 Prove that the following points are vertices of a parallelogram: $A(-4,-1)$, $B(0,-2)$, $C(6,1)$, $D(2,2)$.

17 Given $A(-3,8)$, find the coordinates of the point B such that $M(5,-10)$ is the midpoint of AB.

18 Given $A(5,-8)$ and $B(-6,2)$, find the point on AB that is three-fourths of the way from A to B.

19 Given $A(-4,-3)$ and $B(6,1)$, prove that $P(5,-11)$ is on the perpendicular bisector of AB.

20 Given $A(-4,-3)$ and $B(6,1)$, find a formula which expresses the fact that $P(x,y)$ is on the perpendicular bisector of AB.

21 Find a formula which expresses the fact that $P(x,y)$ is a distance 5 from the origin. Describe the totality of all such points.

22 If r is a positive real number, find a formula that states that $P(x,y)$ is a distance r from a fixed point $C(h,k)$. Describe the totality of all such points.

23 Find all points on the y-axis that are 6 units from the point $(5,3)$.

24 Find all points on the x-axis that are 5 units from $(-2,4)$.

25 Let S denote the set of points of the form $(2x,x)$ where x is a real number. Find the point in S that is in the third quadrant and is a distance 5 from the point $(1,3)$.

26 Let S denote the set of points of the form (x,x) where x is a real number. Find all points in S that are a distance 3 from the point $(-2,1)$.

27 For what values of a is the distance between $(a,3)$ and $(5,2a)$ greater than $\sqrt{26}$?

28 Given the points $A(-2,0)$ and $B(2,0)$, find a formula not containing radicals that expresses the fact that the sum of the distances from $P(x,y)$ to A and to B, respectively, is 5.

29 Prove that the midpoint of the hypotenuse of any right triangle is equidistant from the vertices. (*Hint:* Label the vertices of the triangle $O(0,0)$, $A(a,0)$, and $B(0,b)$.)

30 Prove that the diagonals of any parallelogram bisect each other. (*Hint:* Label three of the vertices of the parallelogram $O(0,0)$, $A(a,b)$, and $C(0,c)$.)

1.5   GRAPHS

If W is a set of ordered pairs, then we may speak of the point $P(x, y)$ in a coordinate plane which corresponds to the ordered pair (x, y) in W. The **graph** of W is the set of all points which correspond to the ordered pairs in W. The phrase "sketch the graph of W" means to illustrate the significant features of the graph geometrically on a coordinate plane, as illustrated in the next example.

Example 1   Sketch the graph of $W = \{(x, y): -1 < x \le 4, 2 \le y < 3\}$.

Solution   The set notation describing W may be translated "the set of all ordered pairs (x, y) such that $-1 < x \le 4$ and $2 \le y < 3$." Thus a point $P(x, y)$ is on the graph of W if and only if the abscissa x is greater than -1 and less than or equal to 4, while the ordinate y is greater than or equal to 2 and less than 3. Hence the graph of W consists of all points inside the shaded rectangle in Figure 1.24, together with the points on the lower and right-hand sides of the rectangle.

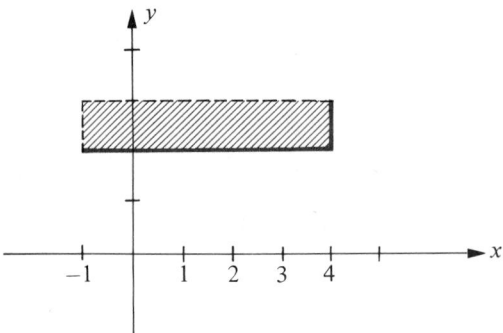

Figure 1.24. $\{(x, y): -1 < x \le 4, 2 \le y < 3\}$

Example 2   Sketch the graph of $W = \{(x, y): y = 2x - 1\}$.

Solution   We begin by finding points with coordinates of the form (x, y) where the ordered pair (x, y) is in W. It is convenient to list these coordinates in tabular form as shown below, where for each real number x the corresponding value for y is $2x - 1$.

x	-2	-1	0	1	2	3
y	-5	-3	-1	1	3	5

After plotting, it appears that the points with these coordinates all lie on a line and we sketch the graph accordingly (see Figure 1.25). Ordinarily, the few points we have plotted would not be enough to illustrate the graph; however, in this elementary case we can be reasonably sure that the graph is a line. In the next chapter it will be proved that our conjecture is correct.

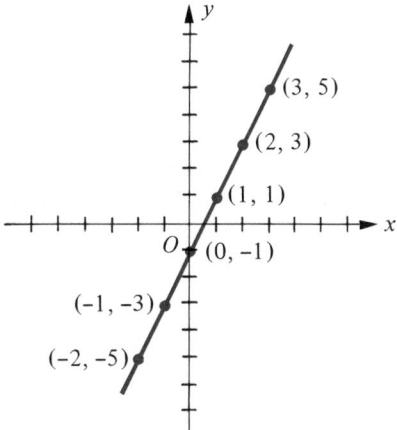

Figure 1.25   ∎

It is impossible to sketch the entire graph in Example 2 since x may be assigned values which are numerically as large as desired. Nevertheless, we often call a drawing of the type given in Figure 1.25 *the graph of W* or *a sketch of the graph*, where it is understood that the drawing is only a device for visualizing the actual graph, and the line does not terminate as shown in the figure. In general, when sketching a graph, you should illustrate enough of the graph so that the remaining parts are evident.

The graph in Example 2 is determined by the equation $y = 2x - 1$ in the sense that for every real number x, the equation can be used to find a number y such that (x, y) is in W. Given an equation in x and y, we say that an ordered pair (a, b) is a **solution** of the equation if equality is obtained when a is substituted for x and b for y. For example, $(2, 3)$ is a solution of $y = 2x - 1$ since substitution of 2 for x and 3 for y leads to $3 = 4 - 1$, or $3 = 3$. Two equations in x and y are said to be **equivalent** if they have exactly the same solutions. The solutions of an equation in x and y determine a set S of ordered pairs, and we define the **graph of the equation** as the graph of S. Notice that the solutions of the equation $y = 2x - 1$ are the pairs (a, b) such that $b = 2a - 1$, and hence the solutions are identical with the set W given in Example 2. Consequently, the graph of the equation $y = 2x - 1$ is the same as the graph of W sketched in Figure 1.25.

For most of the equations we shall encounter in this chapter the technique used for sketching the graph will consist of plotting a sufficient number of points until some pattern emerges, and then sketching the graph accordingly. This is obviously a crude (and often inaccurate) way to arrive at the graph; however, it is a method often employed at the beginning of elementary courses. As we progress through this text, techniques will be introduced which will enable us to sketch a variety of graphs without plotting many points. In order to give accurate descriptions of graphs when complicated expressions are involved, it is usually necessary to employ more advanced mathematical tools of the types introduced in the study of calculus.

Example 3   Sketch the graph of the equation $y = x^2$.

Solution   In order to obtain the graph, it is necessary to plot more points than in the previous example. Substituting for x in the given equation gives us the following table.

x	-3	$-\frac{5}{2}$	-2	$-\frac{3}{2}$	-1	$-\frac{1}{2}$	0	$\frac{1}{2}$	1	$\frac{3}{2}$	2	$\frac{5}{2}$	3
y	9	$\frac{25}{4}$	4	$\frac{9}{4}$	1	$\frac{1}{4}$	0	$\frac{1}{4}$	1	$\frac{9}{4}$	4	$\frac{25}{4}$	9

Larger numerical values of x produce even larger values of y. For example, the points $(3,9)$, $(4,16)$, $(5,25)$, and $(6,36)$ are on the graph, as are $(-3,9)$, $(-4,16)$, $(-5,25)$, and $(-6,36)$. Plotting the points given in the table and drawing a smooth curve through these points, we obtain the sketch in Figure 1.26, where we have labeled only points with integer coordinates.

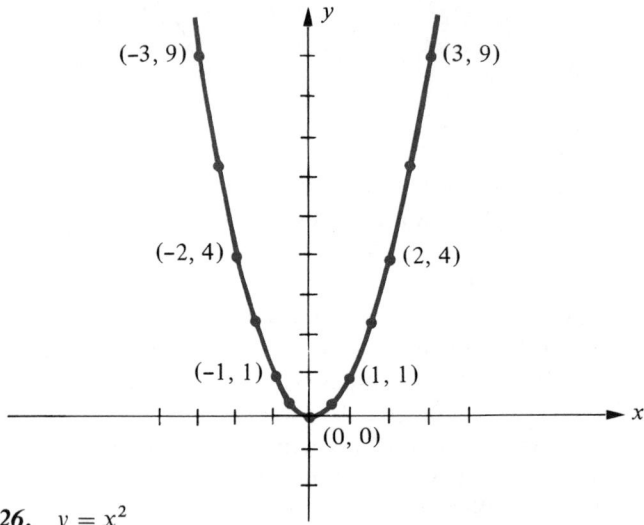

Figure 1.26. $y = x^2$

The graph in Example 3 is called a **parabola**. The lowest point $(0,0)$ is called the **vertex** of the parabola, and we say that the parabola **opens upward**. If the graph were inverted, as would be the case for $y = -x^2$, then the parabola **opens downward**. The y-axis is called the **axis of the parabola**.

If the coordinate plane in Figure 1.26 is folded along the y-axis, then the graph which lies in the left half of the plane coincides with that in the right half. We say that **the graph is symmetric with respect to the y-axis**. As in (i) of Figure 1.27, a graph is symmetric with respect to the y-axis provided that the point $(-x, y)$ is on the graph whenever (x, y) is on the graph. Similarly, as in (ii) of Figure 1.27, **a graph is symmetric with respect to the x-axis** if, whenever a point (x, y) is on the graph, the point $(x, -y)$ is also on the graph. In the latter case, if we fold the coordinate plane along the x-axis, that part of the graph which lies above the x-axis will coincide with the part which lies below. The previous remarks give us the following useful result.

Tests for Symmetry with Respect to an Axis

(i)   The graph of an equation is symmetric with respect to the y-axis if substitution of $-x$ for x does not change the solutions of the equation.

(ii)   The graph of an equation is symmetric with respect to the x-axis if substitution of $-y$ for y does not change the solutions of the equation.

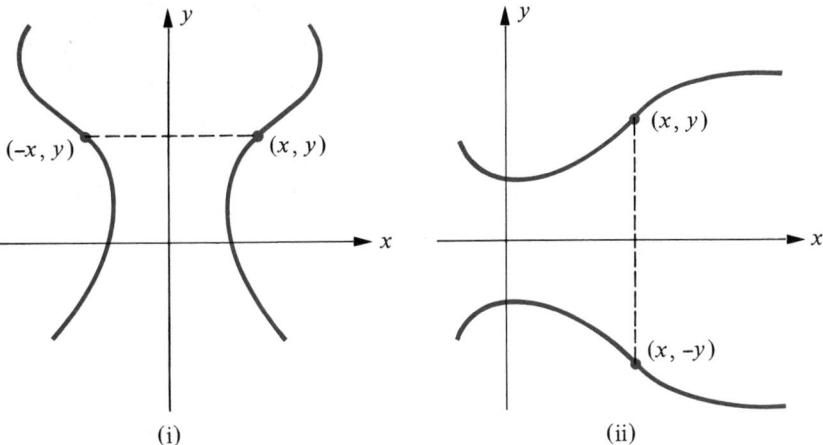

Figure 1.27

If, in the equation of Example 3, we substitute $-x$ for x, we obtain $y = (-x)^2$, which is the same as $y = x^2$. Hence the graph is symmetric with respect to the y-axis.

If symmetry with respect to an axis exists, then it is sufficient to determine the graph in half of the coordinate plane, since the remainder of the graph is a mirror image, or reflection, of that half.

Example 4 Sketch the graph of $y^2 = x$.

Solution Since substitution of $-y$ for y does not change the equation, the graph is symmetric with respect to the x-axis (Why?). It is sufficient, therefore, to plot points with nonnegative ordinates and then reflect through the x-axis. Since $y^2 = x$, the ordinates of points above the x-axis are given by $y = \sqrt{x}$. Coordinates of some points on the graph are tabulated below. A portion of the graph is sketched in Figure 1.28. The graph is a parabola with vertex at the origin and which opens to the right. In this case the x-axis is the axis of the parabola.

x	0	1	2	3	4	9
y	0	1	$\sqrt{2}$	$\sqrt{3}$	2	3

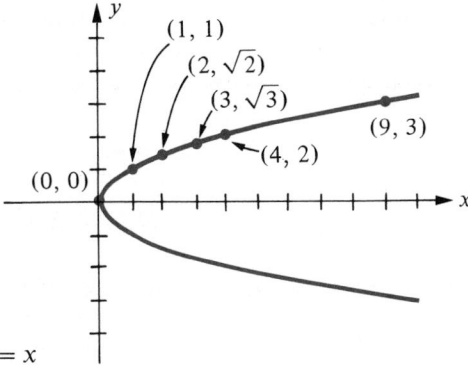

Figure 1.28. $y^2 = x$

Another type of symmetry which certain graphs possess is called **symmetry with respect to the origin**. In this situation, whenever a point (x, y) is on the graph, then $(-x, -y)$ is also on the graph, as illustrated in Figure 1.29. Evidently, we have the following result.

Test for Symmetry with Respect to the Origin

The graph of an equation is symmetric with respect to the origin if the simultaneous substitution of $-x$ for x and $-y$ for y does not change the solutions of the equation.

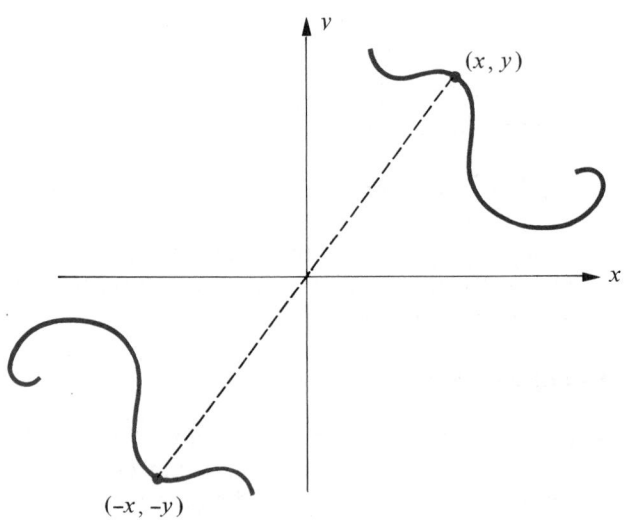

Figure 1.29

Example 5   Sketch the graph of the equation $y = x^3$.

Solution   If we substitute $-x$ for x and $-y$ for y, then

$$-y = (-x)^3 = -x^3.$$

Multiplying both sides by -1, we see that the latter equation has the same solutions as the given equation $y = x^3$. Hence the graph is symmetric with respect to the origin. The following table lists some points on the graph.

x	0	$\frac{1}{4}$	$\frac{1}{2}$	$\frac{3}{4}$	1	$\frac{3}{2}$	2
y	0	$\frac{1}{64}$	$\frac{1}{8}$	$\frac{27}{64}$	1	$\frac{27}{8}$	8

By symmetry (or substitution) we see that the points $(-\frac{1}{4}, -\frac{1}{64})$, $(-\frac{1}{2}, -\frac{1}{8})$, etc., are on the graph. Plotting points leads to the graph in Figure 1.30.

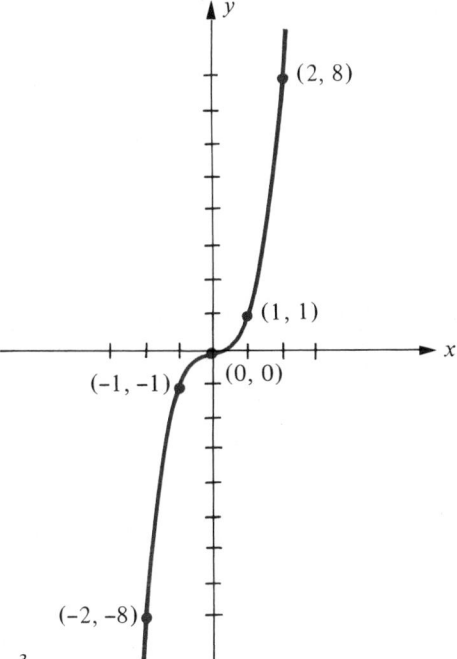

Figure 1.30. $y = x^3$ ■

Given a geometric figure in a coordinate plane, it is sometimes possible to find its equation, in the sense that the figure is the graph of the equation. We shall demonstrate how this can be accomplished for circles.

If $C(h, k)$ is a point in a coordinate plane, then a circle in the plane with center C and radius r may be defined as the collection of all points in the plane that are r units from C. If $P(x, y)$ is an arbitrary point in the plane, then, as illustrated in (i) of Figure 1.31, P is on the circle if and only if $d(C, P) = r$.

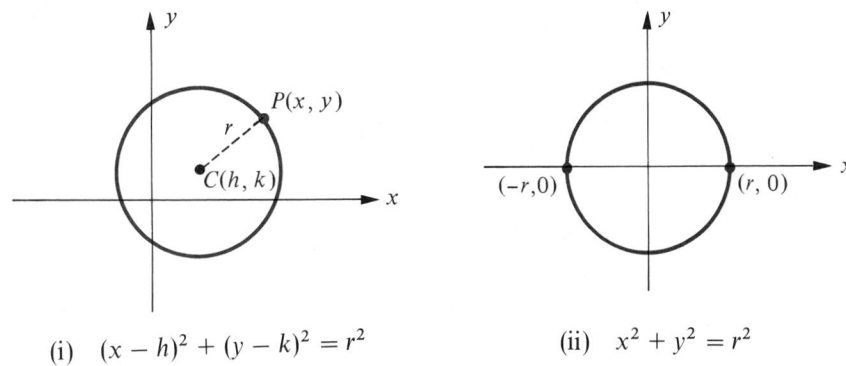

(i) $(x - h)^2 + (y - k)^2 = r^2$ (ii) $x^2 + y^2 = r^2$

Figure 1.31

Using the Distance Formula gives us the equation

$$\sqrt{(x - h)^2 + (y - k)^2} = r.$$

The equivalent equation

$$(x - h)^2 + (y - k)^2 = r^2, \quad r > 0$$

is called the **standard equation of a circle of radius r and center (h, k)**. If the center is at the origin, then $h = 0$ and $k = 0$, and this equation reduces to

$$x^2 + y^2 = r^2$$

which is an equation of a circle of radius r with center at the origin (see (ii) of Figure 1.31). If we next let $r = 1$, the graph is a **unit circle** with center at the origin. A point $P(x, y)$ is on this unit circle if and only if $x^2 + y^2 = 1$.

Example 6   Find an equation of the circle having center $C(-2, 3)$ and containing the point $D(4, 5)$.

Solution   The circle is illustrated in Figure 1.32. Since D is on the circle, the radius r

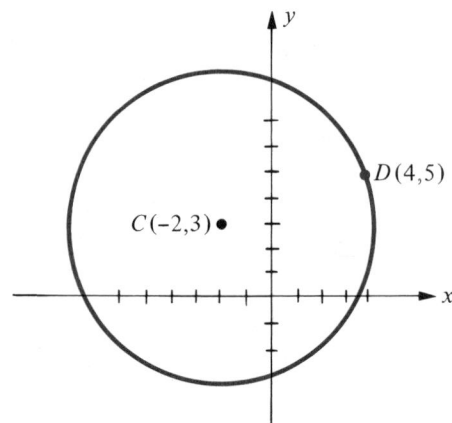

Figure 1.32

is $d(C, D)$. By the Distance Formula,

$$r = d(C, D) = \sqrt{(-2 - 4)^2 + (3 - 5)^2} = \sqrt{36 + 4} = \sqrt{40}.$$

Using the standard equation of a circle with $h = -2$ and $k = 3$, we obtain

$$(x + 2)^2 + (y - 3)^2 = 40. \qquad \blacksquare$$

EXERCISES 1.5

In each of Exercises 1–8 sketch the graph of the given set W of ordered pairs.

1    $W = \{(x, y): x = 4\}$

2    $W = \{(x, y): y = -3\}$

3    $W = \{(x, y): xy < 0\}$

4    $W = \{(x, y): xy = 0\}$

5 $W = \{(x, y): |x| < 2, |y| > 1\}$

6 $W = \{(x, y): |x| > 1, |y| \le 2\}$

7 $W = \{(x, y): |x - 2| < 3, |y| \le 1\}$

8 $W = \{(x, y): |x - 1| < 5, |y + 1| < 2\}$

In each of Exercises 9–28 sketch the graph of the equation after plotting a sufficient number of points.

9 $y = 3x + 1$ 10 $y = 4x - 3$

11 $y = -2x + 3$ 12 $y = 2 - 3x$

13 $2y + 3x + 6 = 0$ 14 $3y - 2x = 6$

15 $y = 2x^2$ 16 $y = -x^2$

17 $y = 2x^2 - 1$ 18 $y = -x^2 + 2$

19 $4y = x^2$ 20 $3y + x^2 = 0$

21 $y = -\frac{1}{2}x^3$ 22 $y = \frac{1}{2}x^3$

23 $y = x^3 - 2$ 24 $y = 2 - x^3$

25 $y = \sqrt{x}$ 26 $y = \sqrt{x} - 1$

27 $y = \sqrt{-x}$ 28 $y = \sqrt{x - 1}$

Describe the graphs in Exercises 29–36.

29 $x^2 + y^2 = 16$ 30 $x^2 + y^2 = 25$

31 $9x^2 + 9y^2 = 1$ 32 $2x^2 + 2y^2 = 1$

33 $(x - 2)^2 + (y + 1)^2 = 4$

34 $(x + 3)^2 + (y - 4)^2 = 1$

35 $x^2 + (y - 3)^2 = 9$ 36 $(x + 5)^2 + y^2 = 16$

In each of Exercises 37–46 find an equation of a circle satisfying the stated conditions.

37 Center $C(3, -2)$, radius 4

38 Center $C(-5, 2)$, radius 5

39 Center $C(\frac{1}{2}, -\frac{3}{2})$, radius 2

40 Center $C(\frac{1}{3}, 0)$, radius $\sqrt{3}$

41 Center at the origin, passing through $P(-3, 5)$

42 Center $C(-4, 6)$, passing through $P(1, 2)$

43 Center $C(-4, 2)$, tangent to the x-axis

44 Center $C(3, -5)$, tangent to the y-axis

45 Endpoints of a diameter $A(4, -3)$ and $B(-2, 7)$

46 Tangent to both axes, center in the first quadrant, radius 2

1.6 REVIEW

Concepts

Define or explain each of the following.

1 The Commutative Properties of real numbers

2 The Associative Properties

3 The Distributive Properties

4 Rational and irrational numbers

5 The integers

6 Prime number

7 Coordinate line

8 A number a is greater than a number b.

9 A number a is less than a number b.

10 The absolute value of a real number

11 The distance between points on a coordinate line

12 Variable

13 Domain of a variable

14 Solution of an equation

15 Root of an equation

16 Identity

17 Conditional equation

18 Properties of inequalities

19   Solution of an inequality

20   Equivalent inequalities

21   Open interval

22   Closed interval

23   Half-open interval

24   Infinite interval

25   Ordered pair

26   Rectangular coordinate system in a plane

27   Coordinate axes

28   Quadrants

29   The abscissa and ordinate of a point

30   The Distance Formula

31   The Midpoint Formula

32   The graph of an equation in x and y

33   Tests for symmetry

34   The standard equation of a circle

35   Unit circle

Exercises

1   Express each of the following as a rational number with least positive numerator.

(a) $\left(\dfrac{2}{3}\right)\left(-\dfrac{5}{8}\right)$ (b) $\dfrac{3}{4}+\dfrac{6}{5}$

(c) $\dfrac{5}{8}-\dfrac{6}{7}$ (d) $\dfrac{3}{4}\div\dfrac{6}{5}$

2   Replace the $\square$ with either $<$, $>$ or $=$.

(a) $-0.1\ \square\ -0.01$ (b) $\sqrt{9}\ \square\ -3$

(c) $1/6\ \square\ 0.166$

3   Express in terms of inequalities:

(a) x is negative.

(b) a is between 1/2 and 1/3.

(c)   The absolute value of x is not greater than 4.

4   Rewrite without using the absolute value symbol:

(a) $|-7|$ (b) $|-5|/(-5)$

(c) $|3^{-1}-2^{-1}|$

5   If points A, B, and C on a coordinate line have coordinates $-8, 4$, and -3, respectively, find the following:

(a) $d(A,C)$ (b) $d(C,A)$

(c) $d(B,C)$

Solve the inequalities in Exercises 6–18.

6   $10-7x<4+2x$

7   $-\dfrac{1}{2}<\dfrac{2x+3}{5}<\dfrac{3}{2}$

8   $(3x-1)(10x+4)\geq(6x-5)(5x-7)$

9   $\dfrac{6}{10x+3}<0$ 10 $|4x+7|<21$

11   $|16-3x|\geq 5$ 12 $2<|x-6|<4$

13   $10x^2+11x-6>0$

14   $x^2-3x\leq 10$ 15 $\dfrac{3}{2x+3}<\dfrac{1}{x-2}$

16   $\dfrac{x+1}{x^2-25}\leq 0$ 17 $x^3>x^2$

18   $(x^2-x)(x^2-5x+6)<0$

19   Plot the points $A(3,1)$, $B(-5,-3)$, and $C(4,-1)$ and prove that they are vertices of a right triangle. What is the area of the triangle?

20   Given the points $P(-5,9)$ and $Q(-8,-7)$,

(a)   find the midpoint of the segment PQ;

(b)   find a point T such that Q is the midpoint of PT.

21   Describe the set of all points (x,y) in a coordinate plane such that $xy<0$.

22   Find an equation of the circle with center $C(7,-4)$ and passing through the point $Q(-3,3)$.

Sketch the graphs of the equations in Exercises 23–30.

23   $2y+5x-8=0$ 24 $x=3y+4$

25   $x+5=0$ 26 $2y-7=0$

27   $y=\sqrt{1-x}$ 28 $y-x^2=1$

29 $y^2 - x = 1$

30 $(x + 2)^2 + (y - 8)^2 = 4$

33 $W = \{(x, y) : x^2 + y^2 < 1\}$

34 $W = \{(x, y) : x^2 + y^2 = 0\}$

In each of Exercises 31–34 sketch the graph of W.

31 $W = \{(x, y) : x > 0\}$ **32** $W = \{(x, y) : y > x\}$

2
FUNCTIONS

One of the most useful concepts in mathematics is that of function. *Indeed, it is safe to say that without the notion of function, little progress could be made in mathematics or in any area of science. We begin this chapter by introducing basic definitions associated with functions and their graphs. Next we make a detailed analysis of the* general linear function. *The chapter concludes with a discussion of* composite *and* inverse *functions.*

2.1   DEFINITION OF FUNCTION

The notion of **correspondence** is encountered frequently in everyday life. For example, to each book in a library there corresponds the number of pages in the book. As another example, to each human being there corresponds a birth date. To cite a third example, if the temperature of the air is recorded throughout a day, then at each instant of time there is a corresponding temperature.

The examples of correspondences we have given involve two sets X and Y. In our first example, X denotes the set of books in a library and Y the set of positive integers. For each book x in X there corresponds a positive integer y, namely the number of pages in the book. In the second example, if we let X denote the set of all human beings and Y the set of all possible dates, then to each person x in X there corresponds a birth date y.

We sometimes represent correspondences by diagrams of the type shown in Figure 2.1, where the sets X and Y are represented by points within regions in a plane. The curved arrow indicates that the element y of Y corresponds to the element x of X. We have pictured X and Y as different sets. However, X and Y may have elements in common. As a matter of fact, we often have $X = Y$.

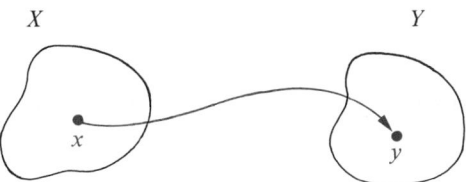

Figure 2.1

Our examples indicate that to each x in X there corresponds *one and only one* y in Y; that is y *is unique* for a given x. However, the same element of Y may correspond to different elements of X. For example, two different books may have the same number of pages, two different people may have the same birthday, and so on.

In most of our work X and Y will be sets of numbers. To illustrate, let X and Y both denote the set $\mathbb{R}$ of real numbers, and to each real number x let us assign its square x^2. Thus, to 3 we assign 9, to -5 we assign 25, and to $\sqrt{2}$ the number 2. This gives us a correspondence from $\mathbb{R}$ to $\mathbb{R}$.

All the examples of correspondences we have given are *functions*, as defined below.

Definition

> A **function** f from a set X to a set Y is a correspondence that assigns to each element x of X a unique element y of Y. The element y is called the **image** of x under f and is denoted by $f(x)$. The set X is called the **domain** of the function. The **range** of the function consists of all images of elements of X.

The notation $f(x)$ used for the element of Y which corresponds to x is usually read "f of x." We also call $f(x)$ the **value** of f at x. In terms of the pictorial representation given earlier, we may now sketch a diagram as in Figure 2.2. The curved arrows indicate that the elements $f(x), f(w), f(z)$, and $f(a)$ of Y correspond to the elements x, w, z, and a of X. Let us repeat the important fact that to each x in X there is assigned precisely one image $f(x)$ in Y; however, different elements of X, such as w and z in Figure 2.2, may have the same image in Y.

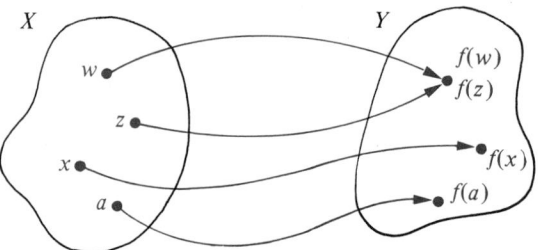

Figure 2.2

Beginning students are sometimes confused by the symbols f and $f(x)$. Remember that f is used to represent the function. It is neither in X nor in Y. However, $f(x)$ is an element of Y, namely the element which f assigns to x.

Two functions f and g from X to Y are said to be **equal**, written

$$f = g \quad \text{if and only if} \quad f(x) = g(x)$$

for every x in X.

Example 1 Let f be the function with domain $\mathbb{R}$ such that $f(x) = x^2$ for every x in $\mathbb{R}$. Find $f(-6), f(\sqrt{3})$, and $f(a)$, where a is any real number. What is the range of f?

Solution Values of f (or images under f) may be found by substituting for x in the equation $f(x) = x^2$. Thus

$$f(-6) = (-6)^2 = 36, \quad f(\sqrt{3}) = (\sqrt{3})^2 = 3 \quad \text{and} \quad f(a) = a^2.$$

If T denotes the range of f, then by definition T consists of all numbers of the form $f(a)$ where a is in $\mathbb{R}$. Hence T is the set of all squares a^2, where a is a real number. Since the square of any real number is nonnegative, T is contained in the set of all nonnegative real numbers. Moreover, every nonnegative real number c is an image under f, since $f(\sqrt{c}) = (\sqrt{c})^2 = c$. Hence the range of f is the set of all nonnegative real numbers. ■

If a function is defined as in the preceding example, the symbol used for the variable is immaterial; that is, expressions such as $f(x) = x^2$, $f(s) = s^2$, $f(t) = t^2$, etc., all define the same function f. This is true because if a is any number in the domain of f, then the same image a^2 is obtained no matter which expression is employed.

Example 2   Let X denote the set of nonnegative real numbers and let f be the function from X to $\mathbb{R}$ defined by $f(x) = \sqrt{x} + 1$ for every x in X. Find $f(4)$ and $f(\pi)$. If b and c are in X, find $f(b + c)$ and $f(b) + f(c)$.

Solution   As in Example 1, finding images under f is simply a matter of substituting the appropriate number for x in the expression for $f(x)$. Thus

$$f(4) = \sqrt{4} + 1 = 2 + 1 = 3$$
$$f(\pi) = \sqrt{\pi} + 1$$
$$f(b + c) = \sqrt{b + c} + 1$$
$$f(b) + f(c) = (\sqrt{b} + 1) + (\sqrt{c} + 1) = \sqrt{b} + \sqrt{c} + 2.$$ ■

In certain branches of mathematics, such as calculus, it is important to carry out manipulations of the type given in the next example.

Example 3   Suppose $f(x) = x^2 + 3x - 2$ for every real number x. If a and t are real numbers, and $t \neq 0$, find

$$\frac{f(a + t) - f(a)}{t}.$$

Solution   We have

$$f(a + t) = (a + t)^2 + 3(a + t) - 2$$
$$= (a^2 + 2at + t^2) + (3a + 3t) - 2$$

and $$f(a) = a^2 + 3a - 2.$$

Hence

$$\frac{f(a + t) - f(a)}{t} = \frac{(a^2 + 2at + t^2) + (3a + 3t) - 2 - (a^2 + 3a - 2)}{t}$$

$$= \frac{2at + t^2 + 3t}{t}$$

$$= 2a + t + 3.$$ ■

Occasionally one of the notations

$$X \xrightarrow{f} Y, \quad f:X \to Y \quad \text{or} \quad f:x \to f(x)$$

is used to signify that f is a function from X to Y. It is not unusual in this event to say f *maps* X *into* Y or f *maps* x *into* $f(x)$. If f is the function in Example 1, then f maps x into x^2 and we may write $f:x \to x^2$.

Many formulas which occur in mathematics and the sciences determine functions. As an illustration, the formula $A = \pi r^2$ for the area A of a circle of radius r assigns to each positive real number r a unique value of A. This determines a function f, where $f(r) = \pi r^2$, and we may write $A = f(r)$. The letter r, which represents an arbitrary number from the domain of f, is often called an **independent variable**. The letter A, which represents a number from the range of f, is called a **dependent variable**, since its value depends on the number assigned to r.

If two variables r and A are related in this manner, it is customary to use the phrase *A is a function of r*. To cite another example, if an automobile travels at a uniform rate of 50 miles per hour, then the distance d (miles) traveled in time t (hours) is given by $d = 50t$ and hence the distance d is a function of time t.

We have seen that different elements in the domain of a function may have the same image. If images are always different, then, as in the next definition, the function is called one-to-one.

Definition

> A function f from X to Y is a **one-to-one function** if, whenever $a \neq b$ in X, then $f(a) \neq f(b)$ in Y.

If f is one-to-one, then each $f(x)$ in the range is the image of *precisely one* x in X. The function illustrated in Figure 2.2 is not one-to-one since two different elements w and z of X have the same image in Y. A one-to-one function is often called a **one-to-one correspondence**. To illustrate, the association between real numbers and points on a coordinate line is an example of a one-to-one correspondence.

Example 4
   (a) If $f(x) = 3x + 2$, where x is real, prove that f is one-to-one.
   (b) If $g(x) = x^2 + 5$, where x is real, prove that g is not one-to-one.

Solution
   (a) If $a \neq b$, then $3a \neq 3b$ and hence $3a + 2 \neq 3b + 2$, or $f(a) \neq f(b)$. Hence f is one-to-one.

   (b) The function g is not one-to-one since different numbers in the domain may have the same image. For example, although $-1 \neq 1$, both $g(-1)$ and $g(1)$ are equal to 6. ∎

The next two definitions introduce terms which are used to describe certain types of functions.

Definition

> The **identity function** f on a set X is defined by $f(x) = x$ for every x in X.

Note that if f is the identity function on X, then every x in X is mapped into itself.

Definition

> A function f from X to Y is a **constant function** if there is some (fixed) element c in Y such that $f(x) = c$ for every x in X.

The diagram in Figure 2.3 illustrates the fact that if f is a constant function, then every arrow from X terminates at the same element in Y.

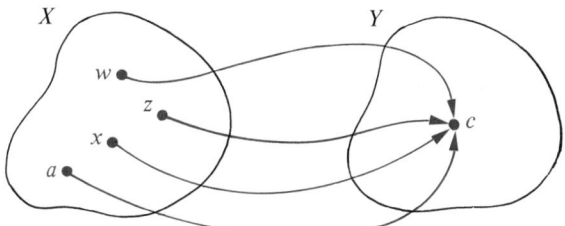

Figure 2.3

In the remainder of our work, unless otherwise specified, the phrase f *is a function* will mean that the domain and range are sets of real numbers. If a function is defined by means of some expression, as in Examples 1–3, and the domain X is not stated explicitly, then X is considered to be the totality of real numbers for which the given expression is meaningful. To illustrate, if $f(x) = \sqrt{x}/(x - 1)$, then the domain is assumed to be the set of nonnegative real numbers different from 1. If x is in the domain, we sometimes say that f **is defined at** x, or that $f(x)$ **exists**, or that f **takes on the value** $f(x)$. If a subset S is contained in the domain, we often say that f **is defined on** S. The terminology f **is undefined at** x means that x is not in the domain of f.

The concept of ordered pair can be used to obtain an alternate approach to functions. We first observe that a function f from X to Y determines the following set W of ordered pairs:

$$W = \{(x, f(x)) : x \text{ is in } X\}.$$

Thus W is the totality of ordered pairs for which the first number is in X and the second number is the image of the first. In Example 2, where $f(x) = \sqrt{x} + 1$, W consists of all pairs of the form $(x, \sqrt{x} + 1)$, where x is a nonnegative real number. It is important to note that for each x there is exactly one ordered pair (x, y) in W having x in the first position.

Conversely, if we begin with a set W of ordered pairs such that each x in X appears exactly once in the first position of an ordered pair, and numbers from Y appear in the second position, then W determines a function from X to Y. Specifically, for any x in X there is a unique pair (x, y) in W, and by letting y correspond to x, we obtain a function from X to Y.

It follows from the preceding discussion that the statement given below could also be used as a definition of function. We prefer, however, to think of it as an alternate approach to this concept.

**Alternate
Definition
of a Function**

> A function with domain X is a set W of ordered pairs such that for each x in X, there is exactly one ordered pair (x, y) in W having x in the first position.

In terms of the preceding definition, the function f of Example 1, where $f(x) = x^2$, is the set of all ordered pairs of the form (x, x^2). Similarly, the ordered pairs $(x, x^2 + 3x - 2)$ determine the function of Example 3, where we had $f(x) = x^2 + 3x - 2$.

EXERCISES 2.1

1 If $f(x) = 2x^2 - 3x + 4$, find $f(1), f(-1), f(0)$, and $f(2)$.

2 If $f(x) = x^3 + 5x^2 - 1$, find $f(2), f(-2), f(0)$, and $f(-1)$.

3 If $f(x) = \sqrt{x - 1} + 2x$, find $f(1), f(3), f(5)$, and $f(10)$.

4 If $f(x) = \dfrac{x}{x - 2}$, find $f(1), f(3), f(-2)$, and $f(0)$.

In Exercises 5–8 find each of the following, where a, b, and h are real numbers:

(a) $f(a)$ (b) $f(-a)$

(c) $-f(a)$ (d) $f(a + h)$

(e) $f(a) + f(h)$

(f) $\dfrac{f(a + h) - f(a)}{h}$ provided $h \neq 0$

5 $f(x) = 5x - 2$ **6** $f(x) = 3 - 4x$

7 $f(x) = 2x^2 - x + 3$ **8** $f(x) = x^3 - 2x$

In Exercises 9–12 find the following:

(a) $g(1/a)$ (b) $1/g(a)$ (c) $g(a^2)$

(d) $(g(a))^2$ (e) $g(\sqrt{a})$ (f) $\sqrt{g(a)}$

9 $g(x) = 3x^2$ **10** $g(x) = 3x - 8$

11 $g(x) = \dfrac{2x}{x^2 + 1}$ **12** $g(x) = \dfrac{x^2}{x + 1}$

In each of Exercises 13–20 find the largest subset of $\mathbb{R}$ that can serve as the domain of the function f.

13 $f(x) = \sqrt{3x - 5}$ **14** $f(x) = \sqrt{7 - 2x}$

15 $f(x) = \sqrt{4 - x^2}$ **16** $f(x) = \sqrt{x^2 - 9}$

17 $f(x) = \dfrac{x + 1}{x^3 - 9x}$

18 $f(x) = \dfrac{4x + 7}{6x^2 + 13x - 5}$

19 $f(x) = \dfrac{\sqrt{x}}{2x^2 - 11x + 12}$

20 $f(x) = \dfrac{x^3 - 1}{x^2 - 1}$

In each of Exercises 21–26 find the number that maps into 4. If $a > 0$, what number maps into a? Find the range of f.

21 $f(x) = 7x - 5$ **22** $f(x) = 3x$

23 $f(x) = \sqrt{x - 3}$ **24** $f(x) = 1/x$

25 $f(x) = x^3$ **26** $f(x) = \sqrt[3]{x - 4}$

In each of Exercises 27–34 determine if the function f is one-to-one.

27 $f(x) = 2x + 9$ **28** $f(x) = 1/(7x + 9)$

29 $f(x) = 5 - 3x^2$ **30** $f(x) = 2x^2 - x - 3$

31 $f(x) = \sqrt{x}$ **32** $f(x) = x^3$

33 $f(x) = |x|$ **34** $f(x) = 4$

A function f with domain X is termed (i) **even** if $f(-a) = f(a)$ for every a in X, or (ii) **odd** if $f(-a) = -f(a)$ for every a in X. In each of Exercises 35–44 determine whether f is even, odd, or neither even nor odd.

35 $f(x) = 3x^3 - 4x$ **36** $f(x) = 7x^4 - x^2 + 7$

37 $f(x) = 9 - 5x^2$ **38** $f(x) = 2x^5 - 4x^3$

39 $f(x) = 2$ **40** $f(x) = 2x^3 + x^2$

41 $f(x) = 2x^2 - 3x + 4$

42 $f(x) = \sqrt{x^2 + 1}$

43 $f(x) = \sqrt[3]{x^3 - 4}$ **44** $f(x) = |x| + 5$

45 If f is an even function, prove that the graph of f is symmetric with respect to the y-axis.

46 If f is an odd function, prove that the graph of f is symmetric with respect to the origin.

47 Find a formula which expresses the radius r of a circle as a function of its circumference C. If the circumference of *any* circle is increased by 12 inches, determine how much the radius increases.

48 Find a formula which expresses the volume of a cube as a function of its surface area. Find the volume if the surface area is 36 square inches.

49 An open box is to be made from a rectangular piece of cardboard having dimensions 20 inches by 30 inches by cutting out identical squares of area x^2 from each corner and turning up the sides. Express the volume V of the box as a function of x.

50 Find a formula which expresses the area A of an equilateral triangle as a function of the length s of a side.

51 Express the perimeter P of a square as a function of its area A.

52 Express the surface area S of a sphere as a function of its volume V.

53 A weather balloon is released at 1:00 P.M. and rises vertically at a rate of 2 meters per second. An observer is situated 100 meters from a point on the ground directly below the balloon. If t denotes the time (in seconds) after 1:00 P.M., express the distance d between the balloon and the observer as a function of t.

54 Two ships leave port at 9:00 A.M., one sailing south at a rate of 16 mph and the other west at a rate of 20 mph. If t denotes the time (in hours) after 9:00 A.M., express the distance d between the ships as a function of t.

55 A manufacturer sells a certain article to dealers at a rate of $20 each if less than 50 are ordered. If 50 or more are ordered (up to 600), the price per article is reduced at a rate of 2 cents times the number ordered. Let A denote the amount of money received when x articles are ordered. Express A as a function of x.

56 A travel agency offers a tour at a cost of $30 per person if not more than 60 take the tour. If more than 60 take the tour the cost is to be reduced 10 cents for each person in excess of 60. Let A denote the total cost if x people go on the tour. Express A as a function of x.

In each of Exercises 57–64 determine whether the set W of ordered pairs is a function in the sense of the alternate definition of function.

57 $W = \{(x, y): 2y = x^2 + 5\}$

58 $W = \{(x, y): x = 3y + 2\}$

59 $W = \{(x, y): x^2 + y^2 = 4\}$

60 $W = \{(x, y): y^2 - x^2 = 1\}$

61 $W = \{(x, y): y = 3\}$

62 $W = \{(x, y): x = y\}$

63 $W = \{(x, y): xy = 0\}$

64 $W = \{(x, y): x + y = 0\}$

2.2 GRAPHS OF FUNCTIONS

Graphs, or more precisely, *sketches of graphs*, are often used to describe the variation of physical quantities. For example, a scientist may use Figure 2.4 to indicate the temperature T of a certain solution at various times t during an experiment. The sketch shows that the temperature increased gradually from time $t = 0$ to time $t = 5$, was steady between $t = 5$ and $t = 8$, and then decreased rapidly from $t = 8$ to $t = 9$. A visual aid of this type reveals the behavior of T more clearly than a long table of numerical values.

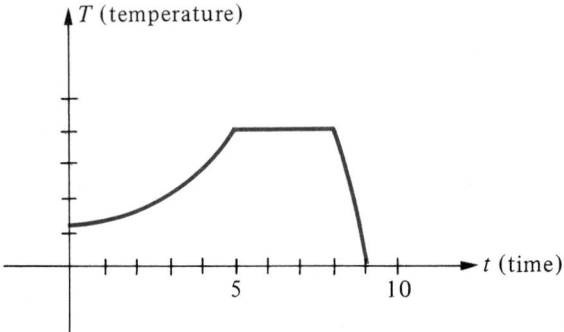

Figure 2.4

In like manner, Figure 2.5 could represent the variation in atmospheric pressure y as the altitude x varies. One can see at a glance the manner in which the pressure decreases as the altitude increases.

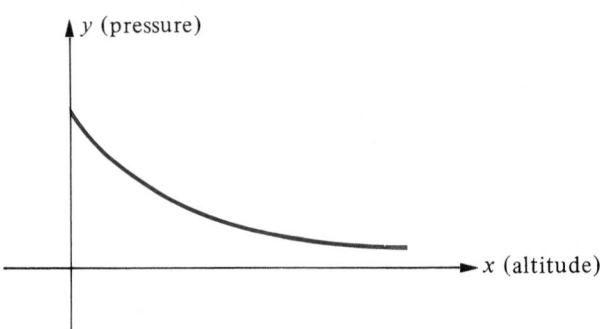

Figure 2.5

If f is a function, we often use a sketch of a graph to exhibit the behavior of $f(x)$ as x varies through the domain of f. Such geometric representations of functions are often more enlightening than algebraic descriptions of the types considered in the examples of the previous section. By definition, the **graph of a function** f is the set of all points $(x, f(x))$ in a coordinate plane, where x is in the domain of f. The phrase "sketch the graph" has the same meaning as in Chapter One. The graph of f can also be described as the set of all points $P(x, y)$ such that $y = f(x)$. Thus the graph of f is the same as the graph of the equation $y = f(x)$, and if $P(a, b)$ is on the graph of f, then the ordinate b is the functional value $f(a)$ as illustrated in Figure 2.6. It is important to note that since there is a unique $f(a)$ for each a in the domain, there is only *one* point on the graph with abscissa a. Thus every vertical line pierces the graph of a function in at most one point. Consequently, for graphs of functions it is impossible to obtain a sketch such as that shown in Figure 1.28, where some vertical lines intersect the graph in more than one point.

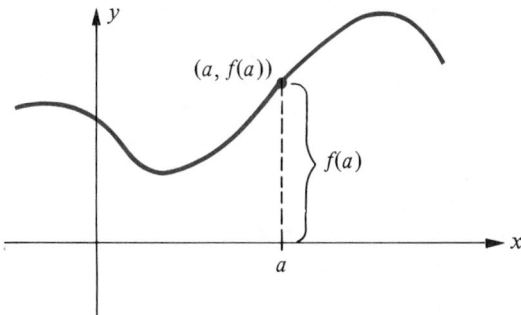

Figure 2.6

Example 1   Sketch the graph of f if $f(x) = 2x + 3$.

Solution   The graph consists of all points $(x, f(x))$, where $f(x) = 2x + 3$. The following table lists coordinates of several points on the graph, where $y = f(x)$.

x	-3	-2	-1	0	1	2
y	-3	-1	1	3	5	7

Plotting, it appears that the points lie on a straight line, and we sketch the graph as in Figure 2.7. (In the next section we shall *prove* that the graph is a straight line.) The graph of f is the same as the graph of the equation $y = 2x + 3$.

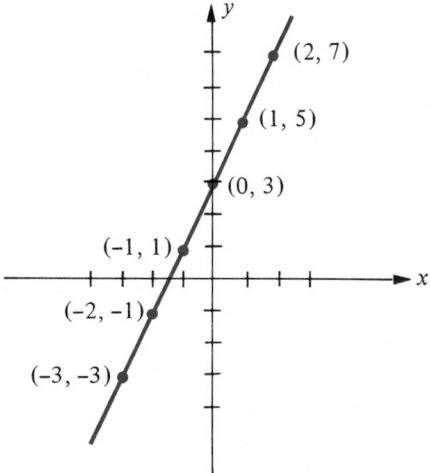

Figure 2.7. $f(x) = 2x + 3$

It is useful to determine the points at which the graph of a function f intersects the x-axis. The abscissas of these points are called the **x-intercepts** of the graph and are found by locating all points with zero ordinates, that is, all points $(x, f(x))$ such that $f(x) = 0$. In Example 1 the x-intercept is $-3/2$, since that number is the solution of $2x + 3 = 0$. A number a such that $f(a) = 0$ is also called a **zero of the function** f.

If the number 0 is in the domain of f, then $f(0)$ is called the **y-intercept** of the graph of f. It is the ordinate of the point at which the graph intersects the y-axis. The graph of a function can have at most one y-intercept. The y-intercept in Example 1 is 3.

If f is the function in Example 1 and if $x_1 < x_2$, then $2x_1 + 3 < 2x_2 + 3$; that is, $f(x_1) < f(x_2)$. This means that, as abscissas of points increase, ordinates also increase. The function f is then said to be *increasing*. If f is increasing, then the graph rises as x increases. For certain functions, we have $f(x_1) > f(x_2)$ whenever $x_1 < x_2$. In this case the graph of f falls as x increases and the function is called a *decreasing* function. In general we shall speak of functions which increase or decrease on certain subsets of their domains, as in the following definition. Functions which neither increase nor decrease are called *constant*. The following definition summarizes these remarks.

Definition

> If S is a subset of the domain of a function f, then
>
> (i) f is **increasing** on S if $f(x_1) < f(x_2)$ whenever $x_1 < x_2$ in S.
>
> (ii) f is **decreasing** on S if $f(x_1) > f(x_2)$ whenever $x_1 < x_2$ in S.
>
> (iii) f is **constant** on S if $f(x_1) = f(x_2)$ for every x_1, x_2 in S.

If we regard Figure 2.4 as the graph of a function f, then f is increasing on the closed interval $[0, 5]$, is constant on the closed interval $[5, 8]$, and is decreasing on $[8, 9]$. Figure 2.5 represents a function which is decreasing throughout its domain.

Example 2   Sketch the graph of f if $f(x) = x^2 - 3$.

Solution   We list coordinates $(x, f(x))$ of some points on the graph of f in tabular form, as shown below.

x	-3	-2	-1	0	1	2	3
$f(x)$	6	1	-2	-3	-2	1	6

The x-intercepts are the solutions of the equation $f(x) = 0$, that is, of $x^2 - 3 = 0$. These are $\pm\sqrt{3}$. The y-intercept is $f(0) = -3$. Plotting the points given by the table and using the x-intercepts leads to the sketch in Figure 2.8. Compare the graph of f with that of the equation $y = x^2$ sketched in Figure 1.26. We could have shortened our work somewhat by observing that since $(-x)^2 = x^2$, the graph of $y = x^2 - 3$ is symmetric with respect to the y-axis.

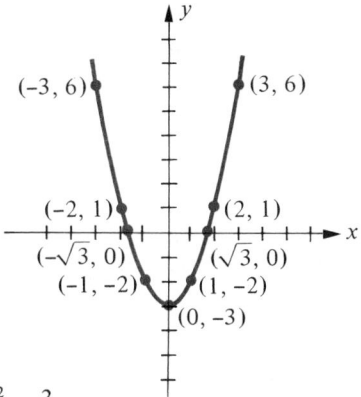

Figure 2.8. $f(x) = x^2 - 3$ ∎

In arriving at the sketch in Figure 2.8 we assumed that f is decreasing on the interval $(-\infty, 0]$ and increasing on the interval $[0, \infty)$. These facts can be readily proved. For example, consider any two positive real numbers x_1 and x_2 such that $x_1 < x_2$. Multiplying both sides of the inequality by x_1 and x_2, respectively, leads to

$$x_1^2 < x_1 x_2 \quad \text{and} \quad x_1 x_2 < x_2^2.$$

Applying properties of inequalities, together with the definition of f, gives us

$$x_1^2 < x_2^2$$
$$x_1^2 - 3 < x_2^2 - 3$$
$$f(x_1) < f(x_2).$$

This shows that f is increasing on the set of positive real numbers. If, in the previous discussion, $x_1 = 0$ and $x_1 < x_2$, then again $x_1^2 < x_2^2$ and $f(x_1) < f(x_2)$. Thus f is increasing on the interval $[0, \infty)$. It can be shown in similar fashion that f is decreasing on $(-\infty, 0]$. It follows that $f(x)$ takes on its least value at $x = 0$. This smallest value, -3, is called the **minimum value** of f. The corresponding point $(0, -3)$ is the lowest point on the graph. Clearly, $f(x)$ does not attain a **maximum value**, that is, a *largest* value. In the future we shall not justify our graphs in this way, but instead rely on sufficient point plotting to determine where functions are increasing or decreasing.

Example 3 Sketch the graph of f if $f(x) = |x|$.

Solution If $x \geq 0$, then $f(x) = x$ and we obtain the set of all points (x, x) on the graph of f. Some special cases are $(0, 0)$, $(1, 1)$, $(2, 2)$, $(3, 3)$, and $(4, 4)$. Negative values of x give rise to the following table:

x	-1	-2	-3	-4
$f(x)$	1	2	3	4

More generally, we obtain all points of the form $(-a, a)$ where $a > 0$. Plotting points leads to the sketch shown in Figure 2.9. As in Example 2, this function decreases on $(-\infty, 0]$ and increases on $[0, \infty)$. Note that the graph is symmetric with respect to the y-axis.

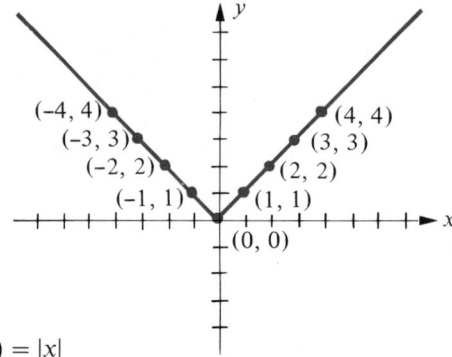

Figure 2.9. $f(x) = |x|$

■

Example 4   Sketch the graph of f if $f(x) = \sqrt{x - 1}$.

Solution   The domain of f does not include values of x such that $x - 1 < 0$, since $f(x)$ is not a real number in this case. Consequently, there are no points (x, y) on the graph with $x < 1$. Moreover, no part of the graph lies below the x-axis. (Why?) The following table lists some points $(x, f(x))$ on the graph.

x	1	2	3	4	5	6
$f(x)$	0	1	$\sqrt{2}$	$\sqrt{3}$	2	$\sqrt{5}$

Plotting points leads to the sketch shown in Figure 2.10. The function is increasing throughout its domain. The x-intercept is 1 and there is no y-intercept.

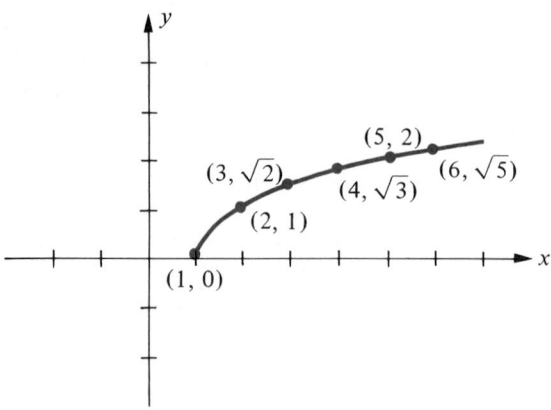

Figure 2.10. $f(x) = \sqrt{x - 1}$

■

Example 5   Sketch the graph of a constant function with domain $\mathbb{R}$.

Solution   By definition, f is a constant function if there is some (fixed) real number c such that $f(x) = c$ for every x. The graph of f consists of all points with coordinates (x, c), where x is any real number. In particular, this includes $(-1, c), (0, c),$ and $(2, c)$. Since all the ordinates equal c, the graph is a line parallel to the x-axis with y-intercept c. A sketch for the case $c > 0$ is shown in Figure 2.11.

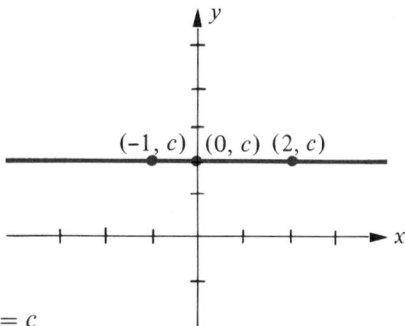

Figure 2.11. $f(x) = c$ ■

Comparing Figures 2.8 and 1.26, we see that the graph of $f(x) = x^2 - 3$ can be obtained by lowering the graph of $y = x^2$ a distance of 3 units, that is, by decreasing the ordinate of each point by 3. Generally, if we know the graph of *any* function f, and if c is a positive real number, then the graph of $y = f(x) + c$ can be obtained by raising the graph of $y = f(x)$ a distance c. To obtain the graph of $y = f(x) - c$ we lower the graph of f a distance c.

Example 6   If $f(x) = x^2 + c$, sketch the graph of f if $c = 4$, $c = 1$, and $c = -2$.

Solution   We shall sketch all three graphs on the same coordinate axes. The graph of $y = x^2$ was sketched in Figure 1.26, and for reference we have represented it by dashes in Figure 2.12. To find the graph of $f(x) = x^2 + 4$ we simply increase the

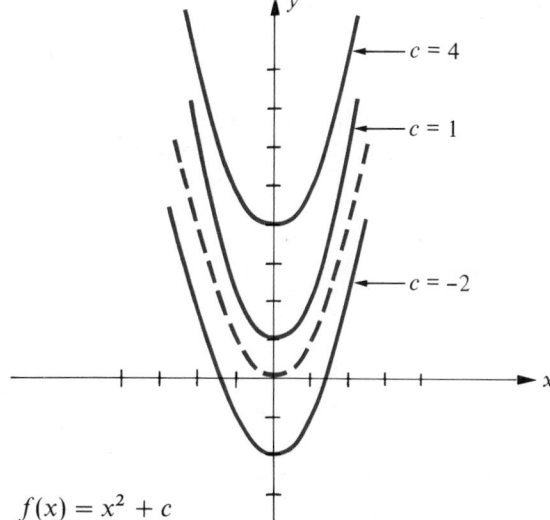

Figure 2.12. $f(x) = x^2 + c$

ordinate of each point on the graph of $y = x^2$ by 4, as shown in the figure. Similarly, for $c = 1$, we increase ordinates by 1, whereas, for $c = -2$, we decrease ordinates by 2. Note that each function is decreasing on the interval $(-\infty, 0]$ and increasing on $[0, \infty)$. Each graph is a parabola symmetric with respect to the y-axis. In order to get the correct position it is advisable to plot several points on each graph.   ■

To obtain the graph of $y = cf(x)$ we may *multiply* the ordinates of points on the graph of $y = f(x)$ by c. For example, if $y = 2f(x)$, we double ordinates, or if $y = \frac{1}{2}f(x)$, we multiply each ordinate by 1/2.

Example 7   If $f(x) = x^2$, sketch the graphs of $y = cf(x)$ if $c = 4$, $c = 1/4$, and $c = -1$.

Solution   We wish to sketch the graphs of the equations

$$y = 4x^2, \quad y = \tfrac{1}{4}x^2, \quad \text{and} \quad y = -x^2.$$

To obtain the graph of $y = 4x^2$ we could refer to the graph of $y = x^2$ in Figure 1.26 and multiply the ordinate of each point by 4. This gives us a narrower parabola which is sharper at the vertex, as illustrated in (i) of Figure 2.13. In order to get the correct position, several points such as $(0, 0)$, $(1/2, 1)$, and $(1, 4)$ should be plotted.

Similarly, for the graph of $y = (1/4)x^2$, we multiply ordinates of points on the graph of $y = x^2$ by 1/4. This gives us a wider parabola which is flatter at the vertex, as shown in (ii) of Figure 2.13.

Finally, for $y = -x^2$, we multiply ordinates in Figure 1.26 by -1. This amounts to reflecting that graph through the x-axis, as shown in (iii) of Figure 2.13.

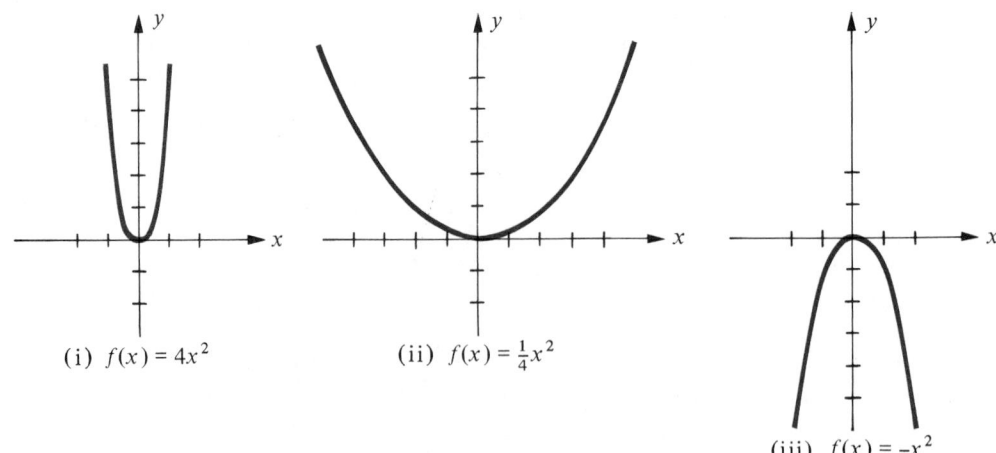

(i) $f(x) = 4x^2$ (ii) $f(x) = \tfrac{1}{4}x^2$ (iii) $f(x) = -x^2$

Figure 2.13   ■

Sometimes functions are described in terms of several expressions, as in the next example.

Example 8 Sketch the graph of the function f which is defined as follows:

$$f(x) = \begin{cases} 2x + 3 & \text{if } x < 0 \\ x^2 & \text{if } 0 \le x < 2 \\ 1 & \text{if } x \ge 2. \end{cases}$$

Solution   If $x < 0$, then $f(x) = 2x + 3$. This means that when x is negative, the expression $2x + 3$ should be used to find functional values. Consequently, if $x < 0$, then the graph of f coincides with the graph in Figure 2.7 and we sketch that portion of the graph to the left of the y-axis, as indicated in Figure 2.14.

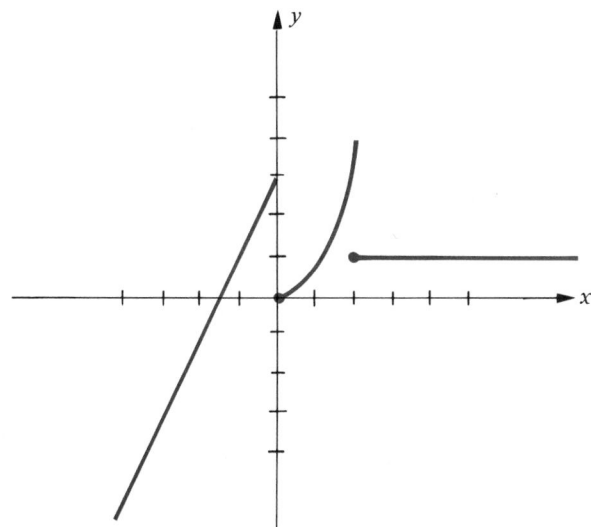

Figure 2.14

If $0 \le x < 2$, we use x^2 to find functional values of f, and therefore this part of the graph of f coincides with the graph of the equation $y = x^2$. We then sketch the part of the graph of f between $x = 0$ and $x = 2$, as indicated in Figure 2.14.

Finally, if $x \ge 2$, the graph of f coincides with the constant function having values equal to 1. That part of f is the horizontal half-line illustrated in Figure 2.14. ■

EXERCISES 2.2

In Exercises 1–20, sketch the graph of f and state where f is increasing, decreasing, or constant.

1   $f(x) = 5x$

2   $f(x) = -3x$

3   $f(x) = -4x + 2$

4   $f(x) = 4x - 2$

5   $f(x) = 3 - x^2$

6   $f(x) = 4x^2 + 1$

7   $f(x) = 2x^2 - 4$

8   $f(x) = \frac{1}{100}x^2$

9   $f(x) = \sqrt{x + 4}$

10   $f(x) = \sqrt{4 - x}$

11   $f(x) = \sqrt{x} + 2$

12   $f(x) = 2 - \sqrt{x}$

13   $f(x) = 4/x$

14   $f(x) = 1/x^2$

15   $f(x) = |x - 2|$

16   $f(x) = |x + 2|$

17   $f(x) = |x| - 2$

18   $f(x) = |x| + 2$

19   $f(x) = \dfrac{x}{|x|}$

20   $f(x) = x + |x|$

In Exercises 21–26, sketch the graph of the function f for the stated values of c.

21 $f(x) = 3x + c; \quad c = 0, c = 2, c = -1$

22 $f(x) = -2x + c; \quad c = 0, c = 1, c = -3$

23 $f(x) = x^3 + c; \quad c = 0, c = 1, c = -2$

24 $f(x) = -x^3 + c; \quad c = 0, c = 2, c = -1$

25 $f(x) = cx + 2; \quad c = 1, c = 2, c = -1/2$

26 $f(x) = cx - 1; \quad c = 3, c = 1/2, c = -2$

In each of Exercises 27–32 sketch the graph of f.

27 $f(x) = \begin{cases} 2 & \text{if } x < 0 \\ -1 & \text{if } x \geq 0 \end{cases}$

28 $f(x) = \begin{cases} -1 & \text{if } x \text{ is an integer} \\ 1 & \text{if } x \text{ is not an integer} \end{cases}$

29 $f(x) = \begin{cases} 3 & \text{if } x < -3 \\ -x & \text{if } -3 \leq x \leq 3 \\ -3 & \text{if } x > 3 \end{cases}$

30 $f(x) = \begin{cases} x & \text{if } x < 0 \\ -2 & \text{if } 0 \leq x < 1 \\ x^2 & \text{if } x \geq 1 \end{cases}$

31 $f(x) = \begin{cases} x^2 & \text{if } x \leq -1 \\ x^3 & \text{if } |x| < 1 \\ 2x & \text{if } x \geq 1 \end{cases}$

32 $f(x) = \begin{cases} x & \text{if } x \leq 1 \\ -x^2 & \text{if } 1 < x < 2 \\ x & \text{if } x \geq 2 \end{cases}$

33 If x is any real number, then there exist consecutive integers n and $n + 1$ such that $n \leq x < n + 1$. Let f be the function defined as follows: if x is a real number and $n \leq x < n + 1$, then $f(x) = n$. Sketch the graph of f. (The function f is called the **greatest integer function**.)

34 Explain why the graph of the equation $x^2 + y^2 = 1$ is not the graph of a function.

35 Sketch the graph of the identity function with domain $\mathbb{R}$.

36 Define a function f whose graph is the upper half of a circle with center at the origin and radius 1.

37 Define a function f whose graph is the lower half of a circle with center at the origin and radius 1.

38 If a function f is increasing throughout its domain, prove that f is one-to-one.

39 If a function f is decreasing throughout its domain, prove that f is one-to-one.

40 Prove that a function f is one-to-one if and only if every horizontal line pierces the graph of f in at most one point.

2.3 LINEAR FUNCTIONS

The following type of function is very important in mathematics and its applications.

Definition

> A function f is a **linear function** if
> $$f(x) = ax + b$$
> where a and b are real numbers and $a \neq 0$.

The reason for the term "linear" is that the graph of f is a line, as we shall see later in this section.

Let us begin by introducing several fundamental concepts pertaining to lines. All lines referred to are considered to be in some fixed coordinate plane.

Definition

If l is a line which is not parallel to the y-axis, and if $P_1(x_1, y_1)$ and $P_2(x_2, y_2)$ are distinct points on l, then the **slope m** of l is given by

$$m = \frac{y_2 - y_1}{x_2 - x_1}.$$

If l is parallel to the y-axis, then the slope is not defined.

The numerator $y_2 - y_1$ in the formula for m is sometimes called the **rise** from P_1 to P_2. It measures the vertical change in direction in proceeding from P_1 to P_2 and may be positive, negative, or zero. The denominator $x_2 - x_1$ is called the **run** from P_1 to P_2. It measures the amount of horizontal change in going from P_1 to P_2. The run may be positive or negative, but is never zero because l is not parallel to the y-axis. Using this terminology, we could write

$$\text{slope of } l = \frac{\text{rise from } P_1 \text{ to } P_2}{\text{run from } P_1 \text{ to } P_2}.$$

In finding the slope of a line it is immaterial which point is labeled P_1 and which is labeled P_2, since

$$\frac{y_2 - y_1}{x_2 - x_1} = \frac{y_1 - y_2}{x_1 - x_2}.$$

Consequently, we may as well assume that the points are labeled so that $x_1 < x_2$, as in Figure 2.15. In this event, $x_2 - x_1 > 0$, and hence the slope is positive, negative, or zero, depending on whether $y_2 > y_1, y_2 < y_1$, or $y_2 = y_1$. The slope of the line shown in (i) of Figure 2.15 is positive, whereas the slope of the line shown in (ii) of the figure is negative. The slope is zero if and only if the line is horizontal.

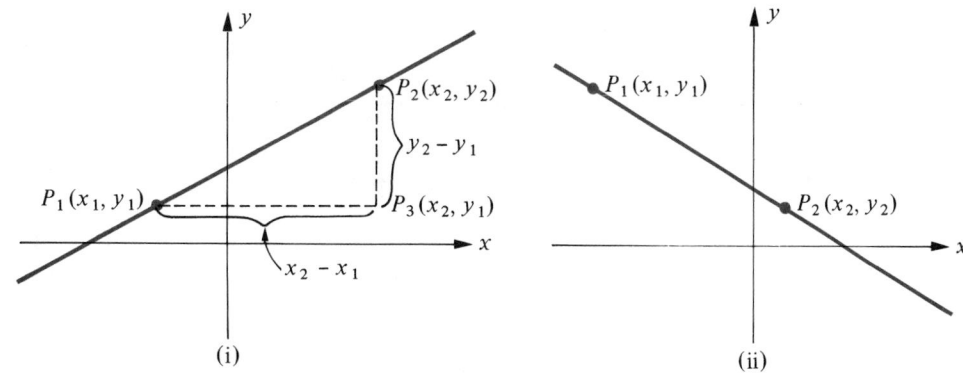

(i)   (ii)

Figure 2.15

Example 1 Sketch the lines through the following pairs of points and find their slopes.
(a) $A(-1, 4)$ and $B(3, 2)$
(b) $A(2, 5)$ and $B(-2, -1)$
(c) $A(4, 3)$ and $B(-2, 3)$
(d) $A(4, -1)$ and $B(4, 4)$.

Solution   The lines are sketched in Figure 2.16. Using the definition of slope gives us

(a) $m = \dfrac{2 - 4}{3 - (-1)} = \dfrac{-2}{4} = -\dfrac{1}{2}$

(b) $m = \dfrac{5 - (-1)}{2 - (-2)} = \dfrac{6}{4} = \dfrac{3}{2}$

(c) $m = \dfrac{3 - 3}{-2 - 4} = \dfrac{0}{-6} = 0$

(d) The slope is undefined since the line is vertical. This is also seen by noting that if the formula for m is used, the denominator is zero.

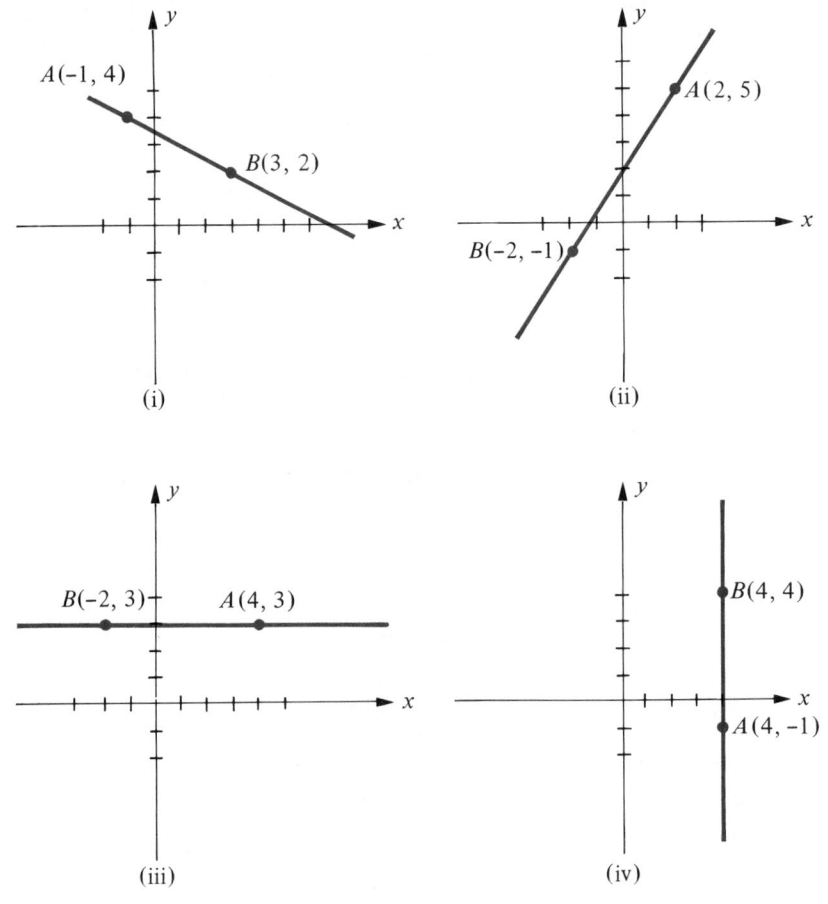

Figure 2.16   ■

It is important to note that the definition of slope is independent of the two points that are chosen on l, for if other points $P_1'(x_1', y_1')$ and $P_2'(x_2', y_2')$ are used, then, as in Figure 2.17, the triangle with vertices P_1', P_2', and $P_3'(x_2', y_1')$ is similar to the triangle with vertices P_1, P_2, $P_3(x_2, y_1)$. Since the ratios of corresponding sides are equal, it follows that

$$\frac{y_2 - y_1}{x_2 - x_1} = \frac{y_2' - y_1'}{x_2' - x_1'}.$$

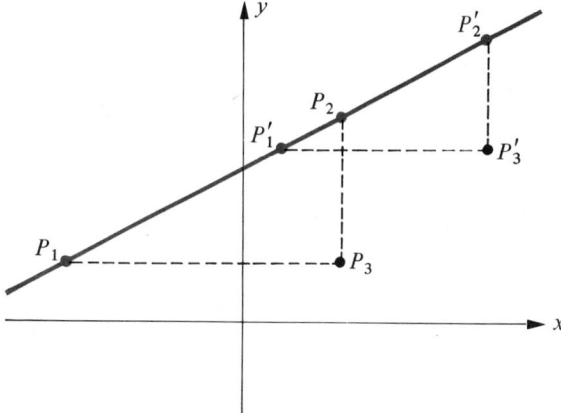

Figure 2.17

Example 2   Construct a line through $P(2, 1)$ that has slope (a) $5/3$;   (b) $-5/3$.

Solution   If the slope of a line is a/b, where b is positive, then for every change of b units in the horizontal direction, the line rises or falls $|a|$ units, depending on whether a is positive or negative, respectively. If $P(2, 1)$ is on the line and $m = 5/3$, we can obtain another point on the line by starting at P and moving 3 units to the right and 5 units upward. This gives us the point $Q(5, 6)$ and the line is determined (see (i) of Figure 2.18). Similarly, if $m = -5/3$, we move 3 units to the right and 5 units downward, obtaining $Q(5, -4)$ as in (ii) of Figure 2.18.

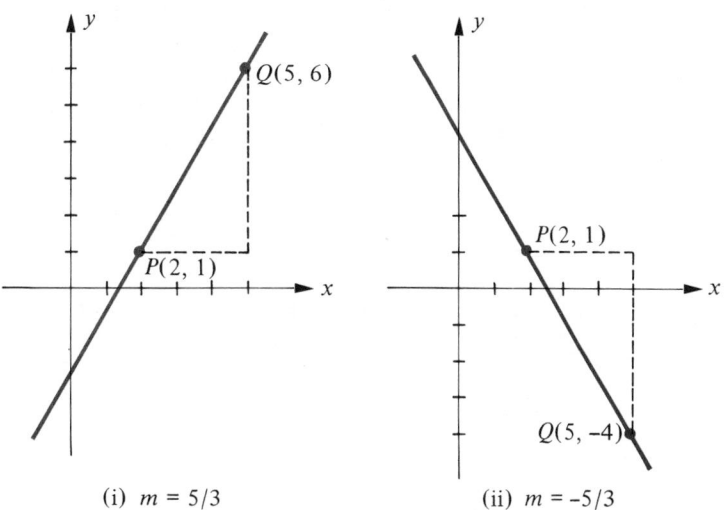

(i) $m = 5/3$ (ii) $m = -5/3$

Figure 2.18

The equation $y = b$, where b is a real number, may be considered as an equation in two variables x and y, since we can write it in the form

$$(0x) + y = b.$$

Some typical solutions of the equation are $(-2, b)$, $(1, b)$, and $(3, b)$. Evidently, all solutions of the equation consist of pairs of the form (x, b), where x may have any value and b is fixed. It follows that the graph of $y = b$ is a line parallel to the x-axis with y-intercept b. This was to be expected since the graph is the same as the graph of the constant function f, where $f(x) = b$ (see Figure 2.11). Conversely, every horizontal line is the graph of an equation of the form $y = b$. A similar argument can be used to show that the graph of the equation $x = a$ is a line parallel to the y-axis with x-intercept a. The graphs of these lines are illustrated in Figure 2.19.

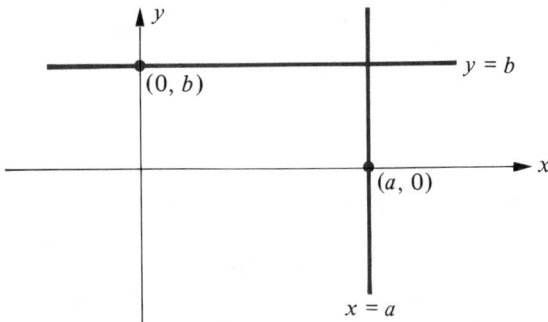

Figure 2.19

Let us now find an equation of a line l through a point $P_1(x_1, y_1)$ with slope m (only one such line exists). If $P(x, y)$ is any point with $x \neq x_1$ (see Figure 2.20), then P is on l if and only if the slope of the line through P_1 and P is m, that is, if and only if

$$\frac{y - y_1}{x - x_1} = m.$$

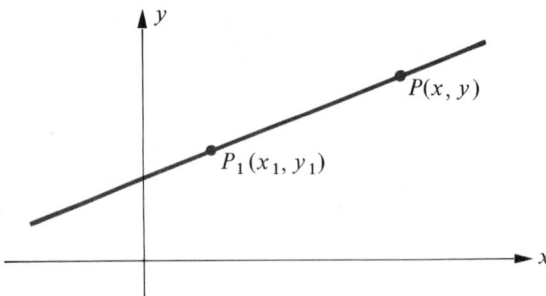

Figure 2.20

This equation may be written in the form

$$y - y_1 = m(x - x_1).$$

Note that (x_1, y_1) is also a solution of the latter equation and hence the points on l are precisely the points which correspond to the solutions. This equation for l is referred to as the **point-slope form**. Our discussion may be summarized as follows:

Point-Slope Form for the Equation of a Line

> An equation for the line through the point $P(x_1, y_1)$ with slope m is
>
> $$y - y_1 = m(x - x_1).$$

Example 3   Find an equation of the line through the points $A(1, 7)$ and $B(-3, 2)$.

Solution   The slope m of the line is

$$m = \frac{7 - 2}{1 - (-3)} = \frac{5}{4}.$$

Using the coordinates of A in the Point-Slope Form gives us

$$y - 7 = \frac{5}{4}(x - 1)$$

which is equivalent to

$$4y - 28 = 5x - 5 \quad \text{or} \quad 5x - 4y + 23 = 0.$$

The same equation would have been obtained if the coordinates of point B had been substituted in the Point-Slope Form.   ■

The Point-Slope Form may be rewritten as $y = mx - mx_1 + y_1$, which is of the form

$$y = mx + b$$

where $b = -mx_1 + y_1$. The real number b is the y-intercept of the graph, as may be seen by setting $x = 0$. Since the equation $y = mx + b$ displays the slope m and y-intercept b of l, it is called the **slope-intercept form** for the equation of a line. Conversely, given an equation of that form, we may change it to

$$y - b = m(x - 0).$$

Comparing with the Point-Slope Form, we see that the graph is a line with slope m and passing through the point $(0, b)$. This gives us the next result.

Slope-Intercept Form for the Equation of a Line

> The graph of the equation $y = mx + b$ is a line having slope m and y-intercept b.

The work we have done shows that every line is the graph of an equation of the form

$$ax + by + c = 0$$

where a, b, and c are real numbers, and a and b are not both zero. We call such an equation a **linear equation** in x and y. Let us show, conversely, that the graph of $ax + by + c = 0$, where a and b are not both zero is always a line. On the one hand, if $b \neq 0$, we may solve for y, obtaining,

$$y = (-a/b)x + (-c/b)$$

which, by the Slope-Intercept Form, is an equation of a line with slope $-a/b$ and y-intercept $-c/b$. On the other hand, if $b = 0$ but $a \neq 0$, then we may solve for x, obtaining $x = -c/a$, which is the equation of a line parallel to the y-axis with x-intercept $-c/a$. This establishes the following important theorem.

Theorem

> The graph of a linear equation $ax + by + c = 0$ is a line and, conversely, every line is the graph of a linear equation.

For simplicity, we shall often use the terminology *the line $ax + by + c = 0$* instead of the more accurate phrase *the line with equation $ax + by + c = 0$.*

Example 4   Sketch the graph of $2x - 5y = 8$.

Solution   We know from the previous theorem that the graph is a line and hence it is sufficient to find two points on the graph. Let us find the x- and y-intercepts. Substituting $y = 0$ in the given equation, we obtain the x-intercept 4. Substituting $x = 0$, we see that the y-intercept is $-8/5$. This leads to the graph in Figure 2.21.

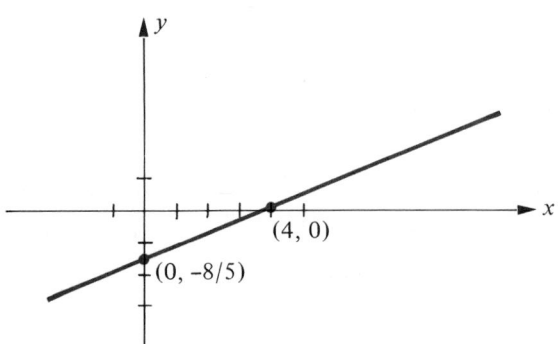

Figure 2.21. $2x - 5y = 8$

Another method of solution is to express the given equation in slope-intercept form. To do this we begin by isolating the term involving y on one side of the equals sign, obtaining

$$5y = 2x - 8.$$

Next, dividing both sides by 5 gives us

$$y = \frac{2}{5}x + \left(\frac{-8}{5}\right)$$

which is of the form $y = mx + b$. Hence the slope is $m = 2/5$ and the y-intercept is $b = -8/5$. We may then sketch a line through the point $(0, -8/5)$ with slope $2/5$. ∎

It can be shown geometrically that *two nonvertical lines are parallel if and only if they have the same slope.* We shall use this fact in the next example.

Example 5 Find an equation of a line through the point $(5, -7)$ which is parallel to the line $6x + 3y - 4 = 0$.

Solution Let us express the given equation in slope-intercept form. We begin by writing

$$3y = -6x + 4$$

and then divide both sides by 3, obtaining

$$y = -2x + \frac{4}{3}.$$

The last equation is in Slope-Intercept Form with $m = -2$, and hence the slope is -2. From the preceding remark, the required line also has slope -2. Applying the Point-Slope Form gives us

$$y + 7 = -2(x - 5)$$

or equivalently,

$$y + 7 = -2x + 10 \quad \text{or} \quad 2x + y - 3 = 0. \qquad ∎$$

The results obtained in this section may be used to prove the following important theorem.

Theorem

| The graph of a linear function is a line. |

Proof If f is linear, then we may write

$$f(x) = mx + b$$

where $m \neq 0$. Since the graph of f is the same as the graph of the equation $y = mx + b$, the conclusion follows at once from the preceding discussion.

EXERCISES 2.3

In each of Exercises 1–6 plot the points A and B and find the slope of the line through A and B.

1 $A(-3, 8), B(-2, 16)$ **2** $A(7, -1), B(-4, 9)$

3 $A(\frac{1}{3}, \frac{2}{3}), B(-\frac{3}{2}, \frac{1}{2})$ **4** $A(-5, -2), B(6, 4)$

5 $A(5, 0), B(-3, 2)$ **6** $A(0, -4), B(8, 3)$

7 Use slopes to show that $A(-4, -1)$, $B(2, 3)$, $C(1, -3)$, and $D(7, 1)$ are vertices of a parallelogram.

8 Use slopes to show that $A(-3, -9)$, $B(1, -1)$, and $C(4, 5)$ lie on a straight line.

In each of Exercises 9–22 find an equation for the line satisfying the given conditions.

9 Through $A(5, -2)$; slope $1/3$

10 Through $B(-3, 7)$; slope -2

11 Through $A(-8, -3)$ and $B(2, -5)$

12 Through $P(4, -1)$ and $Q(-6, 3)$

13 Slope -4, y-intercept 7

14 Slope $3/4$, x-intercept 5

15 x-intercept -3, y-intercept 7

16 Through $A(-7, -8)$; y-intercept 0

17 Through $A(7, -3)$, parallel to the (a) x-axis;   (b) y-axis

18 Through $A(-8, 2)$, perpendicular to the (a) y-axis;   (b) x-axis

19 Bisecting the second and fourth quadrants

20 Coinciding with the y-axis

21 Through $A(-4, 10)$, parallel to the line through $B(0, 5)$ and $C(-8, -8)$

22 Through $P(3/2, -1/4)$, parallel to the line with equation $2x + 4y = 5$

23 Find equations for the medians of the triangle with vertices $A(-4, -3)$, $B(2, 3)$, and $C(5, -1)$.

24 If $A(x_1, y_1)$, $B(x_2, y_2)$, $C(x_3, y_3)$, and $D(x_4, y_4)$ are vertices of an arbitrary quadrilateral, prove that the line segments joining the midpoints of adjacent sides form a parallelogram.

In each of Exercises 25–34 use the slope-intercept form to find the slope and y-intercept of the line with the given equation and sketch the graph.

25 $2x - 5y + 10 = 0$ **26** $2y - 3x = 4$

27 $3x + 5y = 0$ **28** $6x = 1 - 2y$

29 $y + 2 = 0$ **30** $3y = 8$

31 $5x = 20 - 6y$ **32** $9x - 4y = 0$

33 $y = 0$ **34** $x = (2/3)y + 4$

35 Find a real number k such that the point $P(-1, 2)$ is on the line $kx + 2y - 7 = 0$.

36 Find a real number k such that the line $5x + ky - 3 = 0$ has y-intercept -5.

37 Sketch the graph of f if $f(x) = -5x - 2$.

38 Sketch the graph of f if $f(x) = 4x - 3$.

39 Given $f(x) = ax + 3$, sketch the graph of f if:
(a) $a = 0$ (b) $a = 2$
(c) $a = 5$ (d) $a = -3$

40 Given $f(x) = -5x + b$, sketch the graph of f if:
(a) $b = 0$ (b) $b = 2$
(c) $b = 5$ (d) $b = -3$

41 If $f(x) = 3x + 7$ and $g(x) = -2x + 5$, find the abscissa of a point which is on the graphs of both f and g.

42 If the graph of a linear function f contains the points $(2, 7)$ and $(0, 1)$, find $f(x)$.

43 If f is a linear function such that $f(0) = -2$ and $f(-1) = 4$, find $f(x)$.

44 If the graph of a linear function f is parallel to the line $3x + 5y - 2 = 0$ and has x-intercept -4, find $f(x)$.

45 If $a > 0$, prove that the linear function f defined by $f(x) = ax + b$ is an increasing function throughout its domain. If $a < 0$, prove that f is decreasing throughout its domain.

46 Prove that the graph of the equation $ax + by = 0$, where a and b are not both zero, is a line passing through the origin.

47 If a line l has nonzero x- and y-intercepts a and b, respectively, prove that an equation for l is

$$\frac{x}{a} + \frac{y}{b} = 1.$$

(This is called the **intercept form** for the equation of a line.) Express the equation $4x - 2y = 6$ in intercept form.

48 Prove that an equation of the line through $P_1(x_1, y_1)$ and $P_2(x_2, y_2)$ is

$$(y - y_1)(x_2 - x_1) = (y_2 - y_1)(x - x_1).$$

(This is called the **two-point form** for the equation of a line.) Use the two-point form to find an equation of the line through $A(7, -1)$ and $B(4, 6)$.

49 Find all values of r such that the slope of the line through the points $(r, 4)$ and $(1, 3 - 2r)$ is less than 5.

50 Find all values of t such that the slope of the line through $(t, 3t + 1)$ and $(1 - 2t, t)$ is greater than 4.

2.4   COMPOSITE AND INVERSE FUNCTIONS

We shall now describe an important method of using two functions f and g to obtain a third function. Suppose X, Y, and Z are sets of real numbers, and let f be a function from X to Y and g a function from Y to Z. In terms of the arrow notation introduced in Section 2.1 we have

$$X \xrightarrow{f} Y \xrightarrow{g} Z$$

that is, f maps X into Y and g maps Y into Z. A function from X to Z may be defined in a natural way. For every x in X, the number $f(x)$ is in Y. Since the domain of g is Y, we may then find the image of $f(x)$ under g. Of course, this element of Z is written as $g(f(x))$. By associating $g(f(x))$ with x, we obtain a function from X to Z called the *composite function* of g by f. This is illustrated pictorially in Figure 2.22, where the dashes indicate the correspondence we have defined from X to Z. We sometimes use an operational symbol $\circ$ and denote the latter function by $g \circ f$. The following definition summarizes our discussion.

Definition

> If f is a function from X to Y and g is a function from Y to Z, then the **composite function** $g \circ f$ is the function from X to Z defined by
>
> $$(g \circ f)(x) = g(f(x))$$
>
> for every x in X.

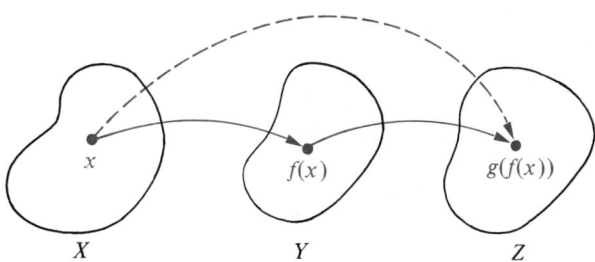

X Y Z

Figure 2.22

Example 1   If $f(x) = x^3$ and $g(x) = 5x^2 + 2x + 1$, find $(g \circ f)(x)$.

Solution   By definition,

$$(g \circ f)(x) = g(f(x)) = g(x^3).$$

Since $g(x^3)$ means that x^3 should be substituted for x in the expression for $g(x)$, we have

$$g(x^3) = 5(x^3)^2 + 2(x^3) + 1.$$

Consequently, $(g \circ f)(x) = 5x^6 + 2x^3 + 1.$ ∎

In applying the definition of $g \circ f$, it is not essential that the domain of g be all of Y, but merely that it *contain* the range of f. In certain cases we may wish to restrict x to some subset of X so that $f(x)$ is in the domain of g as illustrated in the next example.

Example 2   If the functions f and g are defined by $f(x) = x - 2$ and $g(x) = 5x + \sqrt{x}$, find $(g \circ f)(x)$.

Solution   Formal substitutions give us the following:

$$
\begin{aligned}
(g \circ f)(x) &= g(f(x)) && \text{(definition of } g \circ f) \\
&= g(x - 2) && \text{(definition of } f) \\
&= 5(x - 2) + \sqrt{x - 2} && \text{(definition of } g) \\
&= 5x - 10 + \sqrt{x - 2} && \text{(simplifying).}
\end{aligned}
$$

The domain X of f is the set of all real numbers; however, the last equality implies that $(g \circ f)(x)$ is a real number only if $x \geq 2$. Thus, when working with the composite function $g \circ f$, we must restrict x to the interval $[2, \infty)$.   ∎

Given f and g, it may also be possible to find $f(g(x))$. In this case we first obtain the image of x under g and then apply f to $g(x)$. This gives us a function called the *composite function* of f by g and denoted by $f \circ g$. Thus, by definition,

$$\boxed{(f \circ g)(x) = f(g(x)).}$$

Example 3   If $f(x) = x^2 - 1$ and $g(x) = 3x + 5$, find $(f \circ g)(x)$ and $(g \circ f)(x)$.

Solution   We have the following identities:

$$
\begin{aligned}
(f \circ g)(x) &= f(g(x)) && \text{(definition of } f \circ g) \\
&= f(3x + 5) && \text{(definition of } g) \\
&= (3x + 5)^2 - 1 && \text{(definition of } f) \\
&= 9x^2 + 30x + 24 && \text{(simplifying).}
\end{aligned}
$$

Similarly,

$$(g \circ f) = g(f(x)) \qquad \text{(definition of } g \circ f)$$
$$= g(x^2 - 1) \qquad \text{(definition of } f)$$
$$= 3(x^2 - 1) + 5 \qquad \text{(definition of } g)$$
$$= 3x^2 + 2 \qquad \text{(simplifying).} \qquad \blacksquare$$

We see from Example 3 that $f(g(x))$ and $g(f(x))$ are not always the same, that is, $f \circ g \neq g \circ f$. In certain cases it may happen that equality *does* occur. Of major importance is the case in which $f(g(x))$ and $g(f(x))$ are not only identical, but both are equal to x. Needless to say, f and g must be very special functions in order for this to happen. In the following discussion we indicate the manner in which they will be restricted.

Suppose f is a *one-to-one function* with domain X and range Y. As mentioned in Section 2.1 this implies that each element of Y is the image of precisely one element of X. Another way of phrasing this is to say that *each element of Y can be written in one and only one way in the form $f(x)$, where x is in X.* We may then define a function g from Y to X by demanding that

$$g(f(x)) = x \quad \text{for every } x \text{ in } X.$$

This amounts to *reversing* the correspondence given by f. If f is represented geometrically by drawing arrows as in (i) of Figure 2.23, then g can be represented by simply *reversing* these arrows as illustrated in (ii) of the figure. It follows that g is a one-to-one function with domain Y and range X. (See Exercise 38.)

As illustrated in Figure 2.23, if $f(x) = y$, then $x = g(y)$. This means that

$$f(g(y)) = y \quad \text{for every } y \text{ in } Y.$$

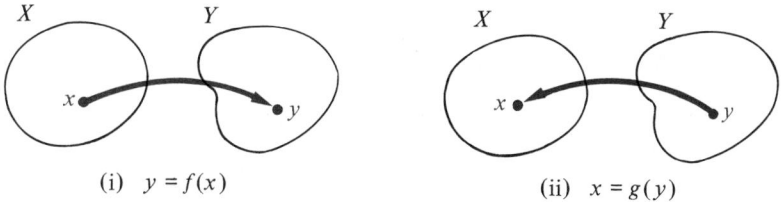

(i) $y = f(x)$ (ii) $x = g(y)$

Figure 2.23

Since the notation used for the variable is immaterial, we may write

$$f(g(x)) = x \quad \text{for every } x \text{ in } Y.$$

The functions f and g are called *inverse functions* of one another, according to the following definition.

Definition

> If f is a one-to-one function with domain X and range Y, then a function g with domain Y and range X is called the **inverse function of** f if
>
> $$f(g(x)) = x \quad \text{for every } x \text{ in } Y \text{ and}$$
> $$g(f(x)) = x \quad \text{for every } x \text{ in } X.$$

There can be only one inverse function of f (see Exercise 39). Moreover, if g is the inverse function of f, then $(g \circ f)(x) = x$, that is, $g \circ f$ is the identity function on X. Similarly, since $(f \circ g)(x) = x$, for all x in Y, $f \circ g$ is the identity function on Y. For this reason the symbol f^{-1} is often used to denote the inverse function of f. Employing this notation,

> $$f^{-1}(f(x)) = x \quad \text{for every } x \text{ in } X$$
> $$f(f^{-1}(x)) = x \quad \text{for every } x \text{ in } Y.$$

The symbol -1 should not be mistaken for an exponent; that is, $f^{-1}(x)$ does not mean $1/f(x)$. The reciprocal $1/f(x)$ may be denoted by $[f(x)]^{-1}$.

Inverse functions are very important in the study of trigonometry. In Chapter Four we shall discuss two other special types of inverse functions. It is important to remember that in order to define the inverse of a function f it is *absolutely essential that f be one-to-one*. The most common examples of one-to-one functions are those which are either increasing or decreasing on their domains, for in this case, if $a \neq b$ in the domain, then $f(a) \neq f(b)$ in the range.

An algebraic method can sometimes be used to find the inverse of a one-to-one function f with domain X and range Y. Given any x in X, its image y in Y may be found by means of the equation $y = f(x)$. In order to determine the inverse function f^{-1}, we wish to *reverse* this procedure, in the sense that given y, the element x may be found. Since x and y are related by means of $y = f(x)$, it follows that if the latter equation can be solved for x in terms of y, we may arrive at the inverse function f^{-1}. This technique is illustrated in the following examples.

Example 4   If $f(x) = 3x - 5$ for every real number x, find the inverse function of f.

Solution   The graph of the given linear function f is a line of slope 3 (Why?) and hence f is increasing for all x. It follows that f is a one-to-one function with domain and range $\mathbb{R}$, and hence the inverse function g exists. If we let

$$y = 3x - 5$$

and then solve for x in terms of y, we get

$$x = \frac{y + 5}{3}.$$

The last equation enables us to find x when given y. Letting

$$g(y) = \frac{y + 5}{3}$$

gives us a function g from Y to X that reverses the correspondence determined by f. Since the symbol used for the independent variable is immaterial, we may replace y by x in the expression for g, obtaining

$$g(x) = \frac{x + 5}{3},$$

To verify that g is actually the inverse function of f, we must verify that the two conditions $f(g(x)) = x$ and $g(f(x)) = x$ are fulfilled. Thus

$$f(g(x)) = f\left(\frac{x + 5}{3}\right) \qquad \text{(definition of } g)$$

$$= 3\left(\frac{x + 5}{3}\right) - 5 \qquad \text{(definition of } f)$$

$$= x \qquad \text{(simplifying).}$$

Also,

$$g(f(x)) = g(3x - 5) \qquad \text{(definition of } f)$$

$$= \frac{(3x - 5) + 5}{3} \qquad \text{(definition of } g)$$

$$= x \qquad \text{(simplifying).}$$

This proves that g is the inverse function of f, and we may write

$$f^{-1}(x) = \frac{x + 5}{3}. \qquad \blacksquare$$

Example 5   Find the inverse function of f if the domain X is the interval $[0, \infty)$ and $f(x) = x^2 - 3$ for all x in X.

Solution   The domain has been restricted so that f is increasing and hence is one-to-one. The range of f is the interval $[-3, \infty)$. As in Example 4, we begin by considering the equation

$$y = x^2 - 3.$$

Solving for x gives us $\qquad x = \pm\sqrt{y + 3}.$

Since x is nonnegative, we reject $x = -\sqrt{y + 3}$ and, as in the preceding example, we let

$$g(y) = \sqrt{y + 3}$$

or equivalently, $\qquad g(x) = \sqrt{x + 3}.$

Since $\qquad f(g(x)) = f(\sqrt{x + 3}) = (\sqrt{x + 3})^2 - 3$

$$= (x + 3) - 3 = x$$

and $$g(f(x)) = g(x^2 - 3) = \sqrt{(x^2 - 3) + 3} = x$$

we see that $$f^{-1}(x) = \sqrt{x + 3} \quad \text{where } x \geq -3.$$ ■

There is an interesting relationship between the graphs of a function f and its inverse function f^{-1}. We first note that f maps a into b if and only if f^{-1} maps b into a; that is, $b = f(a)$ means the same thing as $a = f^{-1}(b)$. These equations imply that the point (a, b) is on the graph of f if and only if the point (b, a) is on the graph of f^{-1}. As an illustration, in Example 5 we found that the functions f and f^{-1} given by

$$f(x) = x^2 - 3 \quad \text{and} \quad f^{-1}(x) = \sqrt{x + 3}$$

are inverse functions of one another, provided x is suitably restricted. Some points on the graph of f are $(0, -3), (1, -2), (2, 1)$ and $(3, 6)$. Corresponding points on the graph of f^{-1} are $(-3, 0), (-2, 1), (1, 2)$, and $(6, 3)$. The graphs of f and f^{-1} are sketched on the same coordinate axes in Figure 2.24. If the page is folded along the line l, which bisects quadrants I and III (as indicated by the dashes in the figure), then the graphs of f and f^{-1} coincide. The two graphs are said to be *reflections* of one another through the line l. This is typical of the graph of every function f that has an inverse function f^{-1}.

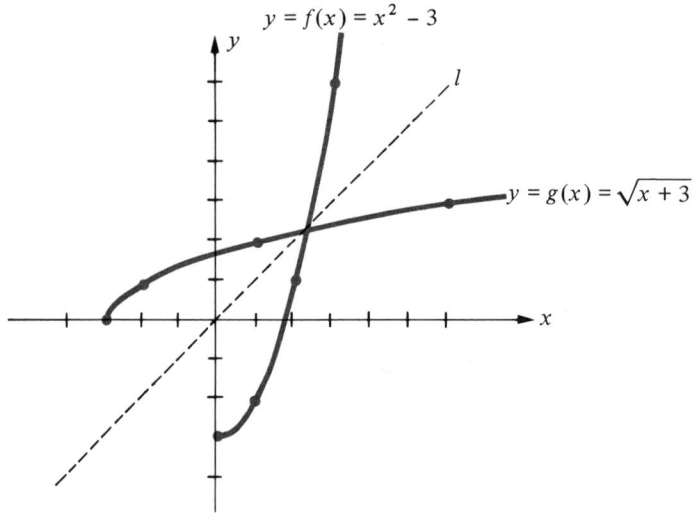

Figure 2.24

EXERCISES 2.4

In Exercises 1–18 find $(f \circ g)(x)$ and $(g \circ f)(x)$.

1 $f(x) = 3x + 2, g(x) = 2x - 1$

2 $f(x) = x + 4, g(x) = 5x - 3$

3 $f(x) = 4x^2 - 5, g(x) = 3x$

4 $f(x) = 7x + 1, g(x) = 2x^2$

5 $f(x) = 3x^2 + 2x, g(x) = 2x - 1$

6 $f(x) = 5x + 7, g(x) = 4 - x^2$

7 $f(x) = x - 1, g(x) = x^3$

8 $f(x) = x^3 + 5x, g(x) = 4x$

9 $f(x) = x^2 + 9x, g(x) = \sqrt{x + 9}$

10 $f(x) = \sqrt[3]{x^2 + 1}, g(x) = x^3 + 1$

11 $f(x) = \dfrac{1}{2x - 5}, g(x) = 1/x^2$

12 $f(x) = \dfrac{x}{x + 1}, g(x) = x - 1$

13 $f(x) = |x|, g(x) = -5$

14 $f(x) = 7, g(x) = 10$

15 $f(x) = x^2, g(x) = 1/x^2$

16 $f(x) = \dfrac{1}{x + 1}, g(x) = x + 1$

17 $f(x) = 2x - 3, g(x) = \dfrac{x + 3}{2}$

18 $f(x) = x^3 - 1, g(x) = \sqrt[3]{x + 1}$

In each of Exercises 19–22 prove that f and g are inverse functions of one another and sketch the graphs of f and g on the same coordinate plane.

19 $f(x) = 7x + 5; g(x) = (x - 5)/7$

20 $f(x) = x^2 - 1, x \geq 0;$
$g(x) = \sqrt{x + 1}, x \geq -1$

21 $f(x) = \sqrt{2x - 4}, x \geq 2;$
$g(x) = (x^2 + 4)/2, x \geq 0$

22 $f(x) = x^3 + 1; g(x) = \sqrt[3]{x - 1}$

In Exercises 23–36 find the inverse function of f.

23 $f(x) = 4x - 3$ **24** $f(x) = 9 - 7x$

25 $f(x) = \dfrac{1}{2x + 5}, x > -5/2$

26 $f(x) = \dfrac{1}{3x - 1}, x > 1/3$

27 $f(x) = 9 - x^2, x \geq 0$

28 $f(x) = 4x^2 + 1, x \geq 0$

29 $f(x) = 5x^3 - 2$ **30** $f(x) = 7 - 2x^3$

31 $f(x) = \sqrt{3x - 5}, x \geq 5/3$

32 $f(x) = \sqrt{4 - x^2}, 0 \leq x \leq 2$

33 $f(x) = \sqrt[3]{x} + 8$ **34** $f(x) = (x^3 + 1)^5$

35 $f(x) = x$ **36** $f(x) = -x$

37 (a) Prove that the linear function f defined by $f(x) = ax + b$, where $a \neq 0$, has an inverse function, and find $f^{-1}(x)$.
  (b) Does a constant function have an inverse? Explain.
  (c) Does an identity function have an inverse? Explain.

38 If f is a one-to-one function with domain X and range Y, prove that f^{-1} is a one-to-one function with domain Y and range X.

39 Prove that a one-to-one function can have at most one inverse function.

40 If $f(x) = ax + b$ and $g(x) = cx + d$, find conditions on c and d in terms of a and b which will guarantee that $f \circ g = g \circ f$. Discuss the case where $a = b = 1$.

2.5 REVIEW

Concepts

Define or discuss each of the following.

1 Function

2 The domain of a function

3 The range of a function

4 The graph of a function

5 Increasing function

6 Decreasing function

7 One-to-one function

8 Identity function

9 Constant function

10 Linear function

11 Slope of a line

12 The point-slope form for the equation of a line

13 The slope-intercept form

14 Linear equation in x and y

15 The composite function of two functions

16 The inverse of a function

Exercises

1 Find the slope of the line through $C(11, -5)$ and $D(-8, 6)$.

2 Prove that the points $A(-3, 1)$, $B(1. -1)$, $C(4, 1)$, and $D(3, 5)$ are vertices of a trapezoid.

3 Find an equation of the line through $A(1/2, -1/3)$ that is parallel to the line $6x + 2y + 5 = 0$.

4 Express the equation $8x + 3y - 24 = 0$ in slope-intercept form.

5 Find an equation of the line having x-intercept -3 and passing through the center of the circle $(x - 2)^2 + (y + 5)^2 = 3$.

In each of Exercises 6–12 sketch the graph of f and determine the intervals in which f is increasing or decreasing.

6 $f(x) = |x + 3|$ **7** $f(x) = \dfrac{1 - 3x}{2}$

8 $f(x) = \sqrt{2 - x}$ **9** $f(x) = 1 - \sqrt{x + 1}$

10 $f(x) = 9 - x^2$ **11** $f(x) = 1000$

12 $f(x) = \begin{cases} x^2 \text{ if } x < 0 \\ 3x \text{ if } 0 \le x < 2 \\ 6 \text{ if } x \ge 2 \end{cases}$

13 If $f(x) = x/\sqrt{x + 3}$ find
(a) $f(1)$ (b) $f(-1)$ (c) $f(0)$
(d) $f(-x)$ (e) $-f(x)$ (f) $f(x^2)$
(g) $(f(x))^2$

14 Find the domain and range of f if
(a) $f(x) = \sqrt{3x - 4}$ (b) $f(x) = 1/(x + 3)^2$

In Exercises 15 and 16 find $(f \circ g)(x)$ and $(g \circ f)(x)$.

15 $f(x) = 2x^2 - 5x + 1, g(x) = 3x + 2$

16 $f(x) = \sqrt{3x + 2}, g(x) = 1/x^2$

In Exercises 17 and 18 find $f^{-1}(x)$ and sketch the graphs of f and f^{-1} on the same coordinate plane.

17 $f(x) = 10 - 15x$

18 $f(x) = 9 - 2x^2, x \le 0$

19 If the altitude and radius of a right circular cylinder are equal, express the volume V as a function of the circumference C of the base.

20 A right circular cylinder is to be inscribed in a cone of altitude 12 cm and base radius 4 cm such that the axes of the cylinder and cone coincide. Express the volume V of the cylinder as a function of its radius r.

3

POLYNOMIAL AND RATIONAL FUNCTIONS

Polynomial functions *occur frequently in both elementary and advanced phases of mathematics. The main objective of this chapter is to study such functions and their algebraic counterpart,* polynomials. *We shall also discuss quotients of polynomial functions, that is,* rational functions.

3.1 THE ALGEBRA OF POLYNOMIALS

If x is a variable, then

$$x^1 = x, \quad x^2 = x \cdot x, \quad x^3 = x \cdot x \cdot x, \quad x^4 = x \cdot x \cdot x \cdot x$$

and, in general, if n is any positive integer,

$$x^n = \underbrace{x \cdot x \cdot \cdots \cdot x}_{n \text{ times}}.$$

The integer n, which specifies the number of times x appears in this product, is called an **exponent**, and x^n is called the ***n*th power** of x. It is also customary to define $x^0 = 1$. Given an expression of the form ax^n, where a is a real number, the number a is referred to as the **coefficient** of x^n. It is important to note that *ax^n is not the same as $(ax)^n$.*

Among the most important functions in mathematics are those defined as follows.

Definition

> A function f is a **polynomial function** if
> $$f(x) = a_n x^n + a_{n-1} x^{n-1} + \cdots + a_1 x + a_0$$
> where the coefficients $a_0, a_1, \ldots, a_n$ are real numbers and the exponents are positive integers.

The expression for $f(x)$ in this definition is called a **polynomial in x** (with real coefficients) and, for each k, $a_k x^k$ is called a **term** of the polynomial. In Chapter Two a number a was called a *zero* of a function f if $f(a) = 0$. If f is a polynomial function, then a is also called a zero of the polynomial $f(x)$.

The coefficient a_n of the highest power of x is the **leading coefficient** of $f(x)$ and, if $a_n \neq 0$, we say that $f(x)$ (or f) has **degree** n. If a coefficient a_i is zero, we often abbreviate the polynomial $f(x)$ by deleting the term $a_i x^i$. If *all* the coefficients of a polynomial are zero, it is called the **zero polynomial** and is denoted by 0. It is customary not to assign a degree to the zero polynomial.

If some of the coefficients are negative, then for convenience we often use minus signs between appropriate terms. To illustrate, instead of $3x^2 + (-5)x + (-7)$, we write $3x^2 - 5x - 7$ for this polynomial of degree 2. Polynomials in other variables may also be considered. For example, $\frac{2}{5}z^2 - 3z^7 + 8 - \sqrt{5}z^4$ is a polynomial in z of degree 7. We ordinarily arrange the terms in order of decreasing powers of the variable and write $-3z^7 - \sqrt{5}z^4 + \frac{2}{5}z^2 + 8$.

According to the definition of degree, if $f(x) = c$, where c is a nonzero real number, then $f(x)$ is a polynomial of degree 0. Such polynomials (together with the zero polynomial) are called **constant polynomials**.

A polynomial in x may be thought of as an expression obtained by employing only additions, subtractions, and multiplications involving x. In particular, the expressions

$$\frac{1}{x} + 3x, \quad \frac{x-5}{x^2+2}, \quad 3x^2 + \sqrt{x} - 2$$

are not polynomials since they also involve divisions by variables, or roots involving variables.

In more advanced courses, coefficients of polynomials are often chosen from some mathematical system other than the set of real numbers. However, *throughout this chapter the terminology "polynomial" (or "polynomial function") will always refer to polynomials with real coefficients*.

By inserting terms with zero coefficients if necessary, we can assume that when discussing several polynomials, the same powers of x appear. As an illustration, given the polynomials $f(x) = 4x^3 - 3x + 2$ and $g(x) = x^2 + 5$, we can rewrite them as $f(x) = 4x^3 + 0x^2 - 3x + 2$ and $g(x) = 0x^3 + x^2 + 0x + 5$.

Let us consider any two polynomials $f(x)$ and $g(x)$. By the preceding remarks we may write

$$f(x) = a_n x^n + a_{n-1} x^{n-1} + \cdots + a_1 x + a_0$$
$$g(x) = b_n x^n + b_{n-1} x^{n-1} + \cdots + b_1 x + b_0$$

where for each i, a_i and b_i are real numbers and n is a nonnegative integer. The polynomials $f(x)$ and $g(x)$ are **equal**, and we write $f(x) = g(x)$ if and only if coefficients of like powers of x are the same, that is, $a_0 = b_0, a_1 = b_1, \ldots, a_n = b_n$. To *add* $f(x)$ and $g(x)$, we may add corresponding coefficients. Thus

$$f(x) + g(x) = (a_n + b_n)x^n + \cdots + (a_1 + b_1)x + (a_0 + b_0).$$

Many properties of real numbers are true for polynomials. The commutative law $f(x) + g(x) = g(x) + f(x)$ follows easily, since for each k, $a_k + b_k = b_k + a_k$. Similarly,

the associative law is a consequence of the fact that the associative law is valid for the coefficients. The zero polynomial

$$0 = 0x^n + 0x^{n-1} + \cdots + 0x + 0$$

is the additive identity, since $f(x) + 0 = f(x)$ for every polynomial $f(x)$. The additive inverse $-f(x)$ of $f(x)$ is given by

$$-f(x) = (-a_n)x^n + \cdots + (-a_1)x + (-a_0)$$

since the sum of $f(x)$ and $-f(x)$ is the zero polynomial. The rule for subtracting the polynomials $f(x)$ and $g(x)$ is

$$f(x) - g(x) = (a_n - b_n)x^n + \cdots + (a_1 - b_1)x + (a_0 - b_0).$$

Multiplication is carried out in the usual way, using properties of real numbers. For this operation there is no advantage in assuming that the highest exponent of x is the same for each polynomial. If $f(x)$ has degree n and $g(x)$ has degree m, we may write

$$f(x) = a_n x^n + a_{n-1}x^{n-1} + \cdots + a_1 x + a_0, \qquad a_n \neq 0,$$
$$g(x) = b_m x^m + b_{m-1}x^{m-1} + \cdots + b_1 x + b_0, \qquad b_m \neq 0,$$

and hence

$$f(x)g(x) = a_n b_m x^{n+m} + (a_n b_{m-1} + a_{n-1}b_m)x^{n+m-1} + \cdots + (a_1 b_0 + a_0 b_1)x + a_0 b_0.$$

Note that we have used a law of exponents which states that $x^i x^j = x^{i+j}$ for all positive integers i and j. For example,

$$x^2 x^3 = (x \cdot x)(x \cdot x \cdot x) = x^5 = x^{2+3}$$
$$x^5 x^4 = (x \cdot x \cdot x \cdot x \cdot x)(x \cdot x \cdot x \cdot x) = x^9 = x^{5+4}.$$

If $a_n \neq 0$ and $b_m \neq 0$, then $a_n b_m \neq 0$, and we see from the preceding discussion that $f(x)g(x)$ has degree $n + m$. This provides the following useful result.

Theorem

> The degree of the product of two nonzero polynomials equals the sum of the degrees of the two polynomials.

As an illustration, if $f(x) = 5x^3 - x + 1$ and $g(x) = 2x^2 + 3$, then the term of highest degree in $f(x)g(x)$ is $10x^5$. Hence the degree of the product is 5, which equals the sum of the degrees of the two given polynomials.

Multiplication of polynomials is commutative and associative, and the distributive laws are valid. The zero degree polynomial $f(x) = 1$ is the identity element relative to multiplication.

Theorem

> If $f(x)$ and $g(x)$ are polynomials such that $f(x)g(x) = 0$, then either $f(x) = 0$ or $g(x) = 0$.

Proof   We shall give an indirect proof. Suppose that $f(x)g(x) = 0$ and that *both* $f(x)$ and $g(x)$ are different from the zero polynomial. If $f(x)$ and $g(x)$ have degrees n and m, respectively, then, by the preceding theorem, $f(x)g(x)$ has degree $n + m$. However, the zero polynomial 0 has *no* degree. This contradiction implies that our supposition is false. Thus, if $f(x)g(x) = 0$, then at least one of the polynomials is the zero polynomial.

Example 1   If $f(x) = x^5 - 4x^3 - 5x^2 + 3$ and $g(x) = x^3 - 5x^2 - 2x + 7$, find the polynomials $f(x) + g(x)$ and $f(x) - g(x)$.

Solution   To find $f(x) + g(x)$ we add coefficients of like powers of x. This gives us

$$f(x) + g(x) = x^5 - 3x^3 - 10x^2 - 2x + 10.$$

In order to keep track of the coefficients, it is sometimes convenient to use the following scheme:

$$
\begin{aligned}
f(x) &= x^5 - 4x^3 - 5x^2 + 3 \\
g(x) &= x^3 - 5x^2 - 2x + 7 \\
\hline
f(x) + g(x) &= x^5 - 3x^3 - 10x^2 - 2x + 10
\end{aligned}
$$

Similarly, by subtracting corresponding coefficients, we obtain

$$f(x) - g(x) = x^5 - 5x^3 + 2x - 4. \qquad\blacksquare$$

Example 2   If $f(x) = 2x^3 + 3x - 1$ and $g(x) = x^2 - x + 4$, find the product $f(x)g(x)$.

Solution   The reader should supply reasons for the following steps:

$$
\begin{aligned}
f(x)g(x) &= (2x^3 + 3x - 1)(x^2 - x + 4) \\
&= (2x^3 + 3x - 1)x^2 + (2x^3 + 3x - 1)(-x) \\
&\quad + (2x^3 + 3x - 1)4 \\
&= 2x^5 + 3x^3 - x^2 - 2x^4 - 3x^2 + x \\
&\quad + 8x^3 + 12x - 4 \\
&= 2x^5 - 2x^4 + (3 + 8)x^3 + (-1 - 3)x^2 \\
&\quad + (1 + 12)x - 4 \\
&= 2x^5 - 2x^4 + 11x^3 - 4x^2 + 13x - 4.
\end{aligned}
$$

Sometimes the above work is arranged as follows:

$$
\begin{array}{l}
2x^3 + 3x - 1 \\
\underline{x^2 - x + 4} \\
2x^5 + 3x^3 - x^2 = (2x^3 + 3x - 1)x^2 \\
 - 2x^4 - 3x^2 + x = (2x^3 + 3x - 1)(-x) \\
\underline{ 8x^3 + 12x - 4} = (2x^3 + 3x - 1)4 \\
2x^5 - 2x^4 + 11x^3 - 4x^2 + 13x - 4 = \text{sum of the above.}
\end{array}
$$

$\blacksquare$

If a polynomial is written as a product of other polynomials, then each polynomial in the product is called a **factor** of the original polynomial. For example, since $x^2 - 9 = (x + 3)(x - 3)$, we see that $x + 3$ and $x - 3$ are factors of $x^2 - 9$. Any polynomial has, as a factor, *every* nonzero real number c. As an illustration, given $3x^2 - 5x + 2$ and any nonzero real number c, we can write

$$3x^2 - 5x + 2 = c\left(\frac{3}{c}x^2 - \frac{5}{c}x + \frac{2}{c}\right).$$

A factor c of this type is called a **trivial factor**. We shall be interested primarily in **nontrivial factors** of polynomials; that is, factors which contain polynomials of degree greater than zero.

An integer $a > 1$ is prime if it cannot be written as a product of two positive integers greater than 1. In similar fashion, if S denotes a set of numbers, then a polynomial with coefficients in S is said to be **prime**, or **irreducible** over S, if it cannot be written as a product of two polynomials of positive degree with coefficients in S. Note that a polynomial may be irreducible over one set S but not over another. For example, $x^2 - 2$ is irreducible over the rational numbers since it cannot be expressed as a product of two polynomials of positive degree which have rational coefficients. If we allow the factors to have *real* coefficients, then $x^2 - 2$ is not prime, since

$$x^2 - 2 = (x + \sqrt{2})(x - \sqrt{2}).$$

Similarly, $x^2 + 1$ is irreducible over the real numbers but, as we shall see in Chapter Seven, not over the complex numbers. Every polynomial $ax + b$ of degree 1 is irreducible, for if $ax + b = f(x)g(x)$ where $f(x)$ and $g(x)$ are polynomials, then the degree of $f(x)g(x)$, that is, the sum of the degrees of $f(x)$ and $g(x)$, is 1. This implies that one of the latter polynomials has degree 0.

Example 3 Express the polynomial $6x^2 - 7x - 3$ as a product of irreducible polynomials.

Solution If we write $6x^2 - 7x - 3 = (ax + b)(cx + d)$, then the product of a and c is 6; the product of b and d is -3; and $ad + bc = -7$. Trying various possibilities, we arrive at the factorization

$$6x^2 - 7x - 3 = (2x - 3)(3x + 1). \qquad \blacksquare$$

It is usually difficult to factor polynomials of degree greater than 2. In simple cases the following formulas may be useful. Each can be verified by multiplication.

$a^2 - b^2 = (a + b)(a - b)$	(Difference of two squares)
$a^3 - b^3 = (a - b)(a^2 + ab + b^2)$	(Difference of two cubes)
$a^3 + b^3 = (a + b)(a^2 - ab + b^2)$	(Sum of two cubes)

Example 4 Express each of the following as a product of polynomials which are irreducible over the set of integers:

(a) $16x^4 - 81$

(b) $x^3 - 8$

(c) $3x^3 + 2x^2 - 12x - 8$

Solution

(a) We apply the difference of two squares formula twice, as follows:

$$16x^4 - 81 = (4x^2)^2 - 9^2$$
$$= (4x^2 + 9)(4x^2 - 9)$$
$$= (4x^2 + 9)[(2x)^2 - 3^2]$$
$$= (4x^2 + 9)(2x + 3)(2x - 3).$$

(b) Using the difference of two cubes formula, with $a = x$ and $b = 2$, we obtain

$$x^3 - 8 = x^3 - 2^3$$
$$= (x - 2)(x^2 + 2x + 4).$$

By trying all possibilities we can show that $x^2 + 2x + 4$ cannot be expressed as a product of two first-degree polynomials with integer coefficients.

(c) We first employ the distributive law and then the difference of two squares formula, as follows:

$$3x^3 + 2x^2 - 12x - 8 = x^2(3x + 2) - 4(3x + 2)$$
$$= (x^2 - 4)(3x + 2)$$
$$= (x + 2)(x - 2)(3x + 2). \qquad ■$$

The technique used in part (c) of Example 4 is called **factorization by grouping**.

EXERCISES 3.1

In Exercises 1–6 perform the indicated operations and find the degree of the resulting polynomial.

1 $(2x^3 + 4x^2 + 3x + 7) + (x^4 - 5x^3 + x - 2)$

2 $(2x^5 - 3x^2 + 2) + (-2x^4 + x^2 - 4x + 5)$

3 $(x^3 + 5x^2 - 7x + 2) - (x^3 + 5x^2 + x + 2)$

4 $(2x^4 - 3x^2 + 1) - (3x^4 + x + 1)$

5 $(2x^2 - 5x + 7)(3x - 5)$

6 $(3x^3 - 2x^2 + 1)(8x - 1)$

In Exercises 7–12 find $f(x) + g(x)$, $f(x) - g(x)$, and $f(x)g(x)$.

7 $f(x) = 2x^3 - x + 5$, $g(x) = x^2 + x + 2$

8 $f(x) = 4x^2 - 1$, $g(x) = x^3 + 6x + 1$

9 $f(x) = 7x^4 + x^2 - 1$, $g(x) = 7x^4 - x^3 + 4x$

10 $f(x) = -x^5 + 5$, $g(x) = x^5 - 5$

11 $f(x) = 3$, $g(x) = x$

12 $f(x) = 0$, $g(x) = 2x^2 - 7x + 4$

13 Without multiplying, find the degree of the product of the polynomials $2x^4 - 3x + 7$ and $3x^3 + 8x^2$.

14 If $f(x)$ and $g(x)$ are polynomials in x of degree 5, does it follow that $f(x) + g(x)$ has degree 5? Explain.

15 Give examples of polynomials $f(x)$ and $g(x)$ of degree 3 such that the degree of $f(x) + g(x)$ is
(a) 3   (b) 2   (c) 1   (d) 0

16 If $f(x)$, $g(x)$, and $h(x)$ are polynomials such that $f(x)g(x)h(x) = 0$, prove that at least one of the polynomials is 0.

17 If $f(x)$ and $g(x)$ are polynomials of degrees n and m, respectively, and $n > m$, prove that the degree of $f(x) + g(x)$ is n. What can be said if $n = m$?

18 Prove that the only polynomials that have multiplicative inverses are polynomials of degree 0.

Express each of the following as a product of polynomials which are irreducible over the set of integers.

19 $3x^2 + 10x - 8$ 20 $10x^2 + 29x - 21$

21 $12x^3 + 21x^2 - 45x$ 22 $6x^4 - 5x^3 - 6x^2$

23 $25x^2 - 9$ 24 $49 - 36x^2$

25 $27x^3 - 1$ 26 $64x^3 + 125$

27 $x^4 - 13x^2 + 36$ 28 $3x^4 - 11x^2 - 4$

29 $x^2 + 2x + 1$ 30 $x^2 - 6x + 9$

31 $x^2 + x + 1$ 32 $x^2 + 16$

33 $x^3 + x^2 + x + 1$ 34 $2x^3 - x^2 - 2x + 1$

35 $x^8 - 10x^4 + 9$ 36 $x^6 - 64$

37 $x^6 - 1$ 38 $(x^2 - 25)^3$

39 $4x^3 - 8x^2 - 9x + 18$

40 $x^6 - x^4$

3.2 QUADRATIC FUNCTIONS

In the preceding chapter we proved that the graph of a polynomial function of degree 1 is a line. In this section we shall investigate graphs of polynomial functions of degree 2. As stated in the next definition, such functions are called *quadratic functions*.

Definition

> A function f is a **quadratic function** if
> $$f(x) = ax^2 + bx + c$$
> where a, b, and c are real numbers and $a \neq 0$.

If $b = c = 0$ in the preceding definition, then $f(x) = ax^2$ and, as indicated in Chapter Two, the graph is a parabola with vertex at the origin, opening upward if $a > 0$ or downward if $a < 0$ (see Figure 2.13). If $b = 0$ and $c \neq 0$, then

$$f(x) = ax^2 + c$$

and the graph is a parabola with vertex at the point $(0, c)$ on the y-axis. Some typical graphs are illustrated in Figure 3.1.

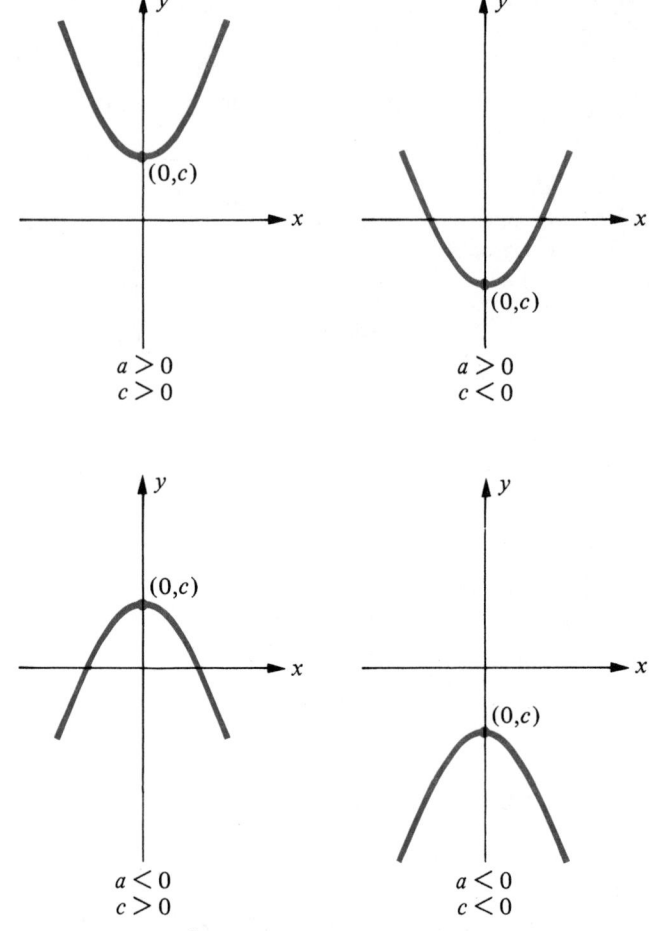

Figure 3.1. $f(x) = ax^2 + c$

Example 1   Sketch the graph of $f(x) = -\frac{1}{2}x^2 + 4$.

Solution   The graph of $f(x) = -\frac{1}{2}x^2$ is similar to the graph of $f(x) = -x^2$ sketched in (iii) of Figure 2.13 but is somewhat wider. The graph of $f(x) = -\frac{1}{2}x^2 + 4$ may be found by raising the graph of $f(x) = -\frac{1}{2}x^2$ four units. To help with the sketch we may plot several points from the following table and use symmetry with respect to the y-axis (see Figure 3.2).

x	0	1	2	$\sqrt{8}$	3
$f(x)$	4	7/2	2	0	$-1/2$

$\blacksquare$

If $f(x) = ax^2 + bx + c$ and $b \neq 0$, then a technique called **completing the square** can be used to change the form of $f(x)$ to

$$f(x) = a(x - h)^2 + k$$

where h and k are real numbers. The right-hand side of this equation may be written $a(x')^2 + k$, where $x' = x - h$ and, as we shall see later, the graph will be similar to that of $f(x) = ax^2 + k$. The process of completing the square is illustrated in the next example.

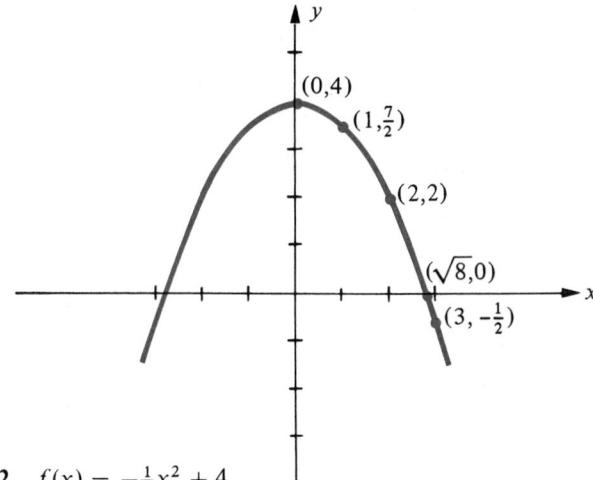

Figure 3.2. $f(x) = -\frac{1}{2}x^2 + 4$

Example 2 Express $f(x)$ in the form $a(x - h)^2 + k$ if:
(a) $f(x) = x^2 - 6x - 5$,   (b) $f(x) = 3x^2 + 24x + 50$.

Solution

(a) We begin by writing

$$f(x) = (x^2 - 6x \qquad) - 5.$$

We next complete the square by determining the number to add to the expression within the parentheses so that it takes on the form $(x - h)^2$ for some h. It is not difficult to show that *the desired number is always the square of one-half the coefficient of* x. In the present example the coefficient of x is -6, and consequently, the square of one-half of -6 is $(-3)^2$, or 9. Adding 9 to $x^2 - 6x$ gives $x^2 - 6x + 9$ or $(x - 3)^2$. Of course, if we add 9 to $f(x)$, then we must also subtract 9, so that $f(x)$ does not change. Accordingly, we obtain

$$\begin{aligned} f(x) &= (x^2 - 6x \qquad) - 5 \\ &= (x^2 - 6x + 9) - 5 - 9 \\ &= (x - 3)^2 - 14 \end{aligned}$$

which has the desired form with $a = 1$, $h = 3$, and $k = -14$.

(b) Given $f(x) = 3x^2 + 24x + 50$, we begin by removing a factor 3 from the first two terms, as follows:

$$f(x) = 3(x^2 + 8x \qquad) + 50.$$

As in part (a), to complete the square for the expression in parentheses, we add the square of one-half the coefficient of x, namely $(8/2)^2$, or 16. However, if we add 16 to the expression within parentheses, then because of the factor 3 we are actually adding 48 to $f(x)$. Hence we must compensate by subtracting 48, as follows:

$$f(x) = 3(x^2 + 8x \quad) + 50$$
$$= 3(x^2 + 8x + 16) + 50 - 48$$
$$= 3(x + 4)^2 + 2$$

which has the desired form with $a = 3$, $h = -4$, and $k = 2$.   ■

Let us compare the two equations

$$\text{(i)}\quad f(x) = a(x - h)^2 + k \quad \text{and} \quad \text{(ii)}\quad y = ax^2 + k.$$

The graph of f in (i) can be obtained by shifting the graph of (ii) either to the right or left, depending on the sign of h. This is true because the ordinate of the point with abscissa x_1 on the graph of (i) is the same as the ordinate of the point with abscissa $x_1 - h$ on the graph of (ii). It follows that the graph of (i) and, therefore, the graph of a quadratic function, is a parabola. If $a > 0$, the parabola opens upward and the vertex is obtained by finding the smallest value of $f(x)$ in (i). Since $(x - h)^2 \geq 0$, this smallest value occurs if $(x - h)^2 = 0$, that is, if $x = h$. The ordinate of the point with abscissa h is

$$f(h) = a(h - h)^2 + k = k$$

and hence the vertex is the point (h, k). If $a < 0$ the vertex is again (h, k), but the parabola opens downward.

If we let $y = f(x)$ in (ii), we obtain $y = a(x - h)^2 + k$, or equivalently,

$$\boxed{y - k = a(x - h)^2.}$$

This is called the **standard equation of a parabola** with vertex (h, k) and axis parallel to the y-axis.

Example 3   Sketch the graph of $f(x) = 2x^2 - 6x + 4$.

Solution   The graph of f is the same as the graph of the equation $y = 2x^2 - 6x + 4$. From the preceding discussion we know that the graph is a parabola, and we may obtain the standard equation by completing the square as follows:

$$y = 2x^2 - 6x + 4$$
$$= 2(x^2 - 3x \quad) + 4$$
$$= 2\left(x^2 - 3x + \frac{9}{4}\right) + \left(4 - \frac{9}{2}\right)$$
$$= 2\left(x - \frac{3}{2}\right)^2 - \frac{1}{2}$$

If we rewrite the last equation as

$$y + \frac{1}{2} = 2\left(x - \frac{3}{2}\right)^2$$

we obtain the standard equation of a parabola with $h = 3/2$ and $k = -1/2$. Consequently, the vertex (h, k) of the parabola is $(3/2, -1/2)$. Since $a = 2 > 0$, the parabola opens upward. The y-intercept is $f(0) = 4$. To find the x-intercepts we solve $2x^2 - 6x + 4 = 0$ or the equivalent equation $(2x - 2)(x - 2) = 0$, obtaining $x = 2$ and $x = 1$. The vertex, together with the x- and y-intercepts, are sufficient for a reasonably accurate sketch (see Figure 3.3).

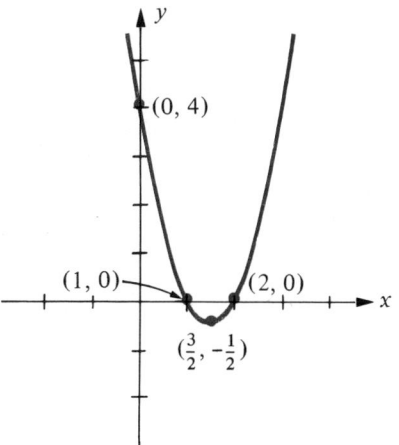

Figure 3.3. $f(x) = 2x^2 - 6x + 4$ ■

Example 4   Sketch the graph of the equation $y = 8 - 2x - x^2$.

Solution   We know from the discussion in this section that the graph is a parabola. Let us find the vertex by completing the square as follows:

$$\begin{aligned} y &= -x^2 - 2x + 8 \\ &= -(x^2 + 2x) + 8 \\ &= -(x^2 + 2x + 1) + 8 + 1 \\ &= -(x + 1)^2 + 9. \end{aligned}$$

Writing $$y - 9 = -(x + 1)^2$$

gives us the standard equation of a parabola with $h = -1$ and $k = 9$, and hence the vertex is $(-1, 9)$. Since $a = -1 < 0$, the parabola opens downward. To find the x-intercepts we solve the equation $8 - 2x - x^2 = 0$. The reader may verify that the roots of this equation are $x = -4$ and $x = 2$. The y-intercept is 8. Using this information gives us the sketch in Figure 3.4.

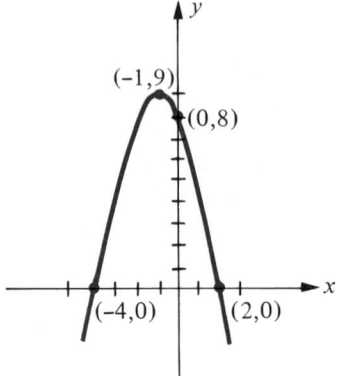

Figure 3.4. $f(x) = 8 - 2x - x^2$ ■

Let us conclude this section with a brief analysis of the general equation $f(x) = ax^2 + bx + c$, where $a > 0$. We begin by writing

$$f(x) = a\left(x^2 + \frac{b}{a}x \right) + c.$$

Next we complete the square by adding $(b/2a)^2$ to the expression within parentheses, as follows:

$$f(x) = a\left(x^2 + \frac{b}{a}x + \frac{b^2}{4a^2}\right) + \left(c - \frac{b^2}{4a}\right).$$

Note that if $b^2/4a^2$ is added *inside* the parentheses, then because of the factor a on the *outside* we have actually added $b^2/4a$ to $f(x)$, and therefore we must compensate by subtracting $b^2/4a$ from c. The preceding equality may be written

$$f(x) = a\left(x + \frac{b}{2a}\right)^2 + \left(\frac{4ac - b^2}{4a}\right).$$

Observe that substituting numbers for x has no effect on the fixed term $(4ac - b^2)/4a$. Consequently, in order to determine where f is increasing or decreasing, it is sufficient to study the expression $(x + b/2a)^2$. Since $(x + b/2a)^2 = 0$ if $x = -b/2a$, and since $(x + b/2a)^2 > 0$ if $x \neq -b/2a$, it follows that $f(x)$ has a minimum value at $x = -b/2a$. This gives us the abscissa of the vertex of the parabola illustrated in Figure 3.5. The ordinate is $f(-b/2a)$, or $(4ac - b^2)/4a$.

It can be shown that f is decreasing on $\{x : x < -b/2a\}$ and increasing for $x > -b/2a$, as illustrated graphically in Figure 3.5. If $a < 0$, then again the abscissa of the vertex is $-b/2a$, but the parabola opens downward.

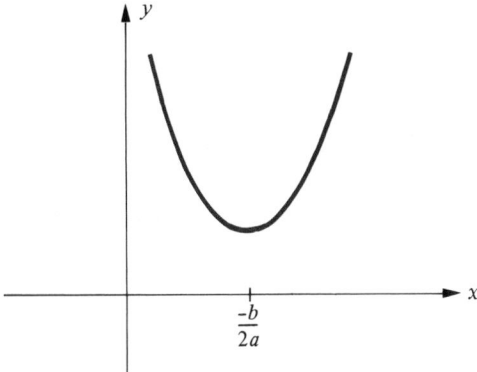

Figure 3.5. $f(x) = ax^2 + bx + c, \quad a > 0$

If we wish to find the x-intercepts of the graph, it is necessary to solve the equation $f(x) = 0$. Using the last expression for $f(x)$ displayed on page 84 leads to the following equations.

$$a\left(x + \frac{b}{2a}\right)^2 + \left(\frac{4ac - b^2}{4a}\right) = 0$$

$$\left(x + \frac{b}{2a}\right)^2 = \frac{b^2 - 4ac}{4a^2}$$

$$x + \frac{b}{2a} = \pm \frac{\sqrt{b^2 - 4ac}}{2a}$$

$$\boxed{x = \frac{-b \pm \sqrt{b^2 - 4ac}}{2a}}$$

The last equation is known as the **quadratic formula** and is used in algebra to find the roots of the quadratic equation $ax^2 + bx + c = 0$. If $b^2 - 4ac > 0$, there are two real and unequal roots and the graph has two x-intercepts. If $b^2 - 4ac = 0$ there is one (double) root and the graph is tangent to the x-axis. Finally, if $b^2 - 4ac < 0$ there are no x-intercepts. We have illustrated the three cases in Figure 3.6 under the assumption that $a > 0$. A similar situation occurs if $a < 0$, but in this case the parabolas open downward.

Parabolas (and hence quadratic functions) are very useful in applications of mathematics to the physical world. It can be shown that if a projectile is fired and is acted upon only by the force of gravity—that is, if air resistance and other outside factors are inconsequential—then the path of the projectile is a parabola. Properties of parabolas are used in the design of mirrors for telescopes and searchlights. They are also employed in the design of field microphones used in television broadcasts of football games. These are only a few of many physical applications.

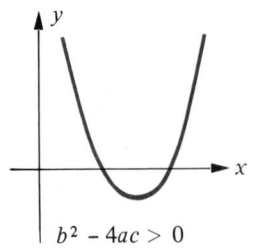 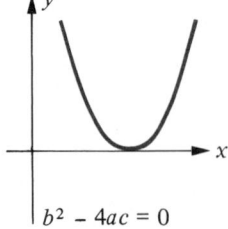 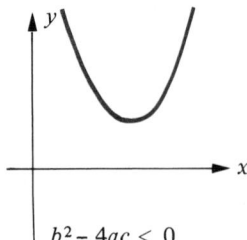

$b^2 - 4ac > 0$ $b^2 - 4ac = 0$ $b^2 - 4ac < 0$

Figure 3.6

EXERCISES 3.2

1 Given $f(x) = ax^2 + 2$, sketch the graph of f if:
(a) $a = 2$ (b) $a = 5$
(c) $a = \frac{1}{2}$ (d) $a = -3$

2 Given $f(x) = 4x^2 + c$, sketch the graph of f if:
(a) $c = 2$ (b) $c = 5$
(c) $c = \frac{1}{2}$ (d) $c = -3$

In each of Exercises 3–6 sketch the graph of f and find the vertex.

3 $f(x) = 4x^2 - 9$ **4** $f(x) = 2x^2 + 3$

5 $f(x) = 9 - 4x^2$ **6** $f(x) = 16 - 9x^2$

In Exercises 7–14 use the quadratic formula to find all real numbers a such that $f(a) = 0$.

7 $f(x) = 4x^2 - 11x - 3$

8 $f(x) = 25x^2 + 10x + 1$

9 $f(x) = 9x^2 - 12x + 4$

10 $f(x) = 10x^2 + x - 21$

11 $f(x) = 2x^2 + x - 5$

12 $f(x) = 2x^2 + 4x - 3$

13 $f(x) = \dfrac{2x + 1}{x - 3} - \dfrac{x - 1}{x + 2}$

14 $f(x) = 9x^2 + 13x$

In Exercises 15–20 use the technique of completing the square to express $f(x)$ in the form $a(x - h)^2 + k$.

15 $f(x) = 2x^2 - 16x + 23$

16 $f(x) = 3x^2 - 12x + 7$

17 $f(x) = -5x^2 - 10x + 3$

18 $f(x) = -2x^2 - 12x - 12$

19 $f(x) = 4x^2 - 4x + 2$

20 $f(x) = 9x^2 + 12x - 8$

In each of Exercises 21–30 sketch the graph of f. (Find the vertex by completing the square.)

21 $f(x) = x^2 + 5x + 4$

22 $f(x) = x^2 - 6x$

23 $f(x) = 8x - 12 - x^2$

24 $f(x) = 10 + 3x - x^2$

25 $f(x) = x^2 + x + 3$

26 $f(x) = x^2 + 2x + 5$

27 $f(x) = 3x^2 - 12x + 16$

28 $f(x) = 2x^2 + 12x + 13$

29 $f(x) = -5x^2 - 10x - 4$

30 $f(x) = -3x^2 + 24x - 42$

31 Find a real number k such that the graph of $y = x^2 + kx + 3$ contains the point $P(-1, -2)$.

32 Prove that a parabola which has an equation of the form $y = ax^2 + c$ has its vertex on the y-axis.

33 A projectile is fired straight upward with an initial velocity of 640 ft/sec. The distance s (in feet) above the ground after t seconds is given by $s = -16t^2 + 640t$. When will the projectile be 1600 feet above the ground? When will it hit the ground?

34 If a projectile is fired upward from a height of s_0 feet above the ground with an initial velocity of v_0 feet per second, then its height above the ground at time t is given by $s = -16t^2 + v_0t + s_0$. Solve for t in terms of s, v_0, and s_0. When will the projectile hit the ground? Express your answer in terms of v_0 and s_0.

35 Prove that the product of two linear functions is a quadratic function.

36 If f is a linear function and g is a quadratic function, prove that $f \circ g$ and $g \circ f$ are quadratic functions.

37 If $f(x) = x^2 + kx - 5 + 3k$, find two different values for k such that f has only one zero.

38 If $f(x) = 3x^2 + kx - 4k$, find all values of k such that f has no real zeros.

39 If one zero of $f(x) = kx^2 + 3x + k$ is -2, find the other zero.

40 If $f(x) = ax^2 + bx + c$, where a, b, and c are rational numbers, can f have one rational and one irrational zero? Explain.

3.3 GRAPHS OF POLYNOMIAL FUNCTIONS OF DEGREE GREATER THAN 2

Recall that f is a polynomial function of degree n if

$$f(x) = a_nx^n + a_{n-1}x^{n-1} + \cdots + a_1x + a_0$$

where $a_n \neq 0$. If all the coefficients except a_n are zero, then we may write

$$f(x) = ax^n, \quad \text{where } a = a_n \neq 0.$$

We know from our previous work that if $n = 1$, the graph of f is a line passing through the origin, whereas, if $n = 2$, the graph is a parabola with vertex at the origin. The special case in which $a = 1$ and $n = 3$, that is, $f(x) = x^3$, is sketched in Figure 1.30. Several other illustrations with $n = 3$ are given in the next example.

Example 1 Sketch the graph of f if   (a) $f(x) = \frac{1}{2}x^3$,   (b) $f(x) = -\frac{1}{2}x^3$.

Solution

(a) The table below gives several points on the graph.

x	0	$\frac{1}{2}$	1	$\frac{3}{2}$	2	$\frac{5}{2}$
$f(x)$	0	$\frac{1}{16}$	$\frac{1}{2}$	$\frac{27}{16}$	4	$\frac{125}{16} \approx 7.8$

The graph of f is symmetric with respect to the origin (Why?) and hence the points $(-\frac{1}{2}, -\frac{1}{16})$, $(-1, -\frac{1}{2})$, etc., are also on the graph. The graph is sketched in (i) of Figure 3.7.

(b) If $f(x) = -\frac{1}{2}x^3$, the graph can be obtained from that in part (a) by multiplying all ordinates by -1. This amounts to reflecting the graph through the x-axis, as shown in (ii) of Figure 3.7.

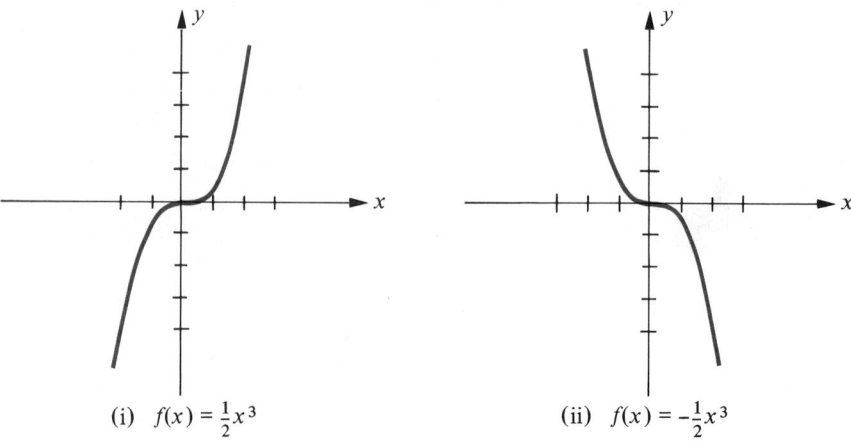

(i) $f(x) = \frac{1}{2}x^3$ (ii) $f(x) = -\frac{1}{2}x^3$

Figure 3.7   ■

In general, if $f(x) = ax^3$, then increasing the numerical value of the coefficient a causes the graph to rise or fall more sharply. For example, if $f(x) = 10x^3$, then $f(1) = 10$, $f(2) = 80$, and $f(-2) = -80$. The effect of using a larger exponent and holding a fixed as, for example, in $f(x) = \frac{1}{2}x^5$, also leads to a graph which rises or falls more rapidly.

If n is an *even* integer, then the graph of $f(x) = ax^n$ is symmetric with respect to the y-axis. (Why?) Two illustrations (with $a = 1$) are given in (i) and (ii) of Figure 3.8. Note that as the exponent increases, the graph becomes much flatter at the origin. It also rises (or falls) more rapidly if we let x increase through positive (or negative) values. Using a negative coefficient such as -1 inverts the graphs.

As usual, the graph of $f(x) = ax^n + c$, where $c \neq 0$, can be found by raising or lowering the graph of $f(x) = ax^n$. For example, if $f(x) = 1 - x^6$, we obtain the sketch in (iii) of Figure 3.8.

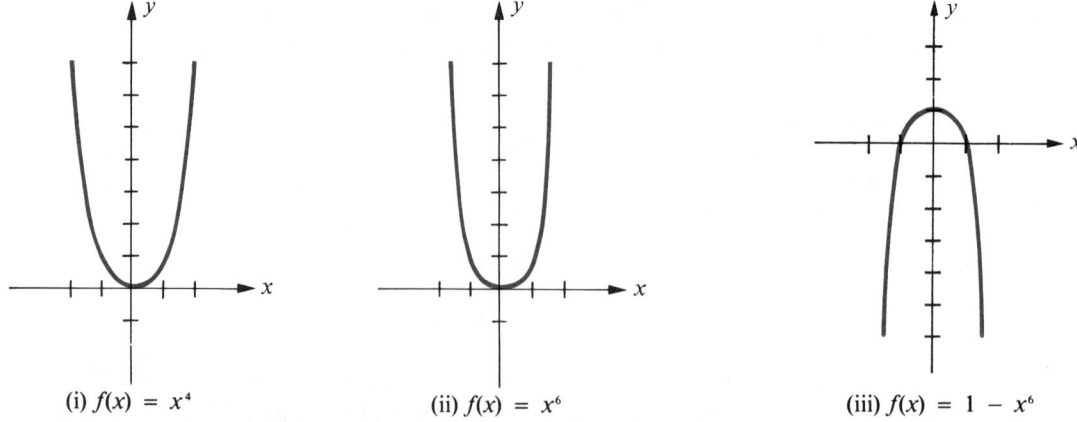

(i) $f(x) = x^4$ (ii) $f(x) = x^6$ (iii) $f(x) = 1 - x^6$

Figure 3.8

A complete analysis of graphs of polynomial functions of degree greater than 2 requires methods studied in calculus. As the degree increases, the graphs usually become more complicated. However, they always have a smooth appearance with a number of "hills" and "valleys," as illustrated in Figure 3.9. This number may be large when the degree is large, but it may also be small, as illustrated in Figures 3.7 and 3.8.

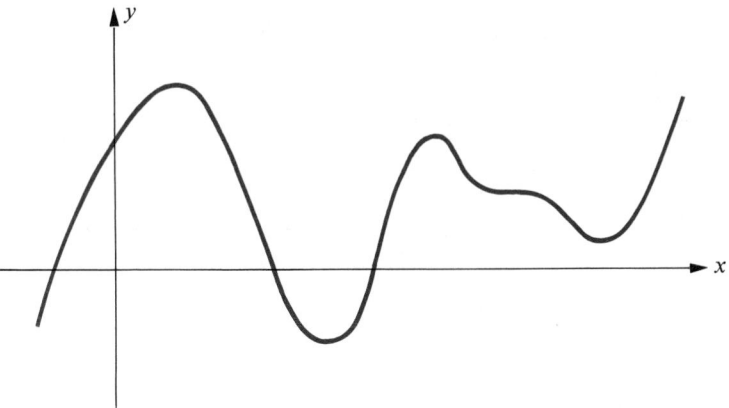

Figure 3.9

If $f(x)$ can be expressed as a product of first-degree polynomials, then a rough sketch of the graph can be obtained easily, as shown in Examples 2 and 3.

Example 2 If $f(x) = x^3 + x^2 - 4x - 4$, determine all values of x such that $f(x) > 0$, and all values of x such that $f(x) < 0$. Use that information as an aid in sketching the graph of f.

Solution The geometric significance of the stated inequalities is that the graph lies above the x-axis for values of x such that $f(x) > 0$, whereas the graph lies below the x-axis if $f(x) < 0$. We may factor $f(x)$ by grouping terms as follows:

$$f(x) = (x^3 + x^2) - (4x + 4)$$
$$= x^2(x + 1) - 4(x + 1)$$
$$= (x^2 - 4)(x + 1)$$
$$= (x + 2)(x - 2)(x + 1).$$

It follows that $f(x)$ is zero if x equals -2, -1, or 2. These are the x-intercepts of the graph. The corresponding points on the graph divide the x-axis into four parts, as illustrated in Figure 3.10. We next consider values of x such that

(a) $x < -2$ (b) $-2 < x < -1$ (c) $-1 < x < 2$ (d) $2 < x$.

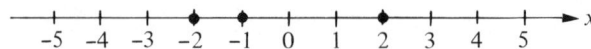

Figure 3.10

The signs of $f(x)$ for the values of x indicated in (a)–(d) may be determined by investigating the signs of the three factors $x + 2$, $x + 1$, and $x - 2$. The diagram in Figure 3.11 displays the signs of these factors for values of x corresponding to each of the four inequalities listed in (a)–(d).

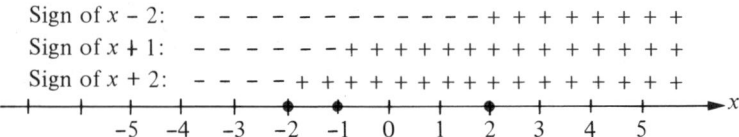

Figure 3.11

We see from the diagram that all three factors are negative if $x < -2$. Since $f(x)$ is the product of the three factors, it follows that $f(x)$ is negative if $x < -2$. Similarly, if $-2 < x < -1$, then two factors are negative and one is positive and hence $f(x)$ is positive. If $-1 < x < 2$, then only one factor is negative and, therefore, $f(x)$ is negative. Finally, if $x > 2$, then all factors are positive and hence so is $f(x)$. The following table summarizes these remarks.

Domain of x	Sign of $f(x)$	Position of graph
$x < -2$	$-$	Below x-axis
$-2 < x < -1$	$+$	Above x-axis
$-1 < x < 2$	$-$	Below x-axis
$x > 2$	$+$	Above x-axis

Coordinates of several points on the graph are listed below.

x	-3	$-\frac{3}{2}$	0	1	$\frac{5}{2}$
$f(x)$	-10	$\frac{7}{8}$	-4	-6	$\frac{63}{8} \approx 7.9$

Using the information provided by the last two tables gives us the graph in Figure 3.12. In order to find the high and low points on the graph it is necessary to use methods developed in calculus.

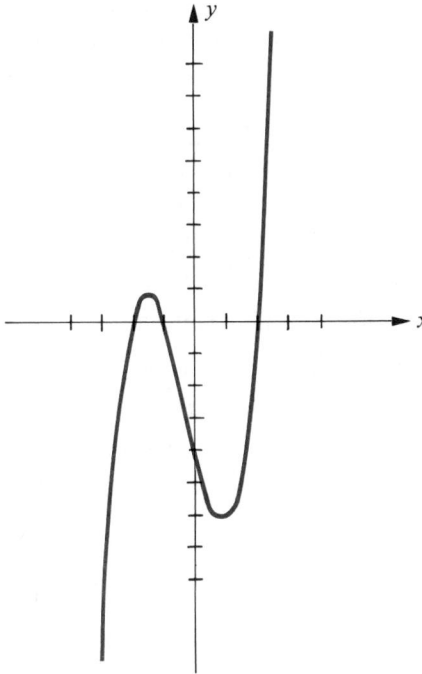

Figure 3.12. $f(x) = x^3 + x^2 - 4x - 4$ ∎

The graph of every polynomial function of degree 3 has an S-shaped appearance similar to that shown in Figure 3.12 or an inverted version of that graph if the coefficient of x^3 is negative. However, sometimes there may be only one x-intercept, or the S may be elongated, as in Figure 3.7.

Example 3   If $f(x) = x^4 - 4x^3 + 3x^2$, determine all values of x such that $f(x) > 0$, and all x such that $f(x) < 0$. Sketch the graph of f.

Solution   We begin by factoring $f(x)$. Thus

$$f(x) = x^2(x^2 - 4x + 3)$$
$$= x^2(x - 1)(x - 3)$$

The x-intercepts of the graph are 0, 1, and 3. (Why?) We next plot the corresponding points on a coordinate line and display the signs of the factors as in the solution of the preceding example (see Figure 3.13). Using the fact that $f(x)$ is

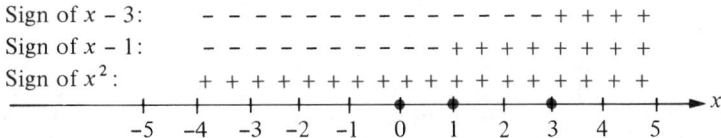

Figure 3.13

positive if two factors are negative and one is positive, or if all three factors are positive, we arrive at the following table.

Domain of x	Sign of $f(x)$	Position of graph
$x < 0$	$+$	Above x-axis
$0 < x < 1$	$+$	Above x-axis
$1 < x < 3$	$-$	Below x-axis
$x > 3$	$+$	Above x-axis

Making use of the above information and plotting several points gives us the sketch in Figure 3.14.

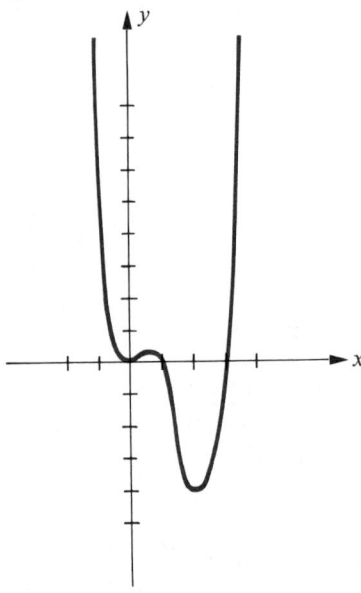

Figure 3.14. $f(x) = x^4 - 4x^3 + 3x^2$ ∎

If $f(x)$ cannot be factored, we may always employ the technique of plotting many points and fitting a smooth curve to the resulting configuration. This is, of course, an extremely tedious procedure. In arriving at such graphs (and also the graphs in the preceding examples) we use a property of polynomial functions called **continuity**. Continuity, which is studied extensively in calculus, implies that a small change in x produces a small change in the functional value $f(x)$. The next theorem specifies another important property of polynomial functions. The proof requires advanced mathematical methods.

The Intermediate Value Theorem for Polynomial Functions

If f is a polynomial function and $f(a) \neq f(b)$, where $a < b$, then f takes on every value between $f(a)$ and $f(b)$ in the interval $[a, b]$.

The Intermediate Value Theorem states that if w is any number between $f(a)$ and $f(b)$, then there is a number c between a and b such that $f(c) = w$. If the graph of the polynomial function f is regarded as extending in an unbroken manner from the point $(a, f(a))$ to the point $(b, f(b))$, as illustrated in Figure 3.15, then for any number w between $f(a)$ and $f(b)$ it appears that a horizontal line with y-intercept w should intersect the graph in at least one point P. The abscissa c of P is a number such that $f(c) = w$.

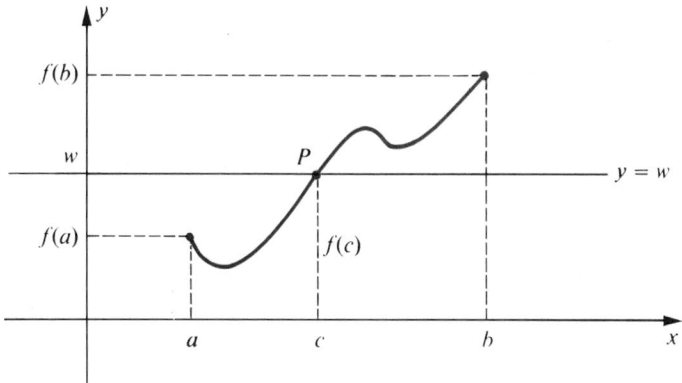

Figure 3.15

A corollary to the Intermediate Value Theorem is that if $f(a)$ and $f(b)$ have opposite signs, then there is a number c between a and b such that $f(c) = 0$; that is, f has a zero at c. Geometrically, this implies that if the point $(a, f(a))$ on the graph of a polynomial function lies below the x-axis, and the point $(b, f(b))$ lies above the x-axis, or vice versa, then the graph crosses the x-axis at some point $(c, 0)$, where $a < c < b$. Actually, there may be *more* than one zero of f between a and b. The Intermediate Value Theorem guarantees that there is at *least* one zero.

Example 4   Show that $f(x) = x^5 + 2x^4 - 6x^3 + 2x - 3$ has a zero between 1 and 2.

Solution   Substitution for x gives us

$$f(1) = 1 + 2 - 6 + 2 - 3 = -4$$
$$f(2) = 32 + 32 - 48 + 4 - 3 = 17.$$

Since $f(1)$ and $f(2)$ have opposite signs, it follows from the preceding discussion that $f(c) = 0$ for some real number c between 1 and 2.   ∎

Example 4 illustrates a possible scheme for locating zeros of polynomial functions. Additional techniques will be introduced in subsequent sections.

EXERCISES 3.3

1 If $f(x) = ax^3 + 2$, sketch the graph of f if:
   (a) $a = 2$ (b) $a = 4$
   (c) $a = \frac{1}{4}$ (d) $a = -2$

2 If $f(x) = 2x^3 + c$, sketch the graph of f if:
   (a) $c = 2$ (b) $c = 4$
   (c) $c = \frac{1}{4}$ (d) $c = -2$

In each of Exercises 3–18 determine all x such that $f(x) > 0$, and all x such that $f(x) < 0$. Sketch the graph of f.

3 $f(x) = \frac{1}{2}x^3 - 4$

4 $f(x) = -\frac{1}{4}x^3 - 16$

5 $f(x) = \frac{1}{8}x^4 + 2$

6 $f(x) = 1 - x^5$

7 $f(x) = x^3 - 9x$

8 $f(x) = 16x - x^3$

9 $f(x) = -x^3 - x^2 + 2x$

10 $f(x) = x^3 + x^2 - 12x$

11 $f(x) = (x + 4)(x - 1)(x - 5)$

12 $f(x) = (x + 2)(x - 3)(x - 4)$

13 $f(x) = x^4 - 16$

14 $f(x) = 16 - x^4$

15 $f(x) = -x^4 - 3x^2 + 4$

16 $f(x) = x^4 - 7x^2 - 18$

17 $f(x) = x(x - 2)(x + 1)(x + 3)$

18 $f(x) = x(x + 1)^2(x - 3)(x - 5)$

19 Find a number k such that the graph of $f(x) = 3x^3 - kx^2 + x - 5k$ contains the point $(-1, 4)$.

20 If one zero of $f(x) = x^3 - 2x^2 - 16x + 16k$ is 2, find two other zeros

In Exercises 21–26 show that f has a zero between a and b.

21 $f(x) = x^3 - 4x^2 + 3x - 2$; $a = 3$, $b = 4$

22 $f(x) = 2x^3 + 5x^2 - 3$; $a = -3$, $b = -2$

23 $f(x) = -x^4 + 3x^3 - 2x + 1$; $a = 2$, $b = 3$

24 $f(x) = 2x^4 + 3x - 2$; $a = \frac{1}{2}$, $b = \frac{3}{4}$

25 $f(x) = x^5 + x^3 + x^2 + x + 1$; $a = -\frac{3}{2}$, $b = -1$

26 $f(x) = x^4 + 2x^3 - 5x^2 + 1$; $a = 1.33$, $b = 1.34$

3.4 PROPERTIES OF DIVISION

If a polynomial $g(x)$ is a factor of a polynomial $f(x)$, we often say that $f(x)$ is **divisible** by $g(x)$. For example, $x^2 - 25$ is divisible by $x - 5$ and by $x + 5$. A division process may also be introduced if $g(x)$ is *not* a factor of $f(x)$. The process is similar to that used in the set of integers. For example, the integer 24 has positive factors 1, 2, 3, 4, 6, 8, 12, and 24; that is, 24 is *divisible* by those numbers. A different situation exists if we divide 24 by nonfactors such as 5, 7, 15, and 32. In such cases a process called *long division* is used which yields a *quotient* and *remainder*. The reader undoubtedly remembers the process from elementary arithmetic where, for example, the work involved in dividing 4126 by

23 can be arranged as follows:

$$
\begin{array}{r}
179 \\
23 \overline{\smash{\big)}\ 4126} \\
23 \\
\hline
182 \\
161 \\
\hline
216 \\
207 \\
\hline
9
\end{array}
$$

Here the number 179 is called the *quotient* and 9 the *remainder*. To complete the terminology, 23 is called the *divisor* and 4126 the *dividend*. The remainder should always be less than the divisor for, otherwise, the quotient can be increased. The above result is sometimes written as follows:

$$\frac{4126}{23} = 179 + \frac{9}{23}.$$

Multiplying by 23 gives us

$$4126 = (23)(179) + 9.$$

The above form is very useful for theoretical purposes and can be generalized to arbitrary integers. Specifically, it can be shown that if a and b are integers with $b > 0$, then there exist unique integers q and r such that

$$a = bq + r$$

where $0 \le r < b$. The integer q is called the **quotient** and r the **remainder** in the division of a by b.

A similar discussion can be given for polynomials. For example, the polynomial $x^4 - 16$ is divisible by $x^2 - 4$, $x^2 + 4$, $x + 2$, and $x - 2$; but $x^2 + 3x + 1$ is not a factor of $x^4 - 16$. However, by another process called *long division*, we write

$$
\begin{array}{r}
x^2 - 3x + 8 \\
x^2 + 3x + 1 \overline{\smash{\big)}\ x^4 - 16} \\
x^4 + 3x^3 + x^2 \\
\hline
-3x^3 - x^2 \\
-3x^3 - 9x^2 - 3x \\
\hline
8x^2 + 3x - 16 \\
8x^2 + 24x + 8 \\
\hline
-21x - 24
\end{array}
$$

which yields the quotient $x^2 - 3x + 8$ and the remainder $-21x - 24$. In this division we proceed as indicated until we arrive at a polynomial (the remainder) which is either 0 or has smaller degree than the divisor. We shall assume familiarity with this process and not attempt to justify it. As with integers, the result of this division process is often written as follows:

$$\frac{x^4 - 16}{x^2 + 3x + 1} = (x^2 - 3x + 8) + \left(\frac{-21x - 24}{x^2 + 3x + 1}\right).$$

Multiplying by $x^2 + 3x + 1$,

$$x^4 - 16 = (x^2 + 3x + 1)(x^2 - 3x + 8) + (-21x - 24).$$

This has the same general form $a = bq + r$ that was given above for integers.

The preceding example illustrates the following theorem, which we state without proof.

Division Algorithm for Polynomials

> If $f(x)$ and $g(x)$ are polynomials and if $g(x) \neq 0$, then there exist unique polynomials $q(x)$ and $r(x)$ such that
>
> $$f(x) = g(x)q(x) + r(x)$$
>
> where either $r(x) = 0$ or the degree of $r(x)$ is less than the degree of $g(x)$. The polynomial $q(x)$ is called the **quotient** and $r(x)$ the **remainder** in the division of $f(x)$ by $g(x)$.

An interesting special case occurs if $f(x)$ is divided by a linear polynomial of the form $x - c$, where c is a real number. If $x - c$ is a factor of $f(x)$, then

$$f(x) = (x - c)q(x)$$

for some polynomial $q(x)$; that is, the remainder $r(x)$ is 0. If the remainder is not 0, then its degree is less than the degree of the divisor $x - c$, and hence $r(x)$ must have degree 0. This, in turn, means that the remainder is a nonzero real number. Consequently, in all cases we have

$$f(x) = (x - c)q(x) + d$$

where d is a real number (possibly $d = 0$). If c is substituted for x in $f(x) = (x - c)q(x) + d$, we obtain

$$f(c) = (c - c)q(c) + d$$

which reduces to $f(c) = d$, that is,

$$f(x) = (x - c)q(x) + f(c).$$

We have proved the following theorem.

Remainder Theorem

> If a polynomial $f(x)$ is divided by $x - c$, where c is a real number, then the remainder is $f(c)$.

Example 1   If $f(x) = x^3 - 3x^2 + x + 5$, use the Remainder Theorem to find $f(2)$.

Solution   According to the Remainder Theorem, $f(2)$ is the remainder when $f(x)$ is divided by $x - 2$. By long division,

$$
\begin{array}{r}
x^2 - x - 1 \\
x - 2 \overline{\smash{\big)}\ x^3 - 3x^2 + x + 5} \\
\underline{x^3 - 2x^2} \\
-x^2 + x \\
\underline{-x^2 + 2x} \\
-x + 5 \\
\underline{-x + 2} \\
3
\end{array}
$$

and hence $f(2) = 3$. To check our work we have, by direct substitution, $f(2) = 2^3 - 3(2)^2 + 2 + 5 = 3$.   ■

Factor Theorem

> A polynomial $f(x)$ has a factor $x - c$ if and only if $f(c) = 0$.

Proof   By the Remainder Theorem, $f(x) = (x - c)q(x) + f(c)$ for some quotient $q(x)$. If $f(c) = 0$, then $f(x) = (x - c)q(x)$; that is, $x - c$ is a factor of $f(x)$. Conversely, if $x - c$ is a factor, then the remainder upon division of $f(x)$ by $x - c$ must be 0, and hence by the Remainder Theorem, $f(c) = 0$.

The Factor Theorem is useful for finding factors of polynomials, as illustrated in the next example.

Example 2   Show that $x - 2$ is a factor of the polynomial

$$f(x) = x^3 - 4x^2 + 3x + 2.$$

Solution   Since $f(2) = 8 - 16 + 6 + 2 = 0$, it follows from the Factor Theorem that $x - 2$ is a factor of $f(x)$. Of course, another method of solution would be to divide $f(x)$ by $x - 2$ and show that the remainder is 0. The quotient in the division would be another factor of $f(x)$.   ■

Example 3   Find a polynomial $f(x)$ of degree 3 that has zeros 2, -1, and 3.

Solution   By the Factor Theorem, $f,(x)$ has factors $x - 2$, $x + 1$. and $x - 3$. We may then write

$$f(x) = a(x - 2)(x + 1)(x - 3)$$

where any nonzero value may be assigned to a. If we let $a = 1$ and multiply, we obtain

$$f(x) = x^3 - 4x^2 + x + 6.$$ ■

EXERCISES 3.4

In Exercises 1–6 find the quotient $q(x)$ and the remainder $r(x)$ if $f(x)$ is divided by $g(x)$.

1 $f(x) = x^4 - 2x^3 + x + 4,\ g(x) = x^2 + 3x - 5$

2 $f(x) = 2x^3 + x^2 - 5,\ g(x) = x^2 - 6x$

3 $f(x) = 5x^3 - 7x,\ g(x) = 2x^2 + 3$

4 $f(x) = -3x^4 - x^3 + x^2 - 4x + 3,$
$g(x) = 2x^3 + 4x - 1$

5 $f(x) = 9x^3 + 4x - 1,\ g(x) = x^4 + 7x^2 - 5$

6 $f(x) = 3x - 10,\ g(x) = 7x^2 + 4x - 13$

In Exercises 7–10 use the Remainder Theorem to find $f(c)$.

7 $f(x) = x^4 - 5x^3 + 2x^2 - x + 2,\quad c = 3$

8 $f(x) = 2x^3 - x^2 - 6x - 3,\quad c = 4$

9 $f(x) = 4x^3 - 7x + 9,\quad c = -2$

10 $f(x) = x^3 - 8x + 4,\quad c = \sqrt{2}$

11 Determine k so that $f(x) = x^3 - kx^2 + 3x + 7k$ is divisible by $x + 2$.

12 Determine all values of k such that $f(x) = k^2x^4 - 3kx^2 + 1$ is divisible by $x - 1$.

13 Use the Factor Theorem to show that $x - 3$ is a factor of $f(x) = x^4 - 2x^3 + x^2 - 8x - 12$.

14 Show that $x + 2$ is a factor of $f(x) = x^{10} - 1024$.

15 Prove that $f(x) = x^4 + 2x^2 + 1$ has no factor of the form $x - c$, where c is a real number.

16 Find the remainder if the polynomial $5x^{100} - 6x^{75} + 4x^{50} + 3x^{25} + 2$ is divided by $x + 1$.

17 Use the Factor Theorem to prove that $x - y$ is a factor of $x^n - y^n$ for all positive integers n. If n is even, show that $x + y$ is also a factor of $x^n - y^n$.

18 If n is an odd positive integer, prove that $x + y$ is a factor of $x^n + y^n$.

19 Prove that a polynomial $f(x)$ has a real zero if and only if $f(x)$ has a first-degree polynomial as a factor.

20 Comment on the following "proof" that if $f(x)$ is divided by $x - c$, then the remainder is $f(c)$.

$$
\begin{array}{r}
f \\
x - c\ \overline{\big)\ f(x)} \\
\underline{f(x) - f(c)} \\
f(c)
\end{array}
$$

3.5 SYNTHETIC DIVISION

When applying the Remainder Theorem, it is necessary to divide by polynomials of the form $x - c$. The process referred to as *synthetic division* simplifies the work if divisors are of that form. We shall illustrate the process by means of examples.

If the polynomial $3x^4 - 8x^3 + 9x + 5$ is divided by $x - 2$, we obtain

$$
\begin{array}{r}
3x^3 - 2x^2 - 4x + 1 \\
x - 2\ \overline{\big)\ 3x^4 - 8x^3 + 0x^2 + 9x + 5} \\
\underline{3x^4 - 6x^3} \\
-2x^3 + 0x^2 \\
\underline{-2x^3 + 4x^2} \\
-4x^2 + 9x \\
\underline{-4x^2 + 8x} \\
x + 5 \\
\underline{x - 2} \\
7
\end{array}
$$

where the term $0x^2$ has been inserted in the dividend so that *all* powers of x are accounted for. Since this technique of long division seems to involve a great deal of labor for so simple a problem, we look for a means of simplifying the notation. After arranging the terms which involve like powers of x in vertical columns as above, it is seen that the repeated expressions $3x^4$, $-2x^3$, $-4x^2$, and x may be deleted without too much chance of confusion. Also, it appears unnecessary to "bring down" the terms $0x^2$, $9x$, and 5 from the dividend as indicated. With the elimination of those repetitions, our work takes on this form:

$$
\begin{array}{r}
3x^3 - 2x^2 - 4x + 1 \\[2pt]
x - 2 \overline{\smash{\big)}\ 3x^4 - 8x^3 + 0x^2 + 9x + 5} \\
-6x^3 \\ \hline
-2x^3 \\
4x^2 \\ \hline
-4x^2 \\
8x \\ \hline
x \\
-2 \\ \hline
7
\end{array}
$$

If we take care to keep like powers of x under one another, and if we account for missing terms by means of zero coefficients as above, then some labor can be saved by omitting the symbol x. Doing this in the preceding display, we obtain the following:

$$
\begin{array}{r}
3 \quad -2 \quad -4 \quad \ \ 1 \\[2pt]
1 - 2 \overline{\smash{\big)}\ 3 \quad -8 \quad \ \ 0 \quad \ \ 9 \quad \ \ 5} \\
-6 \\ \hline
-2 \\
4 \\ \hline
-4 \\
8 \\ \hline
1 \\
-2 \\ \hline
7
\end{array}
$$

Since the divisor is a polynomial of the form $x - c$, the two coefficients in the far left position are always $1 - c$. With this in mind, we shall discard the coefficient 1. Moreover, to make our notation more compact, let us move the numbers up in the following way:

$$
\begin{array}{r}
\ \ 3 \quad -2 \quad -4 \quad \ \ 1 \\[2pt]
-2 \overline{\smash{\big)}\ 3 \quad -8 \quad \ \ 0 \quad \ \ 9 \quad \ \ 5} \\
-6 \quad \ \ 4 \quad \ \ 8 \quad -2 \\ \hline
-2 \quad -4 \quad \ \ 1 \quad \ \ 7
\end{array}
$$

If we now insert the leading coefficient 3 in the first position of the last row, the first four numbers of that row are the coefficients 3, -2, -4, and 1 of the quotient, and the final number 7 is the remainder. Since there is no need to write the coefficients of the quotient two times, we discard the first row in our scheme, obtaining

$$
\begin{array}{r|rrrrr}
-2 & 3 & -8 & 0 & 9 & 5 \\
& & -6 & 4 & 8 & -2 \\
\hline
& 3 & -2 & -4 & 1 & 7
\end{array}
$$

where the top line has also been deleted since there is no longer any need for it.

There is a simple way of interpreting the last display. Note that every number in the second row can be obtained by multiplying the number in the third row of the *preceding* column by -2. Moreover, each number in the third row can be found by subtracting the number above it in the second row from the corresponding number in the first row. This suggests a method for carrying out this procedure without actually thinking of the division process. After arranging the terms of the polynomial in decreasing powers of x, we write the coefficients in a row, supplying 0 for any missing term. Next we write $-c$ (in the above case, -2) to the left of this row. Next we bring down the leading coefficient 3 to the third row. Then we multiply that number by -2 to obtain the first number, -6, in the second row. We subtract -6 from -8 to obtain the second number, -2, in the third row, and then we multiply by $-c$ (in our case, -2) to obtain the second number, 4, in the second row. Again we subtract to get the third number, -4, in the third row. This process is continued until the final number in the third row (the remainder) is obtained.

It is possible to avoid the subtractions performed above if the number c is used in place of $-c$ in the far left position of the first row. In this event, when the process above is used, the signs of the elements in the second row are changed, and hence to find elements of the third row, we *add* the number above it in the second row to the corresponding number in the first row. With this change we obtain

$$
\begin{array}{r|rrrrr}
2 & 3 & -8 & 0 & 9 & 5 \\
& & 6 & -4 & -8 & 2 \\
\hline
& 3 & -2 & -4 & 1 & 7
\end{array}
$$

The latter scheme is called the process of **synthetic division**.

Example 1 Use synthetic division to find the quotient and remainder if $2x^4 + 5x^3 - 2x - 8$ is divided by $x + 3$.

Solution Since we are to divide by $x + 3$, the c in the expression $x - c$ is -3. Hence the synthetic division takes this form:

$$
\begin{array}{r|rrrrr}
-3 & 2 & 5 & 0 & -2 & -8 \\
& & -6 & 3 & -9 & 33 \\
\hline
& 2 & -1 & 3 & -11 & 25
\end{array}
$$

The first four numbers in the third row are the coefficients of the quotient and the last number is the remainder. Hence the quotient is $2x^3 - x^2 + 3x - 11$ and the remainder is 25. ■

Synthetic division can be used to find values of polynomial functions, as illustrated in the next example.

Example 2   If $f(x) = 3x^5 - 38x^3 + 5x^2 - 1$, use synthetic division to find $f(4)$.

Solution   By the Remainder Theorem, $f(4)$ is the remainder when $f(x)$ is divided by $x - 4$. Dividing synthetically, we obtain

$$
\begin{array}{r|rrrrrr}
4 & 3 & 0 & -38 & 5 & 0 & -1 \\
  &   & 12 & 48 & 40 & 180 & 720 \\
\hline
  & 3 & 12 & 10 & 45 & 180 & 719 \\
\end{array}
$$

Consequently, $f(4) = 719$.   ∎

Synthetic division may be employed to help find zeros of polynomials. By the method illustrated in the preceding example, $f(c) = 0$ if and only if the remainder in the synthetic division by $x - c$ is 0.

Example 3   Show that -11 is a zero of the polynomial

$$f(x) = x^3 + 8x^2 - 29x + 44.$$

Solution   Dividing synthetically by $x + 11$ gives us

$$
\begin{array}{r|rrrr}
-11 & 1 & 8 & -29 & 44 \\
    &   & -11 & 33 & -44 \\
\hline
    & 1 & -3 & 4 & 0 \\
\end{array}
$$

Thus $f(-11) = 0$.   ∎

The preceding example shows that -11 is a solution of the equation $x^3 + 8x^2 - 29x + 44 = 0$. In the next section we shall use synthetic division in this way to find solutions of equations.

EXERCISES 3.5

In each of Exercises 1–10 use synthetic division to find the quotient and remainder if the first polynomial is divided by the second.

1 $2x^3 - 3x^2 + 4x - 5, \quad x - 2$

2 $3x^3 - 4x^2 - x + 8, \quad x + 4$

3 $x^3 - 8x - 5, \quad x + 3$

4 $5x^3 - 6x^2 + 15, \quad x - 4$

5 $3x^5 + 6x^2 + 7, \quad x + 2$

6 $-2x^4 + 10x - 3, \quad x - 3$

7 $4x^4 - 5x^2 + 1$, $x - \frac{1}{2}$

8 $9x^3 - 6x^2 + 3x - 4$, $x - \frac{1}{3}$

9 $x^n - 1$, $x - 1$, where n is any positive integer

10 $x^n + 1$, $x + 1$, where n is any positive integer

Use synthetic division to solve Exercises 11–16.

11   If $f(x) = x^4 - 4x^3 + x^2 - 3x - 5$, find $f(2)$ and $f(-2)$.

12   If $f(x) = 0.3x^3 + 0.04x - 0.034$, find $f(0.2)$ and $f(-0.2)$.

13   If $f(x) = x^6 - x^5 + x^4 - x^3 + x^2 - x + 1$, find $f(4)$.

14   If $f(x) = 8x^5 - 3x^2 + 7$, find $f(1/2)$.

15   If $f(x) = x^2 + 3x - 5$, find $f(2 + \sqrt{3})$.

16   If $f(x) = x^3 - 3x^2 - 8$, find $f(1 + \sqrt{2})$.

In Exercises 17–20 use synthetic division to show that c is a zero of $f(x)$.

17 $f(x) = 3x^4 + 8x^3 - 2x^2 - 10x + 4$, $c = -2$

18 $f(x) = 4x^3 - 9x^2 - 8x - 3$, $c = 3$

19 $f(x) = 4x^3 - 6x^2 + 8x - 3$, $c = \frac{1}{2}$

20 $f(x) = 27x^4 - 9x^3 + 3x^2 + 6x + 1$, $c = -\frac{1}{3}$

3.6   REAL ZEROS OF POLYNOMIAL FUNCTIONS

If $f(x)$ is a polynomial, and if a is a real number such that $f(a) = 0$, we shall call a a **real zero** of f (or of $f(x)$). In this event a is also a *root*, or *solution*, of the equation $f(x) = 0$. The Factor Theorem indicates that there is a close relationship between the real zeros and the factors of $f(x)$. Indeed, $f(c) = 0$ if and only if $x - c$ is a factor. Unfortunately, except in special cases, zeros of polynomials are very difficult to find. For example, given the polynomial $f(x) = x^5 - 3x^4 + 4x^3 + 4x - 10$, there are no obvious zeros. Moreover, there is no device such as the quadratic formula which can be used to produce the zeros. In spite of the practical difficulty of determining zeros of polynomials, it is possible to make some headway concerning the *theory* of such zeros. The results in this section form the basis for work in what is known as **The Theory of Equations**.

Theorem

> A polynomial function f of degree n has, at most, n distinct real zeros.

Proof

We shall give an indirect proof. Thus suppose $f(x)$ has *more* than n distinct real zeros. Let us choose $n + 1$ of the zeros and label them $c_1, c_2, \ldots, c_n$, and c. From the Factor Theorem,

$$f(x) = (x - c_1)f_1(x)$$

for some polynomial $f_1(x)$. Since $f(x)$ has degree n, the sum of the degrees of $x - c_1$ and $f_1(x)$ must also be n. (Why?). Since $x - c_1$ has degree 1, it follows that the polynomial $f_1(x)$ must have degree $n - 1$.

If we substitute c_2 for x in the last equation and use the fact that $f(c_2) = 0$, we obtain

$$0 = (c_2 - c_1)f_1(c_2).$$

Since c_1 and c_2 are different, $c_2 - c_1 \neq 0$ and hence $f_1(c_2) = 0$; that is c_2 is a zero of f_1. Applying the Factor Theorem to f_1, we may write

$$f_1(x) = (x - c_2)f_2(x)$$

where the polynomial $f_2(x)$ has degree $n - 2$. Using the fact that $f(x) = (x - c_1)\, f_1(x)$ gives us

$$f(x) = (x - c_1)(x - c_2)f_2(x).$$

We continue the process, obtaining a factor $x - c_3$ of $f_2(x)$. Since the degrees of the polynomials $f_i(x)$ decrease by one at each step, and since the sum of the degrees of the polynomials in the factorization must equal n, it follows that after n steps we reach a polynomial $f_n(x)$ of degree 0. Thus $f_n(x) = a$ for some nonzero real number a, and we have

$$f(x) = a(x - c_1)(x - c_2)\cdots(x - c_n).$$

Substituting c for x and using the fact that $f(c) = 0$, we obtain

$$0 = a(c - c_1)(c - c_2)\cdots(c - c_n).$$

However, each factor on the right side is different from zero because $c \neq c_i$ for every i. Since the product of nonzero real numbers cannot equal zero, we have a contradiction, and hence f cannot have more than n distinct real zeros.

The preceding theorem gives us information about the *maximum* number of real zeros of a polynomial function f; however, it says nothing about the *existence* of real zeros. Indeed, many polynomial functions have *no* real zeros. For example, if $f(x) = x^2 + 1$, then $f(c) = c^2 + 1 > 0$ for every real number c.

In some cases a polynomial may have a factorization as a product of linear factors $x - c_i$, where not all of the zeros c_i are distinct. To illustrate, the polynomial $f(x) = x^3 + x^2 - 5x + 3$ has the factorization

$$f(x) = (x + 3)(x - 1)(x - 1).$$

If a factor $x - c$ occurs exactly k times in the factorization of $f(x)$, then c is called a **zero of multiplicity k** of $f(x)$. In the preceding illustration, 1 is a zero of multiplicity 2, and -3 is a zero of multiplicity 1.

As another example, if

$$f(x) = (x - 2)(x - 4)^3(x + 1)^2$$

then f has degree 6 and possesses three distinct zeros, 2, 4, and -1, where 2 has multiplicity 1, 4 has multiplicity 3, and -1 has multiplicity 2.

In Chapter Seven, after *complex numbers* are introduced, it will be shown that every nth degree polynomial $f(x)$ which has real or complex coefficients can be expressed as a product of n (not necessarily distinct) linear factors $x - c_i$, where c_i is a real or complex number. It will then follow that if a zero of multiplicity k is counted as k zeros, every nth degree polynomial has exactly n (real or complex) zeros.

The next theorem may be used to obtain information about the real zeros of a polynomial. In the statement of the theorem it is assumed that the terms of a polynomial are arranged in order of descending powers of x, and that terms with zero coefficients are deleted. In this case we say there is a **variation of sign** in $f(x)$ if two consecutive coefficients have opposite signs. To illustrate, the polynomial

$$f(x) = 2x^5 - 7x^4 + 3x^2 + 6x - 5$$

has three variations in sign, since there is one variation from $2x^5$ to $-7x^4$, a second from $-7x^4$ to $3x^2$, and a third from $6x$ to -5.

The theorem also refers to the variations of sign in $f(-x)$. Using the previous illustration, note that

$$f(-x) = 2(-x)^5 - 7(-x)^4 + 3(-x)^2 + 6(-x) - 5$$

$$= -2x^5 - 7x^4 + 3x^2 - 6x - 5.$$

Hence there are two variations of sign in $f(-x)$, one from $-7x^4$ to $3x^2$, and a second from $3x^2$ to $-6x$.

**Descartes'
Rule of
Signs**

If $f(x)$ is a polynomial with real coefficients, then

(i)   the number of positive real solutions of the equation $f(x) = 0$ either is equal to the number of variations of sign in $f(x)$ or is less than that number by an even integer.

(ii)   the number of negative real solutions of the equation $f(x) = 0$ either is equal to the number of variations of sign in $f(-x)$ or is less than that number by an even integer.

The proof of Descartes' Rule is rather technical and will not be given in this text.

Example 1   Discuss the number of possible positive and negative solutions of the equation

$$2x^5 - 7x^4 + 3x^2 + 6x - 5 = 0.$$

Solution   The polynomial $f(x)$ on the left side of the equation is the same as that given in the illustration preceding the statement of Descartes' Rule. Since there are three variations of sign in $f(x)$, the equation has either three positive real solutions or one positive real solution.

Since $f(-x) = -2x^5 - 7x^4 + 3x^2 - 6x - 5$ has two variations of sign, the given equation has either two negative solutions or no negative solutions.

The following table summarizes the various possiblilties that can occur, where the number of complex solutions is a consequence of the material to be discussed in Chapter Seven. Note that in each case there are five solutions to the equation.

Number of positive solutions	1	1	3	3
Number of negative solutions	0	2	0	2
Number of complex solutions	4	2	2	0

■

Example 2 Discuss the nature of the roots of the equation

$$3x^5 + 4x^3 + 2x - 5 = 0.$$

Solution The polynomial $f(x)$ on the left side of the equation has one variation of sign and hence by (i) of Descartes' Rule of Signs, there is precisely one positive real root. Since

$$f(-x) = 3(-x)^5 + 4(-x)^3 + 2(-x) - 5$$

$$= -3x^5 - 4x^3 - 2x - 5$$

there are no variations of sign in $f(-x)$ and hence by (ii) of Descartes' Rule of Signs, there are no negative real roots. Thus the equation has one real root and four complex roots. Moreover, since $f(0) = -5$ and $f(1) = 4$, the real root is between 0 and 1. (Why?) ■

When applying Descartes' Rule, roots of multiplicity k should be counted as k roots. For example, given $x^2 - 2x + 1 = 0$, the indicated polynomial has two variations of sign and hence there are either two positive real roots or none. The factored form of the equation is $(x - 1)^2 = 0$, and hence the equation has the *double* root 1.

We shall conclude this section with a discussion of *bounds* for the real solutions of an equation $f(x) = 0$, where $f(x)$ is a polynomial. By definition, a real number b is an **upper bound** for the solutions if no solution is greater than b. A real number a is a **lower bound** for the solutions if no solution is less than a. Thus, if r is a real solution of $f(x) = 0$, then $a \leq r \leq b$. Note that upper and lower bounds are not unique, since any number greater than b is also an upper bound, and any number less than a is a lower bound.

If $f(x)$ is a polynomial we may use synthetic division to find upper and lower bounds for the solutions of $f(x) = 0$. Recall that if $f(x)$ is divided synthetically by $x - c$, then the third row which appears in the division process consists of the coefficients of the quotient $q(x)$ together with the remainder $f(c)$. The following theorem indicates how this third row may be used to find upper and lower bounds for the real solutions.

Bounds for Real Zeros of Polynomials

Suppose $f(x)$ is a polynomial with real coefficients and positive leading coefficient, and let $f(x)$ be divided synthetically by $x - c$.

(i) If $c > 0$, and if all numbers in the third row of the division process are either positive or zero, then c is an upper bound for the real solutions of $f(x) = 0$.

(ii) If $c < 0$, and if the numbers in the third row of the division process are alternatively positive or negative (where a 0 in the third row is to be considered either positive or negative), then c is a lower bound for the real solutions of $f(x) = 0$.

A general proof of this theorem can be patterned after the solution given in the next example.

Example 3   Find upper and lower bounds for the real solutions of the equation

$$2x^3 + 5x^2 - 8x - 7 = 0.$$

Solution   If we divide synthetically by $x - 1$ and $x - 2$, we obtain

$$
\begin{array}{r|rrrr}
1 & 2 & 5 & -8 & -7 \\
  &   & 2 & 7 & -1 \\
\hline
  & 2 & 7 & -1 & -8
\end{array}
\qquad
\begin{array}{r|rrrr}
2 & 2 & 5 & -8 & -7 \\
  &   & 4 & 18 & 20 \\
\hline
  & 2 & 9 & 10 & 13
\end{array}
$$

Since all numbers in the third row of the synthetic division by $x - 2$ are positive, it follows from (i) of the preceding theorem that 2 is an upper bound for the solutions of the given equation. This fact is also evident by noting that if we divide by $x - c$, where $c > 2$, then the numbers in the third row are greater than those which appear in the division by $x - 2$. Hence if $f(x)$ is the indicated polynomial, then $f(c) > 0$ if $c > 2$, and there can be no zero greater than 2.

Dividing synthetically by $x - (-3)$ and $x - (-4)$ gives us

$$
\begin{array}{r|rrrr}
-3 & 2 & 5 & -8 & -7 \\
   &   & -6 & 3 & 15 \\
\hline
   & 2 & -1 & -5 & 8
\end{array}
\qquad
\begin{array}{r|rrrr}
-4 & 2 & 5 & -8 & -7 \\
   &   & -8 & 12 & -16 \\
\hline
   & 2 & -3 & 4 & -23
\end{array}
$$

Since the numbers in the third row of the synthetic division by $x - (-4)$ are alternately positive or negative, it follows from (ii) of the preceding theorem that -4 is a lower bound for the solutions. This can also be proved by noting that if we divide by $x - c$, where $c < -4$, then the numbers in the third row are, in absolute value, greater than the absolute values of those appearing in the division by $x - (-4)$. Hence $f(c) < -23$ if $c < -4$, and there can be no real solutions less than -4. ∎

EXERCISES 3.6

1 Find a polynomial of degree 4 such that both 4 and -3 are zeros of multiplicity 2.

2 Find a polynomial of degree 4 such that 2 is a zero of multiplicity 3 and -2 is a zero of multiplicity 1.

3 Find a polynomial of degree 7 such that 1 is a zero of multiplicity 3 and 0 is a zero of multiplicity 4.

4 Find a polynomial of degree 5 such that -1 is a zero of multiplicity 2 and 0 is a zero of multiplicity 3.

In Exercises 5–12 find the zeros of the polynomials and state the multiplicity of each zero.

5 $f(x) = (x + 5)^2(x - 1)$

6 $f(x) = (x + 4)^3(x^2 - 1)$

7 $f(x) = x^5 + x^4 - 5x^3$

8 $f(x) = (x^2 - 25)^2$

9 $f(x) = (x^2 - 9)^2(x^2 - 4)$

10 $f(x) = (x^2 - 7x + 6)^2$

11 $f(x) = (x^2 + 2x - 3)(x + 2)^4$

12 $f(x) = (5x - 8)^5$

13 Show that 3 is a zero of multiplicity 2 of the polynomial $f(x) = x^4 - 6x^3 + 4x^2 + 30x - 45$ and express $f(x)$ as a product of linear factors.

14 Show that 2 is a zero of multiplicity 2 of the polynomial $f(x) = x^4 - 6x^3 + 9x^2 + 4x - 12$ and express $f(x)$ as a product of linear factors.

15 Show that 1 is a zero of multiplicity 3 of $f(x) = x^4 + x^3 - 9x^2 + 11x - 4$ and find the other zero.

16 Show that -1 is a zero of multiplicity 4 of $f(x) = x^6 + 4x^5 + x^4 - 16x^3 - 29x^2 - 20x - 5$ and find the other zeros.

In each of Exercises 17–24 use Descartes' Rule of Signs to determine the number of possible positive and negative solutions of the given equation.

17 $4x^3 - 6x^2 + x - 3 = 0$

18 $5x^3 - 6x - 4 = 0$

19 $4x^3 + 2x^2 + 1 = 0$

20 $3x^3 - 4x^2 + 3x = 0$

21 $3x^4 + 2x^3 - 4x + 2 = 0$

22 $2x^4 - x^3 + x^2 - 3x + 4 = 0$

23 $x^5 + 4x^4 + 3x^3 - 4x + 2 = 0$

24 $2x^6 + 5x^5 + 2x^2 - 3x + 4 = 0$

In each of Exercises 25–30 find the smallest and largest integers that are upper and lower bounds, respectively, for the real solutions of the given equation.

25 $x^3 - 4x^2 - 5x + 7 = 0$

26 $2x^3 - 5x^2 + 4x - 8 = 0$

27 $x^4 - x^3 - 2x^2 + 3x + 6 = 0$

28 $2x^4 - 9x^3 - 8x - 10 = 0$

29 $2x^5 - 13x^3 + 2x - 5 = 0$

30 $3x^5 + 2x^4 - x^3 - 8x^2 - 7 = 0$

31 Let $f(x)$ and $g(x)$ be polynomials of degree not greater than n, where n is a positive integer. Show that if $f(x)$ and $g(x)$ are equal in value for more than n distinct real values of x, then $f(x)$ and $g(x)$ are identical, that is, coefficients of like powers are the same. (*Hint:* Write $f(x)$ and $g(x)$ as in Section 1 and consider $h(x) = f(x) - g(x) = (a_n - b_n)x^n + \cdots + (a_0 - b_0)$. Then show that $h(x)$ has more than n distinct real zeros and conclude that $a_i = b_i$ for all i.

3.7 RATIONAL ZEROS OF POLYNOMIALS

We have already pointed out that it is generally very difficult to find zeros of polynomials of high degree. However, if all the coefficients are integers or rational numbers, there is a method for finding the *rational* zeros, if they exist. The method is a consequence of the following theorem.

Theorem on Rational Zeros

> Suppose that $f(x) = a_n x^n + a_{n-1} x^{n-1} + \cdots + a_1 x + a_0$ is a polynomial with integer coefficients. If c/d is a rational zero of $f(x)$, where c and d have no common prime factors and $c > 0$, then c is a factor of a_0 and d is a factor of a_n.

Proof

Let us show that c is a factor of a_0. If $c = 1$, the theorem follows at once, since 1 is a factor of *any* number. Now suppose that $c \neq 1$. In this case $c/d \neq 1$, for if $c/d = 1$, we obtain $c = d$, and since c and d have no prime factor in common, this implies that $c = d = 1$, a contradiction. Hence in the following discussion we have $c \neq 1$ and $c \neq d$.

Since $f(c/d) = 0$,

$$a_n(c^n/d^n) + a_{n-1}(c^{n-1}/d^{n-1}) + \cdots + a_1(c/d) + a_0 = 0.$$

Multiplying by d^n and then adding $-a_0 d^n$ to both sides, we obtain

$$a_n c^n + a_{n-1} c^{n-1} d + \cdots + a_1 cd^{n-1} = -a_0 d^n$$

or

$$c(a_n c^{n-1} + a_{n-1} c^{n-2} d + \cdots + a_1 d^{n-1}) = -a_0 d^n.$$

This shows that c is a factor of the integer $a_0 d^n$. If c is factored into primes, say $c = p_1 p_2 \cdots p_k$, then each prime p_i is also a factor of $a_0 d^n$. However, by hypothesis, none of the p_i is a factor of d. This implies that each p_i is a factor of a_0, that is, c is a factor of a_0. A similar argument may be used to prove that d is a factor of a_n.

The technique of using the preceding theorem for finding rational solutions of equations with integer coefficients is illustrated in the following example.

Example 1   Find all rational solutions of the equation

$$3x^4 + 14x^3 + 14x^2 - 8x - 8 = 0.$$

Solution   The problem is equivalent to finding the rational zeros of the indicated polynomial. According to the last theorem, if c/d is a rational zero and $c > 0$, then c is a divisor of -8 and d is a divisor of 3. Hence the possible choices for c are 1, 2, 4, and 8, and the choices for d are ± 1 and ± 3. Consequently, any rational roots are included among the numbers ± 1, ± 2, ± 4, ± 8, $\pm \frac{1}{3}$, $\pm \frac{2}{3}$, $\pm \frac{4}{3}$, and $\pm \frac{8}{3}$. It is necessary to check to see which, if any, are zeros. The method of synthetic division is recommended for carrying out this task. The number of possibilities can be reduced by finding upper and lower bounds for the real solutions; however, we shall not do so here. We find that

$$
\begin{array}{r|rrrrr}
-2 & 3 & 14 & 14 & -8 & -8 \\
  &   & -6 & -16 & 4 & 8 \\
\hline
  & 3 & 8 & -2 & -4 & 0
\end{array}
$$

which shows that -2 is a zero. Moreover, the synthetic division provides the coefficients of the quotient in the division of the polynomial by $x + 2$. Hence we have the following factorization of the given polynomial:

$$(x + 2)(3x^3 + 8x^2 - 2x - 4).$$

The remaining solutions of the equation must be zeros of the second factor, and therefore we may use the latter polynomial to check for solutions. Dividing by $x + \frac{2}{3}$ synthetically gives us

$$
\begin{array}{r|rrrr}
-\frac{2}{3} & 3 & 8 & -2 & -4 \\
  &   & -2 & -4 & 4 \\
\hline
  & 3 & 6 & -6 & 0
\end{array}
$$

and therefore $-\frac{2}{3}$ is a zero.

The remaining zeros are solutions of the equation $3x^2 + 6x - 6 = 0$, or equivalently, $x^2 + 2x - 2 = 0$. By the quadratic formula this equation has solutions

$$\frac{-2 \pm \sqrt{4 - 4(-2)}}{2} = \frac{-2 \pm \sqrt{12}}{2} = \frac{-2 \pm 2\sqrt{3}}{2}.$$

Hence the given polynomial has two rational roots, -2 and $-\frac{2}{3}$, and two irrational roots, $-1 + \sqrt{3}$ and $-1 - \sqrt{3}$.   ■

The Theorem on Rational Zeros may also be applied to equations with rational coefficients. We merely multiply both sides of the equation by the least common denominator of all the coefficients to obtain an equation with integral coefficients and then proceed as above.

Example 2   Find all rational solutions of the equation

$$(\tfrac{2}{3})x^4 + (\tfrac{1}{2})x^3 - (\tfrac{5}{4})x^2 - x - (\tfrac{1}{6}) = 0.$$

Solution   Multiplying both sides of the equation by 12 produces the equivalent equation

$$8x^4 + 6x^3 - 15x^2 - 12x - 2 = 0.$$

If c/d is a rational solution, then the choices for c are 1 and 2 and the choices for d are $\pm 1, \pm 2, \pm 4$, and ± 8. Hence the only possible rational roots are $\pm 1, \pm 2, \pm\frac{1}{2}$, $\pm\frac{1}{4}$, and $\pm\frac{1}{8}$. By trial we have

$$
\begin{array}{r|rrrrr}
-\frac{1}{2} & 8 & 6 & -15 & -12 & -2 \\
 & & -4 & -1 & 8 & 2 \\
\hline
 & 8 & 2 & -16 & -4 & 0
\end{array}
$$

and hence $-\frac{1}{2}$ is a solution. Using synthetic division on the coefficients of the quotient, we obtain

$$
\begin{array}{r|rrrr}
-\frac{1}{4} & 8 & 2 & -16 & -4 \\
 & & -2 & 0 & 4 \\
\hline
 & 8 & 0 & -16 & 0
\end{array}
$$

and consequently, $-\frac{1}{4}$ is a solution. The last synthetic division gave us the quotient $8x^2 - 16$. Setting this equal to zero and solving, we obtain $x^2 = 2$, or $x = \pm\sqrt{2}$. Thus the given equation has rational solutions $-\frac{1}{2}, -\frac{1}{4}$ and irrational solutions $\sqrt{2}, -\sqrt{2}$. ∎

The discussion in this section gives no practical information about finding the irrational zeros of polynomials. The examples we have worked are not typical of problems encountered in applications. Indeed, it is not unusual for a polynomial with rational coefficients to have *no* rational zeros. Except in the simplest cases, the best that can be accomplished is to find decimal approximations to the irrational zeros. There exist methods which may be used to approximate irrational zeros to any degree of accuracy. A standard way is to use a technique studied in calculus called Newton's Method. In practice, computers have taken over the task of approximating irrational solutions of equations to a large extent.

If only rough approximations to the real solutions of an equation $f(x) = 0$ are required, then graphical methods are available. For example, we could sketch the graph of $y = f(x)$ and estimate where $y = 0$; that is, we could approximate the x-intercepts. Needless to say, the accuracy of the approximation depends on the care with which the graph is sketched.

EXERCISES 3.7

In Exercises 1–14 find all rational solutions of the given equations. If possible, find all the real solutions of the equations.

1 $2x^3 - 3x^2 - 17x + 30 = 0$

2 $2x^3 - 3x^2 - 7x - 6 = 0$

3 $6x^3 + 11x^2 - 4x - 4 = 0$

4 $6x^3 + 11x^2 - 57x - 20 = 0$

5 $3x^3 + 8x^2 - x - 20 = 0$

6 $12x^3 - x^2 + 7x + 2 = 0$

7 $x^4 + x^3 - 5x^2 - 15x - 18 = 0$

8 $8x^4 + 2x^3 - 7x^2 + 2x - 15 = 0$

9 $(\frac{9}{4})x^4 - (\frac{15}{4})x^3 - 20x^2 + (\frac{11}{2})x + \frac{1}{2} = 0$

10 $(\frac{1}{3})x^4 - x^3 - x^2 + (\frac{13}{3})x - 2 = 0$

11 $x^4 + 3x^3 - 30x^2 - 6x + 56 = 0$

12 $x^4 - 3x^3 - 43x^2 + 9x + 20 = 0$

13 $4x^5 + 12x^4 - 41x^3 - 99x^2 + 10x + 24 = 0$

14 $4x^5 + 24x^4 - 13x^3 - 174x^2 + 9x + 270 = 0$

Prove that the equations in Exercises 15–18 have no rational roots.

15 $2x^4 + 2x^3 + 9x^2 - x - 5 = 0$

16 $3x^4 - 9x^3 - 2x^2 - 15x - 5 = 0$

17 $x^5 - 3x^3 + 4x^2 + x - 2 = 0$

18 $2x^5 + 3x^3 + 7 = 0$

19 If a polynomial of the form $x^n + a_{n-1}x^{n-1} + \cdots + a_1x + a_0$, where each a_i is an integer, has a rational root r, show that r is an integer and is a factor of a_0.

20 Complete the proof of the Theorem on Rational Zeros by showing that d is a factor of a_n.

3.8 RATIONAL FUNCTIONS

A **rational function** is a quotient of two polynomial functions. Thus f is rational if for all x in its domain,

$$f(x) = \frac{g(x)}{h(x)}$$

where $g(x)$ and $h(x)$ are polynomials. Since division by zero is not permissible, the domain of f consists of all real numbers except those for which the denominator $h(x)$ is zero. In order to obtain the graph of f, it is extremely important to examine the behavior of $f(x)$ when x is near a number c such that $h(c) = 0$ and $g(c) \neq 0$.

Example 1 Sketch the graph of f if $f(x) = \frac{1}{x}$.

Solution The function f is rational since $f(x)$ has the form $g(x)/h(x)$, with $g(x) = 1$ and $h(x) = x$. The domain of f is the set of all nonzero real numbers. (Why?) Before constructing a table, let us make some general observations. If x is positive, so is $1/x$, and hence quadrant IV is an excluded region. Quadrant II is also excluded since, if $x < 0$, then $1/x < 0$. If x is close to zero, then $1/x$ is very large numerically. As x increases through positive values, $1/x$ decreases and is close to zero when x is large. The variation of $f(x)$ is brought out in the following table.

x	$\frac{1}{100}$	$\frac{1}{10}$	$\frac{1}{4}$	$\frac{1}{2}$	1	2	4	10	100
$f(x)$	100	10	4	2	1	$\frac{1}{2}$	$\frac{1}{4}$	$\frac{1}{10}$	$\frac{1}{100}$

By plotting some points and using the previous discussion, we obtain the first quadrant part of the graph in Figure 3.16. Since the graph of f is the same as the graph of $y = 1/x$, we may apply a test for symmetry to show that the graph is symmetric with respect to the origin. This fact gives us the third quadrant part of the graph in Figure 3.16.

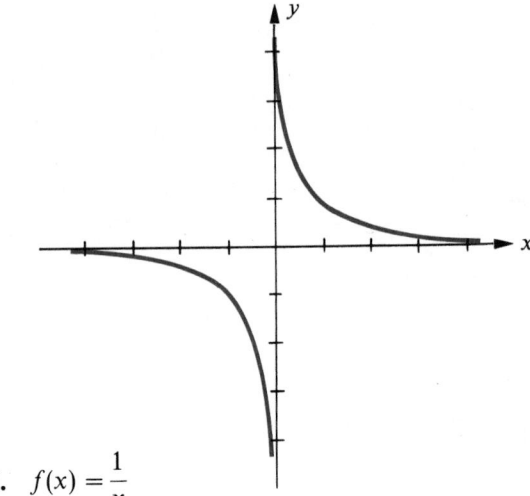

Figure 3.16. $f(x) = \dfrac{1}{x}$ ■

If, in the previous example, we assign values to x which are very close to 0, then $f(x)$ becomes very large numerically. Thus,

$$f(0.0001) = 10,000, \quad f(0.000001) = 1,000,000$$

and so on. Indeed, $f(x)$ can be made as large as desired by taking a positive number x sufficiently close to 0. This is sometimes referred to by the statement:

$f(x)$ increases without bound as x approaches 0 through positive values

and is denoted symbolically by

$$f(x) \to \infty \quad \text{as} \quad x \to 0^+.$$

The phrase "$f(x)$ becomes positively infinite" is sometimes used to describe the variation described above. It is important to remember that the symbol ∞ (read "infinity") does not represent a real number, but is used merely as an abbreviation for certain types of functional behavior.

In like manner, if x is *negative* and close to 0, then $f(x)$ is a numerically large negative number. For example,

$$f(-0.001) = -1,000, \quad f(-0.00001) = -100,000$$

and so on. In this case we say that

$f(x)$ decreases without bound as x approaches 0 through negative values

and write

$$f(x) \to -\infty \quad \text{as} \quad x \to 0^-.$$

The phrase "$f(x)$ becomes negatively infinite" is sometimes used in place of the phrase "decreases without bound." The line $x = 0$ in Figure 3.16, that is, the y-axis, is called a **vertical asymptote** for the graph of f. Vertical asymptotes are common characteristics of graphs of rational functions.

If, in Example 1, we assign very *large* values to x, then $f(x)$ is close to 0. For example,

$$f(1,000) = 0.001 \quad \text{and} \quad f(1,000,000) = 0.000001.$$

Moreover, we can make $f(x)$ as close to 0 as desired by choosing x sufficiently large. In this case we say that

$$f(x) \text{ approaches } 0 \text{ as } x \text{ becomes positively infinite}$$

and denote this fact by writing

$$f(x) \to 0 \quad \text{as} \quad x \to \infty.$$

Similarly, in Example 1,

$$f(x) \text{ approaches } 0 \text{ as } x \text{ becomes negatively infinite,}$$

written

$$f(x) \to 0 \quad \text{as} \quad x \to -\infty.$$

Since the graph gets closer and closer to the line $y = 0$ as $|x|$ increases, we call the x-axis a **horizontal asymptote** for the graph.

Example 2   Sketch the graph of f if $f(x) = 1/x^2$.

Solution   The graph here differs from that in Example 1 since $f(x)$ is never negative. Indeed, we see that

$$f(x) \to \infty \quad \text{as} \quad x \to 0^+$$

and
$$f(x) \to \infty \quad \text{as} \quad x \to 0^-.$$

In addition, the present graph is steeper near the origin and approaches the x-axis more rapidly as x increases. These facts can be seen by comparing the following table with the corresponding table in Example 1.

x	$\frac{1}{100}$	$\frac{1}{10}$	$\frac{1}{4}$	$\frac{1}{2}$	1	2	10	100
$f(x)$	10000	100	16	4	1	$\frac{1}{4}$	$\frac{1}{100}$	$\frac{1}{10000}$

Plotting leads to the first quadrant part of the graph in Figure 3.17. Since the graph is symmetric with respect to the y-axis (Why?), it is unnecessary to tabulate points with negative abscissas. We merely reflect the first quadrant part through the y-

axis. Note that the y-axis is a vertical asymptote and the x-axis is a horizontal asymptote.

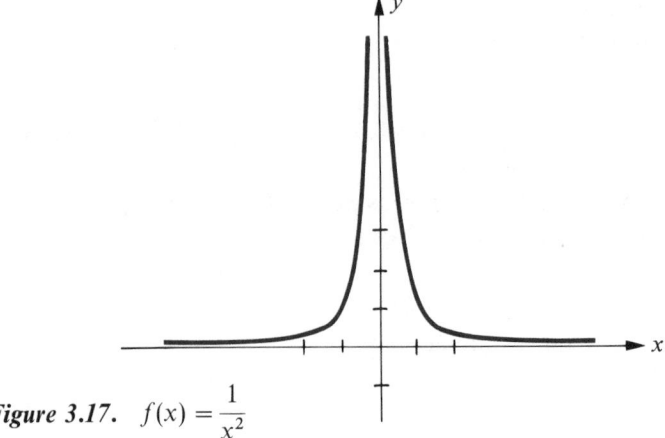

Figure 3.17. $f(x) = \dfrac{1}{x^2}$ ■

Examples 1 and 2 can be generalized to $f(x) = 1/x^n$ where n is *any* positive integer. If n is odd, as for $1/x^3$ and $1/x^5$, the graph has the general appearance shown in Figure 3.16, except that there is a sharper corner near the origin. If n is even, as for $1/x^4$ and $1/x^6$, the graph is similar to that in Figure 3.17, except that the graph again has a sharper corner near the origin.

Let us extend the previous discussion to the case in which x approaches *any* real number a. In particular, the symbol $x \to a^+$ will signify that x approaches a through values *greater* than a, whereas $x \to a^-$ will mean that x approaches a through values *less* than a. Some illustrations of the manner in which a function may increase or decrease without bound, together with the notation used, are shown in Figure 3.18.

Of major importance in sketching the graph of a rational function is the special case

$$f(x) = \frac{1}{(x-a)^n}$$

where n is a positive integer and a is any real number. If n is even, then $(x-a)^n$ is always positive when $x \neq a$ and, therefore, $f(x) \to \infty$ if either $x \to a^-$ or $x \to a^+$, as illustrated in (i) of Figure 3.19. However, if n is odd and if $x < a$, then $x - a < 0$ and hence $(x-a)^n < 0$. In this event $f(x)$ is negative and we see that $f(x) \to -\infty$ as $x \to a^-$ (see (ii) of Figure 3.19).

The remarks made in the preceding paragraph about the variation of $f(x)$ as $x \to a^-$ or $x \to a^+$ are also true if

$$f(x) = \frac{k}{(x-a)^n}$$

where k is any *positive* real number.

To complete the discussion, let us consider what happens if k is *negative*. In this case if n is even, then again $(x-a)^n$ is always positive when $x \neq a$, and hence $f(x)$ is

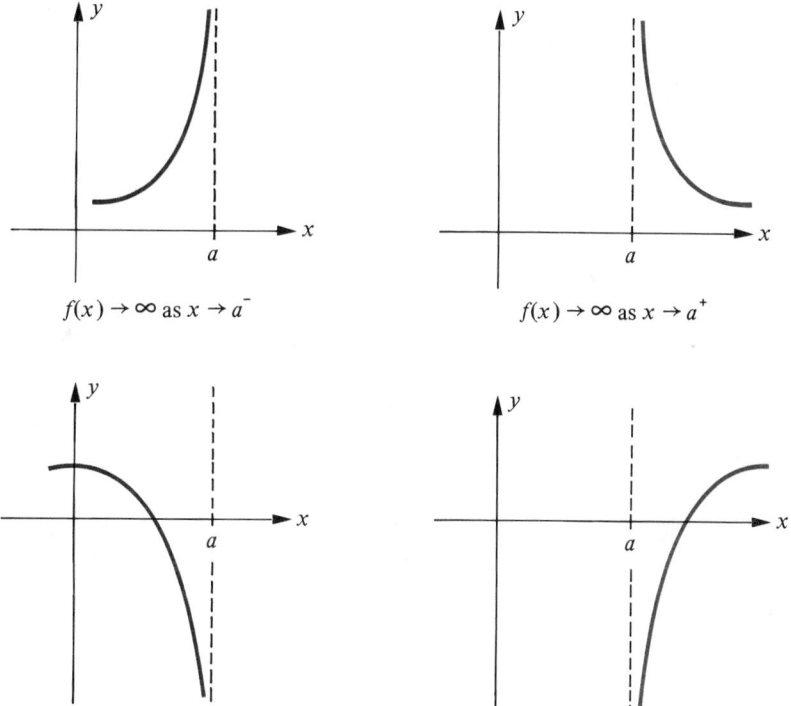

$f(x) \to \infty$ as $x \to a^-$

$f(x) \to \infty$ as $x \to a^+$

$f(x) \to -\infty$ as $x \to a$

$f(x) \to -\infty$ as $x \to a^+$

Figure 3.18

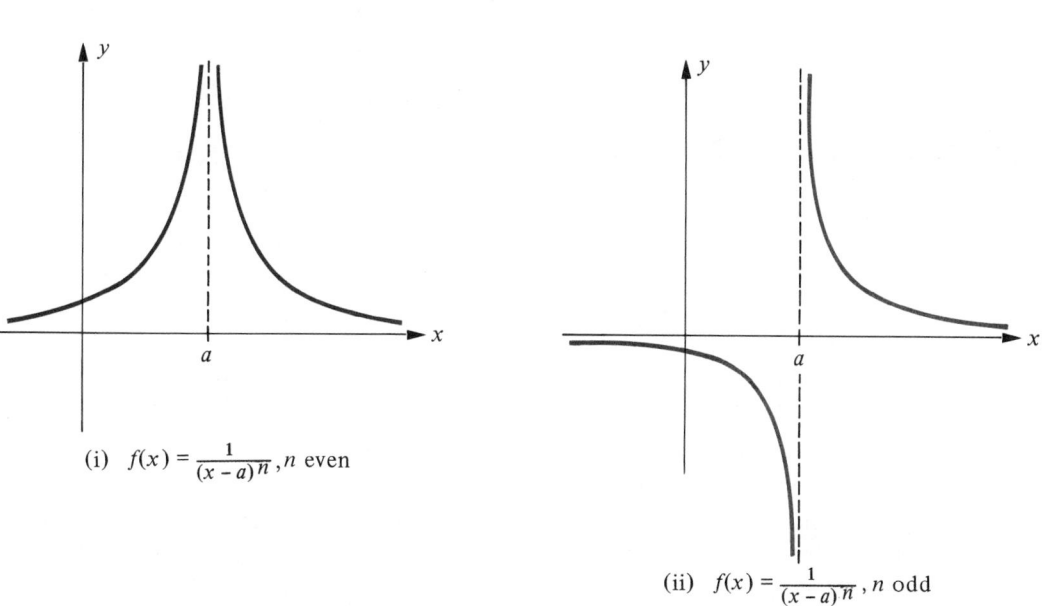

(i) $f(x) = \dfrac{1}{(x-a)^n}$, n even

(ii) $f(x) = \dfrac{1}{(x-a)^n}$, n odd

Figure 3.19

always negative. Consequently, $f(x) \to -\infty$ as $x \to a^+$ or $x \to a^-$. The general shape of the graph can be obtained by *inverting* the graph in (i) of Figure 3.19. Similarly, if n is odd, the shape of the graph may be found by inverting the graph in (ii) of Figure 3.19.

Sketches of graphs of $f(x) = k/(x - a)^n$ for several values of k, a, and n are shown in Figure 3.20. The reader should check each graph by plotting several points and carefully noting the behavior of $f(x)$ if x is near a.

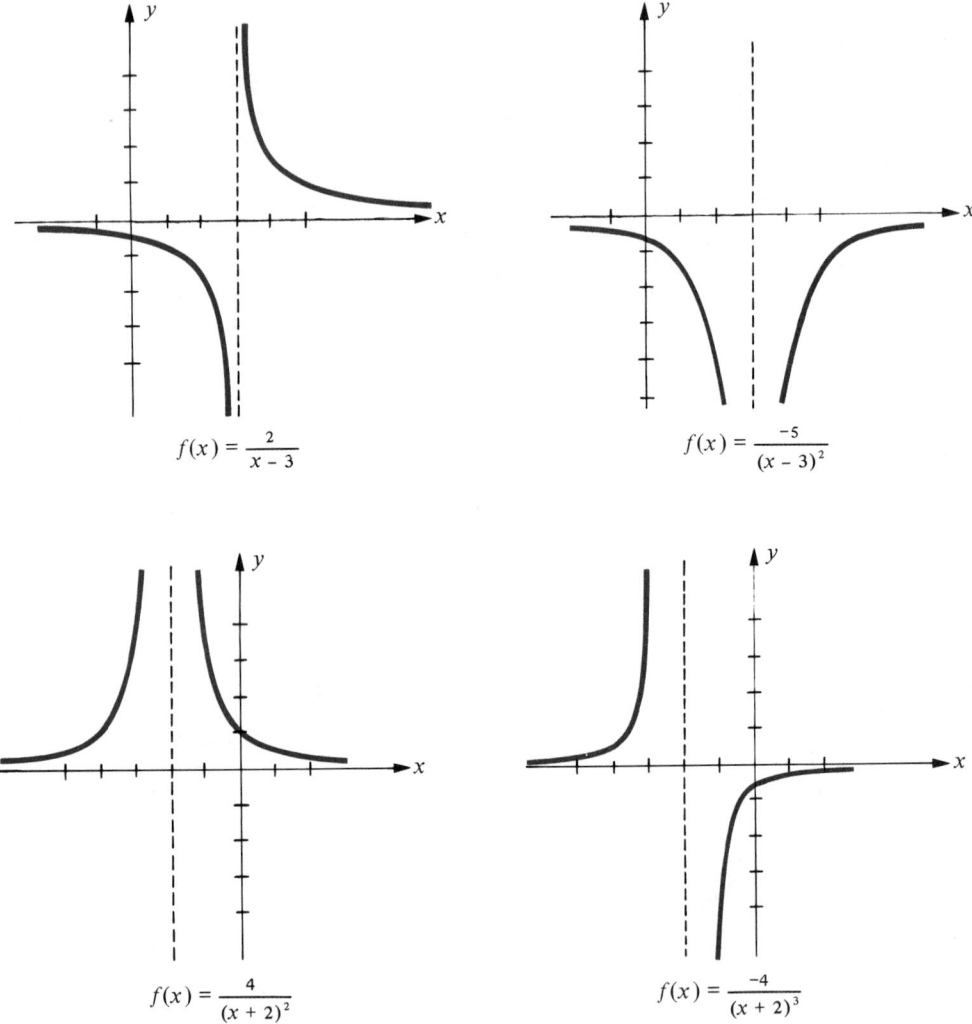

$$f(x) = \frac{2}{x - 3}$$

$$f(x) = \frac{-5}{(x - 3)^2}$$

$$f(x) = \frac{4}{(x + 2)^2}$$

$$f(x) = \frac{-4}{(x + 2)^3}$$

Figure 3.20

In the following examples we consider several other types of rational functions.

Example 3   Sketch the graph of f if $f(x) = \dfrac{x^2}{x^2 - x - 2}$.

Solution   Let us begin by factoring the denominator, obtaining

$$f(x) = \frac{x^2}{(x + 1)(x - 2)}.$$

Since the denominator is zero at 2 and -1, it is necessary to investigate the behavior of $f(x)$ as x approaches either of these two numbers. In order to consider $x \to 2^+$ and $x \to 2^-$, we isolate the factor $x - 2$ as follows:

$$f(x) = \left(\frac{x^2}{x + 1}\right)\left(\frac{1}{x - 2}\right).$$

From the previous discussion we know that

$$\frac{1}{x - 2} \to \infty \quad \text{as} \quad x \to 2^+$$

and

$$\frac{1}{x + 2} \to -\infty \text{ as } x \to 2^-.$$

Since $x^2/(x + 1)$ is positive when x is near 2 and since $f(x)$ is the *product* of $x^2/(x + 1)$ and $1/(x - 2)$, it follows that

$$f(x) \to \infty \quad \text{as} \quad x \to 2^+$$

and

$$f(x) \to -\infty \quad \text{as} \quad x \to 2^-.$$

These facts are illustrated in the partial graph shown in (i) of Figure 3.21

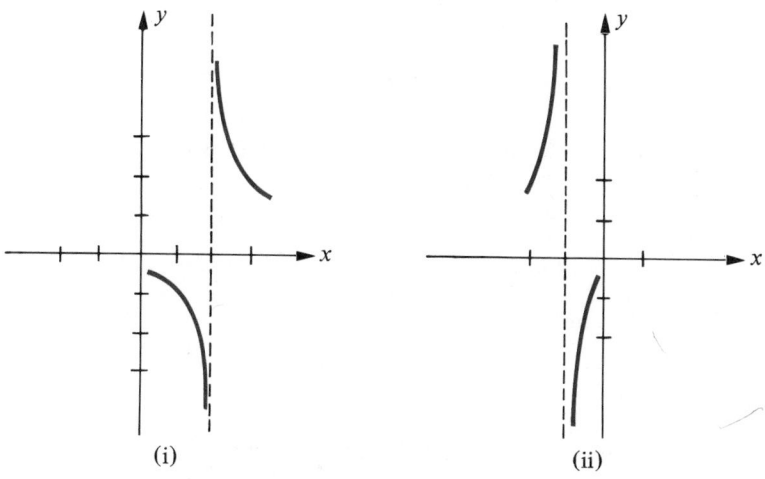

(i)   (ii)

Figure 3.21

In like manner, to determine the variation of $f(x)$ as x approaches -1, we isolate the factor $x + 1$ as follows:

$$f(x) = \left(\frac{x^2}{x-2}\right)\left(\frac{1}{x+1}\right).$$

If $x < -1$, then $x + 1 < 0$ and hence $1/(x + 1)$ is negative. Consequently,

$$\frac{1}{x+1} \to -\infty \quad \text{as} \quad x \to -1^{-}.$$

However, if x is close to -1, the expression $x^2/(x - 2)$ is *negative*, and since $1/(x + 1) \to -\infty$, we see that the *product* of $x^2(x - 2)$ and $1/(x + 1)$ becomes *positively* infinite; that is,

$$f(x) \to \infty \quad \text{as} \quad x \to -1^{-}.$$

Similarly, if $x > -1$, then $x + 1 > 0$. Hence

$$\frac{1}{x+1} \to \infty \quad \text{as} \quad x \to -1^{+}.$$

Again, since $x^2/(x - 2)$ is *negative* when x is near -1, it follows that the product of $x^2/(x - 2)$ and $1/(x + 1)$ becomes *negatively* infinite; that is,

$$f(x) \to -\infty \quad \text{as} \quad x \to -1^{+}.$$

This behavior of $f(x)$ near $x = -1$ is illustrated in (ii) of Figure 3.21.

In order to determine what happens as $x \to \infty$ or $x \to -\infty$, it is helpful to divide numerator and denominator of the given expression for $f(x)$ by x^2, obtaining

$$f(x) = \frac{x^2/x^2}{(x^2 - x - 2)/x^2} = \frac{1}{1 - (1/x) - (2/x^2)}.$$

If we assign very large numerical vlaues to x, the expressions $1/x$ and $2/x^2$ are very close to 0, and hence

$$f(x) \approx \frac{1}{1 - 0 - 0} = 1.$$

This indicates that $\qquad f(x) \to 1 \quad \text{as} \quad x \to \infty$

and $\qquad f(x) \to 1 \quad \text{as} \quad x \to -\infty.$

Using the previous information, together with Figure 3.21, and plotting several points, we obtain the graph in Figure 3.22. Note that the graph has vertical asymptotes $x = -1$ and $x = 2$, and a horizontal asymptote $y = 1$.

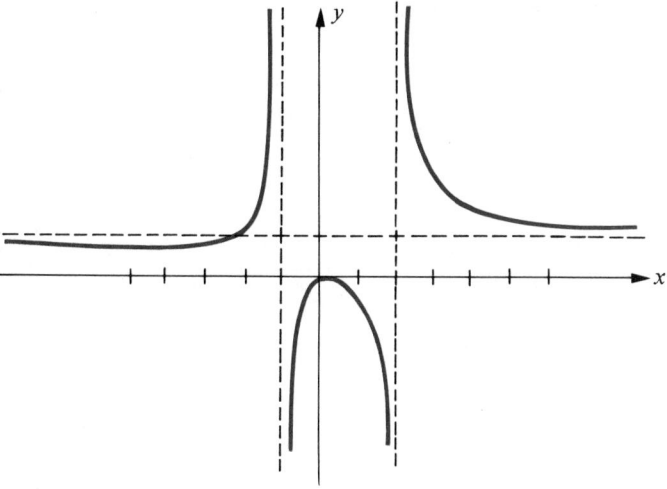

Figure 3.22. $f(x) = \dfrac{x^2}{x^2 - x - 2}$ ■

Example 4   Sketch the graph of f if $f(x) = \dfrac{-x}{x^2 - x - 6}$.

Solution   The factored form of $f(x)$ is

$$f(x) = \frac{-x}{(x + 2)(x - 3)}.$$

The zeros of the denominator are -2 and 3. As in Example 3, we begin by isolating the factor $x - 3$ as follows:

$$f(x) = \left(\frac{-x}{x + 2}\right)\left(\frac{1}{x - 3}\right).$$

Since $\dfrac{1}{x - 3} \to \infty$ as $x \to 3^{+}$

and since $-x/(x + 2)$ is *negative* when x is close to 3, we see that

$$f(x) \to -\infty \quad \text{as} \quad x \to 3^{+}.$$

Similarly, since $\dfrac{1}{x - 3} \to -\infty$ as $x \to 3^{-}$

and since $-x/(x + 2)$ is negative if x is close to 3, we have

$$f(x) \to \infty \quad \text{as} \quad x \to 3^{-}.$$

This behavior is illustrated in (i) of Figure 3.23.

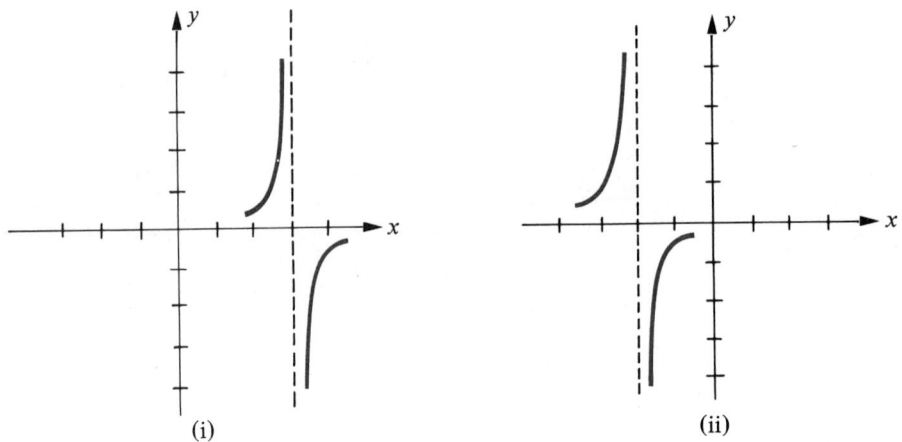

(i)   (ii)

Figure 3.23

In order to investigate the situation near $x = -2$, we write

$$f(x) = \left(\frac{-x}{x-3}\right)\left(\frac{1}{x+2}\right).$$

Since

$$\frac{1}{x+2} \to -\infty \quad \text{as} \quad x \to -2^-$$

and since $-x/(x-3)$ is *negative* if x is close to -2, it follows that

$$f(x) \to \infty \quad \text{as} \quad x \to -2^-$$

It is left to the reader to show that

$$f(x) \to -\infty \quad \text{as} \quad x \to -2^+.$$

The sketch in (ii) of Figure 3.23 illustrates the variation of $f(x)$ near $x = -2$.

To determine what is true if $x \to \infty$ or $x \to -\infty$, we divide numerator and denominator of $f(x)$ by x^2, obtaining

$$f(x) = \frac{(-1/x)}{1 - (1/x) - (6/x^2)}.$$

If x is numerically very large, then the expressions in parentheses are close to 0, and hence

$$f(x) \approx \frac{0}{1 - 0 - 0} = 0.$$

This indicates that $$f(x) \to 0 \quad \text{as} \quad x \to \infty$$

and $$f(x) \to 0 \quad \text{as} \quad x \to -\infty.$$

Using this information, together with Figure 3.23, and plotting several points, gives us the sketch in Figure 3.24. The graph has vertical asymptotes $x = -2$ and $x = 3$ and a horizontal asymptote $y = 0$ (the x-axis).

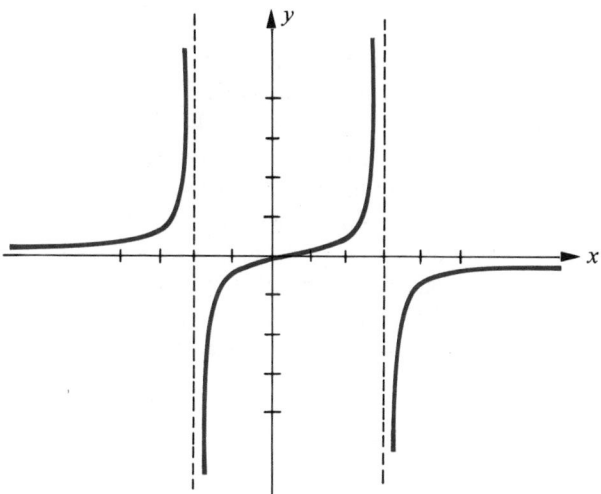

Figure 3.24. $f(x) = \dfrac{-x}{x^2 - x - 6}$ ∎

Example 5   Sketch the graph of f if $f(x) = \dfrac{2x^4}{x^4 + 1}$.

Solution   Since $f(-x) = f(x)$, it follows that the graph is symmetric with respect to the y-axis. Since the denominator of $f(x)$ is never 0, there are no vertical asymptotes. In order to examine what happens to $f(x)$ as $x \to \infty$ or $x \to -\infty$, we divide numerator and denominator by x^4, obtaining

$$f(x) = \frac{2}{1 + (1/x^4)}.$$

Since $$\frac{1}{x^4} \to 0 \quad \text{as} \quad x \to \infty$$

it follows that $$f(x) \to \frac{2}{1 + 0} = 2 \quad \text{as} \quad x \to \infty.$$

Using this fact, plotting several points, and making use of the symmetry with respect to the y-axis, leads to the graph in Figure 3.25. The line $y = 2$ is a horizontal asymptote.

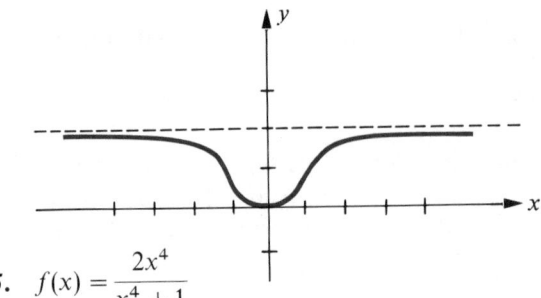

Figure 3.25. $f(x) = \dfrac{2x^4}{x^4 + 1}$ ■

It is worth noting that when we investigated the situation $x \to \infty$ in Examples 3 and 4, we divided numerator and denominator of $f(x)$ by x^2, whereas in Example 5 we divided by x^4. In general, if we wish to investigate $f(x) = g(x)/h(x)$ as $x \to \infty$ or as $x \to -\infty$, and if the degree of $g(x)$ is not higher than the degree of $h(x)$, then we divide numerator and denominator by x^k, where k is the degree of $h(x)$. If the degree of $g(x)$ is higher than the degree of $h(x)$, then it can be shown that $f(x)$ either increases or decreases without bound as $x \to \infty$ or as $x \to -\infty$.

Graphs of rational functions may become very complicated as the degrees of the polynomials in the numerator and denominator increase. For a thorough treatment it is necessary to employ techniques developed in calculus.

EXERCISES 3.8

Sketch the graph of f in each of Exercises 1–20.

1 $f(x) = \dfrac{1}{x - 2}$

2 $f(x) = \dfrac{1}{x + 3}$

3 $f(x) = \dfrac{-2}{x + 4}$

4 $f(x) = \dfrac{-3}{x - 1}$

5 $f(x) = \dfrac{x}{x - 5}$

6 $f(x) = \dfrac{x}{3x + 2}$

7 $f(x) = \dfrac{4}{(x - 1)^2}$

8 $f(x) = \dfrac{-1}{(x + 2)^2}$

9 $f(x) = \dfrac{1}{x^2 - 4}$

10 $f(x) = \dfrac{2}{x^2 + x - 2}$

11 $f(x) = \dfrac{5x}{4 - x^2}$

12 $f(x) = \dfrac{x^2}{x^2 - 4}$

13 $f(x) = \dfrac{x^2}{x^2 - 7x + 10}$

14 $f(x) = \dfrac{x}{x^2 - x - 6}$

15 $f(x) = \dfrac{3x + 2}{x}$

16 $f(x) = \dfrac{x^2 - 4}{x^2}$

17 $f(x) = \dfrac{4}{x^2 + 4}$

18 $f(x) = \dfrac{3x}{x^2 + 1}$

19 $f(x) = \dfrac{1}{x^3 + x^2 - 6x}$

20 $f(x) = \dfrac{x^2 - x}{16 - x^2}$

3.9 PARTIAL FRACTIONS

It is easy to verify that

$$\frac{2}{x^2 - 1} = \frac{1}{x - 1} + \frac{-1}{x + 1}.$$

The expression on the right side of this equation is called *the partial fraction decomposition* of $2/(x^2 - 1)$. In general, if f is *any* rational function, it is theoretically possible to express $f(x)$ as a sum of quotients of polynomials whose denominators involve powers of polynomials of degree not greater than two. This fact has important applications in the study of calculus.

If $g(x)$ and $h(x)$ are polynomials *and the degree of $g(x)$ is less than the degree of $h(x)$*, then it follows from a theorem in algebra that

$$\frac{g(x)}{h(x)} = F_1 + F_2 + \cdots + F_k$$

where each F_i has one of the forms

$$\frac{A}{(px + q)^m} \quad \text{or} \quad \frac{Cx + D}{(ax^2 + bx + c)^n}$$

for some nonnegative integers m and n, and where $ax^2 + bx + c$ is **irreducible**, in the sense that this quadratic polynomial has no real zeros, that is, $b^2 - 4ac < 0$. The sum $F_1 + F_2 + \cdots + F_k$ is called the **partial fraction decomposition** of $g(x)/h(x)$ and each F_i is called a **partial fraction**. We shall not prove this algebraic result but will, instead, give rules for obtaining the decomposition.

To find the partial fraction decomposition of $g(x)/h(x)$ it is *essential* that $g(x)$ have lower degree than $h(x)$. If this is not the case, then long division should be employed to arrive at such an expression. For example, given

$$\frac{x^3 - 6x^2 + 5x - 3}{x^2 - 1}$$

we obtain, by long division,

$$\frac{x^3 - 6x^2 + 5x - 3}{x^2 - 1} = x - 6 + \frac{6x - 9}{x^2 - 1}.$$

The partial fraction decomposition is then found for $(6x - 9)/(x^2 - 1)$.

In order to obtain the decomposition of $g(x)/h(x)$, we begin by expressing the denominator $h(x)$ as a product of factors $px + q$, or irreducible quadratic factors $ax^2 + bx + c$. Repeated factors are then collected so that $h(x)$ is a product of *different* factors of the form $(px + q)^m$ or $(ax^2 + bx + c)^n$, where m and n are nonnegative integers and $ax^2 + bx + c$ is irreducible. We then apply the following rules.

Rule 1.   For each factor of $h(x)$ of the form $(px + q)^m$ where $m \geq 1$, the partial fraction decomposition of $g(x)/h(x)$ contains a sum of m partial fractions of the form

$$\frac{A_1}{px + q} + \frac{A_2}{(px + q)^2} + \cdots + \frac{A_m}{(px + q)^m}$$

where each A_i is a real number.

Rule 2.   For each factor of $h(x)$ of the form $(ax^2 + bx + c)^n$ where $n \geq 1$ and $b^2 - 4ac < 0$, the partial fraction decomposition of $g(x)/h(x)$ contains a sum of n partial fractions of the form

$$\frac{A_1 x + B_1}{ax^2 + bx + c} + \frac{A_2 x + B_2}{(ax^2 + bx + c)^2} + \cdots + \frac{A_n x + B_n}{(ax^2 + bx + c)^n}$$

where, for each i, A_i and B_i are real numbers.

Example 1   Find the partial fraction decomposition of

$$\frac{4x^2 + 13x - 9}{x^3 + 2x^2 - 3x}.$$

Solution   The denominator of the quotient has the factored form $x(x + 3)(x - 1)$. Each of the linear factors is handled according to Rule 1 with $m = 1$. Thus, for the factor x there corresponds a partial fraction of the form A/x. Similarly, for the factors $x + 3$ and $x - 1$, there correspond partial fractions $B/(x + 3)$ and $C/(x - 1)$, respectively. The partial fraction decomposition has the form

$$\frac{4x^2 + 13x - 9}{x(x + 3)(x - 1)} = \frac{A}{x} + \frac{B}{x + 3} + \frac{C}{x - 1}.$$

Multiplying by the lowest common denominator $x(x + 3)(x - 1)$ gives us

$$4x^2 + 13x - 9 = A(x + 3)(x - 1) + Bx(x - 1) + Cx(x + 3).$$

In a case such as this, in which the factors are all linear and nonrepeated, the values for A, B, and C can be found by substituting values for x which make the various factors zero. If we let $x = 0$ in the last equation, then $-9 = -3A$, or $A = 3$. Letting $x = 1$ gives us $8 = 4C$, or $C = 2$. Finally, if $x = -3$, then $-12 = 12B$, or $B = -1$. The partial fraction decomposition is, therefore,

$$\frac{4x^2 + 13x - 9}{x(x + 3)(x - 1)} = \frac{3}{x} + \frac{-1}{x + 3} + \frac{2}{x - 1}.$$

∎

Example 2   Find the partial fraction decomposition of

$$\frac{x^2 + 10x - 36}{x(x - 3)^2}.$$

Solution   By Rule 1, with $m = 1$, there is a partial fraction A/x corresponding to the factor x. Next, applying Rule 1, with $m = 2$, the squared factor $(x - 3)^2$ gives rise to a sum of two partial fractions $B/(x - 3) + C/(x - 3)^2$. Thus the partial fraction decomposition has the form

$$\frac{x^2 + 10x - 36}{x(x - 3)^2} = \frac{A}{x} + \frac{B}{x - 3} + \frac{C}{(x - 3)^2}.$$

Multiplying both sides by $x(x - 3)^2$ gives us

(∗) $$x^2 + 10x - 36 = A(x - 3)^2 + Bx(x - 3) + Cx$$

where the symbol (∗) is used for later identification of this equation. Two of the unknown constants may be determined easily. Letting $x = 3$ in (∗) gives us

$$9 + 30 - 36 = A(0) + B(0) + 3C.$$

This reduces to $3 = 3C$ and hence $C = 1$. In like manner, letting $x = 0$, we get

$$0 + 0 - 36 = A(-3)^2 + B(0) + C(0).$$

Hence $-36 = 9A$, or $A = -4$. The remaining constant can be found by comparing coefficients. If the right side of (∗) is expanded and like powers of x collected we obtain

$$x^2 + 10x - 36 = (A + B)x^2 + (-3A - 3B + C)x + 9A.$$

The coefficient $A + B$ of x^2 on the right must equal the coefficient of x^2 on the left, that is, $A + B = 1$. Since $A = -4$, it follows that $B = 1 - A = 1 - (-4) = 5$. The partial fraction decomposition is, therefore,

$$\frac{x^2 + 10x - 36}{x(x - 3)^2} = \frac{-4}{x} + \frac{5}{x - 3} + \frac{1}{(x - 3)^2}.$$ ∎

Example 3   Find the partial fraction decomposition of

$$\frac{3x^3 - 18x^2 + 29x - 4}{(x + 1)(x - 2)^3}.$$

Solution   By Rule 1, there is a partial fraction of the form $A/(x + 1)$ corresponding to the factor $x + 1$ in the denominator. For the factor $(x - 2)^3$ we apply Rule 1 (with $m = 3$), obtaining a sum of three partial fractions $B/(x - 2)$, $C/(x - 2)^2$, and $D/(x - 2)^3$. Consequently, the decomposition has the form

$$\frac{3x^3 - 18x^2 + 29x - 4}{(x + 1)(x - 2)^3} = \frac{A}{x + 1} + \frac{B}{x - 2} + \frac{C}{(x - 2)^2} + \frac{D}{(x - 2)^3}.$$

Multiplying both sides by $(x + 1)(x - 2)^3$ gives us

$$3x^3 - 18x^2 + 29x - 4$$

(∗)

$$= A(x - 2)^3 + B(x + 1)(x - 2)^2 + C(x + 1)(x - 2) + D(x + 1).$$

Two of the unknown constants may be determined easily. If we let $x = 2$ in (∗) then

$$24 - 72 + 58 - 4 = 3D, \quad 6 = 3D, \quad \text{and} \quad D = 2.$$

Similarly, letting $x = -1$ in (∗) gives us

$$-3 - 18 - 29 - 4 = -27A, \quad -54 = -27A, \quad \text{and} \quad A = 2.$$

The remaining constants may be found by comparing coefficients. If the right side of (∗) is expanded and like powers of x collected, we see that the coefficient of x^3 is $A + B$. This must be equal to the coefficient of x^3 on the left, that is,

$$A + B = 3.$$

Since $A = 2$, it follows that $B = 3 - A = 3 - 2 = 1$. Finally, we compare the constant terms in (∗) by letting $x = 0$. This gives us

$$-4 = -8A + 4B - 2C + D.$$

Substituting the values we have found for A, B, and D leads to

$$-4 = -16 + 4 - 2C + 2$$

which has the solution $C = -3$. The partial fraction decomposition is, therefore,

$$\frac{3x^3 - 18x^2 + 29x - 4}{(x + 1)(x - 2)^3} = \frac{2}{x + 1} + \frac{1}{x - 2} + \frac{-3}{(x - 2)^2} + \frac{2}{(x - 2)^3}. \qquad \blacksquare$$

Example 4 Find the partial fraction decomposition of

$$\frac{x^2 - x - 21}{2x^3 - x^2 + 8x - 4}.$$

Solution The denominator may be factored by grouping as follows:

$$2x^3 - x^2 + 8x - 4 = x^2(2x - 1) + 4(2x - 1) = (x^2 + 4)(2x - 1).$$

Applying Rule 2 to the irreducible quadratic factor $x^2 + 4$, we see that one of the partial fractions has the form $(Ax + B)/(x^2 + 4)$. By Rule 1, there is also a partial fraction $C/(2x - 1)$ corresponding to the factor $2x - 1$. Consequently,

$$\frac{x^2 - x - 21}{2x^3 - x^2 + 8x - 4} = \frac{Ax + B}{x^2 + 4} + \frac{C}{2x - 1}.$$

As in previous examples, this leads to

(∗)
$$x^2 - x - 21 = (Ax + B)(2x - 1) + C(x^2 + 4).$$

Substituting $x = 1/2$, we obtain $(1/4) - (1/2) - 21 = (17/4)C$, which has the solution $C = -5$. The remaining constants may be found by comparing coefficients. Rearranging the right side of (∗) gives us

$$x^2 - x - 21 = (2A + C)x^2 + (-A + 2B)x - B + 4C.$$

Comparing the coefficients of x^2, we see that $2A + C = 1$. Since $C = -5$ it follows that $2A = 6$ or $A = 3$. Similarly, comparing the constant terms, $-B + 4C = -21$ and hence $-B - 20 = -21$ or $B = 1$. Thus the partial fraction decomposition is

$$\frac{x^2 - x - 21}{2x^3 - x^2 + 8x - 4} = \frac{3x + 1}{x^2 + 4} + \frac{-5}{2x - 1}.$$ ∎

Example 5 Find the partial fraction decomposition of

$$\frac{5x^3 - 3x^2 + 7x - 3}{(x^2 + 1)^2}.$$

Solution Applying Rule 2, with $n = 2$,

$$\frac{5x^3 - 3x^2 + 7x - 3}{(x^2 + 1)^2} = \frac{Ax + B}{x^2 + 1} + \frac{Cx + D}{(x^2 + 1)^2}.$$

Multiplying both sides by $(x^2 + 1)^2$ gives us

$$5x^3 - 3x^2 + 7x - 3 = (Ax + B)(x^2 + 1) + Cx + D$$

or $\qquad 5x^3 - 3x^2 + 7x - 3 = Ax^3 + Bx^2 + (A + C)x + (B + D).$

Comparing the coefficients of x^3 and x^2, we obtain $A = 5$ and $B = -3$. From the coefficients of x we see that $A + C = 7$ or $C = 7 - A = 7 - 5 = 2$. Finally, the constant terms give us $B + D = -3$ or $D = -3 - B = -3 - (-3) = 0$. Therefore,

$$\frac{5x^3 - 3x^2 + 7x - 3}{(x^2 + 1)^2} = \frac{5x - 3}{x^2 + 1} + \frac{2x}{(x^2 + 1)^2}.$$ ∎

EXERCISES 3.9

Find the partial fraction decompositions in Exercises 1–24.

1 $\dfrac{8x - 1}{(x - 2)(x + 3)}$ $\qquad$ **2** $\dfrac{x - 29}{(x - 4)(x + 1)}$ $\qquad$ **5** $\dfrac{4x^2 - 15x - 1}{(x - 1)(x + 2)(x - 3)}$

3 $\dfrac{x + 34}{x^2 - 4x - 12}$ $\qquad$ **4** $\dfrac{5x - 12}{x^2 - 4x}$ $\qquad$ **6** $\dfrac{x^2 + 19x + 20}{x(x + 2)(x - 5)}$ $\qquad$ **7** $\dfrac{4x^2 - 5x - 15}{x^3 - 4x^2 - 5x}$

8 $\dfrac{37 - 11x}{(x + 1)(x^2 - 5x + 6)}$

9 $\dfrac{2x + 3}{(x - 1)^2}$

10 $\dfrac{5x^2 - 4}{x^2(x + 2)}$

11 $\dfrac{19x^2 + 50x - 25}{3x^3 - 5x^2}$

12 $\dfrac{10 - x}{x^2 + 10x + 25}$

13 $\dfrac{x^2 - 6}{(x + 2)^2(2x - 1)}$

14 $\dfrac{2x^2 + x}{(x - 1)^2(x + 1)^2}$

15 $\dfrac{3x^3 + 11x^2 + 16x + 5}{x(x + 1)^3}$

16 $\dfrac{4x^3 + 3x^2 + 5x - 2}{x^3(x + 2)}$

17 $\dfrac{x^2 + x - 6}{(x^2 + 1)(x - 1)}$

18 $\dfrac{x^2 - x - 21}{(x^2 + 4)(2x - 1)}$

19 $\dfrac{9x^2 - 3x + 8}{x^3 + 2x}$

20 $\dfrac{2x^3 + 2x^2 + 4x - 3}{x^4 + x^2}$

21 $\dfrac{4x^3 - x^2 + 4x + 2}{(x^2 + 1)^2}$

22 $\dfrac{3x^3 + 13x - 1}{(x^2 + 4)^2}$

23 $\dfrac{2x^4 - 2x^3 + 6x^2 - 5x + 1}{x^3 - x^2 + x - 1}$

24 $\dfrac{x^3}{x^3 - 3x^2 + 9x - 27}$

3.10 REVIEW

Concepts

Define or discuss each of the following concepts.

1 Polynomial function

2 Polynomial

3 Degree of a polynomial function

4 Factor of a polynomial

5 Irreducible polynomial

6 Quadratic function

7 Graph of a quadratic function

8 Graphs of polynomial functions of degree greater than 2

9 Division algorithm for polynomials

10 Remainder Theorem

11 Factor Theorem

12 Synthetic division

13 The multiplicity of a zero of a polynomial

14 Descartes' Rule of Signs

15 Upper and lower bounds for solutions of equations

16 Rational zeros of polynomial functions

17 Rational function

18 Graph of a rational function

19 Vertical and horizontal asymptotes

20 Partial Fractions

Exercises

Perform the operations indicated in Exercises 1–6 and find the degree of the resulting polynomial.

1 $(2x^3 - x^2 + 5x - 7) - (3x^3 + 2x^2 - x + 2)$

2 $(x^4 + 3x^2 - 7) + (x^3 - 2x^2 + x + 1)$

3 $(x^2 - 3x + 4)(x - 5)$

4 $(x^2 - 1)(x^2 + 3x + 2)$

5 $(x + 1)^2 - (3x - 1)^2$

6 $(2x + 1)(3x - 5)(x + 4)$

Express the polynomials in Exercises 7–12 as products of polynomials which are irreducible over the set of integers.

7 $8x^2 + 10x - 3$ **8** $9x^4 + 12x^2 + 4$

9 $3x^4 - 2x^3 + 12x^2 - 8x$

10 $64x^3 + 8$

11 $2x^4 - 32$

12 $3x^4 + 6x^3 - 24x^2$

In each of Exercises 13–16 find the quotient and remainder if $f(x)$ is divided by $g(x)$:

13 $f(x) = 4x^5 - 2x^3 + x + 7, \quad g(x) = x^3 + 2x - 3$

14 $f(x) = 2x^2 + 5x - 3, \quad g(x) = x^3$

15 $f(x) = 6x - 5, \quad g(x) = 2x + 3$

16 $f(x) = 2x^3 + x^2 - 2x + 1, \quad g(x) = x^3$

17 If $f(x) = -2x^4 - 5x^3 + 4x^2 - 9$,
  (a) use the Remainder Theorem to find $f(2)$.
  (b) prove that $x + 3$ is a factor of $f(x)$.

18 Use synthetic division to find the quotient and remainder if $3x^5 - x^2 + 3$ is divided by $x + 2$.

19 Find polynomials of degree 3 with the following zeros:
  (a) 1, 2, 3   (b) $0, \sqrt{2}, \sqrt{3}$

20 Find a polynomial of degree 5 such that:
  (a) -2 is a zero of multiplicity 3, and 0 is a zero of multiplicity 2.
  (b) 0 is a zero of multiplicity 3, and -2 is a zero of multiplicity 2.

21 Show that -2 is a zero of multiplicity 3 of the polynomial $x^5 + 8x^4 + 21x^3 + 14x^2 - 20x - 24$, and express this polynomial as a product of linear factors.

For each equation in Exercises 22 and 23, (a) use Descartes' Rule of Signs to determine the number of possible positive and negative solutions. (b) find the smallest and largest integers that are upper and lower bounds, respectively, of the real solutions.

22 $2x^4 - 4x^3 + 2x^2 - 5x - 7 = 0$

23 $x^5 - 4x^3 + 6x^2 + x + 4 = 0$

Find all rational solutions of the equations in Exercises 24–26.

24 $x^4 - 3x^3 - 15x^2 + 19x + 30 = 0$

25 $30x^3 + 107x^2 - 5x - 42 = 0$

26 $x^4 + 2x^3 + 6x^2 + x + 10 = 0$

Sketch the graphs of the equations in Exercises 27 and 28.

27 $y = x^2 + 6x + 16$ **28** $4x^2 + y = 10$

In each of Exercises 29–38 sketch the graph of f.

29 $f(x) = (x - 4)^2$

30 $f(x) = 12x^2 + 5x - 3$

31 $f(x) = 16x - 4x^2$

32 $f(x) = 2x^3 - 8x^2 + 64x$

33 $f(x) = 2x^2 + x^3 - x^4$

34 $f(x) = \dfrac{-4}{(x + 1)^2}$ **35** $f(x) = \dfrac{3x^2}{16 - x^2}$

36 $f(x) = \dfrac{x}{(x + 5)(x - 1)(x - 4)}$

37 $f(x) = (x + 2)^3$ **38** $f(x) = x^6 - 32$

Find the partial fraction decompositions in Exercises 39 and 40.

39 $\dfrac{4x^2 + 54x + 134}{(x + 3)(x^2 + 4x - 5)}$

40 $\dfrac{x^2 + 14x - 13}{x^3 + 5x^2 + 4x + 9}$

EXPONENTIAL AND LOGARITHMIC FUNCTIONS

Exponential and logarithmic functions are very important in mathematics and its applications. In this chapter we shall define these functions and study some of their properties.

4.1 A REVIEW OF EXPONENTS AND RADICALS

We have had occasion to use the notation a^2 for the real number $a \cdot a$. Similarly, a^3 is used to denote $a \cdot a \cdot a$. In general, if n is any positive integer,

$$a^n = \underbrace{a \cdot a \cdot \cdots \cdot a}_{n \text{ factors}}$$

where n factors, all equal to a, appear on the right-hand side of the equals sign. The positive integer n is called the **exponent** of a in the expression a^n, and a^n is read "a to the nth power" or simply "a to the n." Note that $a^1 = a$. Some numerical examples are

$$\left(\frac{1}{2}\right)^5 = \frac{1}{2} \cdot \frac{1}{2} \cdot \frac{1}{2} \cdot \frac{1}{2} \cdot \frac{1}{2} = \frac{1}{32}$$

$$(-3)^3 = (-3)(-3)(-3) = -27$$

$$(\sqrt{2})^4 = \sqrt{2}\sqrt{2}\sqrt{2}\sqrt{2} = (\sqrt{2})^2(\sqrt{2})^2 = (2)(2) = 4.$$

If $a \neq 0$, the extension to nonpositive exponents is made by defining

$$a^0 = 1 \quad \text{and} \quad a^{-n} = \frac{1}{a^n}.$$

The following rules are indispensable when working with exponents.

Laws of Exponents

If a and b are real numbers and m and n are integers, then

(i) $a^m a^n = a^{m+n}$;   (ii) $(a^m)^n = a^{mn}$;   (iii) $(ab)^n = a^n b^n$;

(iv) $\left(\dfrac{a}{b}\right)^n = \dfrac{a^n}{b^n}$, if $b \neq 0$;   (v) If $a \neq 0$, then $\dfrac{a^m}{a^n} = a^{m-n} = \dfrac{1}{a^{n-m}}$.

Complete proofs of (i) – (v) require the method of mathematical induction discussed in Section 9.1. However, if we are allowed to count an arbitrary number of factors, then it is easy to supply arguments which establish the laws for positive exponents. To prove (i) we have

$$a^m a^n = \underbrace{a \cdot a \cdot \cdots \cdot a}_{m \text{ factors}} \ \underbrace{a \cdot a \cdot \cdots \cdot a}_{n \text{ factors}} .$$

Since the total number of factors a on the right is $m + n$, this expression is equal to a^{m+n}.
Similarly, we could write

$$(a^m)^n = \underbrace{a^m \cdot a^m \cdot \cdots \cdot a^m}_{n \text{ factors } a^m}$$

and count the number of times a appears as a factor on the right-hand side. Since $a^m = a \cdot a \cdot \cdots \cdot a$ where a occurs as a factor m times, and since the number of such groups of m factors is n, the total number of factors is $m \cdot n$. This gives us Law (ii).
Laws (iii) and (iv) can be obtained in similar fashion and are left to the reader. Law (v) is clear if $m = n$. For the case $m > n$, the integer $m - n$ is positive, and we may write

$$\frac{a^m}{a^n} = \frac{a^n a^{m-n}}{a^n} = \frac{a^n}{a^n} \cdot a^{m-n} = 1 \cdot a^{m-n} = a^{m-n}.$$

A similar argument can be used if $n > m$.
We can, of course, use symbols for real numbers other than a and b when applying laws of exponents, as in the following illustrations.

$$x^5 x^6 = x^{5+6} = x^{11} \qquad (y^5)^7 = y^{5 \cdot 7} = y^{35}$$

$$(rs)^7 = r^7 s^7 \qquad \left(\frac{p}{q}\right)^{10} = \frac{p^{10}}{q^{10}}$$

$$\frac{c^8}{c^3} = c^{8-3} = c^5 \qquad \frac{u^3}{u^8} = \frac{1}{u^{8-3}} = \frac{1}{u^5}$$

The Laws of Exponents can be extended to rules such as $a^m a^n a^p = a^{m+n+p}$ and $(abc)^n = a^n b^n c^n$. To simplify statements, we shall assume that in all problems involving exponents, symbols which appear in denominators represent *nonzero* real numbers.

Example 1 Simplify each of the following:

(a) $(3x^3 y^4)(4xy^5)$ (b) $(2a^2 b^3 c)^4$ (c) $\left(\dfrac{2r^3}{s}\right)^2 \left(\dfrac{s}{r^3}\right)^3$

Solutions   We may proceed as follows:

(a)
$$(3x^3y^4)(4xy^5) = (3)(4)x^3xy^4y^5$$
$$= 12x^4y^9$$

(b)
$$(2a^2b^3c)^4 = 2^4(a^2)^4(b^3)^4c^4$$
$$= 16a^8b^{12}c^4$$

(c)
$$\left(\frac{2r^3}{s}\right)^2\left(\frac{s}{r^3}\right)^3 = \left(\frac{2^2r^6}{s^2}\right)\left(\frac{s^3}{r^9}\right)$$
$$= 2^2\left(\frac{r^6}{r^9}\right)\left(\frac{s^3}{s^2}\right)$$
$$= 4\left(\frac{1}{r^3}\right)(s)$$
$$= \frac{4s}{r^3} \qquad\qquad ∎$$

Example 2   Eliminate negative exponents and simplify:

(a) $(a^{-2}b^3)^{-3}$ (b) $\dfrac{8x^3y^{-5}}{4x^{-1}y^2}$

Solutions

(a)
$$(a^{-2}b^3)^{-3} = (a^{-2})^{-3}(b^3)^{-3}$$
$$= a^6b^{-9}$$
$$= a^6 \cdot \frac{1}{b^9} = \frac{a^6}{b^9}$$

(b)
$$\frac{8x^3y^{-5}}{4x^{-1}y^2} = \frac{8}{4}\frac{x^3}{x^{-1}}\frac{y^{-5}}{y^2}$$
$$= 2x^{3-(-1)}y^{-5-2}$$
$$= 2x^4y^{-7}$$
$$= \frac{2x^4}{y^7}$$

Another method for simplifying (b) is to multiply numerator and denominator of the given fraction by y^5x, thereby eliminating negative exponents. Thus

$$\frac{8x^3y^{-5}}{4x^{-1}y^2} = \frac{8x^3y^{-5}}{4x^{-1}y^2}\cdot\frac{y^5x}{y^5x}$$
$$= \frac{8}{4}\frac{x^4y^0}{x^0y^7}$$
$$= 2\frac{x^4(1)}{(1)y^7}$$
$$= \frac{2x^4}{y^7}. \qquad\qquad ∎$$

Roots of real numbers are defined by the statement

$$\sqrt[n]{b} = a \quad \text{if and only if} \quad a^n = b$$

provided that both a and b are nonnegative real numbers and n is a positive integer, or that both a and b are negative and n is an odd positive integer. The number $\sqrt[n]{a}$ is called the **principal nth root** of a. If $n = 2$, it is customary to write $\sqrt{a}$ instead of $\sqrt[2]{a}$ and to call $\sqrt{a}$ the (principal) **square root** of a. The number $\sqrt[3]{a}$ is referred to as the **cube root** of a. For example,

$$\sqrt[5]{\tfrac{1}{32}} = \tfrac{1}{2} \quad \text{since} \quad (\tfrac{1}{2})^5 = \tfrac{1}{32}$$

$$\sqrt[4]{81} = 3 \quad \text{since} \quad 3^4 = 81$$

$$\sqrt[3]{-8} = -2 \quad \text{since} \quad (-2)^3 = -8$$

$$\sqrt{16} = 4 \quad \text{since} \quad 4^2 = 16.$$

Complex numbers (see Chapter Seven) are needed to define $\sqrt[n]{a}$ if $a < 0$ and n is an *even* positive integer, since for all real numbers b, $b^n \geq 0$ whenever n is even. We shall not define such roots at this time.

It is important to observe that if $\sqrt[n]{a}$ exists, it is a *unique* real number. More generally, if $b^n = a$ for a positive integer n, then b is called *an nth root* of a. For example, both 4 and -4 are square roots of 16, since $4^2 = 16$ and also $(-4)^2 = 16$. However, as in Example 1, the *principal* square root of 16 is 4, and we write $\sqrt{16} = 4$. In elementary arithmetic the expression $\sqrt{16} = \pm 4$ is sometimes used to denote *all* square roots of 16. It should be emphasized that this is *not* done in advanced mathematics.

To complete our terminology, the symbol $\sqrt[n]{a}$ is called a **radical**, the number a is called the **radicand**, and n is the **index** of the radical. The symbol $\sqrt{}$ is called a **radical sign**.

If $\sqrt{b} = a$, then $a^2 = b$; that is, $(\sqrt{b})^2 = b$. Similarly, if $\sqrt[3]{b} = a$, then $a^3 = b$, or $(\sqrt[3]{b})^3 = b$. In general, if n is any positive integer, then

$$(\sqrt[n]{b})^n = b.$$

It also follows that if $b > 0$, or if $b < 0$ and n is an odd positive integer, then

$$\sqrt[n]{b^n} = b.$$

For example, $\sqrt{5^2} = 5, \quad \sqrt[3]{(-2)^3} = -2, \quad \sqrt[4]{3^4} = 3.$

The following three laws can be proved, where m and n denote positive integers and where it is assumed that the indicated roots exist.

Laws of Radicals

$$\text{(i)} \quad \sqrt[n]{bc} = \sqrt[n]{b}\sqrt[n]{c}; \qquad \text{(ii)} \quad \sqrt[n]{\frac{b}{c}} = \frac{\sqrt[n]{b}}{\sqrt[n]{c}}; \qquad \text{(iii)} \quad \sqrt[m]{\sqrt[n]{b}} = \sqrt[mn]{b}$$

Example 3   Show that (a) $\sqrt{50} = 5\sqrt{2}$;

(b) $\sqrt[3]{-108} = -3\sqrt[3]{4}$;

(c) $\sqrt[3]{\sqrt{64}} = 2$.

Solutions

(a) $$\sqrt{50} = \sqrt{25 \cdot 2} = \sqrt{25}\sqrt{2} = 5\sqrt{2}$$

(b) $$\sqrt[3]{-108} = \sqrt[3]{(-27)(4)} = \sqrt[3]{-27}\sqrt[3]{4} = -3\sqrt[3]{4}$$

(c)   Applying (iii) with $m = 3$ and $n = 2$,

$$\sqrt[3]{\sqrt{64}} = \sqrt[6]{64} = 2.$$

As a check we have $$\sqrt[3]{\sqrt{64}} = \sqrt[3]{8} = 2.$$ ∎

Example 4   Simplify the following, where all letters denote positive real numbers.

(a) $\sqrt[3]{16x^3y^8z^4}$ (b) $\sqrt{3a^2b^3}\sqrt{6a^5b}$

Solutions

(a) $$\sqrt[3]{16x^3y^8z^4} = \sqrt[3]{(2^3x^3y^6z^3)(2y^2z)}$$
$$= \sqrt[3]{(2xy^2z)^3(2y^2z)}$$
$$= \sqrt[3]{(2xy^2z)^3}\sqrt[3]{2y^2z}$$
$$= 2xy^2z\sqrt[3]{2y^2z}$$

(b) $$\sqrt{3a^2b^3}\sqrt{6a^5b} = \sqrt{18a^7b^4}$$
$$= \sqrt{(9a^6b^4)(2a)}$$
$$= \sqrt{(3a^3b^2)^2(2a)}$$
$$= \sqrt{(3a^3b^2)^2}\sqrt{2a}$$
$$= 3a^3b^2\sqrt{2a}$$ ∎

An important type of simplification involving radicals is that of **rationalizing a denominator**. The process involves beginning with a quotient which contains a radical in the denominator and then multiplying numerator and denominator by some expression so that the resulting denominator contains no radicals. The next example illustrates this technique.

Example 5   Rationalize the denominator in each of the following.

(a) $\dfrac{1}{\sqrt{5}}$ (b) $\sqrt{\dfrac{2}{3}}$ (c) $\sqrt[3]{\dfrac{x}{y}}$

Solutions   We may proceed as follows (supply reasons):

(a)
$$\frac{1}{\sqrt{5}} = \frac{1}{\sqrt{5}} \cdot \frac{\sqrt{5}}{\sqrt{5}} = \frac{\sqrt{5}}{(\sqrt{5})^2} = \frac{\sqrt{5}}{5}$$

(b)
$$\sqrt{\frac{2}{3}} = \sqrt{\frac{2}{3} \cdot \frac{3}{3}} = \sqrt{\frac{6}{3^2}} = \frac{\sqrt{6}}{\sqrt{3^2}} = \frac{\sqrt{6}}{3}$$

(c)
$$\sqrt[3]{\frac{x}{y}} = \sqrt[3]{\frac{x}{y} \cdot \frac{y^2}{y^2}} = \frac{\sqrt[3]{xy^2}}{\sqrt[3]{y^3}} = \frac{\sqrt[3]{xy^2}}{y} \quad \blacksquare$$

The denominators of certain fractional expressions contain sums or differences involving radicals. In some cases the denominators can be rationalized, as illustrated in the next example.

Example 6   Simplify $\dfrac{1}{\sqrt{x} - \sqrt{y}}$.

Solution   Multiplying numerator and denominator by $\sqrt{x} + \sqrt{y}$ we obtain

$$\frac{1}{\sqrt{x} - \sqrt{y}} = \frac{1}{\sqrt{x} - \sqrt{y}} \cdot \frac{\sqrt{x} + \sqrt{y}}{\sqrt{x} + \sqrt{y}}$$

$$= \frac{\sqrt{x} + \sqrt{y}}{(\sqrt{x})^2 - \sqrt{y}\sqrt{x} + \sqrt{x}\sqrt{y} - (\sqrt{y})^2}$$

$$= \frac{\sqrt{x} + \sqrt{y}}{x - y}. \quad \blacksquare$$

Radicals may be used to define rational exponents. First, if n is a positive integer and a is a real number, we define

$$\boxed{a^{1/n} = \sqrt[n]{a}}$$

provided $\sqrt[n]{a}$ is a real number. Next we have the following:

Definition

> If m/n is a rational number, where n is a positive integer, and if b is a real number such that $\sqrt[n]{b}$ exists, then
>
> $$b^{m/n} = (\sqrt[n]{b})^m = \sqrt[n]{b^m}.$$

We may also write
$$b^{m/n} = (b^{1/n})^m = (b^m)^{1/n}.$$

It can be shown that the Laws of Exponents are true for all rational exponents.

Example 7   Simplify:

(a) $(-27)^{2/3}(4)^{-5/2}$

(b) $(4a^{1/3})(2a^{1/2})$

(c) $(r^2 s^6)^{1/3}$

(d) $\left(\dfrac{2x^{2/3}}{y^{1/2}}\right)^2 \left(\dfrac{3x^{-5/6}}{y^{1/3}}\right)$

Solutions   The reader should supply reasons for the following:

(a) $(-27)^{2/3}(4)^{-5/2} = (\sqrt[3]{-27})^2(\sqrt{4})^{-5} = (-3)^2(2)^{-5} = \frac{9}{32}$

(b) $(4a^{1/3})(2a^{1/2}) = 8a^{1/3+1/2} = 8a^{5/6}$

(c) $(r^2s^6)^{1/3} = (r^2)^{1/3}(s^6)^{1/3} = r^{2/3}s^2$

(d) $\left(\dfrac{2x^{2/3}}{y^{1/2}}\right)^2\left(\dfrac{3x^{-5/6}}{y^{1/3}}\right) = \left(\dfrac{4x^{4/3}}{y}\right)\left(\dfrac{3x^{-5/6}}{y^{1/3}}\right) = \dfrac{12x^{1/2}}{y^{4/3}}$ ∎

In the next section we shall extend our work to irrational exponents such as $3^{\sqrt{2}}$, 5^{π}, etc.

In scientific areas it is not unusual to work with numbers that are very large or very small. To simplify matters it is customary to write such numbers in the form $a \cdot 10^n$, where a is expressed as a decimal between 1 and 10 and n is an integer. This is referred to as the **scientific form** for real numbers. In applications it is customary to use $\times$ for the multiplication symbol and write $a \times 10^n$ instead of $a \cdot 10^n$. As an illustration, the distance a ray of light travels in one year is approximately 5,900,000,000,000 miles. This may be written in scientific form as 5.9×10^{12}. The positive exponent 12 indicates that the decimal point should be moved 12 places to the *right*. The notation works equally well for small numbers. To illustrate, it is estimated that the weight of an oxygen molecule is 0.0000000000000000000000053 grams or, in scientific form, 5.3×10^{-23} grams. The negative exponent indicates that the decimal point should be moved 23 places to the *left*. Some other illustrations of scientific form are

$$513 = 5.13 \times 10^2, \qquad 20,700 = 2.07 \times 10^4, \qquad 92,000,000 = 9.2 \times 10^7$$
$$0.000648 = 6.48 \times 10^{-4}, \qquad 0.00000000043 = 4.3 \times 10^{-10}.$$

There are a number of advantages to this notation. It is compact, and it enables the reader to see quickly (without counting zeros) the relative magnitudes of large or small quantities. It may also be used to simplify calculations which involve quantities such as those illustrated in the next example.

Example 8   Calculate $\dfrac{(1,100,000)^2\sqrt{0.00000004}}{(8,000,000,000)^{2/3}}$.

Solution   Writing each number in scientific form and using the Laws of Exponents, we have

$$\frac{(1.1 \times 10^6)^2(4 \times 10^{-8})^{1/2}}{(8 \times 10^9)^{2/3}} = \frac{[(1.1)^2 \times 10^{12}][4^{1/2} \times 10^{-4}]}{8^{2/3} \times 10^6}$$

$$= \frac{[1.21 \times 10^{12}][2 \times 10^{-4}]}{4 \times 10^6}$$

$$= \frac{(1.21)(2)}{4} \times 10^{12-4-6}$$

$$= (0.605) \times 10^2$$

$$= 6.05 \times 10.$$ ∎

Hand-held calculators often employ scientific notation in their display panels. In this case the number 10 used in the scientific notation $a \times 10^n$ is suppressed and only the exponent is shown. For example, if we wanted to find $(4,500,000)^2$ on a typical calculator, we would enter the integer 4,500,000 and press the x^2 (or squaring) key. The display panel would show

which is translated as 2.025×10^{13}. Thus,

$$(4,500,000)^2 = 20,250,000,000,000.$$

Many calculators also allow the entry of a number in scientific form. The owner of a calculator should consult the user's manual for details.

EXERCISES 4.1

Express the numbers in Exercises 1–10 in the form a/b, where a and b are integers.

1 $(-\frac{2}{3})^4$

2 $(-3)^3$

3 $\dfrac{2^{-3}}{3^{-2}}$

4 $\dfrac{2^0 + 0^2}{2 + 0}$

5 $(-2)^3 + 3^{-2}$

6 $(-\frac{3}{2})^4 - 2^{-4}$

7 $16^{-3/4}$

8 $9^{5/2}$

9 $(-0.008)^{2/3}$

10 $(0.008)^{-2/3}$

For Exercises 11–46 eliminate negative exponents and simplify.

11 $(\frac{1}{2}x^4)(16x^5)$

12 $(-3x^{-2})(4x^4)$

13 $\dfrac{(2x^3)(3x^2)}{(x^2)^3}$

14 $\dfrac{(2x^2)^3}{4x^4}$

15 $(\frac{1}{6}a^5)(-3a^2)(4a^7)$

16 $(-4b^3)(\frac{1}{6}b^2)(-9b^4)$

17 $\dfrac{(6x^3)^2}{(2x^2)^3}$

18 $\dfrac{(3y^3)(2y^2)^2}{(y^4)^3}$

19 $(3u^7v^3)(4u^4v^{-5})$

20 $(x^2yz^3)(-2xz^2)(x^3y^{-2})$

21 $(8x^4y^{-3})(\frac{1}{2}x^{-5}y^2)$

22 $\left(\dfrac{4a^2b}{a^3b^2}\right)\left(\dfrac{5a^2b}{2b^4}\right)$

23 $(\frac{1}{3}x^4y^{-3})^{-2}$

24 $(-2xy^2)^5\left(\dfrac{x^7}{8y^3}\right)$

25 $(3y^3)^4(4y^2)^{-3}$

26 $(-3a^2b^{-5})^3$

27 $(-2r^4s^{-3})^{-2}$

28 $(2x^2y^{-5})(6x^{-3}y)(\frac{1}{3}x^{-1}y^3)$

29 $(5x^2y^{-3})(4x^{-5}y^4)$

30 $(-2r^2s)^5(3r^{-1}s^3)^2$

31 $\left(\dfrac{3x^5y^4}{x^0y^{-3}}\right)^2$

32 $(4a^2b)^4\left(\dfrac{-a^3}{2b}\right)^2$

33 $(4a^{3/2})(2a^{1/2})$

34 $(-6x^{7/5})(2x^{8/5})$

35 $(3x^{5/6})(8x^{2/3})$

36 $(8r)^{1/3}(2r^{1/2})$

37 $(27a^6)^{-2/3}$

38 $(25z^4)^{-3/2}$

39 $(8x^{-2/3})x^{1/6}$

40 $(3x^{1/2})(-2x^{5/2})$

41 $\left(\dfrac{-8x^3}{y^{-6}}\right)^{2/3}$

42 $\left(\dfrac{-y^{3/2}}{y^{-1/3}}\right)^3$

43 $\left(\dfrac{x^6}{9y^{-4}}\right)^{-1/2}$

44 $\left(\dfrac{c^{-4}}{16d^8}\right)^{3/4}$

45 $\dfrac{(x^6y^3)^{-1/3}}{(x^4y^2)^{-1/2}}$ **46** $a^{4/3}a^{-3/2}a^{1/6}$

Simplify each of the expressions in Exercises 47–72 (all letters denote positive real numbers).

47 $\sqrt{81}$ **48** $\sqrt[3]{-125}$

49 $\sqrt[5]{-64}$ **50** $\sqrt[4]{256}$

51 $\dfrac{1}{\sqrt[3]{2}}$ **52** $\sqrt{\dfrac{1}{7}}$

53 $\sqrt{9x^{-4}y^6}$ **54** $\sqrt{16a^8b^{-2}}$

55 $\sqrt[3]{8a^6b^{-3}}$ **56** $\sqrt[4]{81r^5s^8}$

57 $\sqrt[3]{\dfrac{54a^7}{b^2}}$ **58** $\sqrt[5]{\dfrac{-96x^7}{y^3}}$

59 $\sqrt{\dfrac{1}{3u^3v}}$ **60** $\sqrt[3]{\dfrac{1}{4x^5y^2}}$

61 $\sqrt[4]{(3x^5y^{-2})^4}$ **62** $\sqrt[6]{(2u^{-3}v^4)^6}$

63 $\sqrt[5]{\dfrac{8x^3}{y^4}}\sqrt[5]{\dfrac{4x^4}{y^2}}$ **64** $\sqrt{5xy^7}\sqrt{10x^3y^3}$

65 $\sqrt[3]{3t^4v^2}\sqrt[3]{-9t^{-1}v^4}$

66 $\sqrt[3]{(2r-s)^3}$ **67** $\dfrac{1}{\sqrt{a}+\sqrt{b}}$

68 $\dfrac{\sqrt{a}+\sqrt{b}}{\sqrt{a}-\sqrt{b}}$ **69** $\dfrac{1}{\sqrt{x}}+\dfrac{1}{\sqrt{y}}$

70 $\dfrac{r}{\sqrt{s}}-\dfrac{s}{\sqrt[3]{r}}$ **71** $\dfrac{c}{1-\sqrt{c}}$

72 $\dfrac{3}{\sqrt{t+4}}$

73 The mass of a hydrogen atom is approximately 0.00000000000000000000000017 grams. Express this number in scientific form.

74 The mass of an electron is approximately 9.1×10^{-31} kilograms. Express this number in decimal form.

Express the numbers in Exercises 75 and 76 in scientific form.

75 (a) 427,000   (b) 0.000000098
     (c) 810,000,000

76 (a) 85,200   (b) 0.0000055
     (c) 24,900,000

Express the numbers in Exercises 77 and 78 in decimal form.

77 (a) 8.3×10^5 (b) 2.9×10^{-12}
     (c) 5.63×10^8

78 (a) 2.3×10^7 (b) 7.01×10^{-9}
     (c) 1.23×10^{10}

Use scientific form to find the numbers in Exercises 79–84.

79 (2,100,000,000)(0.0000000033)

80 (840,000,000)(0.00000021)

81 $\dfrac{\sqrt{81,000,000}}{(0.0000002)^3}$

82 $\dfrac{(300,000)^4}{(8,000,000)^{2/3}}$

83 $\dfrac{\sqrt[5]{(0.00243)^2}(20,000)^3}{\sqrt[3]{8,000,000}}$

84 $\dfrac{(160,000)^{3/2}(12,100,000)}{(0.0000011)^2}$

85 In astronomy, distances to stars are measured in light years, where 1 light year is the distance a ray of light travels in one year. If the speed of light is 186,000 miles per second, approximate 1 light year.

86 (a) It is estimated that the Milky Way galaxy contains 100 billion stars. Express this number in scientific form.

     (b) The diameter d of the Milky Way galaxy is estimated as 100,000 light years. Express d in miles (refer to Exercise 85).

4.2   EXPONENTIAL FUNCTIONS

Throughout this section the letter a will denote a positive real number. In Section 4.1 we defined a^r, where r is any rational number. It is also possible to define a unique real number a^x for every real number x (rational or irrational) in such a way that the Laws of Exponents remain valid. To illustrate, given a number such as a^π, we may use the nonterminating decimal representation $3.14159\ldots$ for π and consider the numbers a^3, $a^{3.1}, a^{3.14}, a^{3.141}, a^{3.1415}, \ldots$. We might expect that each successive power gets closer to a^π. This is precisely what happens if a^x is properly defined. However, the definition requires deeper concepts than are available to us and, consequently, it is better to leave it for a more advanced mathematics course such as calculus.

Although definitions and proofs are omitted, we shall assume, henceforth, that formulas such as $a^x a^y = a^{x+y}$, $(a^x)^y = a^{xy}$, and so on, are valid for all real numbers x and y.

Before continuing the discussion of real exponents, we shall establish several results about rational exponents. Let us first show that if $a > 1$, then $a^r > 1$ for every positive rational number r. If $a > 1$, then multiplying both sides by a we obtain $a^2 > a$, and hence $a^2 > a > 1$. Multiplying by a again we obtain $a^3 > a^2 > a > 1$. Continuing, we see that $a^n > 1$ for every positive integer n. (A complete proof requires the method of mathematical induction discussed in Section 9.1.) Similarly, if $0 < a < 1$, it follows that $a^n < 1$ for every positive integer n. Next suppose that $a > 1$ and $r = p/q$, where p and q are positive integers. If it were true that $a^{p/q} \le 1$, then from the previous discussion, $(a^{p/q})^q \le 1$, or $a^p \le 1$, which contradicts the fact that $a^n > 1$ for every positive integer n. Consequently, $a^r > 1$ for every positive rational number r. We may use this fact to prove the following theorem.

Theorem

> If $a > 1$ and if r and s are rational numbers such that $r < s$, then $a^r < a^s$.

Proof

If $r < s$, then $s - r$ is a positive rational number, and from the previous discussion, $a^{s-r} > 1$. Multiplying both sides by a^r we obtain

$$(a^{s-r})a^r > 1 \cdot a^r \quad \text{or} \quad a^s > a^r$$

which is what we wished to prove.

Since to each real number x there corresponds a unique real number a^x, we can define a function as follows.

Definition

> If $a > 0$, then the **exponential function f with base a** is defined by
>
> $$f(x) = a^x$$
>
> where x is any real number.

It is possible to extend the theorem preceding this definition to the case of *real* exponents r and s. Specifically, if $a > 1$ and x_1 and x_2 are real numbers such that $x_1 < x_2$, then it can be shown that $a^{x_1} < a^{x_2}$, that is, $f(x_1) < f(x_2)$. This means that if

$a > 1$, then the exponential function f with base a is increasing for all real numbers. It can also be shown that if $0 < a < 1$, then f is decreasing for all real numbers.

Example 1 Sketch the graph of f if $f(x) = 2^x$.

Solution Coordinates of some points on the graph are listed in the following table.

x	-3	-2	-1	0	1	2	3	4
2^x	$\frac{1}{8}$	$\frac{1}{4}$	$\frac{1}{2}$	1	2	4	8	16

Plotting points and using the fact that f is increasing, gives us the sketch in Figure 4.1.

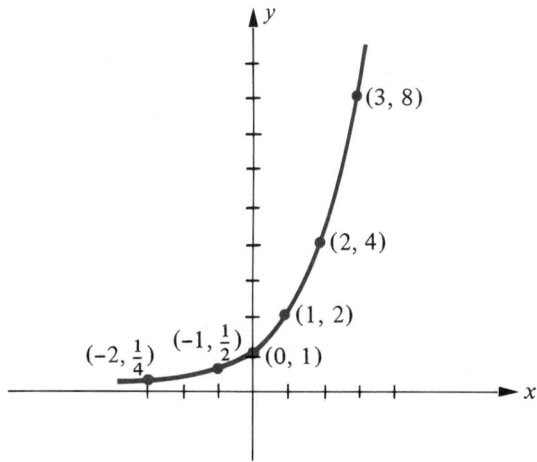

Figure 4.1. $f(x) = 2^x$ ■

The graph in Figure 4.1 is typical of the exponential function with base a if $a > 1$. Since $a^0 = 1$, the y-intercept is always 1. Observe that as x decreases through negative values, the graph approaches the x-axis but never intersects it, since $a^x > 0$ for all x. This means that the x-axis is a **horizontal asymptote** for the graph. As x increases through positive values the graph rises very rapidly. Indeed, if we begin with $x = 0$ and consider successive unit changes in x, then the corresponding changes in y are 1, 2, 4, 8, 16, 32, 64, and so on. This type of variation is very common in nature and is characteristic of the **exponential law of growth**. At the end of this section we shall give several practical illustrations of this behavior.

Example 2 If $f(x) = (3/2)^x$ and $g(x) = 3^x$, sketch the graphs of f and g on the same coordinate plane.

Solution   The following table displays coordinates of several points on the graphs.

x	-2	-1	0	1	2	3	4
$\left(\frac{3}{2}\right)^x$	$\frac{4}{9}$	$\frac{2}{3}$	1	$\frac{3}{2}$	$\frac{9}{4}$	$\frac{27}{8}$	$\frac{81}{16}$
3^x	$\frac{1}{9}$	$\frac{1}{3}$	1	3	9	27	81

With the aid of these points we obtain Figure 4.2, where dashes have been used for the graph of f in order to distinguish it from the graph of g.

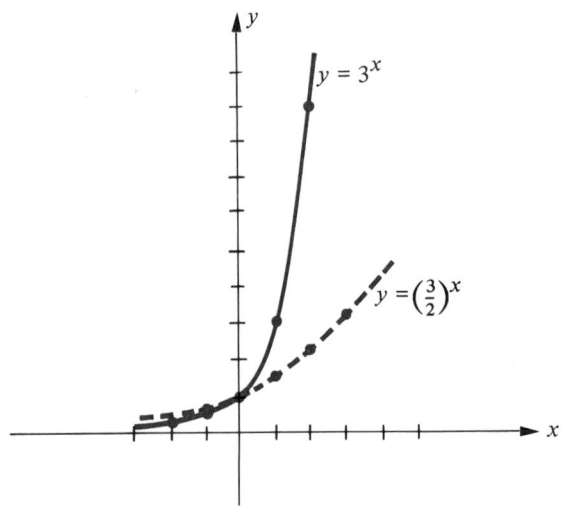

Figure 4.2   ■

Example 2 brings out the fact that if $1 < a < b$, then $a^x < b^x$ for positive values of x and $b^x < a^x$ for negative values of x. In particular, since $\frac{3}{2} < 2 < 3$, this tells us that the graph in Example 1 lies between the graphs of the functions in Example 2.

Example 3   Sketch the graph of the equation $y = \left(\frac{1}{2}\right)^x$.

Solution   Some points on the graph may be obtained from the following table.

x	-3	-2	-1	0	1	2	3
$\left(\frac{1}{2}\right)^x$	8	4	2	1	$\frac{1}{2}$	$\frac{1}{4}$	$\frac{1}{8}$

The graph is sketched in Figure 4.3. Since $\left(\frac{1}{2}\right)^x = 2^{-x}$, the graph is the same as the graph of the equation $y = 2^{-x}$.

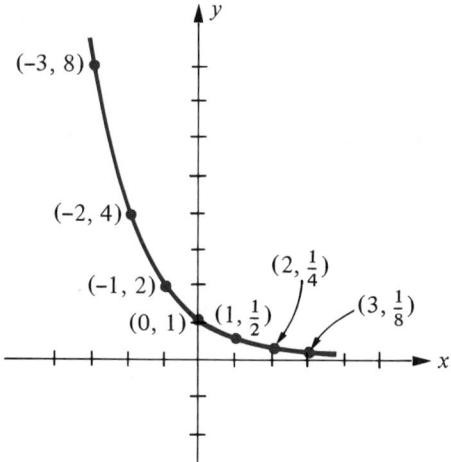

Figure 4.3. $y = (\frac{1}{2})^x = 2^{-x}$ ∎

In advanced mathematics and applications it is often necessary to consider a function f such that $f(x) = a^p$, where p is some expression in x. We do not intend to study such functions in detail; however, let us consider one example.

Example 4 Sketch the graph of f if $f(x) = 2^{-x^2}$.

Solution Since $f(x) = 1/2^{x^2}$, it follows that if x increases numerically, the corresponding point $(x, f(x))$ on the graph approaches the x-axis. Thus the x-axis is a horizontal asymptote for the graph. The maximum value of $f(x)$ occurs at $x = 0$. Tabulating some coordinates of points on the graph, we obtain the following table.

x	-2	-1	0	1	2
$f(x)$	$\frac{1}{16}$	$\frac{1}{2}$	1	$\frac{1}{2}$	$\frac{1}{16}$

The graph is sketched in Figure 4.4. Note the symmetry with respect to the y-axis. Functions of this type arise in the study of the branch of mathematics called *probability*.

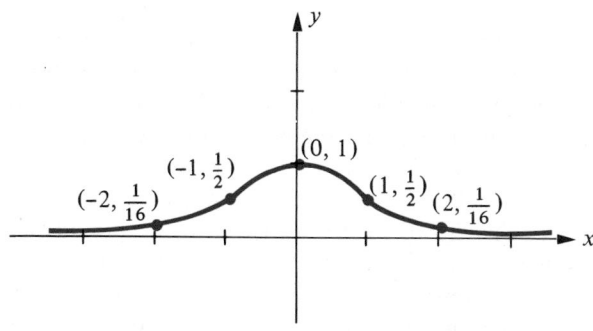

Figure 4.4. $y = 2^{-x^2}$ ∎

The variation of many physical entities can be described by means of exponential functions. One of the most common examples occurs in the growth of certain populations. As an illustration, it might be observed experimentally that the number of bacteria in a culture doubles every hour. If 1,000 bacteria are present at the start of the experiment, then the experimenter would obtain the readings listed below, where t is the time in hours and $f(t)$ is the bacteria count at time t.

t	0	1	2	3	4
$f(t)$	1,000	2,000	4,000	8,000	16,000

It appears that $f(t) = (1,000)2^t$. This formula makes it possible to predict the number of bacteria present at any time t. For example, at $t = 1.5$ we have

$$f(t) = (1,000)2^{3/2} = 1,000\sqrt{2^3} = 1,000\sqrt{8} \approx 2,828.$$

Exponential growth patterns of this type may also be observed in some human and animal populations.

Compound interest provides another illustration of exponential growth. If a sum of money P (called the **principal**) is invested at a simple interest rate of i per cent, then the interest at the end of one interest period is Pi. For example, if $P = \$100$ and i is 8% per year, then the interest at the end of one year is $\$100(0.08)$, or $\$8$. If the interest is reinvested at the end of this period then the new principal is

$$P + Pi, \quad \text{or} \quad P(1 + i).$$

Note that to find the new principal we multiply the original principal by $(1 + i)$. In the preceding illustration the new principal is $\$100(1.08)$, or $\$108$.

If another time period elapses, then the new principal may be found by multiplying $P(1 + i)$ by $(1 + i)$. Thus the principal after two time periods is $P(1 + i)^2$. If we reinvest and continue this process, the principal after three periods is $P(1 + i)^3$; after four it is $P(1 + i)^4$; and in general, after n time periods the principal P_n is given by

$$\boxed{P_n = P(1 + i)^n.}$$

Interest accumulated in this way is called **compound interest**. We see that the principal is given in terms of an exponential function whose base is $1 + i$ and exponent is n. The time period may vary, being measured in years, months, weeks, days, or any other suitable unit of time. When this formula for P_n is employed, it must be remembered that i is the interest rate per time period. For example, if the rate is stated as 9% *per year compounded monthly*, then the rate per month is $\frac{9}{12}\%$, or equivalently, $\frac{3}{4}\%$. In this case, $i = 0.75\% = 0.0075$ and n is the number of months. The sketch in Figure 4.5 illustrates the growth of $\$100$ invested at this rate over a period of 15 years. We have connected the principal amounts by a smooth curve in order to indicate the growth during this time. Due to the complexity of the base, the use of the formula for P_n is extremely tedious unless tables or calculators are available. For those with calculators, several problems on compound interest have been included in the exercises.

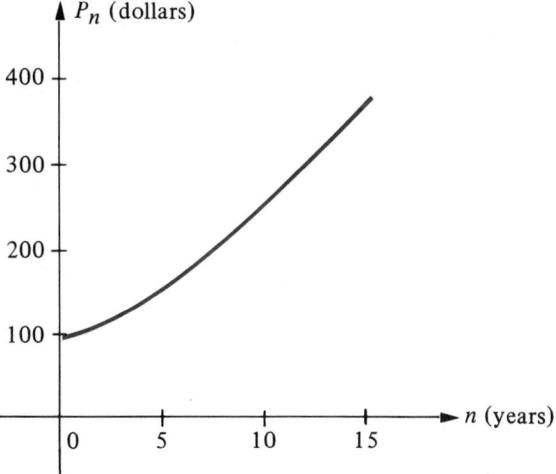

Figure 4.5. *Compound interest:* $P_n = 100(1.0075)^n$

Certain physical quantities may *decrease* exponentially. In such cases the base a of the exponential function is between 0 and 1. One of the most common examples is the decay of a radioactive substance. As an illustration, the polonium isotope ^{210}Po has a half-life of approximately 140 days; that is, given any amount, one-half of it will disintegrate in 140 days. If there is initially 20 mg present, then the following table indicates the amount remaining after various intervals of time.

t (days)	0	140	280	420	560
Amount remaining	20	10	5	2.5	1.25

The sketch in Figure 4.6 illustrates the exponential nature of the disintegration.

The behavior of an electrical condenser can be used to illustrate exponential decay. If the condenser is allowed to discharge, the initial rate of discharge is relatively high, but it then tapers off as in the preceding example of radioactive decay.

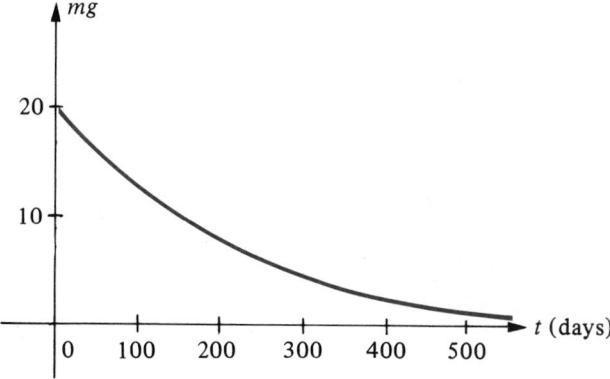

Figure 4.6. *Decay of polonium*

In calculus and its applications a certain irrational number, denoted by e, is often used as the base of an exponential function. To five decimal places,

$$e \approx 2.71828.$$

The base e arises naturally in the discussion of many physical phenomena. For this reason the function f defined by $f(x) = e^x$ is called the **natural exponential function.** Since $2 < e < 3$, the graph of f lies "between" the graphs of the equations $y = 2^x$ and $y = 3^x$. Many hand-held calculators have an e^x key which can be used to approximate values of the natural exponential function. There may also be a key marked y^x that can be used for any positive base y. For those who have access to calculators, we have included some exercises involving e^x and y^x.

EXERCISES 4.2

In each of Exercises 1–24 sketch the graph of the function f.

1   $f(x) = 4^x$

2   $f(x) = 5^x$

3   $f(x) = 10^x$

4   $f(x) = 8^x$

5   $f(x) = 3^{-x}$

6   $f(x) = 4^{-x}$

7   $f(x) = -2^x$

8   $f(x) = -3^x$

9   $f(x) = 4 - 2^{-x}$

10   $f(x) = 2 + 3^{-x}$

11   $f(x) = (2/3)^x$

12   $f(x) = (3/4)^{-x}$

13   $f(x) = (5/2)^{-x}$

14   $f(x) = (4/3)^x$

15   $f(x) = 2^{|x|}$

16   $f(x) = 2^{-|x|}$

17   $f(x) = 2^{x+3}$

18   $f(x) = 3^{x+2}$

19   $f(x) = 2^{3-x}$

20   $f(x) = 3^{-2-x}$

21   $f(x) = 3^{1-x^2}$

22   $f(x) = 2^{-(x+1)^2}$

23   $f(x) = 3^x + 3^{-x}$

24   $f(x) = 3^x - 3^{-x}$

25   If $f(x) = 2^x$ and $g(x) = x^2$, illustrate the difference in the rate of growth of f and g for $x \geq 0$ by sketching the graphs of both functions on the same coordinate plane.

26   Repeat Exercise 25 if $f(x) = 2^{-x}$ and $g(x) = x^{-2}$.

27   The half-life of radium is 1,600 years; that is, given any quantity, one-half of it will disinte-

grate in 1,600 years. If the initial amount is q_0 milligrams, then it can be shown that the quantity $q(t)$ remaining after t years is given by $q(t) = q_0 2^{kt}$. Find k.

28   The number of bacteria in a certain culture at time t is given by $Q(t) = 2(3^t)$, where t is measured in hours and $Q(t)$ in thousands. What is the initial number of bacteria? What is the number after 10 minutes? After 30 minutes? After 1 hour?

29   If \$1,000 is invested at a rate of 12 % per year compounded monthly, what is the principal after 1 month? 2 months? 6 months? 1 year?

30   If 10 grams of salt are added to a certain quantity of water, then the amount $q(t)$ which is undissolved after t minutes is given by $q(t) = 10(\frac{4}{5})^t$. Sketch a graph which shows the value $q(t)$ at any time from $t = 0$ to $t = 10$.

31   Why was $a < 0$ ruled out in the discussion of a^x?

32   Prove that if $0 < a < 1$ and r and s are rational numbers such that $r < s$, then $a^r > a^s$.

33   How does the graph of $y = a^x$ compare with the graph of $y = -a^x$?

34   If $a > 1$, how does the graph of $y = a^x$ compare with the graph of $y = a^{-x}$?

CALCULATOR EXERCISES 4.2

Solve Exercises 1–4 by using the compound interest formula $P_n = P(1 + i)^n$.

1 If $1,000 is invested at an interest rate of 6% per year compounded quarterly, find the principal at the end of
   (a) one year.   (b) two years.
   (c) five years.   (d) ten years.
   (*Hint*: $i = \frac{1}{4}(6\%) = 1.5\%$ and, at the end of one year, $n = 4$.)

2 Rework Exercise 1 if the interest rate is 6% per year compounded monthly.

3 If $10,000 is invested at a rate of 9% per year compounded semi-annually, how long will it take for the principal to exceed
   (a) $15,000? (b) $20,000? (c) $30,000?

4 A certain department store requires its credit card customers to pay interest at the rate of 18% per year compounded monthly on any unpaid bills. If a man buys a $500 television set on credit and then makes no payments for one year, how much does he owe?

5 Sketch the graph of f if $f(x)$ is given by:

   (a) e^x (b) e^{-x}

   (c) $\dfrac{e^x + e^{-x}}{2}$ (d) $\dfrac{2}{e^x + e^{-x}}$

   (e) $\dfrac{e^x - e^{-x}}{2}$ (f) $\dfrac{2}{e^x - e^{-x}}$

6 Sketch the graph of the equation $y = e^{-x^2/2}$.

7 Sketch the graph of $y = x(2^x)$.

8 If $f(x) = x^2(5\sqrt{x})$, find $f(1.01) - f(1)$.

9 If $f(x) = (1 + x)^{1/x}$, approximate the following to four decimal places:

$$f(1), \quad f(0.1), \quad f(0.01), \quad f(0.001),$$
$$f(0.0001), \quad f(0.00001).$$

Compare your answers with the value of $e \approx 2.71828$ and arrive at a conjecture.

10 The population of a certain city is increasing at the rate of 5% per year and the present population is 500,000. Using methods developed in calculus, it is estimated that the population after t more years will be approximately $500,000e^{0.05t}$. Estimate the population after   (a) 10 years.   (b) 20 years.   (c) 40 years.

11 Under certain conditions the atmospheric pressure p at altitude h ft is given by

$$p = 29e^{-0.000034h}.$$

What is the pressure at an altitude of 40,000 ft?

4.3 LOGARITHMS

Throughout this section and the next, it is assumed that a is a positive real number different from 1. Let us begin by examining the graph of f, where $f(x) = a^x$ (see Figure 4.7 for the case $a > 1$). It appears that every positive real number u is the ordinate of some point on the graph; that is, there is a number v such that $u = a^v$. Indeed, v is the abscissa of the point where the line $y = u$ intersects the graph. Moreover, since f is increasing throughout its domain, the number u can occur as an ordinate only once. A similar situation exists if $0 < a < 1$. This makes the following theorem plausible.

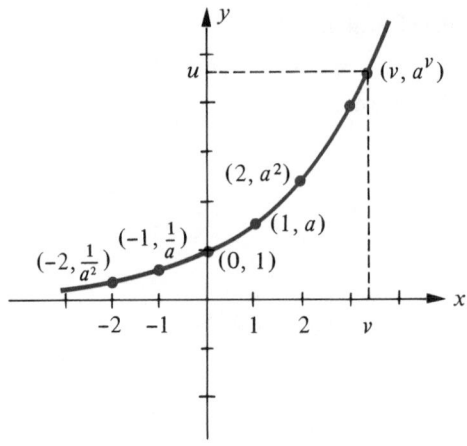

Figure 4.7. $f(x) = a^x, a > 1$

Theorem

> For each positive real number u there is a unique real number v such that
>
> $$a^v = u.$$

A rigorous proof of this theorem requires concepts studied in calculus.

Definition

> If u is any positive real number, then the (unique) exponent v such that $a^v = u$ is called the **logarithm of u with base a** and denoted by $\log_a u$.

A convenient way to memorize this definition is by means of the following statement:

> $$v = \log_a u \quad \text{if and only if} \quad a^v = u.$$

As illustrations we may write

$$3 = \log_2 8 \quad \text{since } 2^3 = 8$$
$$-2 = \log_5 \tfrac{1}{25} \quad \text{since } 5^{-2} = \tfrac{1}{25}$$
$$4 = \log_{10} 10{,}000 \quad \text{since } 10^4 = 10{,}000.$$

The fact that $\log_a u$ is the exponent v such that $a^v = u$ gives us the following important identity:

> $$a^{\log_a u} = u.$$

Since $a^r = a^r$ for every real number r, we may use the definition of logarithm with $v = r$ and $u = a^r$ to obtain

> $$r = \log_a a^r.$$

In particular, letting $r = 1$ we see that $\log_a a = 1$.

Since $a^0 = 1$, it follows from the definition of logarithm that

$$\log_a 1 = 0.$$

Example 1   In each of the following find s.

(a) $s = \log_4 2$ (b) $\log_5 s = 2$ (c) $\log_s 8 = 3$

Solutions

(a) If $s = \log_4 2$, then $4^s = 2$ and hence $s = \frac{1}{2}$.
(b) If $\log_5 s = 2$, then $5^2 = s$ and hence $s = 25$.
(c) If $\log_s 8 = 3$, then $s^3 = 8$ and hence $s = \sqrt[3]{8} = 2$.   ■

Example 2   Solve the equation $\log_4 (5 + x) = 3$.

Solution   If $\log_4 (5 + x) = 3$, then

$$5 + x = 4^3 \quad \text{or} \quad 5 + x = 64.$$

Hence the solution is $x = 59$.   ■

The following laws are fundamental for all work with logarithms, where it is assumed that u and w are positive real numbers.

Laws of Logarithms

(i) $\log_a (uw) = \log_a u + \log_a w$
(ii) $\log_a (u/w) = \log_a u - \log_a w$
(iii) $\log_a (u^c) = c \log_a u$, for every real number c

Proof   To prove (i), we begin by letting

$$r = \log_a u \quad \text{and} \quad s = \log_a w.$$

By the definition of logarithm, this implies that $a^r = u$ and $a^s = w$. Consequently,

$$a^r a^s = uw$$

and hence, by a law of exponents,

$$a^{r+s} = uw.$$

Again using the definition of logarithm, we see that the last equation is equivalent to

$$r + s = \log_a (uw).$$

Substituting for r and s from the first step in the proof gives us

$$\log_a u + \log_a w = \log_a uw.$$

This completes the proof of (i).

To prove (ii) we begin as in the proof of (i), but this time we divide a^r by a^s, obtaining

$$\frac{a^r}{a^s} = \frac{u}{w}, \quad \text{or} \quad a^{r-s} = \frac{u}{w}.$$

The latter equation is equivalent to

$$r - s = \log_a (u/w).$$

Substituting for r and s gives us (ii).

Finally, if c is any real number, then using the same notation as above,

$$(a^r)^c = u^c, \quad \text{or} \quad a^{cr} = u^c.$$

According to the definition of logarithm, the last equality implies that

$$cr = \log_a u^c.$$

Substituting for r, we obtain

$$c\log_a u = \log_a u^c.$$

This proves Law (iii).

The following examples illustrate uses of the Laws of Logarithms.

Example 3   If $\log_a 3 = 0.4771$ and $\log_a 2 = 0.3010$, find each of the following.

(a) $\log_a 6$ (b) $\log_a (\tfrac{3}{2})$ (c) $\log_a \sqrt{2}$ (d) $\dfrac{\log_a 3}{\log_a 2}$

Solutions

(a) Since $6 = 2 \cdot 3$, we may use Law (i) to obtain

$$\log_a 6 = \log_a (2 \cdot 3) = \log_a 2 + \log_a 3$$

$$= 0.4771 + 0.3010 = 0.7781.$$

(b) By Law (ii),

$$\log_a(\tfrac{3}{2}) = \log_a 3 - \log_a 2$$

$$= 0.4771 - 0.3010 = 0.1761.$$

(c) Using Law (iii),

$$\log_a \sqrt{2} = \log_a 2^{1/2} = (\tfrac{1}{2}) \log_a 2$$
$$= (\tfrac{1}{2})(0.3010) = 0.1505.$$

(d) There is no law of logarithms which allows us to simplify $(\log_a 3)/(\log_a 2)$. Consequently, we *divide* 0.4771 by 0.3010, obtaining the approximation 1.585. It is important to notice the difference between this problem and the one stated in part (b). ■

Example 4   Solve each of the following equations.

(a) $\log_5 (2x + 3) = \log_5 11 + \log_5 3$

(b) $\log_4 (x + 6) - \log_4 10 = \log_4 (x - 1) - \log_4 2$

Solutions

(a)   Using Law (i), the given equation may be written as

$$\log_5 (2x + 3) = \log_5 (11 \cdot 3) = \log_5 33.$$

Consequently, $2x + 3 = 33$, or $2x = 30$, and therefore the solution is $x = 15$.

(b)   The given equation is equivalent to

$$\log_4 (x + 6) - \log_4 (x - 1) = \log_4 10 - \log_4 2.$$

Applying Law (ii) gives us

$$\log_4 \left(\frac{x + 6}{x - 1} \right) = \log_4 \frac{10}{2} = \log_4 5$$

and hence

$$\frac{x + 6}{x - 1} = 5.$$

The last equation implies that

$$x + 6 = 5x - 5 \quad \text{or} \quad 4x = 11.$$

Thus the solution is $x = 11/4$. ■

Extraneous solutions sometimes occur in the process of solving logarithmic equations, as illustrated in the next example.

Example 5   Solve the equation $2 \log_7 x = \log_7 36$.

Solution   Applying Law (iii), we obtain $2\log_7 x = \log_7 x^2$, and substitution in the given equation leads to

$$\log_7 x^2 = \log_7 36.$$

Consequently, $x^2 = 36$ and hence either $x = 6$ or $x = -6$. However, $x = -6$ is not a solution of the original equation since x must be positive in order for $\log_7 x$ to exist. Thus there is only one solution, $x = 6$.   ■

Sometimes it is necessary to *change the base* of a logarithm by expressing $\log_b u$ in terms of $\log_a u$, for some positive real number b different from 1. This can be accomplished as follows. We begin with the equivalent equations

$$v = \log_b u \quad \text{and} \quad b^v = u.$$

Taking the logarithm, base a, of both sides of the second equation gives us

$$\log_a b^v = \log_a u.$$

Applying Law (iii), we get

$$v \log_a b = \log_a u.$$

Solving for v (that is, $\log_b u$), we see that

$$\boxed{\log_b u = \frac{\log_a u}{\log_a b}.}$$

The most important special case of this formula is obtained by letting $u = a$. Since $\log_a a = 1$, this gives us

$$\boxed{\log_b a = \frac{1}{\log_a b}.}$$

The Laws of Logarithms are often used as in the next example.

Example 6   Express $\log_a \dfrac{x^3\sqrt{y}}{z^2}$ in terms of the logarithms of x, y, and z.

Solution   Using the three Laws of Logarithms,

$$\log_a \frac{x^3\sqrt{y}}{z^2} = \log_a (x^3\sqrt{y}) - \log_a z^2$$

$$= \log_a x^3 + \log_a \sqrt{y} - \log_a z^2$$

$$= 3\log_a x + (\tfrac{1}{2})\log_a y - 2\log_a z.$$ ■

The procedure in Example 6 can also be reversed; that is, beginning with the final equation, we can retrace our steps to obtain the original expression (see Exercises 67–70).

As a final remark, note that there is no general law for expressing $\log_a (u + w)$ in terms of simpler logarithms. It is evident that it does not always equal $\log_a u + \log_a w$, since the latter equals $\log_a (uw)$.

EXERCISES 4.3

Change the equations in Exercises 1–8 to logarithmic form.

1 $4^3 = 64$ **2** $3^5 = 243$

3 $2^7 = 128$ **4** $5^3 = 125$

5 $10^{-3} = 0.001$ **6** $10^{-2} = 0.01$

7 $t^r = s$ **8** $v^w = u$

Change the equations in Exercises 9–16 to exponential form.

9 $\log_{10} 1000 = 3$ **10** $\log_3 81 = 4$

11 $\log_3 \dfrac{1}{243} = -5$ **12** $\log_4 \dfrac{1}{64} = -3$

13 $\log_7 1 = 0$ **14** $\log_9 1 = 0$

15 $\log_t r = p$ **16** $\log_v w = q$

Find the numbers in Exercises 17–30.

17 $\log_4 (1/16)$ **18** $\log_2 32$

19 $\log_{10} 100$ **20** $\log_8 64$

21 $10^{\log_{10} 5}$ **22** $\log_{10} 0.0001$

23 $\log_7 \sqrt[3]{7}$ **24** $10^{2 \log_{10} 3}$

25 $\log_{10} (1/10)$ **26** $\log_{10} 100{,}000$

27 $\log_{1/2} 8$ **28** $\log_3 (1/81)$

29 $3^{4 \log_3 2}$ **30** $\log_6 \sqrt[5]{6}$

Given $\log_a 7 = 0.8$ and $\log_a 3 = 0.5$, change the expressions in Exercises 31–40 to decimal form.

31 $\log_a (7/3)$ **32** $\log_a (3/7)$

33 $\log_a 7^3$ **34** $\log_a (3^7)$

35 $\log_a (1/\sqrt{3})$ **36** $\log_a \sqrt[4]{3}$

37 $\log_a 63$ **38** $(\log_a 7)/(\log_a 3)$

39 $\log_7 a$ **40** $(\log_3 a)(\log_a 3)$

Find the solutions of the equations in Exercises 41–58.

41 $\log_3 (x - 4) = 2$ **42** $\log_2 (x - 5) = 4$

43 $\log_9 x = 3/2$ **44** $\log_4 x = -3/2$

45 $\log_5 x^2 = -2$ **46** $\log_{10} x^2 = -4$

47 $\log_2 (x^2 - 5x + 14) = 3$

48 $\log_2 (x^2 - 10x + 18) = 1$

49 $\log_x 7 = 3$

50 $5^{\log_5 x} = 10$

51 $\log_6 (2x - 3) = \log_6 12 - \log_6 3$

52 $2 \log_3 x = 3 \log_2 5$

53 $\log_2 x - \log_2 (x + 1) = 3 \log_2 4$

54 $\log_5 x + \log_5 (x + 6) = \tfrac{1}{2} \log_5 9$

55 $\log_{10} x^2 = \log_{10} x$

56 $\log_x 10 = 10$

57 $\tfrac{1}{2} \log_5 (x - 2) = 3 \log_5 2 - \tfrac{3}{2} \log_5 (x - 2)$

58 $\log_{10} 5^x = \log_3 1$

In each of Exercises 59–66, express the logarithm in terms of logarithms of x, y, and z.

59 $\log_a \dfrac{x^2 y}{z^3}$ **60** $\log_a \dfrac{x^3 y^2}{z^5}$

61 $\log_a \dfrac{\sqrt{xz^2}}{y^4}$

62 $\log_a x \sqrt[3]{\dfrac{y^2}{z^4}}$

63 $\log_a \sqrt[3]{\dfrac{x^2}{yz^5}}$

64 $\log_a \dfrac{\sqrt{xy^6}}{\sqrt[3]{z^2}}$

65 $\log_a \sqrt{x\sqrt{yz^3}}$

66 $\log_a \sqrt[3]{x^2 y \sqrt{z}}$

In each of Exercises 67–70 write the expression as one logarithm.

67 $2\log_a x + \frac{1}{3}\log_a (x - 2) - 5\log_a (2x + 3)$

68 $5\log_a x - \frac{1}{2}\log_a (3x - 4) + 3\log_a (5x + 1)$

69 $\log_a (y^2 x^3) - 2\log_a x\sqrt[3]{y} + 3\log_a \left(\dfrac{x}{y}\right)$

70 $2\log_a \dfrac{y^3}{x} - 3\log y + \frac{1}{2}\log_a x^4 y^2$

4.4 LOGARITHMIC FUNCTIONS

We shall now use the concept of logarithm to introduce a new function whose domain is the set of positive real numbers.

Definition

> The function f defined by
>
> $$f(x) = \log_a x$$
>
> for all positive real numbers x is called the **logarithmic function with base a**.

The graph of f is the same as the graph of the equation $y = \log_a x$ which, by the definition of logarithm, is equivalent to

$$x = a^y.$$

In order to find some pairs which are solutions of the preceding equation, we may substitute for y and find the corresponding values of x, as illustrated in the following table.

y	-3	-2	-1	0	1	2	3
x	$\dfrac{1}{a^3}$	$\dfrac{1}{a^2}$	$\dfrac{1}{a}$	1	a	a^2	a^3

If $a > 1$, we obtain the sketch in Figure 4.8. In this case f is an increasing function throughout its domain. If $0 < a < 1$, then the graph has the general shape shown in Figure 4.9, and hence f is a decreasing function. Note that for every a under consideration, the region to the left of the y-axis is excluded. There is no y-intercept and the x-intercept is 1.

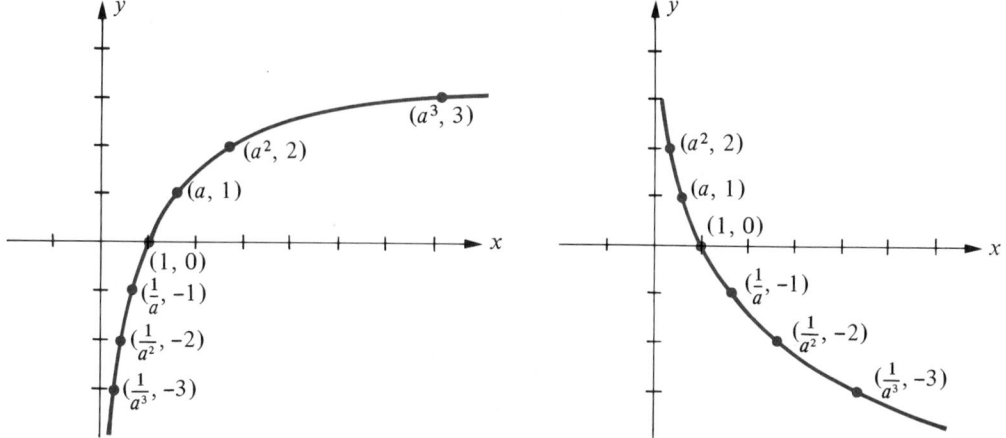

Figure 4.8. $f(x) = \log_a x, a > 1$ **Figure 4.9.** $f(x) = \log_a x, 0 < a < 1$

Functions defined by expressions of the form $\log_a p$, where p is some expression in x, often occur in mathematics and its applications. Functions of this type are classified as members of the logarithmic family; however, the graphs may differ from those sketched in Figures 4.8 and 4.9, as is illustrated in the following examples.

Example 1   Sketch the graph of f if $f(x) = \log_3(-x)$, $x < 0$.

Solution   If $x < 0$, then $-x > 0$, and hence $\log_3(-x)$ is defined. We wish to sketch the graph of the equation $y = \log_3(-x)$, or equivalently, $3^y = -x$. The following table displays coordinates of some points on the graph, which is sketched in Figure 4.10.

y	-2	-1	0	1	2
x	$-\frac{1}{9}$	$-\frac{1}{3}$	-1	-3	-9

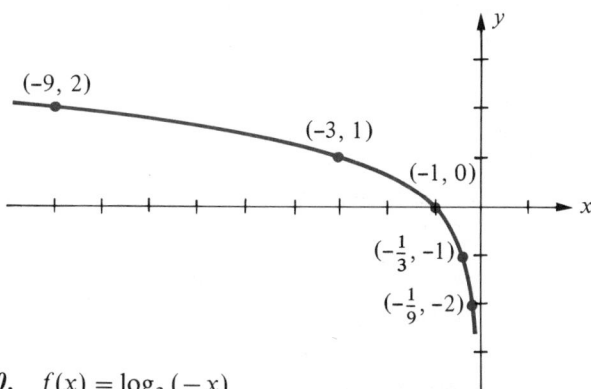

Figure 4.10. $f(x) = \log_3(-x)$

Example 2   Sketch the graph of the equation $y = \log_3 |x|,\ x \neq 0$.

Solution   Since $|x| > 0$ for all $x \neq 0$, there are points on the graph corresponding to negative values of x as well as to positive values. If $x > 0$, then $|x| = x$ and hence to the right of the y-axis the graph coincides with the graph of $y = \log_3 x$, or equivalently, $x = 3^y$. If $x < 0$, then $|x| = -x$ and the graph is the same as that of $y = \log_3(-x)$ (see Example 1). The graph is sketched in Figure 4.11. Note that the graph is symmetric with respect to the y-axis.

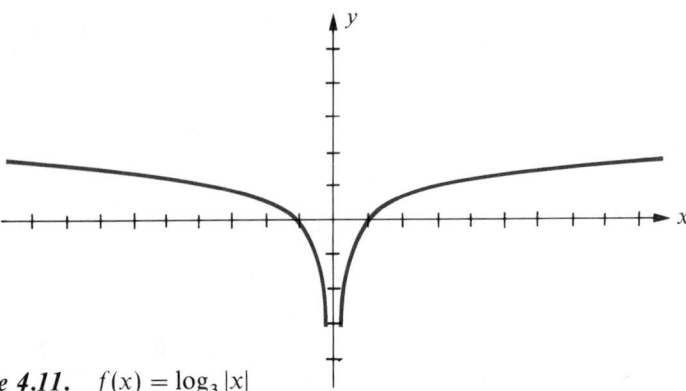

Figure 4.11. $f(x) = \log_3 |x|$ ■

There is a close relationship between exponential and logarithmic functions. This is to be expected since the logarithmic function was defined in terms of the exponential function. A precise description of the relationship is stated in the following theorem.

Theorem

> The exponential and logarithmic functions with base a are inverse functions of one another.

Proof   If $f(x) = a^x$ and $g(x) = \log_a x$, then, according to the definition of inverse function (see Section 2.4), we must show that

$$f(g(x)) = x \quad \text{and} \quad g(f(x)) = x.$$

For every positive real number x,

$$f(g(x)) = a^{g(x)}$$
$$= a^{\log_a x}$$
$$= x.$$

For every real number x,

$$g(f(x)) = \log_a f(x)$$
$$= \log_a a^x$$
$$= x.$$

This proves the theorem.

Logarithmic functions occur frequently in applications. Indeed, if two variables u and v are related so that u is an exponential function of v, then v is a logarithmic function of u. As a specific example, if \$1.00 is invested at a rate of 9 % per year compounded monthly, then from the discussion in Section 4.2, the principal P after n interest periods is

$$P = (1.0075)^n.$$

The logarithmic form of this equation is

$$n = \log_{1.0075} P.$$

Thus, the number of periods n required for \$1.00 to grow to an amount P is related logarithmically to P.

Another example we considered in Section 4.2 was that of population growth. In particular, the equation for the number N of bacteria in a certain culture after t hours was

$$N = (1000)2^t \quad \text{or} \quad 2^t = N/1000.$$

Changing to logarithmic form, we obtain

$$t = \log_2 (N/1000).$$

Hence the time t is a logarithmic function of N.

In Section 4.2 we defined the natural exponential function f by means of the equation $f(x) = e^x$. The inverse of f, that is, the logarithmic function with base e, is called the **natural logarithmic function**. We use **ln** x as an abbreviation for $\log_e x$ and refer to it as the **natural logarithm of x**. Since $e \approx 3$, the graph of $y = \ln x$ is similar in appearance to the graph of $y = \log_3 x$. For natural logarithms the Laws of Logarithms are written

$$\ln (uv) = \ln u + \ln v, \qquad \ln \frac{u}{v} = \ln u - \ln v, \qquad \ln u^c = c \ln u.$$

Example 3 According to Newton's Law of Cooling, the rate at which an object cools is directly proportional to the difference in temperature between the object and the surrounding medium. Newton's Law can be used to show that under certain conditions the temperature T of an object is given by

$$T = 75e^{-2t}$$

where t is time. Express t as a function of T.

Solution The given equation may be rewritten

$$e^{-2t} = T/75.$$

Using logarithms with base e yields

$$-2t = \log_e (T/75) = \ln (T/75).$$

Consequently,

$$t = -\tfrac{1}{2}\ln{(T/75)} \quad \text{or} \quad t = -\tfrac{1}{2}[\ln T - \ln 75]. \qquad \blacksquare$$

We shall not have many occasions to work with natural logarithms in this text. For computational purposes the base 10 is used. Base 10 gives us the system of common logarithms discussed in subsequent sections.

Many hand-held calculators have a key labeled $\ln x$ which can be used to approximate values of the natural logarithmic function. Several exercises are included at the end of this section for those who have a calculator of that type.

EXERCISES 4.4

Sketch the graph of f in each of Exercises 1–10.

1 (a) $f(x) = \log_2 x$ (b) $f(x) = \log_4 x$

2 (a) $f(x) = \log_5 x$ (b) $f(x) = \log_{10} x$

3 (a) $f(x) = \log_3 x$
   (b) $f(x) = \log_3 (x - 2)$

4 (a) $f(x) = \log_2 (-x)$
   (b) $f(x) = \log_2 (3 - x)$

5 (a) $f(x) = \log_3 (3x)$ (b) $f(x) = 3 \log_3 x$

6 (a) $f(x) = \log_2 (x^2)$ (b) $f(x) = \log_2 (x^3)$

7 (a) $f(x) = \log_2 \sqrt{x}$ (b) $f(x) = \log_2 \sqrt[3]{x}$

8 (a) $f(x) = \log_2 |x|$ (b) $f(x) = |\log_2 x|$

9 (a) $f(x) = \log_3 (1/x)$
   (b) $f(x) = 1/(\log_3 x)$

10 (a) $f(x) = \log_3 (2 + x)$
    (b) $f(x) = 2 + \log_3 x$

11 What is the geometric relationship between the graphs of $y = \log_a x$ and $y = a^x$? Why does this relationship exist?

12 Describe the relationship of the graph of $y = \ln x$ to the graphs of $y = \log_2 x$ and $y = \log_3 x$.

13 If one begins with q_0 milligrams of pure radium, then the amount q remaining after t years is given by the formula

$$q = q_0(2)^{-t/1600}.$$

Use logarithms with base 2 to solve for t in terms of q and q_0.

14 The number of bacteria in a certain culture at time t is given by

$$N = 10^4(3)^t.$$

Use logarithms with base 3 to solve for t in terms of N.

15 An electrical condenser with initial charge Q_0 is allowed to discharge. After t seconds the charge Q is given by

$$Q = Q_0 e^{kt}$$

where k is a constant of proportionality. Use natural logarithms to solve for t in terms of Q_0, Q, and k.

16 Under certain conditions the atmospheric pressure p at altitude h is given by

$$p = 29e^{-0.000034h}.$$

Use natural logarithms to solve for h as a function of p.

17 The loudness of sound, as experienced by the human ear, is based upon intensity levels. A formula used for finding the intensity level α which corresponds to a sound intensity I is

$$\alpha = 10\log_{10}(I/I_0) \text{ decibels}$$

where I_0 is a special value of I agreed to be the weakest sound that can be detected by the ear under certain conditions. Find α if:

(a) I is 10 times as great as I_0.
(b) I is 1,000 times as great as I_0.
(c) I is 10,000 times as great as I_0. (This is the intensity level of the average voice.)

CALCULATOR EXERCISES 4.4

1 Sketch the graph of
(a) $y = \ln x$ (b) $y = \ln|x|$ (c) $y = |\ln x|$

2 Sketch the graph of f if $f(x) = x^2 \ln x$.

3 Verify that $\ln e^x = x$ and $e^{\ln x} = x$ for the following values of x:
(a) 1 (b) 3.4 (c) 0.056 (d) 8.143

4 Refer to Calculator Exercise 10 of Section 4.2. For what value of t will the population of the city be 1,000,000?

5 Letting $a = e$ in the formula $\log_b u = (\log_a u)/(\log_a b)$ leads to

$$\log_b u = \frac{\ln u}{\ln b}.$$

Use this formula to approximate
(a) $\log_3 4$ (b) $\log_5(1.64)$ (c) $\log_7\sqrt{2}$
(d) $\log_{10} 382$ (e) $\log_{10}(47/62)$

6 The LOG key on a calculator refers to the logarithmic function with base 10. Use a calculator as an aid to sketching the graph of f if $f(x) = \log_{10} x$.

4.5 COMMON LOGARITHMS

Logarithms with base 10 are useful for certain numerical computations. It is customary to refer to such logarithms as **common logarithms** and to use the symbol **log x** as an abbreviation for $\log_{10} x$.

Base 10 is used in computational work because every positive real number can be written in the scientific form $c \cdot 10^k$, where $1 \le c < 10$ and k is an integer (see Section 4.1). For example,

$$513 = (5.13)10^2, \quad 2375 = (2.375)10^3, \quad 720000 = (7.2)10^5,$$
$$0.641 = (6.41)10^{-1}, \quad 0.00000438 = (4.38)10^{-6}, \quad 4.601 = (4.601)10^0.$$

If x is any positive real number and we write

$$x = c \cdot 10^k$$

where $1 \le c < 10$ and k is an integer, then applying (iii) of the Laws of Logarithms gives us

$$\log x = \log c + \log 10^k.$$

Since $\log 10^k = k$ (Why?) we see that

$$\log x = \log c + k.$$

The last equation tells us that to find $\log x$ for any positive real number x it is sufficient to know the logarithms of numbers between 1 and 10. The number $\log c$, where $1 \le c < 10$, is called the **mantissa**, and the integer k is called the **characteristic** of $\log x$.

If $1 \le c < 10$, then since $\log x$ increases as x increases,

$$\log 1 \le \log c < \log 10,$$

or equivalently, $$0 \le \log c < 1.$$

Hence the mantissa of a logarithm is a number between 0 and 1. When working numerical problems, it is usually necessary to approximate logarithms. For example, it can be shown that

$$\log 2 = 0.3010299957\ldots$$

where the decimal is nonrepeating and nonterminating. We shall round off such logarithms to four decimal places and write

$$\log 2 \approx 0.3010.$$

If a number between 0 and 1 is written as a finite decimal, it is sometimes referred to as a **decimal fraction**. Thus the equation $\log x = \log c + k$ implies that if x is any positive real number, then *log x may be approximated by the sum of a positive decimal fraction (the mantissa) and an integer k (the characteristic)*. We shall refer to this representation as the **standard form** for $\log x$.

Logarithms of many of the numbers between 1 and 10 have been calculated. Table 2 contains four-decimal-place approximations for logarithms of numbers between 1.00 and 9.99 at intervals of 0.01. This table can be used to find the logarithm of any three-digit number to four-decimal-place accuracy. There exist far more extensive tables which provide logarithms of many additional numbers to much greater accuracy than four decimal places. The use of Table 2 is illustrated in the following examples.

Example 1   Find each of the following.

    (a) $\log 43.6$ (b) $\log 43{,}600$ (c) $\log 0.0436$

Solutions

    (a) Since $43.6 = (4.36)10^1$, the characteristic of $\log 43.6$ is 1. Referring to Table 2, we find that the mantissa of $\log 4.36$ may be approximated by 0.6395. Hence, as in the preceding discussion,

$$\log 43.6 \approx 0.6395 + 1 = 1.6395.$$

    (b) Since $43{,}600 = (4.36)10^4$, the mantissa is the same as in part (a); however, the characteristic is 4. Consequently,

$$\log 43{,}600 \approx 0.6395 + 4 = 4.6395.$$

(c) If we write $0.0436 = (4.36)10^{-2}$, then

$$\log 0.0436 = \log 4.36 + (-2).$$

Hence $\qquad \log 0.0436 \approx 0.6395 + (-2).$

We could subtract 2 from 0.6395 and obtain

$$\log 0.0436 \approx -1.3605$$

but this is not standard form, since $-1.3605 = -0.3605 + (-1)$, a number in which the decimal fraction is *negative*. A common error is to write $0.6395 + (-2)$ as -2.6395. This is incorrect since $-2.6395 = -0.6395 + (-2)$, which is not the same as $0.6395 + (-2)$. ∎

If a logarithm has a negative characteristic, it is customary either to leave it in standard form or to rewrite the logarithm, keeping the decimal part positive. To illustrate the latter technique, let us add and subtract 8 on the right side of the equation

$$\log 0.0436 \approx 0.6395 + (-2).$$

This gives us

$$\log 0.0436 \approx 0.6395 + (8 - 8) + (-2)$$

or $\qquad \log 0.0436 \approx 8.6395 - 10.$

We could also write

$$\log 0.0436 \approx 18.6395 - 20 = 43.6395 - 45$$

and so on, as long as the *integral part* of the logarithm is -2.

Example 2 Find each of the following.
(a) $\log (0.00652)^2$ (b) $\log (0.00652)^{-2}$ (c) $\log (0.00652)^{1/2}$

Solutions
(a) By (iii) of the Laws of Logarithms,

$$\log (0.00652)^2 = 2 \log 0.00652.$$

Since $0.00652 = (6.52)10^{-3}$,

$$\log 0.00652 = \log 6.52 + (-3).$$

Referring to Table 2, we see that $\log 6.52$ is approximately 0.8142 and, therefore,

$$\log 0.00652 \approx 0.8142 + (-3).$$

Hence
$$\log (0.00652)^2 = 2 \log 0.00652$$
$$\approx 2[0.8142 + (-3)]$$
$$= 1.6284 + (-6).$$

The standard form is $0.6284 + (-5)$.

(b) Again using Law (iii) and the value for $\log 0.00652$ found in part (a),

$$\log (0.00652)^{-2} = -2 \log 0.00652$$
$$\approx -2[0.8142 + (-3)]$$
$$= -1.6284 + 6.$$

It is important to note that -1.6284 means $-0.6284 + (-1)$ and, consequently, the decimal part is negative. To obtain the standard form, we may write

$$-1.6284 + 6 = 6.0000 - 1.6284$$
$$= 4.3716.$$

This shows that the mantissa is 0.3716 and the characteristic is 4.

(c) By Law (iii),

$$\log (0.00652)^{1/2} = (\tfrac{1}{2}) \log 0.00652$$
$$\approx \tfrac{1}{2}[0.8142 + (-3)].$$

If we multiply by 1/2, the standard form is not obtained since neither number in the resulting sum is the characteristic. In order to avoid this, we may adjust the expression within brackets by adding and subtracting a suitable number. If we use 1 in this way, we obtain

$$\log (0.00652)^{1/2} \approx \tfrac{1}{2}[1.8142 + (-4)]$$
$$= 0.9071 + (-2)$$

which is in standard form. We could also have added and subtracted a number other than 1. For example,

$$\tfrac{1}{2}[0.8142 + (-3)] = \tfrac{1}{2}[17.8142 + (-20)]$$
$$= 8.9071 + (-10). \qquad \blacksquare$$

Table 2 can be used to find an approximation to x if $\log x$ is given, as illustrated in the following example.

Example 3 Find a decimal approximation to x if $\log x$ is as follows.

(a) $\log x = 1.7959$ (b) $\log x = -3.5918$

Solutions

(a) The mantissa 0.7959 determines the sequence of digits in x and the characteristic determines the position of the decimal point. Referring to the *body* of Table 2, we see that the mantissa 0.7959 is the logarithm of 6.25. Since the characteristic is 1, x lies between 10 and 100. Consequently, $x \approx 62.5$.

(b) In order to find x from Table 2, $\log x$ must be written in standard form. In order to change $\log x = -3.5918$ to standard form, we may add and subtract 4, obtaining

$$\log x = (4 - 3.5918) - 4$$
$$= 0.4082 - 4.$$

Referring to Table 2, we see that the mantissa 0.4082 is the logarithm of 2.56. Since the characteristic of $\log x$ is -4, it follows that $x \approx 0.000256$. ∎

In each of the previous examples the mantissas appeared in Table 2. If this is not the case, then x can be approximated by the method of interpolation discussed in Section 4.7.

If a hand-held calculator with a log key is used to determine common logarithms, then the standard form for $\log x$ is obtained only if $x \geq 1$. For example, to find $\log 463$ on a typical calculator, we enter 463 and press the log key, obtaining the standard form

However, if we find $\log 0.0463$ in similar fashion, then the following number appears on the display panel:

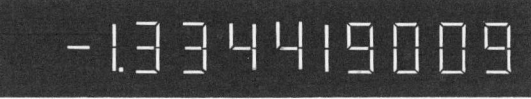

This is not the standard form for the logarithm, but is similar to that which occurred in the solutions to Example 1(c) and Example 3(b). To find the standard form we could add 2 to the logarithm (using a calculator) obtaining 0.665580991 and then append -2. Thus

$$\log 0.0463 \approx 0.665580991 - 2.$$

EXERCISES 4.5

In Exercises 1–16 use Table 2 and laws of logarithms to approximate the common logarithms of the given numbers. If a calculator with a log key is available, compare your answers with those obtained by means of the calculator.

1 347; 0.00347; 3.47

2 86.2; 8,620; 0.862

3 0.54; 540; 540,000

4 208; 2.08; 20,800

5 60.2; 0.0000602; 602

6 5; 0.5; 0.0005

7 $(44.9)^2$; $(44.9)^{1/2}$; $(44.9)^{-2}$

8 $(1810)^4$; $(1810)^{40}$; $(1810)^{1/4}$

9 $(0.943)^3$; $(0.943)^{-3}$; $(0.943)^{1/3}$

10 $(0.017)^{10}$; $10^{0.017}$; $10^{1.43}$

11 $(638)(17.3)$

12 $\dfrac{(2.73)(78.5)}{621}$

13 $\dfrac{(47.4)^3}{(29.5)^2}$

14 $\dfrac{(897)^4}{\sqrt{17.8}}$

15 $\sqrt[3]{20.6}(371)^3$

16 $\dfrac{(0.0048)^{10}}{\sqrt{0.29}}$

In Exercises 17–30 use Table 2 to find a decimal approximation to x. If a suitable calculator is available, compare your approximations with those obtained by means of the calculator.

17 $\log x = 3.6274$

18 $\log x = 1.8965$

19 $\log x = 0.9469$

20 $\log x = 0.5729$

21 $\log x = 5.2095$

22 $\log x = 6.7300 - 10$

23 $\log x = 9.7348 - 10$

24 $\log x = 7.6739 - 10$

25 $\log x = 8.8306 - 10$

26 $\log x = 4.9680$

27 $\log x = 2.2765$

28 $\log x = 3.0043$

29 $\log x = -1.6253$

30 $\log x = -2.2118$

31 Chemists use a number denoted by pH to describe quantitatively the acidity or basicity of solutions. By definition,

$$pH = -\log[H^+]$$

where $[H^+]$ is the hydrogen ion concentration in moles per liter. Given the following, with the indicated $[H^+]$, approximate the pH of each substance.

(a) vinegar: $[H^+] \approx 6.3 \times 10^{-3}$
(b) carrots: $[H^+] \approx 1.0 \times 10^{-5}$
(c) sea water: $[H^+] \approx 5.0 \times 10^{-9}$

32 Approximate the hydrogen ion concentration $[H^+]$ in each of the following substances. (Refer to Exercise 31 and use the closest entry in Table 2 to approximate $[H^+]$.)
(a) apples: pH ≈ 3.0 (b) beer: pH ≈ 4.2
(c) milk: pH ≈ 6.6

33 A solution is considered acidic if $[H^+] > 10^{-7}$ or basic if $[H^+] < 10^{-7}$. What are the corresponding inequalities involving pH?

34 Many solutions have a pH between 1 and 14. What is the corresponding range of the hydrogen ion content $[H^+]$?

35 The sound intensity level formula considered in Exercise 17 of Section 4.4 may be written

$$\alpha = 10\log(I/I_0).$$

Find α if I is 285,000 times as great as I_0.

36 A sound intensity level of 140 decibels produces pain in the average human ear. Approximately how many times greater than I_0 must I be in order for α to reach this level? (Refer to Exercise 35 and use the nearest entry in Table 2 to find the approximation.)

4.6 EXPONENTIAL AND LOGARITHMIC EQUATIONS

The variables in certain equations appear as exponents or logarithms, as illustrated in the following examples.

Example 1   Solve the equation $3^x = 21$.

Solution   Taking the common logarithm of both sides and using (iii) of the Laws of Logarithms, we obtain

$$\log(3^x) = \log 21$$
$$x \log 3 = \log 21$$
$$x = \frac{\log 21}{\log 3}.$$

If an approximation is desired, we may use Table 2 to obtain

$$x \approx \frac{1.3222}{0.4771} \approx 2.77$$

where the last number is obtained by dividing 1.3222 by 0.4771. A partial check on the solution is to note that since $3^2 = 9$ and $3^3 = 27$, the number x such that $3^x = 21$ should lie between 2 and 3, somewhat closer to 3 than to 2.   ∎

Example 2   Solve $5^{2x+1} = 6^{x-2}$.

Solution   Taking the common logarithm of both sides and using (iii) of the Laws of Logarithms, we obtain

$$(2x + 1)\log 5 = (x - 2)\log 6.$$

We may now solve for x as follows:

$$2x \log 5 + \log 5 = x \log 6 - 2 \log 6$$
$$2x \log 5 - x \log 6 = -\log 5 - 2 \log 6$$
$$x(2 \log 5 - \log 6) = -(\log 5 + \log 6^2)$$
$$x = \frac{-(\log 5 + \log 36)}{2 \log 5 - \log 6}$$
$$= \frac{-\log(5 \cdot 36)}{\log 5^2 - \log 6}$$
$$x = \frac{-\log 180}{\log(25/6)}.$$

If an approximation to the solution is desired, we may proceed as in Example 1.   ∎

Example 3 Solve the equation $\log(5x - 1) - \log(x - 3) = 2$.

Solution The given equation may be written as

$$\log\frac{5x - 1}{x - 3} = 2.$$

Using the definition of logarithm with $a = 10$ gives us

$$\frac{5x - 1}{x - 3} = 10^2.$$

Consequently,

$$5x - 1 = 10^2(x - 3) = 100x - 300, \quad \text{or} \quad 299 = 95x.$$

Hence

$$x = \frac{299}{95}. \qquad \blacksquare$$

Example 4 Solve the equation $\dfrac{5^x - 5^{-x}}{2} = 3$.

Solution Multiplying both sides of the given equation first by 2 and then by 5^x gives us

$$5^x - 5^{-x} = 6$$
$$5^{2x} - 1 = 6(5^x)$$

which may be written $(5^x)^2 - 6(5^x) - 1 = 0.$

Letting $u = 5^x$ gives us the quadratic equation

$$u^2 - 6u - 1 = 0$$

in the variable u. Applying the Quadratic Formula,

$$u = \frac{6 \pm \sqrt{36 + 4}}{2} = 3 \pm \sqrt{10},$$

that is, $5^x = 3 \pm \sqrt{10}$. Since 5^x is never negative, the number $3 - \sqrt{10}$ must be discarded; therefore,

$$5^x = 3 + \sqrt{10}.$$

Taking the common logarithm of both sides and using (iii) of the Laws of Logarithms,

$$x \log 5 = \log(3 + \sqrt{10}) \quad \text{or} \quad x = \frac{\log(3 + \sqrt{10})}{\log 5}.$$

To obtain an approximate solution we may write $3 + \sqrt{10} \approx 6.16$ and use Table 2. This gives us

$$x \approx \frac{\log 6.16}{\log 5} \approx \frac{0.7896}{0.6990} \approx 1.13. \qquad \blacksquare$$

EXERCISES 4.6

Find the solutions of the equations in Exercises 1–20.

1 $10^x = 7$

2 $5^x = 8$

3 $4^x = 3$

4 $10^x = 6$

5 $3^{4-x} = 5$

6 $(1/3)^x = 100$

7 $3^{x+4} = 2^{1-3x}$

8 $4^{2x+3} = 5^{x-2}$

9 $2^{-x} = 8$

10 $2^{-x^2} = 5$

11 $\log x = 1 - \log(x - 3)$

12 $\log(5x + 1) = 2 + \log(2x - 3)$

13 $\log(x^2 + 4) - \log(x + 2) = 3 + \log(x - 2)$

14 $\log(x - 4) - \log(3x - 10) = \log(1/x)$

15 $\log(x^2) = (\log x)^2$

16 $\log\sqrt{x} = \sqrt{\log x}$

17 $\log(\log x) = 2$

18 $\log\sqrt{x^3 - 9} = 2$

19 $x^{\sqrt{\log x}} = 10^8$

20 $\log(x^3) = (\log x)^3$

In Exercises 21–24 solve for x in terms of y.

21 $y = \dfrac{10^x + 10^{-x}}{2}$

22 $y = \dfrac{10^x - 10^{-x}}{2}$

23 $y = \dfrac{10^x - 10^{-x}}{10^x + 10^{-x}}$

24 $y = \dfrac{1}{10^x - 10^{-x}}$

In Exercises 25–28 use natural logarithms to solve for x in terms of y.

25 $y = \dfrac{e^x - e^{-x}}{2}$

26 $y = \dfrac{e^x + e^{-x}}{2}$

27 $y = \dfrac{e^x - e^{-x}}{e^x + e^{-x}}$

28 $y = \dfrac{e^x + e^{-x}}{e^x - e^{-x}}$

29 The current i in a certain electrical circuit is given by

$$i = \frac{E}{R}(1 - e^{-Rt/L}).$$

Use natural logarithms to solve for t in terms of the remaining symbols.

30 If a sum of money P_0 is invested at an interest rate of $100r$ per cent per year, compounded m times per year, then the principal P at the end of t years is given by

$$P = P_0\left(1 + \frac{r}{m}\right)^{mt}.$$

Solve for t in terms of the other symbols.

31 In Exercise 17 of Section 4.4 we considered the sound intensity level formula

$$\alpha = 10\log(I/I_0).$$

(a) Solve for I in terms of α and I_0.
(b) Show that a one-decibel rise in the intensity level corresponds to a 26 % increase in the intensity.

32 The chemical formula pH $= -\log[\mathrm{H}^+]$ was introduced in Exercise 31 of Section 4.5. Solve for $[\mathrm{H}^+]$ in terms of pH.

CALCULATOR EXERCISES 4.6

Use natural logarithms to approximate the solutions of the equations in Exercises 1–3.

1 $5^x = 7$ (*Hint:* $\ln 5^x = \ln 7$, or $x \ln 5 = \ln 7$)

2 $4^{-x} = 10$

3 $e^{-x^2} = 0.163$

4 If $10,000 is invested at an interest rate of 8% per year compounded quarterly, when will the principal exceed $20,000? (*Hint:* Use Exercise 30.)

5 How long will it take money to double if it is invested at a rate of 6% per year compounded monthly?

4.7 LINEAR INTERPOLATION

The only logarithms that can be found *directly* from Table 2 are logarithms of numbers that contain at most three nonzero digits. If *four* nonzero digits are involved, then it is possible to obtain an approximation by using the method of linear interpolation described in this section. The terminology *linear interpolation* is used because, as we shall see, the method is based upon approximating portions of the graph of $y = \log x$ by line segments. It should be pointed out that there are quicker, more mechanical, ways to find logarithms accurately. Indeed, if a hand-held calculator is available, it is usually only necessary to press a few keys to obtain accuracy to many decimal places. Nonetheless, linear interpolation is a very useful skill, since it can be applied to any tabular data in which entries vary in a manner similar to that of a linear function, at least over small portions of the table. Consequently, as you proceed through this section keep in mind that you are learning a process that can be used with many tables other than those for logarithms.

In order to illustrate the process of linear interpolation, and at the same time give some justification for it, let us consider the specific example $\log 12.64$. Since the logarithmic function with base 10 is increasing, this number lies between $\log 12.60 \approx 1.1004$ and $\log 12.70 \approx 1.1038$. Examining the graph of $y = \log x$, we have the situation shown in Figure 4.12, where we have distorted the units on the x- and y-axes and also the portion of the graph shown. A more accurate drawing would indicate that the graph of $y = \log x$ is much closer to the line segment joining $P(12.60, 1.1004)$ to $Q(12.70, 1.1038)$ than is shown in the figure. Since $\log 12.64$ is the ordinate of the point on the graph having abscissa 12.64, it can be approximated by the ordinate of the point with abscissa 12.64 on the *line segment PQ*. Referring to Figure 4.12, we see that the latter ordinate is $1.1004 + d$. The number d can be approximated by using similar triangles. Referring to Figure 4.13, where the graph of $y = \log x$ has been deleted, we may form the following proportion:

$$\frac{d}{0.0034} = \frac{0.04}{0.1}.$$

Hence

$$d = \frac{(0.04)(0.0034)}{0.1} = 0.00136.$$

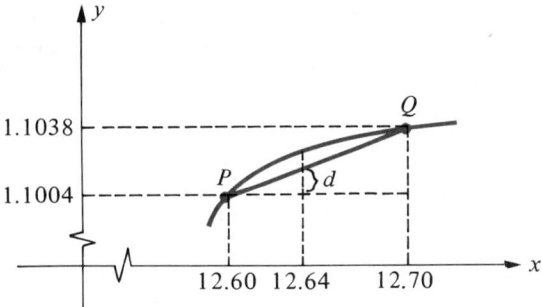

Figure 4.12

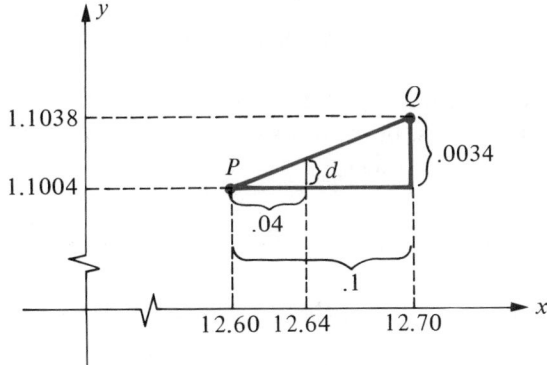

Figure 4.13

When using this technique, we always round off decimals to the same number of places as appear in the body of the table. Consequently, $d \approx 0.0014$ and

$$\log 12.64 \approx 1.1004 + 0.0014 = 1.1018.$$

Hereafter we shall not sketch a graph when interpolating. Instead we shall use the scheme illustrated in the next example.

Example 1 Approximate $\log 572.6$.

Solution It is convenient to arrange our work as follows:

$$1.0 \left\{ 0.6 \left\{ \begin{array}{l} \log 572.0 \approx 2.7574 \\ \log 572.6 = ? \end{array} \right\} d \\ \log 573.0 \approx 2.7582 \end{array} \right\} 0.0008$$

where we have indicated differences by appropriate symbols alongside of the braces. This leads to the proportion

$$\frac{d}{0.0008} = \frac{0.6}{1.0} = \frac{6}{10} \quad \text{or} \quad d = \left(\frac{6}{10} \right)(0.0008) = 0.00048 \approx 0.0005.$$

Hence $\log 572.6 \approx 2.7574 + 0.0005 = 2.7579.$

Another way of working this type of problem is to reason that since 572.6 is $\frac{6}{10}$ of the way from 572.0 to 573.0, then log 572.6 is (approximately) $\frac{6}{10}$ of the way from 2.7574 to 2.7582. Hence

$$\log 572.6 \approx 2.7574 + (\tfrac{6}{10})(0.0008) \approx 2.7574 + 0.0005 = 2.7579. \qquad \blacksquare$$

Example 2   Approximate $\log 0.003678$.

Solution   We begin by arranging our work as in the solution of Example 1. Thus,

$$10 \left\{ \begin{array}{l} 8 \left\{ \begin{array}{l} \log 0.003670 \approx 0.5647 + (-3) \\ \log 0.003678 = \ ? \end{array} \right\} d \\[2mm] \quad \log 0.003680 \approx 0.5658 + (-3) \end{array} \right\} 0.0011$$

Since we are only interested in ratios, we have used the numbers 8 and 10 on the left side because their ratio is the same as the ratio of 0.000008 to 0.000010. This leads to the proportion

$$\frac{d}{0.0011} = \frac{8}{10} = 0.8 \quad \text{or} \quad d = (0.0011)(0.8) = 0.00088 \approx 0.0009.$$

Hence, $$\log 0.003678 \approx [0.5647 + (-3)] + 0.0009$$
$$= 0.5656 + (-3). \qquad \blacksquare$$

If a number x is written in the scientific form $x = c \cdot 10^k$, where $1 \le c < 10$, then before using Table 2 to find $\log x$ by interpolation, c should be rounded off to three decimal places. Another way of saying this is that x should be rounded off to four **significant figures**. Some examples will help to clarify the procedure. If $x = 36.4635$, we round off to 36.46 before approximating $\log x$. The number 684,279 should be rounded off to 684,300. For a decimal such as 0.096202 we use 0.09620. The reason for doing this is that Table 2 does not guarantee more than four-digit accuracy, since the mantissas which appear in it are approximations. This means that if *more* than four-digit accuracy is required in a problem, then Table 2 cannot be used. If, in more extensive tables, the logarithm of a number containing n digits can be found directly, then interpolation is allowed for numbers involving $n + 1$ digits and numbers should be rounded off accordingly.

The method of interpolation can also be used to find x when we are given $\log x$. If we use Table 2, then x may be found to four significant figures. In this case we are given the *ordinate* of a point on the graph of $y = \log x$ and are asked to find the *abscissa*. A geometric argument similar to the one given earlier can be used to justify the procedure illustrated in the next example.

Example 3   Find x to four significant figures if $\log x = 1.7949$.

Solution   The mantissa 0.7949 does not appear in Table 2, but it can be isolated between adjacent entries, namely the mantissas corresponding to 6.230 and 6.240.

We shall arrange our work as follows:

$$0.1 \left\{ \begin{array}{l} r \left\{ \begin{array}{l} \log 62.30 \approx 1.7945 \\ \log x \quad = 1.7949 \end{array} \right\} 0.0004 \\[2ex] \log 62.40 \approx 1.7952 \end{array} \right\} 0.0007.$$

This leads to the proportion

$$\frac{r}{0.1} = \frac{0.0004}{0.0007} = \frac{4}{7} \quad \text{or} \quad r = (0.1)\left(\frac{4}{7}\right) \approx 0.06.$$

Hence, $\qquad\qquad\qquad x \approx 62.30 + 0.06 = 62.36.$ ■

If we are given $\log x$, then the number x is called the **antilogarithm** of $\log x$. In Example 3 the antilogarithm of $\log x = 1.7949$ is $x \approx 62.36$. Sometimes the notation antilog $(1.7949) \approx 62.36$ is used.

EXERCISES 4.7

Use linear interpolation to approximate the common logarithms of the numbers in Exercises 1–20. If a calculator with a log key is available, compare your approximation with those obtained by means of the calculator.

1	25.48	2	421.6
3	5363	4	0.3817
5	0.001259	6	69,450
7	123,400	8	0.0212
9	0.7786	10	1.203
11	384.7	12	54.44
13	0.9462	14	7259
15	66,590	16	0.001428
17	0.04321	18	300,100
19	3.003	20	1.236

In Exercises 21–40 use linear interpolation to approximate x. If a suitable calculator is available, compare your approximations with those obtained by means of the calculator.

21	$\log x = 1.4437$	22	$\log x = 3.7455$
23	$\log x = 4.6931$	24	$\log x = 0.5883$
25	$\log x = 9.1664 - 10$	26	$\log x = 8.3902 - 10$
27	$\log x = 3.8153 - 6$	28	$\log x = 5.9306 - 9$
29	$\log x = 2.3705$	30	$\log x = 4.2867$
31	$\log x = 0.1358$	32	$\log x = 0.0194$
33	$\log x = 8.9752 - 10$	34	$\log x = 2.4979 - 5$
35	$\log x = 5.0409$	36	$\log x = 1.3796$
37	$\log x = -2.8712$	38	$\log x = -1.8164$
39	$\log x = -0.6123$	40	$\log x = -3.1426$

4.8 COMPUTATIONS WITH LOGARITHMS

The importance of logarithms for numerical computations has diminished in recent years because of the development of computers and hand-held calculators. However, since mechanical devices are not always available, it is worthwhile to have some familiarity with the use of logarithms for solving arithmetic problems. At the same time, practice in working numerical problems leads to a deeper understanding of logarithms and logarithmic functions. The following examples illustrate some computational techniques.

Example 1 Approximate $N = \dfrac{(59700)(0.0163)}{41.7}$.

Solution   Using laws of logarithms and Table 2 leads to

$$\begin{aligned}
\log N &= \log 59700 + \log 0.0163 - \log 41.7 \\
&\approx 4.7760 + (0.2122 - 2) - (1.6201) \\
&= 4.9882 - 3.6201 \\
&= 1.3681.
\end{aligned}$$

Referring to Table 2 for the antilogarithm we have, to three significant figures,

$$N \approx 23.3. \qquad\qquad \blacksquare$$

Example 2 Approximate $N = \sqrt[3]{56.11}$ to four significant figures.

Solution   Writing $N = (56.11)^{1/3}$ and using (iii) of the Laws of Logarithms gives us

$$\log N = \tfrac{1}{3}\log 56.11.$$

To find $\log 56.11$ we interpolate from Table 2 as follows:

$$10\left\{1\begin{cases} \log 56.10 \approx 1.7490 \\ \log 56.11 = ? \\ \log 56.20 \approx 1.7497 \end{cases}\right\}d\right\}0.0007$$

$$\frac{d}{0.0007} = \frac{1}{10}$$

$$d = 0.00007 \approx 0.0001.$$

Hence, $\log 56.11 \approx 1.7490 + 0.0001 = 1.7491$

and therefore, $\log N \approx \tfrac{1}{3}(1.7491) \approx 0.5830.$

The antilogarithm may be found by interpolation from Table 2 as follows

$$0.01 \left\{ r \left\{ \begin{matrix} \log 3.820 \approx 0.5821 \\ \log N \quad \approx 0.5830 \end{matrix} \right\} 9 \\ \log 3.830 \approx 0.5832 \end{matrix} \right\} 11$$

$$\frac{r}{0.01} = \frac{9}{11}$$

$$r = \frac{9}{11}(0.01) \approx 0.008.$$

Consequently, $\qquad N \approx 3.820 + 0.008 = 3.828.$ ∎

If we were interested in only *three* significant figures, then the interpolations in Example 2 could have been avoided. In the remaining examples we shall, for simplicity, work only with three-digit numbers.

Example 3 Approximate $N = \dfrac{(1.32)^{10}}{\sqrt[5]{0.0268}}$.

Solution Using laws of logarithms and Table 2,

$$\begin{aligned} \log N &= 10 \log 1.32 - \tfrac{1}{5} \log 0.0268 \\ &\approx 10(0.1206) - \tfrac{1}{5}(3.4281 - 5) \\ &= 1.206 - 0.6856 + 1 \\ &= 1.5204. \end{aligned}$$

Finding the antilogarithm (to three significant figures), we obtain

$$N \approx 33.1.$$ ∎

Example 4 Find x if $x^{2.1} = 6.5$.

Solution Taking the common logarithm of both sides and using (iii) of the Laws of Logarithms gives us

$$(2.1)\log x = \log 6.5.$$

Hence, $\qquad \qquad \log x = \dfrac{\log 6.5}{2.1}$

and by the definition of logarithm,

$$x = 10^{(\log 6.5)/(2.1)}.$$

If an approximation to x is desired, then from the second equation above we have

$$\log x = \frac{\log 6.5}{2.1} \approx \frac{0.8129}{2.1} \approx 0.3871.$$

Using Table 2, the antilogarithm (to three significant figures) is

$$x \approx 2.44. \qquad \blacksquare$$

Example 5   Approximate $N = \dfrac{69.3 + \sqrt[3]{56.1}}{\log 807}$.

Solution   Since we have no formula for the logarithm of a sum, the two terms in the numerator must be *added* before the logarithm can be found. From Example 2, $\sqrt[3]{56.1} \approx 3.8$ and hence the numerator is approximately 73.1. Using Table 2 we see that

$$\log 807 \approx 2.9069 \approx 2.91.$$

Hence the given expression may be approximated by

$$N \approx \frac{73.1}{2.91}.$$

It is now an easy matter to find N, either by using logarithms or by long division. It is left to the reader to verify that $N \approx 25.1$. $\blacksquare$

EXERCISES 4.8

Use logarithms to approximate the numbers in Exercises 1–20 to three significant figures. If a calculator is available, compare your answers with those obtained by means of the calculator.

1   $(638)(57.2)$

2   $(0.0178)(0.00729)$

3   $\dfrac{35,900}{8,430}$

4   $\dfrac{3.14}{63.2}$

5   $(4.21)^{10}$

6   $(0.712)^6$

7   $\sqrt[5]{0.517}$

8   $\sqrt[4]{2.23}$

9   $\dfrac{(26.7)^3(1.48)}{(67.4)^2}$

10   $\dfrac{(3.04)^3}{\sqrt{1.12}(86.6)}$

11   $\sqrt{1.65\,\sqrt[3]{(70.8)^2}}$

12   $\left[\dfrac{(11.1)^3}{\sqrt[5]{4.17}}\right]^{-1/2}$

13   $\sqrt{\dfrac{0.563}{(0.105)^3}}$

14   $\sqrt[5]{\dfrac{(124)^2}{(9.83)^3}}$

15   $10^{-3.14}$

16   $(100)^{0.523}$

17   $(4.12)^{0.220}$

18   $\sqrt{8.46\,\sqrt{3.07}}$

19   $\dfrac{\log 37.4}{\log 6.19}$

20   $\dfrac{56.8 + \log(7.13)}{\sqrt[10]{4.42}}$

Use interpolation in Table 2 to approximate the numbers in Exercises 21–26 to four significant figures.

21   $(2.461)^5$

22   $1/(33.89)^4$

23   $\sqrt[10]{0.5138}$

24   $\sqrt[5]{(17.04)^2}$

25   $(5.375)^{2/3}$

26   $(1776)^{11}$

27   The area A of a triangle with sides a, b, and c may be calculated from the formula $A = \sqrt{s(s-a)(s-b)(s-c)}$, where s is one-half the perimeter. Use logarithms to approximate the area of a triangle with sides 12.6, 18.2, and 14.1.

28   The volume V of a right circular cone of altitude h and radius of base r is $V = \frac{1}{3}\pi r^2 h$. Use logarithms to approximate the volume of a cone of radius 2.43 cm and altitude 7.28 cm.

29 The formula used in physics to approximate the period T (seconds) of a simple pendulum of length L (feet) is $T = 2\pi\sqrt{L/(32.2)}$. Approximate the period of a pendulum 33 inches long.

30 The pressure p (pounds per cubic foot) and volume v (cubic feet) of a certain gas are related by the formula $pv^{1.4} = 600$. Approximate the pressure if $v = 8.22$ cubic feet.

4.9 REVIEW

Concepts

Define or discuss the following.

1 The exponential function with base a

2 The natural exponential function

3 The logarithm of u with base a

4 The natural logarithmic function

5 The laws of logarithms

6 The logarithmic function with base a

7 Common logarithms

8 Mantissa

9 Characteristic

10 Linear interpolation

Exercises

Find the numbers in Exercises 1–6.

1 $\log_2 (1/16)$ **2** $\log_5 \sqrt[3]{5}$

3 $6^{\log_6 4}$ **4** $10^{3\log 2}$

5 $\log 1,000,000$ **6** $\ln e$

In Exercises 7–16 sketch the graph of f.

7 $f(x) = 3^{x+2}$ **8** $f(x) = (3/5)^x$

9 $f(x) = (3/2)^{-x}$ **10** $f(x) = 3^{-2x}$

11 $f(x) = 3^{-x^2}$ **12** $f(x) = 1 - 3^{-x}$

13 $f(x) = \log_6 x$ **14** $f(x) = \log_3 (x^2)$

15 $f(x) = 2\log_3 x$ **16** $f(x) = \log_2 (x + 4)$

Find the solutions of the equations in Exercises 17–24.

17 $\log_8 (x - 5) = 2/3$

18 $\log_4 (x + 1) = 2 + \log_4 (3x - 2)$

19 $2\log_3 (x + 3) - \log_3 (x + 1) = 3\log_3 2$

20 $\log \sqrt[4]{x + 1} = 1/2$

21 $2^{5-x} = 6$ **22** $3^{x^2} = 7$

23 $2^{5x+3} = 3^{2x+1}$ **24** $5^{\log_5 (x+1)} = 3$

25 Express $\log x^4 \sqrt[3]{y^2/z}$ in terms of logarithms of x, y, and z.

26 Express $\log (x^2/y^3) + 4\log y - 6\log \sqrt{xy}$ as one logarithm.

Solve the equations in Exercises 27 and 28 for x in terms of y.

27 $y = \dfrac{10^x + 10^{-x}}{10^x - 10^{-x}}$ **28** $y = \dfrac{1}{10^x + 10^{-x}}$

Use linear interpolation to approximate the logarithms in Exercises 29–32.

29 $\log 47.82$

30 $\log 0.001347$

31 $\log 300,600$

32 $\log 0.2143$

Use linear interpolation to approximate x in Exercises 33–36.

33 $\log x = 2.4995$

34 $\log x = 1.5045$

35 $\log x = 8.7970 - 10$ **36** $\log x = -1.3146$

Use logarithms to approximate the numbers in Exercises 37–40.

37 $\dfrac{(38.2)^3}{\sqrt{4.67}}$

38 $\sqrt[5]{\dfrac{21.8}{62.2}}$

39 $(5.16)^{2.1}$

40 $(1.01)^{44}$

5

THE TRIGONOMETRIC FUNCTIONS

Trigonometry originated over 2,000 years ago, when the Greeks developed precise methods for measuring angles and sides of triangles. Although such measurements are still important in certain branches of science and engineering, there are many modern applications of trigonometry in which the notion of angle either is secondary or does not enter into the picture. Of primary importance are properties of the so-called trigonometric functions, *whose domains are sets of real numbers. For this reason, these functions are introduced in Section 5.2 without reference to angles. However, since the classical techniques are also important, our modern definition is followed almost immediately, in Section 5.5, by a more traditional approach involving angles and ratios. As a consequence, we have available, very early in our work, two different, but equivalent, formulations of the trigonometric functions. Students should strive to become comfortable using both approaches, since each has certain advantages. Graphs involving the trigonometric functions are considered in Sections 5.6–5.8. The chapter concludes with some modern and ancient applications of trigonometry to problems involving right triangles.*

5.1 THE UNIT CIRCLE *U*

Let U be a unit circle, that is, a circle of radius 1, with center at the origin of a rectangular coordinate system. Thus U is the graph of the equation $x^2 + y^2 = 1$. If A is the point with coordinates $(1, 0)$ and P is any other point on U, then proceeding along U in a *counterclockwise direction* from A (see Figure 5.1), there is a unique positive real number t called the **length of the arc** $\widehat{AP}$. Since the circumference of U is 2π, we have $0 < t < 2\pi$. If we let $t = 0$ when $P = A$, then with each point P on U there is associated precisely one real number in the half-open interval $[0, 2\pi)$.

Conversely, it can be shown that for each real number t in the half-open interval $[0, 2\pi)$ there is one and only one point P on U such that the length of $\widehat{AP}$, obtained by proceeding in the counterclockwise direction from A to P, is t. This establishes a one-to-

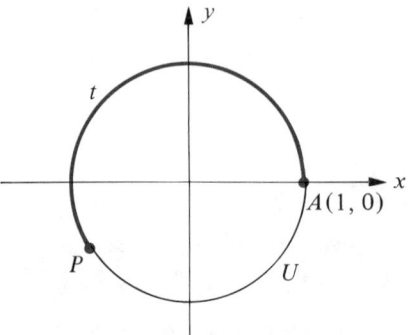

Figure 5.1

one correspondence between the real numbers in the interval $[0, 2\pi)$ and the points on U. Since the position of P depends on t, we shall use functional notation and denote it by $P(t)$. In certain applications a (variable) point is regarded as moving along U, and we refer to the point as *traveling*, or *traversing*, a *distance* t from A to $P(t)$.

For most values of t it is very difficult to determine the rectangular coordinates of the point $P(t)$ on U. Three simple cases are illustrated in Figure 5.2. In Case (i) we note that since the circumference of the unit circle is 2π, the number $\pi/2$ is one-fourth the circumference, and hence the point $P(\pi/2)$ has coordinates $(0, 1)$. Similarly, in Case (ii), π is one-half the circumference, and hence $P(\pi)$ may be found by starting at A and traveling half-way around U to the point $(-1, 0)$. Finally, in Case (iii), $P(3\pi/2)$ can be located by traversing three-fourths of the circumference (in the counterclockwise direction) to the point $(0, -1)$.

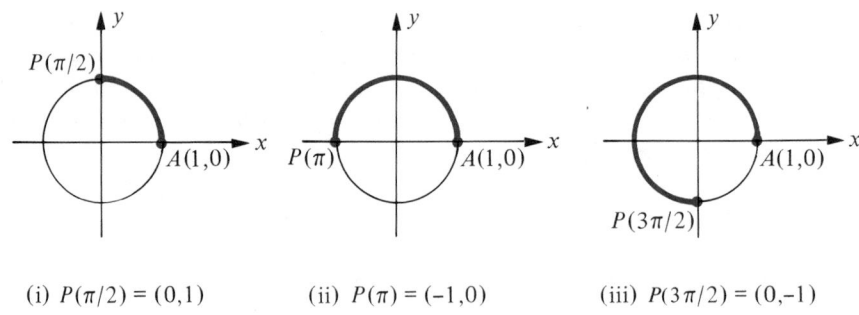

(i) $P(\pi/2) = (0,1)$ (ii) $P(\pi) = (-1,0)$ (iii) $P(3\pi/2) = (0,-1)$

Figure 5.2

A more difficult problem involving the location of $P(t)$ is illustrated in the next example.

Example 1   Find the rectangular coordinates of $P(\pi/4)$, $P(3\pi/4)$, $P(5\pi/4)$, and $P(7\pi/4)$.

Solution   The rectangular coordinates of the point $P(\pi/4)$ can be found by noting that $\pi/4$ is one-half of $\pi/2$, and consequently, $P(\pi/4)$ is in the first quadrant and bisects the circular arc from $(1, 0)$ to $(0, 1)$. It follows that $P(\pi/4)$ lies on the line $y = x$

and hence has coordinates (c, c) for some positive real number c (see Figure 5.3). Since (c, c) is a point on the unit circle $x^2 + y^2 = 1$, we must have $c^2 + c^2 = 1$, or $2c^2 = 1$. Thus $c = \sqrt{1/2} = \sqrt{2}/2$ and

$$P(\pi/4) = (\sqrt{2}/2, \sqrt{2}/2).$$

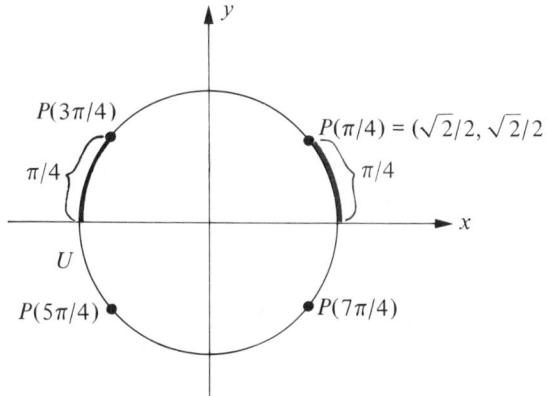

Figure 5.3

Since $3\pi/4 = \pi - (\pi/4)$, we see that $P(3\pi/4)$ and $P(\pi/4)$ are symmetric with respect to the y-axis, as illustrated in Figure 5.3. It follows that

$$P(3\pi/4) = (-\sqrt{2}/2, \sqrt{2}/2).$$

Again using symmetry, it is evident from the figure that

$$P\left(\frac{5\pi}{4}\right) = \left(-\frac{\sqrt{2}}{2}, -\frac{\sqrt{2}}{2}\right) \quad \text{and} \quad P\left(\frac{7\pi}{4}\right) = \left(\frac{\sqrt{2}}{2}, -\frac{\sqrt{2}}{2}\right). \qquad \blacksquare$$

Geometric arguments can be used to find the rectangular coordinates of several other special points on U, such as $P(\pi/6)$ and $P(\pi/3)$; however, it is impossible for us to cope with values of t such as $2/3, \sqrt{5}, 4.627$, or for that matter, integer values such as $1, 2, 3$, and 4. The reason for the difficulty is that techniques studied in calculus are required to handle all possibilities. Fortunately, in our future work it will be unnecessary to calculate the exact coordinates (x, y) of $P(t)$ for every value of t.

The preceding discussion can be extended so that to *every* real number there corresponds exactly one point on U. Specifically, to associate a point $P(t)$ with any *positive* real number t, we proceed along U in a counterclockwise direction from A, *making several revolutions*, if necessary. For example, if $t = 5\pi/2$, we first note that

$$\frac{5\pi}{2} = \frac{\pi}{2} + 2\pi.$$

Thus the point $P(5\pi/2)$ may be found by first traversing the distance $\pi/2$ from A to the point $(0, 1)$, and then making one complete counterclockwise revolution through the distance 2π. Consequently, $P(5\pi/2) = (0, 1)$. Similarly, to locate $P(9\pi/2)$, we note that

$$\frac{9\pi}{2} = \frac{\pi}{2} + 4\pi$$

and hence *two* complete revolutions should be made after reaching the point $(0, 1)$. This gives us $P(9\pi/2) = (0, 1)$. In general, if t_1 and t_2 are positive real numbers such that $t_2 = t_1 + 2\pi n$ for some positive integer n, then $P(t_2) = P(t_1)$, since the circumference of U is 2π.

To associate a point on U with each *negative* real number t, we start at A and proceed a distance $|t|$ in a *clockwise* direction along U. The three special cases $t = -\pi/2$, $t = -\pi$, and $t = -7\pi/4$ are illustrated in Figure 5.4 The reader should compare the third case with Example 1. Of course, for real numbers less than -2π, it is necessary to make more than one revolution (in the clockwise direction) along U.

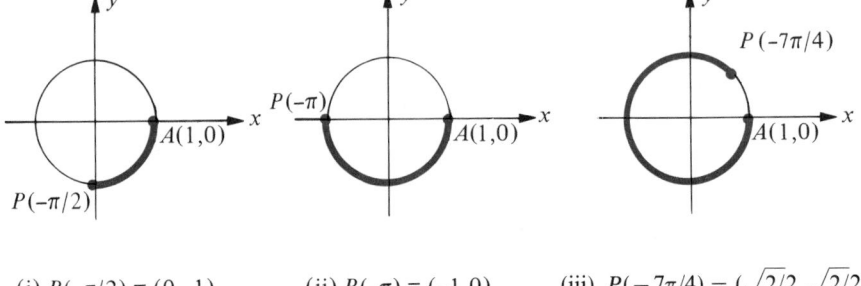

(i) $P(-\pi/2) = (0, -1)$ (ii) $P(-\pi) = (-1, 0)$ (iii) $P(-7\pi/4) = (\sqrt{2}/2, \sqrt{2}/2)$

Figure 5.4

We have demonstrated that to each real number t there is associated a unique point $P(t)$ on the unit circle U. We shall refer to $P(t)$ as **the point on U that corresponds to t**. Since the circumference of U is 2π, it follows that two real numbers t_1 and t_2 are associated with the same point if and only if $t_2 = t_1 + 2\pi n$ for some integer n. This gives us the important formula

$$P(t) = P(t + 2\pi n)$$

for every real number t and every integer n. Several uses of this formula are given in the next two examples.

Example 2   Find the rectangular coordinates of $P(27\pi/4)$.

Solution   Since $27/4 = 6 + (3/4)$ we see that

$$\frac{27\pi}{4} = 6\pi + \frac{3\pi}{4}.$$

Using the formula preceding this example (with $n = 3$),

$$P\left(\frac{27\pi}{4}\right) = P\left(\frac{3\pi}{4} + 6\pi\right) = P\left(\frac{3\pi}{4}\right).$$

Finally, referring to Example 1, we obtain $P(27\pi/4) = (-\sqrt{2}/2, \sqrt{2}/2)$.   ■

Example 3   Find the quadrant which contains $P(21)$.

Solution   From the discussion in this section, $P(21) = P(21 + 2\pi n)$ for every integer n. By choosing n negative we can find a number of the form $21 + 2\pi n$ which is in the interval $[0,2\pi)$. The quadrant may then be readily determined. In particular, if we let $n = -3$, then

$$21 - 6\pi \approx 21 - 18.85 = 2.15.$$

Since $\pi/2 < 2.15 < \pi$, we see that

$$\frac{\pi}{2} < 21 - 6\pi < \pi.$$

It follows that $P(21)$ is in the second quadrant.   ■

EXERCISES 5.1

In Exercises 1–10 find the rectangular coordinates of the given points.

1 $P(5\pi)$ **2** $P(-3\pi)$

3 $P(7\pi/2)$ **4** $P(5\pi/2)$

5 $P(-6\pi)$ **6** $P(11\pi/2)$

7 $P(37\pi)$ **8** $P(99\pi)$

9 $P(-9\pi/2)$ **10** $P(1000000\pi)$

Find the rectangular coordinates of the points in Exercises 11–16. (*Hint:* See Figure 5.3.)

11 $P(-3\pi/4)$ **12** $P(-\pi/4)$

13 $P(-9\pi/4)$ **14** $P(-5\pi/4)$

15 $P(11\pi/4)$ **16** $P(9\pi/4)$

If $P(t)$ has the rectangular coordinates given in each of Exercises 17–24, find (a) $P(t + \pi)$, (b) $P(t - \pi)$, (c) $P(-t)$, (d) $P(-t - \pi)$.

17 $(3/5, 4/5)$ **18** $(4/5, -3/5)$

19 $(-8/17, 15/17)$ **20** $(-15/17, -8/17)$

21 $(-1, 0)$ **22** $(0, -1)$

23 $(a/\sqrt{a^2 + b^2}, b/\sqrt{a^2 + b^2})$ where $a^2 + b^2 \neq 0$

24 $(a, \sqrt{1 - a^2})$

In each of Exercises 25–28, approximate the y-coordinate of $P(t)$ if $P(t)$ is in the indicated quadrant and has the given x-coordinate.

25 I, $x = 1/2$ **26** II, $x = -\sqrt{3}/2$

27 III, $x = -2/3$ **28** IV, $x = 0.1$

In each of Exercises 29–32, approximate the x-coordinate of $P(t)$ if $P(t)$ is in the indicated quadrant and has the given y-coordinate.

29 II, $y = \sqrt{3}/2$ **30** III, $y = -1/2$

31 IV, $y = -0.01$ **32** II, $y = 0.9$

CALCULATOR EXERCISES 5.1

Find the quadrants containing the points in Exercises 1–4.

1 (a) $P(43)$ (b) $P(-28)$

2 (a) $P(23.64)$ (b) $P(3.141591)$

3 (a) $P(1776)$ (b) $P(\sqrt[4]{782}\pi)$

4 (a) $P(-2001)$ (b) $P(6\sqrt{2} + 5\sqrt{7})$

5 Approximate the y-coordinate of $P(t)$ if $P(t)$ is in the fourth quadrant and has x-coordinate 0.9703.

6 Approximate the x-coordinate of $P(t)$ if $P(t)$ is in the third quadrant and has y-coordinate -0.6561.

5.2 THE TRIGONOMETRIC FUNCTIONS

In the preceding section we introduced a technique for associating, with each real number t, a unique point $P(t)$ on the unit circle U. The rectangular coordinates (x, y) of $P(t)$ can be used to define the six **trigonometric** or (**circular**) **functions**. These functions are referred to as the **sine, cosine, tangent, cotangent, secant**, and **cosecant functions**, and are designated by the symbols **sin, cos, tan, cot, sec**, and **csc**, respectively. If t is a real number, then the real number which the sine function associates with t will be denoted by either sin (t) or sin t, and similarly for the other five functions.

Definition of the Trigonometric Functions

For any real number t, let $P(t)$ be the point on the unit circle U that corresponds to t. If the rectangular coordinates of $P(t)$ are (x, y), then

$$\sin t = y \qquad\qquad \csc t = \frac{1}{y} \quad (\text{if } y \neq 0)$$

$$\cos t = x \qquad\qquad \sec t = \frac{1}{x} \quad (\text{if } x \neq 0)$$

$$\tan t = \frac{y}{x} \quad (\text{if } x \neq 0) \quad \cot t = \frac{x}{y} \quad (\text{if } y \neq 0).$$

In order to use this definition to find the values of trigonometric functions, it is necessary to determine the rectangular coordinates (x, y) of the point $P(t)$ on U and then substitute for x and y in the appropriate formulas. In later sections we shall introduce other techniques for finding functional values. The domains and ranges of the trigonometric functions will be discussed in Section 5.3.

Example 1 Find the values of the trigonometric functions corresponding to the number $t = \pi/4$.

Solution   Referring to Figure 5.3 we see that $P(\pi/4) = (\sqrt{2}/2, \sqrt{2}/2)$. Letting

$x = \sqrt{2}/2$ and $y = \sqrt{2}/2$ in the last definition gives us

$$\sin\frac{\pi}{4} = \frac{\sqrt{2}}{2} \qquad\qquad \csc\frac{\pi}{4} = \frac{2}{\sqrt{2}} = \sqrt{2}$$

$$\cos\frac{\pi}{4} = \frac{\sqrt{2}}{2} \qquad\qquad \sec\frac{\pi}{4} = \frac{2}{\sqrt{2}} = \sqrt{2}$$

$$\tan\frac{\pi}{4} = \frac{\sqrt{2}/2}{\sqrt{2}/2} = 1 \quad \cot\frac{\pi}{4} = \frac{\sqrt{2}/2}{\sqrt{2}/2} = 1. \qquad\qquad ∎$$

Before considering additional functional values, we shall discuss several important relationships which exist among the trigonometric functions. The formulas listed next are, without doubt, the most important identities in trigonometry, because they may be used to simplify and unify many different aspects of the subject. Since the formulas are true for every allowable value of t, and are part of the foundation for work in trigonometry, they are called the *Fundamental Identities*. Every student should carefully memorize these identities before proceeding to the next section of this text.

Three of the Fundamental Identities involve squares such as $(\sin t)^2$ and $(\cos t)^2$. In general, if n is an integer different from -1, then powers such as $(\cos t)^n$ are written in the form $\cos^n t$. The symbols $\sin^{-1} t$ and $\cos^{-1} t$ are reserved for inverse trigonometric functions, to be discussed in the next chapter. With this agreement on notation we have, for example,

$$\cos^2 t = (\cos t)^2 = (\cos t)(\cos t)$$
$$\tan^3 t = (\tan t)^3 = (\tan t)(\tan t)(\tan t)$$
$$\sec^4 t = (\sec t)^4 = (\sec t)(\sec t)(\sec t)(\sec t).$$

Let us first list all the fundamental identities and then discuss the proofs. The formulas to follow are true for all values of t in the domains of the indicated functions.

The Fundamental Identities

$$\csc t = \frac{1}{\sin t}, \quad \sec t = \frac{1}{\cos t}, \quad \cot t = \frac{1}{\tan t}$$

$$\tan t = \frac{\sin t}{\cos t}, \quad \cot t = \frac{\cos t}{\sin t}$$

$$\sin^2 t + \cos^2 t = 1, \quad 1 + \tan^2 t = \sec^2 t, \quad 1 + \cot^2 t = \csc^2 t$$

The proofs follow directly from the definition of the trigonometric functions. Thus,

$$\csc t = \frac{1}{y} = \frac{1}{\sin t}, \quad \sec t = \frac{1}{x} = \frac{1}{\cos t}, \quad \cot t = \frac{x}{y} = \frac{1}{(y/x)} = \frac{1}{\tan t},$$

$$\tan t = \frac{y}{x} = \frac{\sin t}{\cos t}, \quad \cot t = \frac{x}{y} = \frac{\cos t}{\sin t},$$

where we assume that no denominator is zero.

If (x, y) is a point on the unit circle U, then

$$y^2 + x^2 = 1.$$

Since $y = \sin t$ and $x = \cos t$, this gives us

$$(\sin t)^2 + (\cos t)^2 = 1$$

or equivalently, $$\sin^2 t + \cos^2 t = 1.$$

If $\cos t \neq 0$, then, dividing both sides of the last equation by $\cos^2 t$, we obtain

$$\frac{\sin^2 t}{\cos^2 t} + 1 = \frac{1}{\cos^2 t}$$

or $$\left(\frac{\sin t}{\cos t}\right)^2 + 1 = \left(\frac{1}{\cos t}\right)^2.$$

Since $\tan t = \sin t / \cos t$ and $\sec t = 1/\cos t$ we see that

$$\tan^2 t + 1 = \sec^2 t.$$

The final fundamental identity is left as an exercise.

Example 2   Find the values of the trigonometric functions at the number t if
(a) $t = 0$;   (b) $t = \pi/2$.

Solutions

(a) The rectangular coordinates of the point $P(0)$ on the unit circle are $(1, 0)$, and hence we take $x = 1$ and $y = 0$ in the definition of the trigonometric functions. This gives us

$$\sin 0 = 0 \quad \text{and} \quad \cos 0 = 1.$$

Next, using two fundamental identities,

$$\tan 0 = \frac{\sin 0}{\cos 0} = \frac{0}{1} = 0, \quad \sec 0 = \frac{1}{\cos 0} = \frac{1}{1} = 1.$$

Of course, we could also find $\tan 0$ and $\sec 0$ by direct substitution for x and y in the definition. The cosecant and cotangent functions are undefined at $t = 0$, since 0 appears in a denominator in the definitions of these functions.

(b) As shown in Figure 5.2, $P(\pi/2) = (0, 1)$, and hence we let $x = 0$ and $y = 1$ in the definition of the trigonometric functions, obtaining

$$\sin(\pi/2) = 1 \quad \text{and} \quad \cos(\pi/2) = 0.$$

Next, applying appropriate fundamental identities,

$$\csc(\pi/2) = \frac{1}{\sin(\pi/2)} = \frac{1}{1} = 1$$

$$\cot(\pi/2) = \frac{\cos(\pi/2)}{\sin(\pi/2)} = \frac{0}{1} = 0.$$

The tangent and secant functions are undefined. (Why?) ∎

It is not difficult to determine the signs of the functional values of the trigonometric functions when $P(t)$ is in various quadrants. For example, if $P(t) = (x, y)$ is in quadrant II, then y is positive, x is negative, and we see from the definition that $\sin t$ and $\csc t$ are positive, whereas the other four functions are negative. The reader should check the remaining quadrants. The table below indicates the signs in all four quadrants.

Signs of the Trigonometric Functions

Quadrant containing $P(t) = (x, y)$	Positive functions	Negative functions
I	All	None
II	sin, csc	cos, sec, tan, cot
III	tan, cot	sin, csc, cos, sec
IV	cos, sec	sin, csc, tan, cot

Example 3 Find the quadrant containing $P(t)$ if both $\sin t < 0$ and $\cos t > 0$.

Solution   Referring to the preceding table, we see than $\sin t < 0$ if $P(t)$ is in quadrants III or IV, and $\cos t > 0$ if $P(t)$ is in quadrants I or IV. Hence, for both conditions to be satisfied, $P(t)$ must be in quadrant IV. ∎

Example 4 If $\sin t = \sqrt{3}/2$ and $\sec t = -2$, find the following:
(a) $\csc t$;   (b) $\cos t$;   (c) $\tan t$.

Solutions
(a) Using a fundamental identity,

$$\csc t = \frac{1}{\sin t} = \frac{1}{(\sqrt{3}/2)} = \frac{2}{\sqrt{3}} = \frac{2\sqrt{3}}{3}.$$

(b) Since $\sec t = 1/\cos t$, we also have $(\cos t)(\sec t) = 1$, and hence

$$\cos t = \frac{1}{\sec t} = \frac{1}{-2} = -\frac{1}{2}.$$

(c)   Since $\sin t = \sqrt{3}/2$ and $\cos t = -1/2$,

$$\tan t = \frac{\sin t}{\cos t} = \frac{\sqrt{3}/2}{-1/2} = -\sqrt{3}. \qquad \blacksquare$$

The identity $\sin^2 t + \cos^2 t = 1$ can be used to express $\sin t$ in terms of $\cos t$, or vice versa. For example, since

$$\sin^2 t = 1 - \cos^2 t$$

we have

$$\sin t = \pm\sqrt{1 - \cos^2 t}$$

where the "+" sign is used if $P(t)$ is in quadrants I or II, and the "−" sign is used if $P(t)$ is in quadrants III or IV. Similarly,

$$\cos t = \pm\sqrt{1 - \sin^2 t}$$

where "+" is used if $P(t)$ is in quadrants I or IV and "−" is used if $P(t)$ is in quadrants II or III. The formulas $1 + \tan^2 t = \sec^2 t$ and $1 + \cot^2 t = \csc^2 t$ can be used in like manner.

Example 5   If $\sin t = 3/5$ and $P(t)$ is in quadrant II, find the values of the other trigonometric functions.

Solution   Since $\cos t$ is negative in quadrant II,

$$\cos t = -\sqrt{1 - \sin^2 t} = -\sqrt{1 - (3/5)^2} = -\sqrt{16/25} = -4/5.$$

Next we see that

$$\tan t = \frac{\sin t}{\cos t} = \frac{3/5}{-4/5} = -\frac{3}{4}.$$

Finally, applying the first three fundamental identities gives us

$$\csc t = \frac{5}{3}, \quad \sec t = -\frac{5}{4}, \quad \cot t = -\frac{4}{3}. \qquad \blacksquare$$

EXERCISES 5.2

1   Prove that $1 + \cot^2 t = \csc^2 t$.

2   Prove that $\sin t = 1/\csc t$.

In each of Exercises 3–6, use fundamental identities to find the values of $\csc t$, $\sec t$, $\tan t$, and $\cot t$.

3 $\sin t = -3/5$, $\cos t = 4/5$

4 $\sin t = 8/17$, $\cos t = -15/17$

5 $\sin t = a/\sqrt{a^2 + b^2}$, $\cos t = b/\sqrt{a^2 + b^2}$
where $a^2 + b^2 \neq 0$

6 $\sin t = \sqrt{5}/5$, $\sec t = \sqrt{5}/2$

In each of Exercises 7–18 find the values of the trigonometric functions corresponding to the number t.

7 $t = -3\pi/2$ 8 $t = -\pi/2$

9 $t = 3\pi/4$ 10 $t = -3\pi/4$

11 $t = -\pi$ 12 $t = 5\pi$

13 $t = 7\pi/2$ 14 $t = 9\pi/2$

15 $t = 200\pi$ 16 $t = 375\pi$

17 $t = -5\pi/4$ 18 $t = 29\pi/4$

In each of Exercises 19–28 find the quadrant containing $P(t)$ if the given conditions are true.

19 $\cos t > 0$ and $\sin t < 0$

20 $\tan t < 0$ and $\cos t > 0$

21 $\sin t < 0$ and $\cot t > 0$

22 $\sec t > 0$ and $\tan t < 0$

23 $\csc t > 0$ and $\sec t < 0$

24 $\csc t > 0$ and $\cot t < 0$

25 $\sec t < 0$ and $\tan t > 0$

26 $\sin t < 0$ and $\sec t > 0$

27 $\cos t > 0$ and $\tan t > 0$

28 $\cos t < 0$ and $\csc t < 0$

In each of Exercises 29–36 use fundamental identities to find the values of all six trigonometric functions if the given conditions are true.

29 $\tan t = -\frac{3}{4}$ and $\sin t > 0$

30 $\cot t = \frac{3}{4}$ and $\cos t < 0$

31 $\sin t = -\frac{5}{13}$ and $\sec t > 0$

32 $\cos t = \frac{1}{2}$ and $\sin t < 0$

33 $\cos t = -\frac{1}{3}$ and $\sin t < 0$

34 $\csc t = 5$ and $\cot t < 0$

35 $\sec t = -4$ and $\csc t > 0$

36 $\sin t = \frac{2}{5}$ and $\cos t < 0$

37 Is there a real number t such that $7 \sin t = 9$? Explain.

38 Is there a real number t such that $3 \csc t = 1$? Explain.

39 If $f(t) = \cos t$ and $g(t) = t/4$, find the following.
    (a) $f(g(\pi))$ (b) $g(f(\pi))$

40 If $f(t) = \tan t$ and $g(t) = t/4$, find the following.
    (a) $f(g(\pi))$ (b) $g(f(\pi))$

41 Show that if $P(t)$ is not on the y-axis, then $(\cos t)(\sec t) = 1$.

42 Show that if $P(t)$ is not on a coordinate axis, then $(\tan t)(\cot t) = 1$.

CALCULATOR EXERCISES 5.2

In each of Exercises 1–6 use fundamental identities to approximate the values of all six trigonometric functions to four significant figures if the two given conditions are true.

1 $\sin t = 0.3145$, $\cos t = -0.9492$

2 $\cos t = 0.7314$, $\sin t = 0.6820$

3 $\cos t = 0.8910$, $\tan t = 0.5095$

4 $\sin t = 0.9744$, $\cot t = 0.2309$

5 $\cos t = -0.1219$, $\csc t > 0$

6 $\sin t = 0.9063$, $\tan t < 0$

5.3   THE VARIATION OF THE TRIGONOMETRIC FUNCTIONS

In this section we shall discuss the domains and ranges of each trigonometric function and make some preliminary observations about the variation of $\sin t$ and $\cos t$ as t varies through the set of real numbers.

The domain of each trigonometric function can be determined by referring to the definition in Section 5.2. If $P(t)$ has coordinates (x, y), then the sine and cosine functions always exist and, therefore, each has domain $\mathbb{R}$. In the definitions of the tangent and secant functions, x appears in the denominator and hence we must exclude values of t for which x is 0, that is, the values of t corresponding to the points $(0, 1)$ and $(0, -1)$ on the unit circle U. It follows that the domain of the tangent and secant functions consists of all numbers *except* those of the form $(\pi/2) + n\pi$, where n is an integer. In particular, we exclude $\pm\pi/2, \pm3\pi/2, \pm5\pi/2$, and so on. As a special case, in (b) of Example 2 of the preceding section we saw that $\tan(\pi/2)$ and $\sec(\pi/2)$ do not exist.

Since $\cot t = x/y$ and $\csc t = 1/y$, the domain of the cotangent and cosecant functions is the set of all real numbers except those numbers t for which the y-coordinate of $P(t)$ is 0. The latter include the numbers, 0, $\pm\pi$, $\pm2\pi$, $\pm3\pi$, and, in general, all numbers of the form $n\pi$, where n is an integer. (Recall that in (a) of Example 2 of the preceding section we found that $\cot 0$ and $\csc 0$ did not exist.) In the future, when we work with $\tan t$, $\cot t$, $\sec t$, and $\csc t$, we will always assume that t is in the appropriate domain, even though this fact will not always be mentioned explicitly.

It was pointed out in Section 5.1 that, except for special values of t, the determination of the point $P(t)$ on U (and hence the values of the trigonometric functions) requires techniques studied in calculus. We can, however, make some general observations. For the time being, let us concentrate on the sine function. Referring to the unit circle in Figure 5.5, we see that if t varies from 0 to $\pi/2$, then the coordinates (x, y) of $P(t)$ vary from $(1, 0)$ to $(0, 1)$. In particular, the y-coordinate, that is, the value of the sine function, increases from 0 to 1. Moreover, this function takes on *all* values between 0 and 1. If we let t vary from $\pi/2$ to π, then the coordinates (x, y) of $P(t)$ vary from $(0, 1)$ to $(-1, 0)$, and hence the sine function, that is, the ordinate y of $P(t)$, decreases from 1 to 0. In similar fashion, we see that if t varies from π to $3\pi/2$, then $\sin t$ decreases from 0 to -1; and as t varies from $3\pi/2$ to 2π, $\sin t$ increases from -1 to 0.

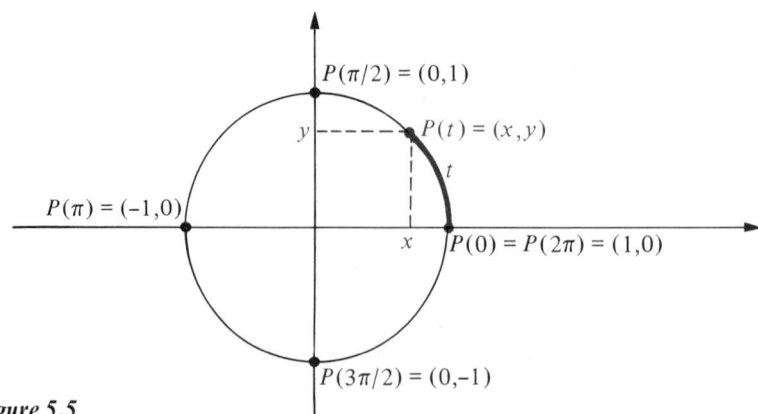

Figure 5.5

The following table partially indicates this behavior of $\sin t$ in the interval $[0, 2\pi]$. For a more complete description we would have to insert many more values of t and $\sin t$.

t	0	$\dfrac{\pi}{2}$	π	$\dfrac{3\pi}{2}$	2π
$\sin t$	0	1	0	-1	0

If we let t vary through the interval $[2\pi, 4\pi]$, then $P(t)$ traces the circle again and the identical pattern for $\sin t$ is repeated, that is,

$$\sin(t + 2\pi) = \sin t$$

for every t in $[0, 2\pi]$. The same is true for other intervals of length 2π. Indeed,

$$\boxed{\sin (t + 2\pi n) = \sin t}$$

since, as we observed in Section 5.1, $P(t) = P(t + 2\pi n)$ for every t and for every integer n. According to the next definition, this repetitive behavior is specified by saying that the sine function is *periodic*.

Definition

> A function f is **periodic** if there exists a positive real number k such that
>
> $$f(t + k) = f(t)$$
>
> for every t in the domain of f. The least such positive real number k, if it exists, is called the **period** of f.

The period of the sine function is 2π (see Exercise 3).

The variation of the cosine function in the interval $[0, 2\pi]$ can be determined by observing the behavior of the x-coordinate of the point $P(t)$ on U as t varies from 0 to 2π. Again referring to Figure 5.5, we see that the cosine function (that is, x) decreases from 1 to 0 in the interval $[0, \pi/2]$, decreases from 0 to -1 in $[\pi/2, \pi]$, increases from -1 to 0 in $[\pi, 3\pi/2]$, and increases from 0 to 1 in $[3\pi/2, 2\pi]$. The pattern is then repeated in successive intervals of length 2π. It follows that the cosine function is periodic with period 2π.

The preceding discussion suggests that the graphs of the sine and cosine functions might have the appearance of those in Figures 5.6 and 5.7, respectively. In the figures we have plotted several points of the form $(t, \sin t)$ and $(t, \cos t)$ using the functional values mentioned earlier. Since each function has period 2π, each graph repeats itself every 2π units along the horizontal axis. These sketches should be considered as very rough approximations, since we have not yet developed techniques for obtaining many intermediate points. Note, however, that we could use Example 1 of the preceding section to get the points $(\pi/4, \sin \pi/4)$ and $(\pi/4, \cos \pi/4)$, that is, $(\pi/4, \sqrt{2}/2)$. We will study the graphs of the trigonometric functions more carefully in Section 5.7.

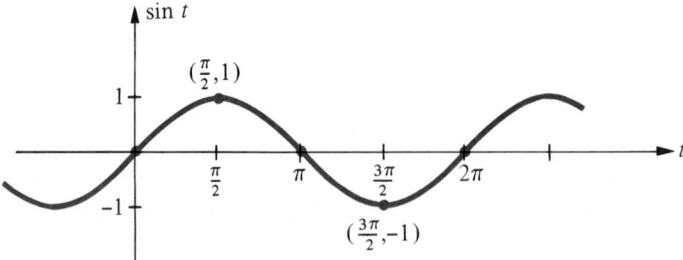

Figure 5.6

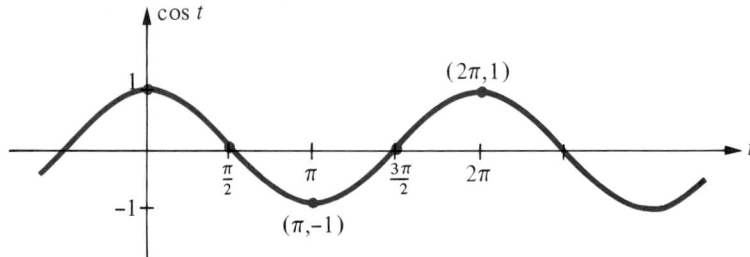

Figure 5.7

It can be shown that the secant and cosecant functions are also periodic with period 2π (see Exercise 6). The variations of these functions will be discussed in Section 5.7. At that time we shall see that the tangent and cotangent functions are periodic with period π.

Finally, let us briefly discuss the range of each trigonometric function. Since the pair (x, y) in the definition of the trigonometric functions gives the coordinates of a point on the unit circle U, we have

$$-1 \le x \le 1 \quad \text{and} \quad -1 \le y \le 1.$$

Since $\sin t = y$ and $\cos t = x$, it follows that the range of both the sine and cosine functions is the set of all numbers in the closed interval $[-1, 1]$.

If x is a nonzero real number such that $-1 \le x \le 1$, then $|x| \le 1$. Multiplying both sides of the last inequality by $1/|x|$, we obtain $1 \le 1/|x|$. Since $\sec t = 1/x$, we see that $1 \le |\sec t|$ for every real number t, and therefore the range of the secant function consists of all real numbers having absolute value greater than or equal to 1. The same is true for the range of the cosecant function.

It can be shown that $\tan t$ and $\cot t$ take on all real values (see Exercise 5). That will also become apparent in Section 5.7, when we study the graphs of the trigonometric functions.

EXERCISES 5.3

In each of Exercises 1 and 2 plot the points $(a, 0)$ on the x-axis if a satisfies both of the given conditions.

1 (a) The tangent and secant functions are undefined at a

    (b) $-4\pi \le a \le 4\pi$

2 (a) The cotangent and cosecant functions are undefined at a.

    (b) $-4\pi \le a \le 4\pi$

3 Prove that the sine function has period 2π. (*Hint:* Assume that there is a positive real number k less than 2π such that $\sin(t + k) = \sin t$ for all t. Arrive at a contradiction by letting $t = 0$).

4 Prove that the cosine function has period 2π.

5 Prove that the range of the tangent function is $\mathbb{R}$ by showing that if a is any real number, then there is a point $P(t)$ on U such that $\tan t = a$. (*Hint:* If $P(t) = (x, y)$ is on U, consider the equation $\tan t = y/x = a$, where $x^2 + y^2 = 1$.)

6 Prove that the cosecant and secant functions each have period 2π. (*Hint:* Use Exercises 3 and 4 and two fundamental identities.)

5.4 ANGLES

The definitions of the trigonometric functions can also be based upon the notion of angles. This more traditional approach is quite common in applications of mathematics and hence should not be obscured by the definitions stated in Section 5.2. Indeed, for a thorough appreciation of the trigonometric functions, it is best to blend the two ideas. In this section we restrict our discussion to angles and their measurement. In Section 5.5 we shall demonstrate how angles can be used to find values of the trigonometric functions.

An angle is often regarded as the geometric configuration formed by two rays or half-lines, l_1 and l_2, having the same initial point O. If A and B are points on l_1 and l_2, respectively (see Figure 5.8), then we may refer to **angle AOB**. The same is true for finite line segments with a common endpoint. For trigonometric purposes it is convenient to regard angle AOB as generated by starting with the fixed ray l_1 with endpoint O and rotating it about O, in a plane, to a position specified by ray l_2. We call l_1 the **initial side**, l_2 the **terminal side**, and O the **vertex** of the angle. The amount or direction of rotation is not restricted in any way. Thus we might let l_1 make several revolutions in either direction about O before coming to the position l_2.

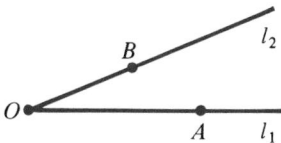

Figure 5.8

If a rectangular coordinate system is introduced, then the **standard position** of an angle is obtained by taking the vertex at the origin and letting l_1 coincide with the positive x-axis. If l_1 is rotated in a counterclockwise direction to position l_2, then the angle is considered **positive**, whereas if l_1 is rotated in a clockwise direction, the angle is **negative**. We often denote angles by lower-case Greek letters and specify the direction of rotation by means of a circular arc or spiral with an arrow attached. Figure 5.9 contains sketches of two positive angles α and β and a negative angle γ. If the terminal side of an angle is in a certain quadrant, we speak of the *angle* as being in that quadrant. In Figure 5.9 α is in quadrant II, β is in quadrant I, and γ is in quadrant III. If the terminal side coincides with a coordinate axis, then the angle is referred to as a **quadrantal angle**. It is important to observe that there are many different angles in standard position which have the same terminal side. Any two such angles are called **coterminal**.

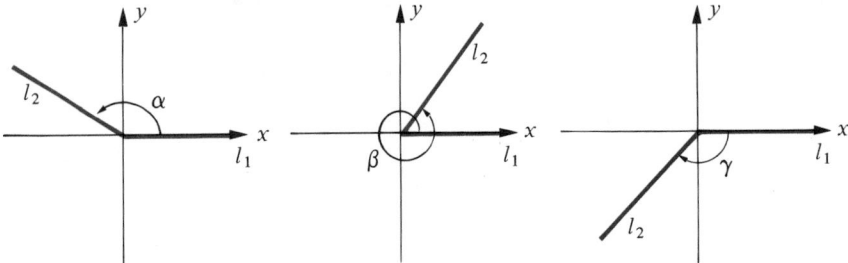

Figure 5.9

We shall next consider the problem of assigning a measure to a given angle. Let U be a unit circle with the center at the origin O of a rectangular coordinate system, and let θ be an angle in standard position. We regard θ as generated by rotating the positive x-axis about O. As the axis rotates to its terminal position, its point of intersection with U travels a certain distance t before arriving at its final position P, as illustrated in Figure 5.10. If t is considered positive for a counterclockwise rotation and negative for a clockwise rotation, then P is precisely the point on U that corresponds to the real number t. A natural way of assigning a measure to θ is to use the number t. When this is done, we say that θ **is an angle of t radians** and we write $\theta = t$ or $\theta = t$ *radians*. In particular, if $\theta = 1$, then θ is an angle that subtends an arc of unit length on the unit circle U. The notation $\theta = -7.5$ means that θ is the angle generated by a clockwise rotation in which the point of intersection of the terminal side of θ with the unit circle U travels 7.5 units. Several angles, measured in radians, are sketched in Figure 5.11.

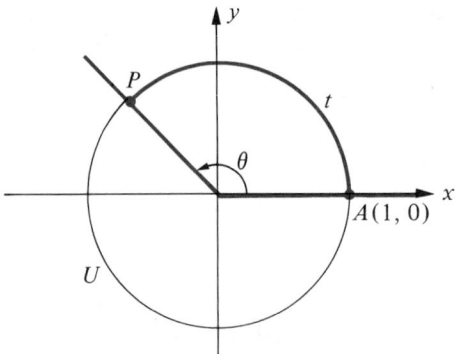

Figure 5.10

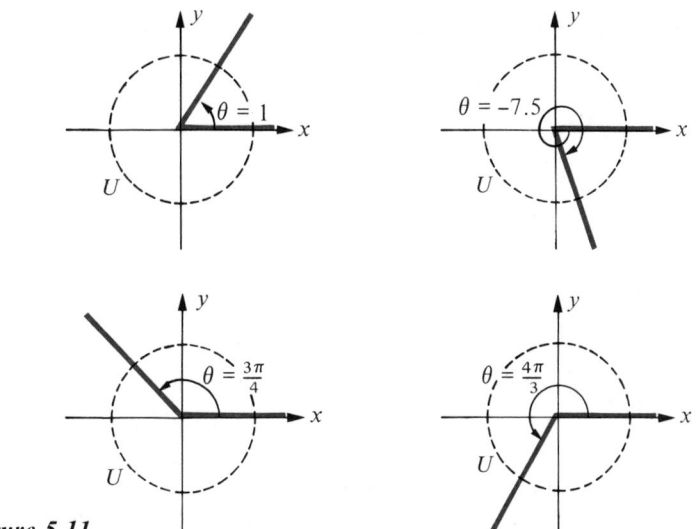

Figure 5.11

The radian measure of an angle can be found by using a circle of *any* radius. In the following discussion, the terminology **central angle** of a circle refers to an angle whose vertex is at the center of the circle. Thus suppose that θ is a central angle of a circle of radius r and that θ subtends an arc of length s, where $0 \le s < 2\pi r$. To find the radian measure of θ let us place θ in standard position on a rectangular coordinate system and superimpose a unit circle U, as shown in Figure 5.12. If t is the length of arc subtended

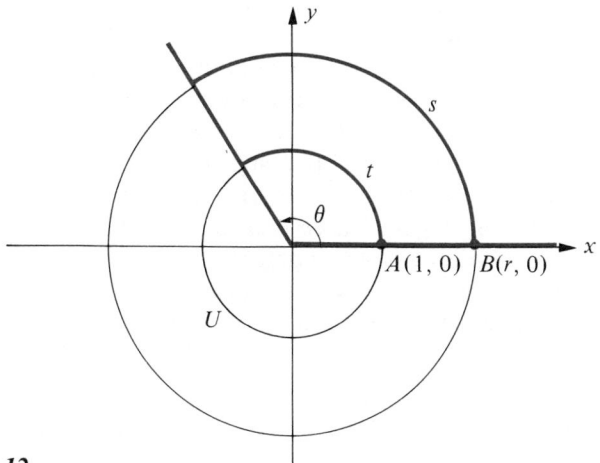

Figure 5.12

by θ on U, then by definition we may write $\theta = t$. From plane geometry, the ratio of the arcs in Figure 5.12 is the same as the ratio of the radii; that is,

$$\frac{t}{s} = \frac{1}{r} \quad \text{or} \quad t = \frac{s}{r}.$$

Substituting θ for t gives us the following result.

> If a central angle θ of a circle of radius r subtends an arc of length s, then the radian measure of θ is given by
>
> $$\theta = \frac{s}{r}.$$

Example 1   A central angle θ subtends an arc 10 cm long on a circle of radius 4 cm. Find the radian measure of θ.

Solution   Substituting in the formula $\theta = s/r$ gives us the radian measure

$$\theta = \frac{10}{4} = 2.5.$$

The formula $\theta = s/r$ indicates that the radian measure of an angle is independent of the size of the circle. For example, if the radius of the circle is $r = 4$ cm and the arc subtended by a central angle θ is 8 cm, then the radian measure is

$$\theta = \frac{8\,\text{cm}}{4\,\text{cm}} = 2.$$ ■

If the radius of the circle is 5 km and the subtended arc is 10 km then

$$\theta = \frac{10\,\text{km}}{5\,\text{km}} = 2.$$

These calculations indicate that the radian measure of an angle is dimensionless, and hence may be regarded as a real number. Indeed, it is for this reason that we usually employ the notation $\theta = t$ instead of $\theta = t$ radians.

Another unit of measurement for angles is the **degree**. If the angle is placed in standard position on a rectangular coordinate system, then an angle of 1 degree is, by definition, the measure of the angle formed by $\frac{1}{360}$ of a complete revolution in the counterclockwise direction. The symbol "°" is used to denote the degrees in the measure of an angle. In Figure 5.13 several angles measured in degrees are shown in standard position on a rectangular coordinate system. It is customary to refer to an angle of measure 90° as a **right angle**. An angle is **acute** if its degree measure is between 0° and 90°. If its measure is between 90° and 180° an angle is **obtuse**.

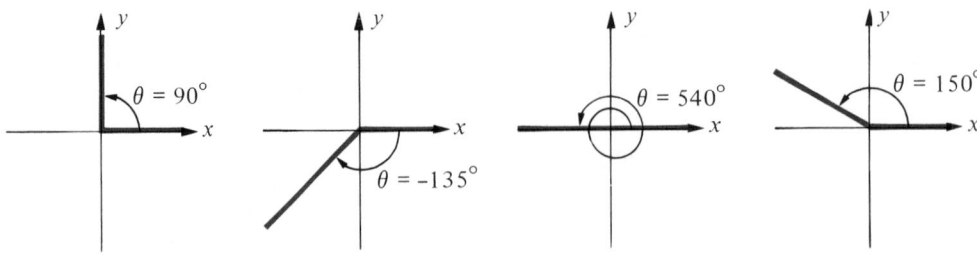

Figure 5.13

If smaller measurements than those afforded by the degree or radian are required, we can use tenths, hundredths, or thousandths of radians or degrees. If degrees are used, then another method is to divide each degree into 60 equal parts called **minutes** (denoted by ′) and each minute into 60 equal parts called **seconds** (denoted by ″). Thus 1′ is $\frac{1}{60}$ of 1°, and 1″ is $\frac{1}{60}$ of 1′. A notation such as $\theta = 73°56′18″$ refers to an angle θ of measure 73 degrees, 56 minutes, and 18 seconds.

It is not difficult to transform angular measure from one system to another. If we consider the angle θ, in standard position, generated by one-half of a complete counterclockwise rotation, then $\theta = 180°$. Using the formula $\theta = s/r$, we see that the radian measure of θ is π. This gives us the basic relation

$$180° = \pi \text{ radians.}$$

Equivalent formulas are

$$1° = \frac{\pi}{180} \text{ radians} \quad \text{and} \quad 1 \text{ radian} = \left(\frac{180}{\pi}\right)°.$$

If we use the approximations $\pi \approx 3.14159$ and calculate $\pi/180$ and $180/\pi$ we obtain

$$1° \approx 0.01745 \text{ radians} \quad \text{and} \quad 1 \text{ radian} \approx 57.296°.$$

Example 2
  (a) Find the radian measure of θ if $\theta = 150°$ and if $\theta = 225°$.
  (b) Find the degree measure of θ if $\theta = 7\pi/4$ and if $\theta = \pi/3$.

Solutions
  (a) Since there are $\pi/180$ radians in each degree, the number of radians in 150° can be found by multiplying 150 by $\pi/180$. Thus

$$150° = 150\left(\frac{\pi}{180}\right) = \frac{5\pi}{6} \text{ radians.}$$

Similarly, $\qquad 225° = 225\left(\frac{\pi}{180}\right) = \frac{5\pi}{4} \text{ radians.}$

  (b) The number of degrees in 1 radian is $180/\pi$. Consequently, to find the number of degrees in $7\pi/4$ radians, we multiply by $180/\pi$, obtaining

$$\frac{7\pi}{4} \text{ radians} = \frac{7\pi}{4}\left(\frac{180}{\pi}\right) = 315°.$$

In like manner, $\qquad \frac{\pi}{3} \text{ radians} = \frac{\pi}{3}\left(\frac{180}{\pi}\right) = 60°.$ ∎

Example 3 If the measure of an angle θ is 3 radians, find the approximate measure of θ in terms of degrees, minutes, and seconds.

Solution   Since 1 radian $\approx 57.296°$, we have

$$3 \text{ radians} \approx 171.888° = 171° + 0.888°.$$

Since there are 60′ in each degree, the number of minutes in 0.888° is 60(0.888), or 53.28′. Hence

$$3 \text{ radians} \approx 171°53.28′.$$

Finally, $0.28′ = (0.28)60″ \approx 17″$. Therefore,

$$3 \text{ radians} \approx 171°53′17″.$$ ■

Many hand-held calculators have a key which will automatically convert the degree measure of an angle to radians, and vice versa. The owner of a calculator should refer to the user's guide for information. Since entries are made in the decimal system, angles which are expressed in terms of degrees and minutes must be converted to decimal form before they are entered into the calculator. For example,

$$67°30′ = 67.5° \quad \text{since} \quad 30′ = (30/60)° = 0.5°$$
$$123°45′ = 123.75° \quad \text{since} \quad 45′ = (45/60)° = 0.75°$$
$$8°13′ \approx 8.21667° \quad \text{since} \quad 13′ = (13/60)° \approx 0.21667°.$$

If seconds are involved, another appropriate decimal must be calculated. To illustrate, since there are 3,600 seconds in each degree, $18″ = (18/3600)° = 0.005′$ and hence

$$67°30′18″ = 67.505°.$$

EXERCISES 5.4

In each of Exercises 1–12 place the angle with the indicated measure in standard position on a rectangular coordinate system, and find the measure of two positive angles and two negative angles which are coterminal with the given angle.

In Exercises 13–24 find the radian measure that corresponds to the given degree measure.

13	150°	**14**	120°
15	−60°	**16**	−135°
17	225°	**18**	210°
19	450°	**20**	630°
21	72°	**22**	54°
23	100°	**24**	95°

1	120°	**2**	240°
3	135°	**4**	315°
5	−30°	**6**	−150°
7	620°	**8**	570°
9	5π/6	**10**	2π/3
11	−π/4	**12**	−5π/4

In Exercises 25–36 find the degree measure that corresponds to the given radian measure.

25	2π/3	**26**	5π/6

27 $11\pi/6$

28 $4\pi/3$

29 $3\pi/4$

30 $11\pi/4$

31 $-7\pi/2$

32 $-5\pi/2$

33 7π

34 9π

35 $\pi/9$

36 $\pi/16$

In Exercises 37–40 find the approximate measure of θ in terms of degrees, minutes, and seconds.

37 $\theta = 2$

38 $\theta = 1.5$

39 $\theta = 5$

40 $\theta = 4$

41 A central angle θ subtends an arc 7 cm long on a circle of radius 4 cm. Approximate the measure of θ in (a) radians; (b) degrees.

42 Answer (a) and (b) of Exercise 41 if θ subtends an arc 3 feet long on a circle of radius 20 inches.

43 Approximate the length of an arc subtended by a central angle of measure 50° on a circle of radius 8 meters.

44 Approximate the length of an arc subtended by a central angle of 2.2 radians on a circle of radius 60 cm.

45 If a central angle θ of measure 20° subtends a circular arc of length 3 km, find the radius of the circle.

46 If a central angle θ of radian measure 4 subtends a circular arc 10 cm long, find the radius of the circle.

47 The distance between two points A and B on the earth is measured along a circle having center C at the center of the earth and radius equal to the distance from C to the surface. If the diameter of the earth is approximately 8,000 miles, approximate the distance between A and B if angle ACB has measure (a) 60°; (b) 45°; (c) 30°; (d) 10°; (e) 1°.

48 Refer to Exercise 47. If angle ACB has measure $1'$, then the distance between A and B is called a **nautical mile**. Approximate the number of ordinary (**statute**) miles in a nautical mile.

49 Refer to Exercise 47. If two points A and B are 500 miles apart, find angle ACB in both degree measure and radian measure.

50 The **angular speed** of a wheel which is rotating at a constant rate is the angle generated by a line segment from the center of the wheel to the circumference, per unit of time. If a wheel of diameter 3 feet is rotating at a rate of 2,400 rpm (revolutions per minute), find the angular speed.

CALCULATOR EXERCISES 5.4

In Exercises 1–6 approximate, to four decimal places, the radian measure that corresponds to the given degree measure of θ, or the degree measure that corresponds to the given radian measure.

1 $\theta = 73°24'$

2 $\theta = 261°37'$

3 $\theta = 482°16'43''$

4 $\theta = \sqrt{317} + \pi$

5 $\theta = \ln(13.6521)$

6 $\theta = e^{\sqrt{\pi}}$

7–9 The diameter of the earth at the equator is estimated as 7,926.41 miles. Rework Exercises 47–49 using that measurement, assuming that points A and B are on the equator.

10 Refer to Exercise 50. If a wheel of diameter 1.683 meters is rotating at a rate of 2,875 rpm, find the angular speed.

5.5 TRIGONOMETRIC FUNCTIONS OF ANGLES

For certain applications it is convenient to change the domain of the trigonometric functions from a set of real numbers to a set of angles. This is very easy to do. If θ is an angle, we merely agree on the values sin θ, cos θ, and so on. The usual way of assigning such values is to use the radian measure of θ, as in the following definition.

**Definition of
Trigonometric
Functions
of Angles**

> If θ is an angle, and if the radian measure of θ is t, then the value of each trigonometric function at θ is its value at the real number t.

We see from the preceding definition that if t is the radian measure of θ, then

$$\sin \theta = \sin t, \quad \cos \theta = \cos t, \quad \tan \theta = \tan t,$$

and likewise for the other functions. For convenience we shall use the terminology *trigonometric functions* regardless of whether angles or real numbers are employed for the domain. To make the unit of angular measure clear we shall use the degree symbol and write sin 65°, tan 150°, and so on, whenever the angle is measured in degrees. Numerals without any symbol attached, such as cos 3 and csc $(\pi/6)$, will indicate that radian measure is being used. This is not in conflict with our previous work where, for example, cos 3 meant the value of the cosine function at the real number 3, since by definition, the cosine of an angle of measure 3 radians is identical with the cosine of the real number 3.

Example 1   Find sin 90°, cos 45°, and tan 720°.

Solution   The angles are shown in standard position in Figure 5.14. The radian measures of the angles may be found as follows:

$$90° = 90\left(\frac{\pi}{180}\right) = \frac{\pi}{2}, \quad 45° = 45\left(\frac{\pi}{180}\right) = \frac{\pi}{4}, \quad 720° = 720\left(\frac{\pi}{180}\right) = 4\pi.$$

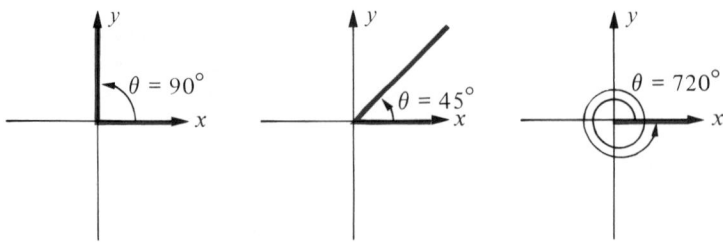

Figure 5.14

Using the definition of trigonometric functions of angles and Example 1 of Section 5.2, we obtain

$$\sin 90° = \sin(\pi/2) = 1$$

$$\cos 45° = \cos(\pi/4) = \sqrt{2}/2$$

$$\tan 720° = \tan 4\pi = \tan 0 = 0.$$ ∎

In the next section we shall introduce techniques for approximating trigonometric functional values corresponding to *any* angle θ.

The values of the trigonometric functions at an angle θ may be determined by means of an arbitrary point on the terminal side of θ. To prove this, let θ be an angle in standard position and let $Q(a, b)$ be any point on the terminal side of θ, where $d(O, Q) = r > 0$. Figure 5.15 illustrates the case in which the terminal side lies in quadrant III; however, our discussion applies to any angle.

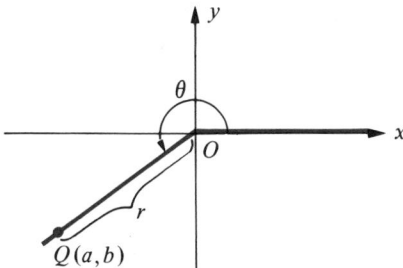

Figure 5.15

The point $Q(a, b)$ is not necessarily a point on the unit circle U, since r may be different from 1. Let $P(x, y)$ be the point on the terminal side of θ such that $d(O, P) = 1$. Hence $P(x, y)$ is on the unit circle U. If t is the radian measure of θ, then by definition we have

$$\sin \theta = \sin t = y$$
$$\cos \theta = \cos t = x.$$

As in Figure 5.16, let us consider vertical lines through Q and P intersecting the x-axis at $Q'(a, 0)$ and $P'(x, 0)$, respectively. Since triangles OQQ' and OPP' are similar, we have

$$\frac{d(P', P)}{d(O, P)} = \frac{d(Q', Q)}{d(O, Q)} \quad \text{or} \quad \frac{|y|}{1} = \frac{|b|}{r}.$$

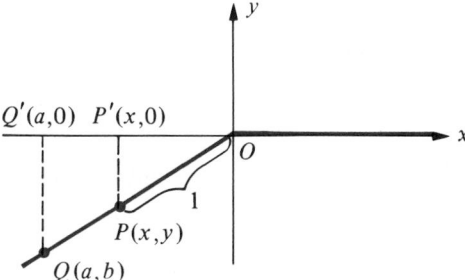

Figure 5.16

Since b and y always have the same sign, this gives us

$$y = \frac{b}{r} \quad \text{and hence} \quad \sin \theta = y = \frac{b}{r}.$$

In similar fashion we obtain

$$\cos \theta = \frac{a}{r}.$$

Using a fundamental identity,

$$\tan \theta = \frac{\sin \theta}{\cos \theta} = \frac{(b/r)}{(a/r)} = \frac{b}{a}.$$

The remaining three functions may be obtained by taking reciprocals. This gives us the following theorem.

Theorem on Trigonometric Functions as Ratios

Let θ be an angle in standard position on a rectangular coordinate system and let $Q(a, b)$ be any point other than O on the terminal side of θ. If $d(O, Q) = r$, then

$$\sin \theta = \frac{b}{r} \qquad\qquad \csc \theta = \frac{r}{b} \quad (\text{if } b \neq 0)$$

$$\cos \theta = \frac{a}{r} \qquad\qquad \sec \theta = \frac{r}{a} \quad (\text{if } a \neq 0)$$

$$\tan \theta = \frac{b}{a} \quad (\text{if } a \neq 0) \qquad \cot \theta = \frac{a}{b} \quad (\text{if } b \neq 0).$$

It can be shown, by using similar triangles, that the formulas given in this theorem are independent of the point $Q(a, b)$ that is chosen on the terminal side of θ. Note that if $r = 1$, then the formulas reduce to the definition of the trigonometric functions given in Section 5.2, with $a = x$ and $b = y$. This theorem has many applications in trigonometry. In Section 5.10 we shall use it to solve problems concerned with right triangles. Another reason for its importance is that it can often be used to obtain values of trigonometric functions. Indeed, note that it is sufficient to find *one* point (other than O) on the terminal side of an angle provided the angle is in standard position on a rectangular coordinate system.

Example 2 If θ is an angle in standard position on a rectangular coordinate system, and if the point $Q(-15, 8)$ is on the terminal side of θ, find the values of the trigonometric functions of θ.

Solution By the Distance Formula, the distance r from the origin O to any point $Q(a, b)$ is $r = \sqrt{(a - 0)^2 + (b - 0)^2} = \sqrt{a^2 + b^2}$. Hence for $Q(-15, 8)$ we have

$$r = \sqrt{(-15)^2 + 8^2} = \sqrt{225 + 64} = \sqrt{289} = 17.$$

Applying the last theorem with $a = -15$, $b = 8$, and $r = 17$, we obtain

$$\sin \theta = \frac{8}{17} \qquad \csc \theta = \frac{17}{8}$$

$$\cos \theta = -\frac{15}{17} \qquad \sec \theta = -\frac{17}{15}$$

$$\tan \theta = -\frac{8}{15} \qquad \cot \theta = -\frac{15}{8}.$$ ∎

Example 3   An angle θ is in standard position on a rectangular coordinate system. If its terminal side is in quadrant III and it lies on the line $y = 3x$, find the values of the trigonometric functions of θ.

Solution   The graph of $y = 3x$ is sketched in Figure 5.17, together with the initial and terminal sides of θ (in color). We begin by choosing a convenient point, say $Q(-1, -3)$, on the terminal side. The distance r from the origin to Q is

$$r = d(O, Q) = \sqrt{(-1)^2 + (-3)^2} = \sqrt{10}.$$

Applying the Theorem on Trigonometric Functions as Ratios with $a = -1$, $b = -3$, and $r = \sqrt{10}$ gives us

$$\sin \theta = \frac{-3}{\sqrt{10}} = \frac{-3\sqrt{10}}{10}, \qquad \csc \theta = \frac{\sqrt{10}}{-3} = -\frac{\sqrt{10}}{3}$$

$$\cos \theta = \frac{-1}{\sqrt{10}} = \frac{-\sqrt{10}}{10}, \qquad \sec \theta = \frac{\sqrt{10}}{-1} = -\sqrt{10}$$

$$\tan \theta = \frac{-3}{-1} = 3, \qquad \cot \theta = \frac{-1}{-3} = \frac{1}{3}.$$

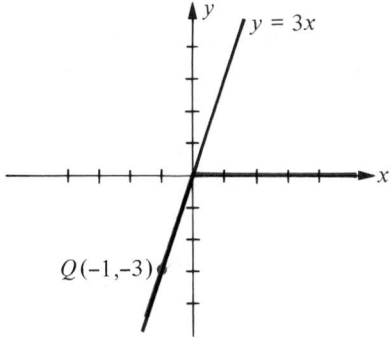

Figure 5.17   ∎

In the proof of the Theorem on Trigonometric Functions as Ratios we considered θ as a nonquadrantal angle; however, the formulas we derived are also true if the terminal side of θ lies on either the x- or y-axis. This is illustrated by the next example.

Example 4   Find the values of the trigonometric functions of θ if $\theta = 270°$.

Solution   If we place θ in standard position, then the terminal side coincides with the negative y-axis. We next choose any point Q on the terminal side of θ. For simplicity we consider $Q(0, -1)$. In this case $r = 1$, $a = 0$, $b = -1$, and hence

$$\sin \theta = \frac{-1}{1} = -1 \qquad \csc \theta = \frac{1}{-1} = -1$$

$$\cos \theta = \frac{0}{1} = 0 \qquad \cot \theta = \frac{0}{-1} = 0.$$

The tangent and secant functions are undefined, since the meaningless expressions $\tan \theta = (-1)/0$ and $\sec \theta = 1/0$ arise when we substitute in the appropriate formulas.   ■

We shall conclude this section by showing that for acute angles, values of the trigonometric functions can be interpreted as ratios of the lengths of the sides of a right triangle. Recall that a triangle is called a **right triangle** if one of its angles is a right angle. If θ is an acute angle, then it can be regarded as an angle of a right triangle and we may refer to the lengths of the **hypotenuse**, the **opposite side**, and the **adjacent side** in the usual way. For convenience we shall use **hyp**, **opp**, and **adj**, respectively, to denote these numbers. Introducing a rectangular coordinate system as in Figure 5.18 we see that the

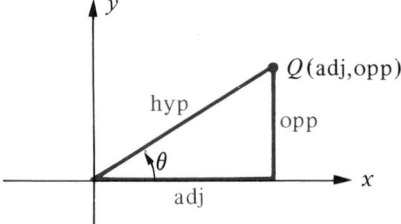

Figure 5.18

lengths of the adjacent side and the opposite side for θ are the abscissa and ordinate, respectively, of a point Q on the terminal side of θ. By the Theorem on Trigonometric Functions as Ratios, we have

$$\sin \theta = \frac{\text{opp}}{\text{hyp}} \qquad \csc \theta = \frac{\text{hyp}}{\text{opp}}$$

$$\cos \theta = \frac{\text{adj}}{\text{hyp}} \qquad \sec \theta = \frac{\text{hyp}}{\text{adj}}$$

$$\tan \theta = \frac{\text{opp}}{\text{adj}} \qquad \cot \theta = \frac{\text{adj}}{\text{opp}}$$

These formulas are very important in work with right triangles and should be memorized. The next example illustrates how they may be used.

Example 5   Find the values of $\sin\theta$, $\cos\theta$, and $\tan\theta$ for the following values of θ:

(a) $\theta = 60°$;   (b) $\theta = 30°$;   (c) $\theta = 45°$.

Solutions   Let us consider an equilateral triangle having sides of length 2. The median from one vertex to the opposite side bisects the angle at that vertex, as illustrated in (i) of Figure 5.19. Using the colored triangle, we obtain the following.

(a) $$\sin 60° = \frac{\sqrt{3}}{2}, \quad \cos 60° = \frac{1}{2}, \quad \tan 60° = \frac{\sqrt{3}}{1} = \sqrt{3}$$

(b) $$\sin 30° = \frac{1}{2}, \quad \cos 30° = \frac{\sqrt{3}}{2}, \quad \tan 30° = \frac{1}{\sqrt{3}} = \frac{\sqrt{3}}{3}$$

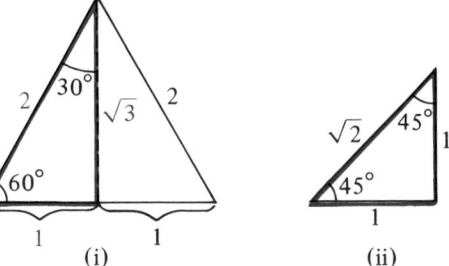

Figure 5.19   (i)   (ii)

(c) To find the functional values for $\theta = 45°$, let us consider an isosceles right triangle whose two equal sides have length 1, as illustrated in (ii) of Figure 5.19. Thus

$$\sin 45° = \frac{1}{\sqrt{2}} = \frac{\sqrt{2}}{2} = \cos 45°, \quad \tan 45° = \frac{1}{1} = 1. \qquad \blacksquare$$

The radian measures corresponding to 30°, 45°, and 60° are $\pi/6$, $\pi/4$, and $\pi/3$, respectively. Since the values of the trigonometric functions at a real number t are the same as those at an angle of t radians, Example 5 provides a convenient method for obtaining these special values. For convenience they are listed in tabular form in Table A on page 204, together with the values at 0 and $\pi/2$. The reader should check each entry (see Exercise 19). A dash indicates that the function is undefined. Two reasons for stressing these special values are (1) they are exact, and (2) they arise frequently in work involving trigonometry. It is a good idea either to memorize Table A or to be able to find the values quickly, using triangles and points on the coordinate axes.

We have now discussed two different approaches to the trigonometric functions. The definition introduced in Section 5.2 emphasizes the fact that the trigonometric functions have domains consisting of real numbers. Such functions are the building blocks for the subject of calculus. In addition, we shall see later that the unit circle approach is useful for discussing graphs and deriving various trigonometric identities. The development in terms of angles and ratios considered in this section has many applications in the sciences and engineering. Students should strive to become proficient in the use of both formulations of the trigonometric functions, since each will reinforce the other and make it easier to master more advanced aspects of trigonometry.

Table A. Special Values of the Trigonometric Functions

t number, radians	t degrees	$\sin t$	$\cos t$	$\tan t$	$\cot t$	$\sec t$	$\csc t$
0	$0°$	0	1	0	—	1	—
$\dfrac{\pi}{6}$	$30°$	$\dfrac{1}{2}$	$\dfrac{\sqrt{3}}{2}$	$\dfrac{\sqrt{3}}{3}$	$\sqrt{3}$	$\dfrac{2\sqrt{3}}{3}$	2
$\dfrac{\pi}{4}$	$45°$	$\dfrac{\sqrt{2}}{2}$	$\dfrac{\sqrt{2}}{2}$	1	1	$\sqrt{2}$	$\sqrt{2}$
$\dfrac{\pi}{3}$	$60°$	$\dfrac{\sqrt{3}}{2}$	$\dfrac{1}{2}$	$\sqrt{3}$	$\dfrac{\sqrt{3}}{3}$	2	$\dfrac{2\sqrt{3}}{3}$
$\dfrac{\pi}{2}$	$90°$	1	0	—	0	—	1

EXERCISES 5.5

In Exercises 1–18 use the Theorem on Trigonometric Functions as Ratios to find the values of the six trigonometric functions of θ if θ is in standard position and satisfies the given condition.

1   The point $P(4, -3)$ is on the terminal side of θ.

2   The point $P(-8, -15)$ is on the terminal side of θ.

3   The point $P(-2, -5)$ is on the terminal side of θ.

4   The point $P(-1, 2)$ is on the terminal side of θ.

5   The terminal side of θ is in quadrant II and lies on the line $y = -4x$.

6   The terminal side of θ is in quadrant IV and lies on the line $3y + 5x = 0$.

7   The terminal side of θ is in quadrant III and is parallel to the line $2y - 7x + 2 = 0$.

8   The terminal side of θ is in quadrant II and is parallel to the line through the points $A(1, 4)$ and $B(3, -2)$.

9   The terminal side of θ is in quadrant I and lies on a line having slope 4/3.

10   The terminal side of θ bisects the third quadrant.

11 $\theta = 450°$ **12** $\theta = 180°$

13 $\theta = 135°$ **14** $\theta = 225°$

15 $\theta = -45°$ **16** $\theta = 405°$

17 $\theta = 360°$ **18** $\theta = 630°$

19   Check each entry in Table A.

20   Prove geometrically that the formulas given in the Theorem on Trigonometric Functions as Ratios are independent of the point $Q(a, b)$ that is chosen on the terminal side of θ.

In Exercises 21–30 find the values of the trigonometric functions of θ if θ is the angle of the pictured right triangle.

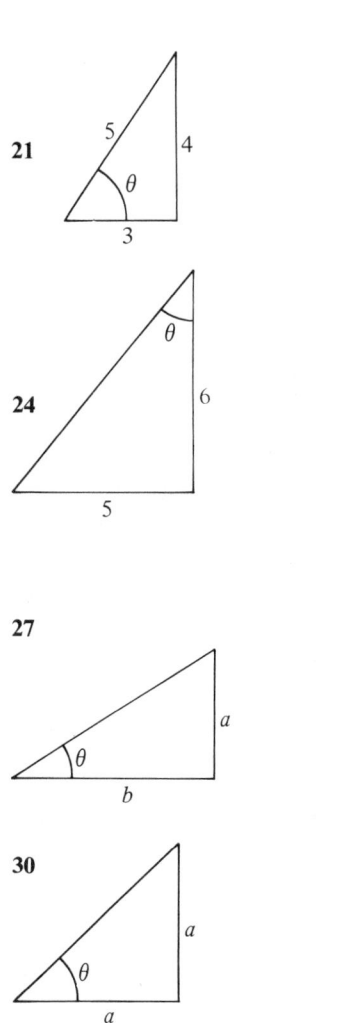

21

22

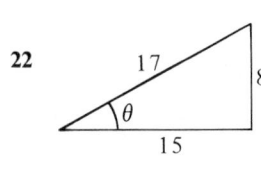

23

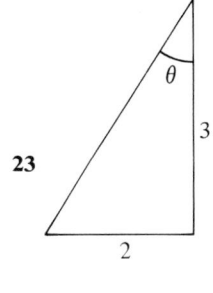

24

25

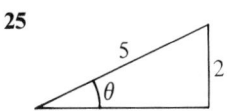

26

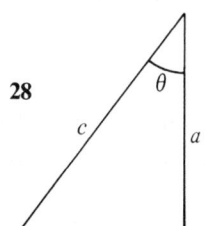

27

28

29

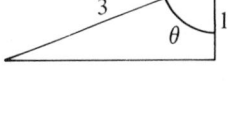

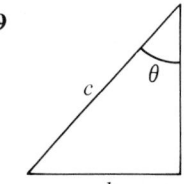

30

CALCULATOR EXERCISES 5.5

1 Use a calculator to find the values of the sine, cosine, and tangent functions at 30°, 45°, and 60°, and compare your answers with the entries in Table A. How do you account for any discrepancies?

2 Find the values of the six trigonometric functions of θ to two-decimal-place accuracy if θ

is in standard position and the point $P(\ln 4.37, \sqrt{9.61}\, e^{1.35})$ is on the terminal side.

3 Find the values of the trigonometric functions of θ to three-decimal-place accuracy if θ is the angle of the right triangle on page 206.

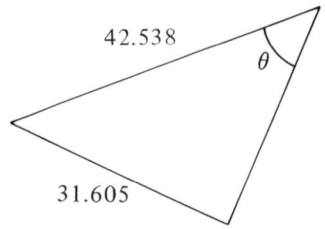

5.6   VALUES OF THE TRIGONOMETRIC FUNCTIONS

Several special values of the trigonometric functions were calculated in previous sections. Let us now turn to the problem of finding *all* values. Since the sine function has period 2π, it is sufficient to know the values of $\sin t$ for $0 \le t \le 2\pi$, because these values are repeated in every t-interval of length 2π. The same is true for the other trigonometric functions. As a matter of fact, the values of any trigonometric function can be determined if its values in the t-interval $[0, \pi/2]$ are known. In order to prove this, suppose that t is any real number and let $P(t)$ be the point on the unit circle U that corresponds to t. We shall make use of the following concept.

Definition

> The shortest arc length t' between $P(t)$ and the x-axis, on the unit circle U, is called the **reference number** associated with t.

Figure 5.20 illustrates arcs of length t' for positions of $P(t)$ in various quadrants. Note that $0 \le t' < \pi/2$ for all values of t.

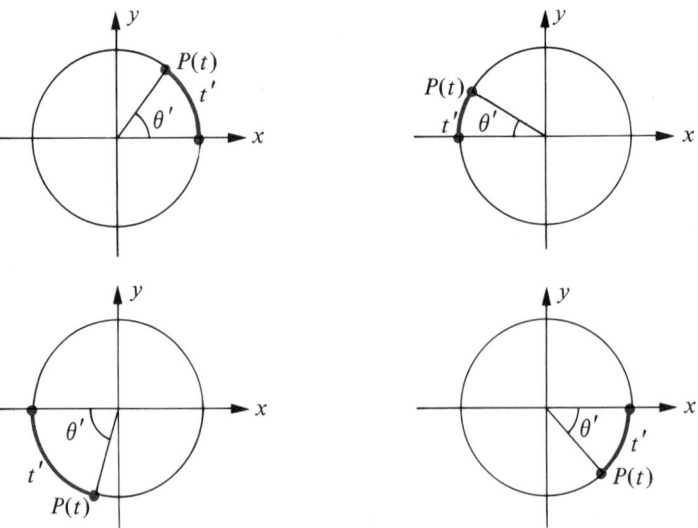

Figure 5.20

If we let θ' denote the angle with vertex O which subtends the arc of length t', as illustrated in Figure 5.20, then θ' is called the *reference angle associated with $P(t)$, or with any angle θ whose terminal side contains $P(t)$.* Thus, if θ is an angle in standard position and θ is not a quadrantal angle, we have the following definition.

Definition

> The **reference angle** associated with θ is the acute angle θ' that the terminal side of θ makes with the positive or negative x-axis.

Referring to Figure 5.20 and using the definition of trigonometric functions of angles stated in Section 5.5, we see that

$$\sin \theta' = \sin t', \quad \cos \theta' = \cos t', \quad \tan \theta = \tan t'.$$

In the remainder of our work we shall use the approximation $\pi \approx 3.1416$ for estimating reference numbers and angles.

Example 1   Approximate the reference number t' if

(a) $t = 2$;   (b) $t = 4$;   (c) $t = 7\pi/4$;   (d) $t = -2\pi/3$.

Solutions   Each point $P(t)$ on the unit circle U, and the magnitude of its reference number t', are indicated in Figure 5.21. Referring to the figure we see that

(a) $t' = \pi - 2 \approx 3.1416 - 2 = 1.1416$

(b) $t' = 4 - \pi \approx 4 - 3.1416 = 0.8584$

(c) $t' = 2\pi - (7\pi/4) = \pi/4$

(d) $t' = \pi - (2\pi/3) = \pi/3$

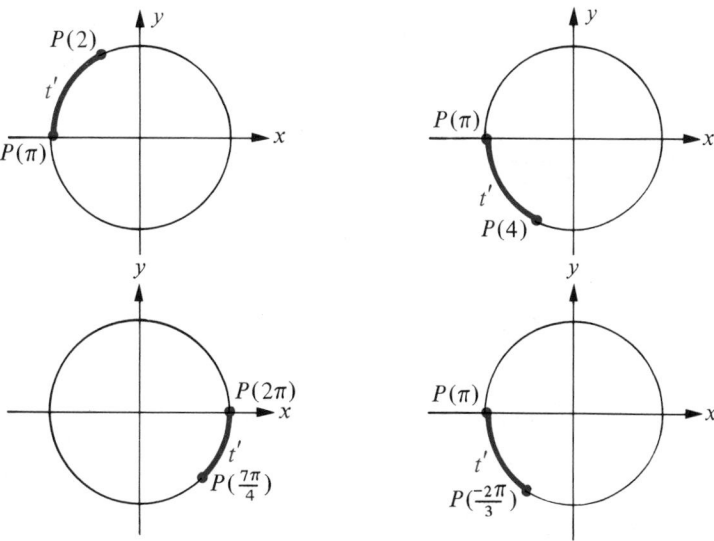

Figure 5.21   ■

Example 2   Sketch the reference angle θ' for each of the following and express the measure of θ' in terms of both radians and degrees.

(a) $\theta = 5\pi/6$ (b) $\theta = 315°$ (c) $\theta = -240°$

Solutions   Each angle θ and its reference angle θ' are sketched in Figure 5.22. Evidently

(a) $\theta' = \pi/6 = 30°$

(b) $\theta' = 45° = \pi/4$

(c) $\theta' = 60° = \pi/3$

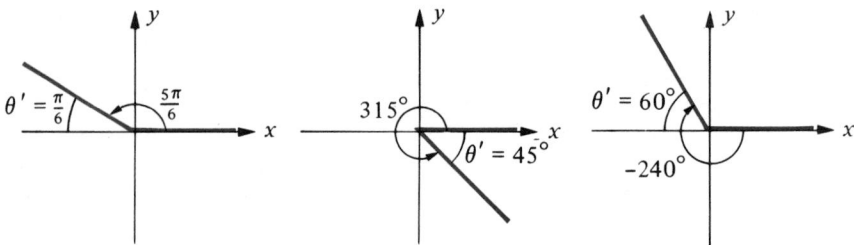

Figure 5.22   ■

Let us now demonstrate how reference numbers or reference angles can be used to find values of the trigonometric functions. If the point $P(t)$ is not on a coordinate axis and t' is the reference number for t, then $0 < t' < \pi/2$. Suppose that $P(t)$ has rectangular coordinates (x, y), so that $P(t) = P(x, y)$. Consider the point $A(1, 0)$ and let $P'(x', y')$ be the point on U in quadrant I such that $\overset{\frown}{AP'} = t'$. Illustrations in which $P(x, y)$ lies in quadrants II, III, or IV are given in Figure 5.23.

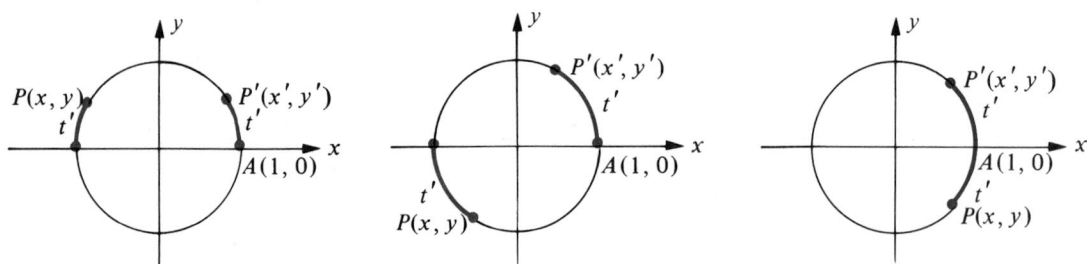

Figure 5.23

We see that in all cases

$$x' = |x| \quad \text{and} \quad y' = |y|$$

and hence we have $|\cos t| = |x| = x' = \cos t'$

$$|\sin t| = |y| = y' = \sin t'.$$

It is easy to show that the absolute value of *every* trigonometric function at *t* is the same as its value at *t'*. For example,

$$|\tan t| = \left|\frac{y}{x}\right| = \frac{|y|}{|x|} = \frac{y'}{x'} = \tan t'.$$

In terms of angles, if θ' is the reference angle of θ, then

$$|\sin \theta| = \sin \theta', \quad |\cos \theta| = \cos \theta'$$

and similarly for the other trigonometric functions. We may, therefore, state the following rule.

Rules for Finding Values of the Trigonometric Functions

> (i) To find the value of a trigonometric function at a number *t*, determine its value for the reference number *t'* associated with *t* and prefix the appropriate sign.
>
> (ii) To find the value of a trigonometric function at an angle θ, determine its value for the reference angle θ' associated with θ and prefix the appropriate sign.

The "appropriate sign" can be determined from the table of signs in Section 5.2.

Example 3 Find $\sin(7\pi/4)$ and $\sec(-7\pi/6)$.

Solution   As indicated in Figure 5.24, the reference numbers of $7\pi/4$ and $-7\pi/6$ are $\pi/4$ and $\pi/6$, respectively. Hence, by the preceding rules and Table A of Section 5.5,

$$\sin \frac{7\pi}{4} = -\sin \frac{\pi}{4} = -\frac{\sqrt{2}}{2}$$

$$\sec \frac{-7\pi}{6} = -\sec \frac{\pi}{6} = -\frac{2\sqrt{3}}{3}.$$

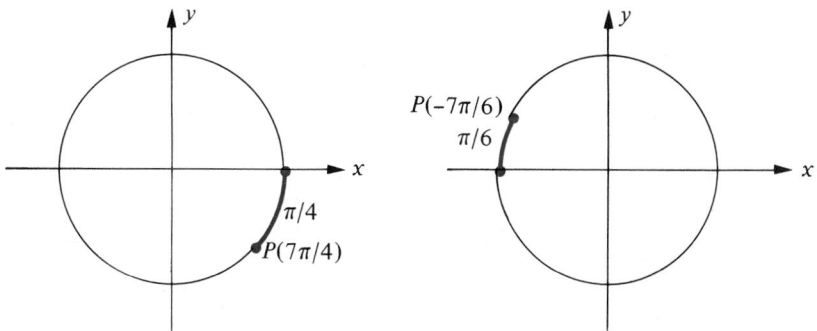

Figure 5.24   ■

Example 4 Find each of the following:

(a) $\sin 150°$;   (b) $\tan 315°$;   (c) $\sec(-240°)$.

Solutions   The angles and their reference angles are shown in Figure 5.25. Using the rules for finding functional values and Table A of Section 5.5 gives us

(a) $$\sin 150° = \sin 30° = 1/2$$
(b) $$\tan 315° = -\tan 45° = -1$$
(c) $$\sec(-240°) = -\sec 60° = -2$$

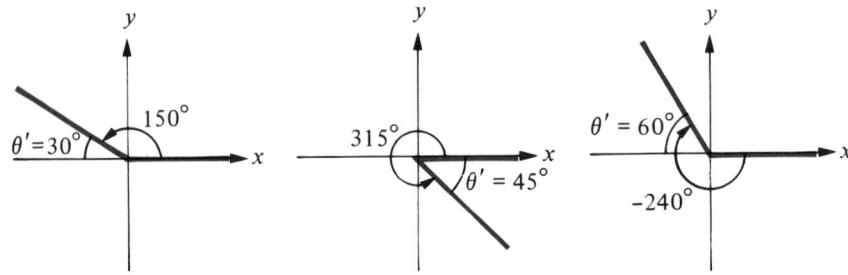

Figure 5.25   ■

By employing advanced techniques it is possible to compute, to any degree of accuracy, all the values of the trigonometric functions in the t-interval $[0, \pi/2]$ or, equivalently, in the degree interval $[0°, 90°]$. Table 3, part of which is reproduced here, gives approximations to such values.

Part of Table 3

t	t degrees	$\sin t$	$\cos t$	$\tan t$	$\cot t$	$\sec t$	$\csc t$		
.4887	**28° 00′**	.4695	.8829	.5317	1.881	1.133	2.130	**62° 00′**	1.0821
.4916	10	.4720	.8816	.5354	1.868	1.134	2.118	50	1.0792
.4945	20	.4746	.8802	.5392	1.855	1.136	2.107	40	1.0763
.4974	30	.4772	.8788	.5430	1.842	1.138	2.096	30	1.0734
.5003	40	.4797	.8774	.5467	1.829	1.140	2.085	20	1.0705
.5032	50	.4823	.8760	.5505	1.816	1.142	2.074	10	1.0676
.5061	**29° 00′**	.4848	.8746	.5543	1.804	1.143	2.063	**61° 00′**	1.0647
.5091	10	.4874	.8732	.5581	1.792	1.145	2.052	50	1.0617
.5120	20	.4899	.8718	.5619	1.780	1.147	2.041	40	1.0588
.5149	30	.4924	.8704	.5658	1.767	1.149	2.031	30	1.0559
.5178	40	.4950	.8689	.5696	1.756	1.151	2.020	20	1.0530
.5207	50	.4975	.8675	.5735	1.744	1.153	2.010	10	1.0501
.5236	**30° 00′**	.5000	.8660	.5774	1.732	1.155	2.000	**60° 00′**	1.0472
⋮	⋮	⋮	⋮	⋮	⋮	⋮	⋮	⋮	⋮
		$\cos t$	$\sin t$	$\cot t$	$\tan t$	$\csc t$	$\sec t$	t degrees	t

The domain of t in Table 3 is from 0 to 1.5708. The latter number is a four-decimal-place approximation to $\pi/2$. Table 3 is arranged so that functional values which correspond to angles in degree measure may be found directly. Angular measures are given in $10'$ intervals from $0°$ to $90°$. The inclusion of the degree columns is the reason that t varies at intervals of approximately 0.0029, since $10' \approx 0.0029$ radians.

To find values of trigonometric functions at a real number t, whenever $0 \le t \le 0.7854 \approx \pi/4$ or $0° \le \theta \le 45°$, the labels at the *top* of the columns in Table 3 should be used. For example,

$$\sin (0.5003) \approx 0.4797 \qquad \tan 28°30' \approx 0.5430$$
$$\cos (0.4945) \approx 0.8802 \qquad \sec 29°00' \approx 1.143$$
$$\cot (0.5120) \approx 1.780 \qquad \csc 29°40' \approx 2.020$$

However, if $0.7854 \le t \le 1.5708$, or if $45° \le \theta \le 90°$, then the labels at the *bottom* of the columns should be employed. For example,

$$\sin (1.0705) \approx 0.8774 \qquad \csc 62°00' \approx 1.133$$
$$\cos (1.0530) \approx 0.4950 \qquad \cot 61°10' \approx 0.5505$$
$$\tan (1.0821) \approx 1.881 \qquad \sec 60°30' \approx 2.031.$$

The reason that the table can be so arranged follows from the fact, to be proved later, that

$$\sin t = \cos \left(\frac{\pi}{2} - t\right), \quad \cot t = \tan \left(\frac{\pi}{2} - t\right), \quad \csc t = \sec \left(\frac{\pi}{2} - t\right)$$

or equivalently

$$\sin \theta = \cos (90° - \theta), \quad \cot \theta = \tan (90° - \theta), \quad \csc \theta = \sec (90° - \theta).$$

In particular,

$$\sin 29° = \cos (90° - 29°) = \cos 61°$$
$$\cot 28°20' = \tan (90° - 28°20') = \tan 61°40'$$

as shown in the table.

If it is necessary to find functional values when t lies *between* numbers given in the table, the method of linear interpolation used for logarithms may be employed. Similarly, given a value such as $\sin t = 0.6371$, we may refer to the body of Table 3 and use linear interpolation, if necessary, to obtain an approximation to t. If t is measured in degrees we round off to the nearest minute.

Example 5 Find approximations for the following:

    (a) $\tan (2.3824)$;   (b) $\cos (0.7)$.

Solutions

    (a) Since $P(2.3824)$ is in quadrant II, the reference number t' is $\pi - 2.3824$ or

$$t' \approx 3.1416 - 2.3824 = 0.7592.$$

Using the rules for finding functional values stated in this section, together with Table 3, we obtain

$$\tan (2.3824) \approx - \tan (0.7592) \approx -0.9490.$$

(b) To find $\cos (0.7)$ we locate the number 0.7000 between successive values of t in Table 3 and interpolate as follows:

$$0.0029 \left\{ 0.0019 \left\{ \begin{array}{l} \cos (0.6981) \approx 0.7660 \\ \cos (0.7000) \approx \quad ? \\ \cos (0.7010) \approx 0.7642 \end{array} \right\} d \right\} 0.0018$$

$$\frac{0.0019}{0.0029} = \frac{d}{0.0018}$$

or

$$d = \frac{19}{29} (0.0018) \approx 0.0012.$$

Hence

$$\cos (0.7000) \approx 0.7660 - 0.0012 = 0.7648.$$

Note that since the cosine function is decreasing in the given interval we must *subtract* d from 0.7660. ∎

Example 6 Approximate the smallest positive real number t such that $\sin t = 0.6635$.

Solution We locate 0.6635 between successive entries in the sine column of Table 3 and interpolate as follows:

$$0.0029 \left\{ d \left\{ \begin{array}{l} \sin (0.7243) \approx 0.6626 \\ \sin t \quad\quad = 0.6635 \\ \sin (0.7272) = 0.6648 \end{array} \right\} 0.0009 \right\} 0.0022$$

$$\frac{d}{0.0029} = \frac{0.0009}{0.0022}$$

or

$$d = \frac{9}{22} (0.0029) \approx 0.0012.$$

Hence

$$t \approx 0.7243 + 0.0012 = 0.7255.$$ ∎

Example 7 Approximate $\tan 155°44'$.

Solution Since the angle is in quadrant II, the reference angle is $180° = 155°44' = 24°16'$ and hence $\tan 155°44' = - \tan 24°16'$ (Why?). We now consult Table 3 and interpolate as follows:

$$10'\left\{ 6'\left\{ \begin{matrix} \tan 24°10' \approx 0.4487 \\ \tan 24°16' = \quad ? \end{matrix} \right\} d \atop \tan 24°20' \approx 0.4522 \right\} 0.0035$$

$$\frac{d}{0.0035} = \frac{6}{10} \quad \text{or} \quad d = \frac{6}{10}(0.0035) \approx 0.0021$$

$$\tan 24°16' \approx 0.4487 + 0.0021 = 0.4508$$

$$\tan 155°44' \approx -0.4508. \qquad \blacksquare$$

Example 8 Approximate $\cos(-117°47')$.

Solution The angle is in quadrant III. The reader should check that the reference angle is $62°13'$. Consequently, $\cos(-117°47') = -\cos 62°13'$. Interpolating, we have

$$10'\left\{ 3'\left\{ \begin{matrix} \cos 62°10' \approx 0.4669 \\ \cos 62°13' = \quad ? \end{matrix} \right\} d \atop \cos 62°20' \approx 0.4643 \right\} 0.0026$$

$$\frac{d}{0.0026} = \frac{3}{10} \quad \text{or} \quad d \approx 0.0008.$$

Since the cosine function is decreasing, we have

$$\cos 62°13' \approx 0.4669 - 0.0008 = 0.4661.$$

Hence, $\cos(-117°47') \approx -0.4661.$ $\qquad \blacksquare$

Example 9 If $\sin \theta = -0.7963$, approximate the degree of all angles θ which are in the interval $[0°, 360°]$.

Solution Let θ' be the reference angle, so that $\sin \theta' = 0.7963$. Interpolating in Table 3,

$$10'\left\{ d\left\{ \begin{matrix} \sin 52°40' \approx 0.7951 \\ \sin \theta' = 0.7963 \end{matrix} \right\} 0.0012 \atop \sin 52°50' \approx 0.7969 \right\} 0.0018$$

$$\frac{d}{10} = \frac{0.0012}{0.0018} \quad \text{or} \quad d \approx 7'$$

$$\theta' \approx 52°47'.$$

Since $\sin \theta$ is negative, θ lies in quadrant III or IV. Using the reference angle $52°47'$ we have

$$\theta \approx 180° + 52°47' = 232°47'$$
$$\theta \approx 360° - 52°47' = 307°13'. \qquad \blacksquare$$

As a final remark, Table 4 can be used to find functional values of t if t is a two-decimal-place approximation of a real number or the radian measure of an angle, and $0 \le t \le 1.57$. For example, referring to the table we see that

$$\sin(0.95) \approx 0.8134$$
$$\cos(1.48) \approx 0.0907$$
$$\cot(0.50) \approx 1.830.$$

Interpolation can be used if t is given to three decimal places. We shall not include exercises which require the use of Table 4, but instead leave it to the discretion of the teacher and student as to what use should be made of that table.

EXERCISES 5.6

In Exercises 1–6 find the reference number t' if t has the given value.

1 (a) $3\pi/4$ (b) $4\pi/3$ (c) $-\pi/6$

2 (a) $5\pi/6$ (b) $2\pi/3$ (c) $-3\pi/4$

3 (a) $9\pi/4$ (b) $7\pi/6$ (c) $-2\pi/3$

4 (a) $8\pi/3$ (b) $7\pi/4$ (c) $-7\pi/6$

5 (a) 1.5 (b) 5

6 (a) 3.5 (b) -4

In Exercises 7–10 find the reference angle θ' if θ has the given measure.

7 (a) $240°$ (b) $340°$ (c) $-110°$

8 (a) $165°$ (b) $275°$ (c) $-202°$

9 (a) $130°40'$ (b) $-405°$ (c) $-260°35'$

10 (a) $335°20'$ (b) $-620°$ (c) $-185°40'$

In Exercises 11–16 find the exact values without the use of tables or calculators.

11 (a) $\sin 2\pi/3$ (b) $\sin 4\pi/3$

12 (a) $\cos 5\pi/6$ (b) $\cos 7\pi/6$

13 (a) $\tan(-5\pi/4)$ (b) $\cot 315°$

14 (a) $\sin 210°$ (b) $\csc(-150°)$

15 (a) $\csc 300°$ (b) $\sec(-120°)$

16 (a) $\tan(-135°)$ (b) $\sec 225°$

In Exercises 17–30 use Table 3 to approximate the given numbers.

17 $\sin(0.2676)$ **18** $\cos(0.3258)$

19 $\tan(0.9948)$ **20** $\sec(0.8988)$

21 $\csc(0.7738)$ **22** $\cot(0.7883)$

23 $\cos 38°30'$ **24** $\sin 73°20'$

25 $\cot 9°10'$ **26** $\csc 43°40'$

27 $\sec 168°50'$ **28** $\tan 207°10'$

29 $\sin 342°20'$ **30** $\cos 432°40'$

In Exercises 31–42 use interpolation in Table 3 to approximate the given numbers.

31 $\sin(0.46)$ **32** $\cos(0.82)$

33 $\tan 3$ **34** $\cot 6$

35 $\sec(1/4)$ **36** $\csc(1.54)$

37 $\cos 37°43'$ **38** $\sin 22°34'$

39 $\cot 62°27'$ **40** $\tan 57°16'$

41 $\csc 16°55'$ **42** $\sec 9°12'$

In Exercises 43–48 use interpolation to approximate the smallest positive number t for which the equality is true.

43 $\cos t = 0.8620$ **44** $\sin t = 0.6612$

45 $\tan t = 4.501$	**46** $\sec t = 3.641$
47 $\csc t = 1.436$	**48** $\cot t = 1.165$

In Exercises 49–56 use interpolation in Table 3 to approximate, to the nearest minute, the degree measure of all angles θ which lie in the interval $[0°, 360°]$.

49 $\sin \theta = 0.3672$	**50** $\cos \theta = 0.8426$
51 $\tan \theta = 0.5042$	**52** $\cot \theta = 1.348$
53 $\cos \theta = 0.3465$	**54** $\csc \theta = 1.219$
55 $\sec \theta = 1.385$	**56** $\sin \theta = 0.7534$

5.7 GRAPHS OF THE TRIGONOMETRIC FUNCTIONS

In our previous work with graphs of functions we used the symbols x and y as labels for the coordinate axes. In the present chapter x has been used primarily for the abscissa of a point on the unit circle U, and hence in this section we shall use the symbol t for the horizontal axis. However, we shall continue to use y to denote the vertical axis. In the ty-coordinate system the graph of the sine function is the same as the graph of the equation $y = \sin t$.

It is not difficult to sketch the graphs of the trigonometric functions. For example, since $-1 \leq \sin t \leq 1$ for every real number t, the graph of the sine function lies between the horizontal lines $y = 1$ and $y = -1$. Moreover, the sine function is periodic with period 2π, and therefore it is sufficient to determine the graph for $0 \leq t \leq 2\pi$, since the same pattern is repeated in intervals of length 2π along the t-axis. In Section 5.3 we discussed the behavior of $\sin t$ in the interval $[0, 2\pi]$ by concentrating on the ordinate y of the point $P(t)$ as $P(t)$ traversed the unit circle U once in the counterclockwise direction (see Figure 5.5). The manner in which $\sin t$ varies in $[0, 2\pi]$ may be briefly summarized as follows.

Variation of t	Variation of $\sin t$
0 to $\pi/2$	0 to 1
$\pi/2$ to π	1 to 0
π to $3\pi/2$	0 to -1
$3\pi/2$ to 2π	-1 to 0

Using these facts we conjectured that the graph of the sine function might look like that in Figure 5.6. This guess may be further justified by calculating some intermediate values of $\sin t$, as indicated in the following table.

t	0	$\dfrac{\pi}{4}$	$\dfrac{\pi}{2}$	$\dfrac{3\pi}{4}$	π	$\dfrac{5\pi}{4}$	$\dfrac{3\pi}{2}$	$\dfrac{7\pi}{4}$	2π
$\sin t$	0	$\dfrac{\sqrt{2}}{2}$	1	$\dfrac{\sqrt{2}}{2}$	0	$\dfrac{-\sqrt{2}}{2}$	-1	$\dfrac{-\sqrt{2}}{2}$	0

To obtain a rough sketch we may plot the points $(t, \sin t)$ listed at the bottom of page 215, draw a smooth curve through them, and extend the configuration to the right and left in periodic fashion. This gives us the portion of the graph shown in Figure 5.26. Of course, the graph does not terminate, but continues indefinitely to the right and left. If greater accuracy is desired, additional points could be plotted using, for example, $\sin \pi/6 = 1/2$, $\sin \pi/3 = \sqrt{3}/2 \approx 0.86$, etc. We could also use Table 3 to obtain many additional points on the graph. We refer to the part of the graph corresponding to the interval $[0, 2\pi]$ as a **sine wave**.

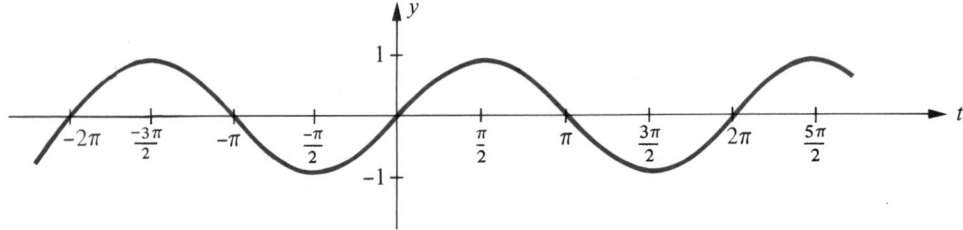

Figure 5.26 $y = \sin t$

The graph of the cosine function can be found in similar fashion, by studying the abscissa x of the point $P(t)$ in Figure 5.5 as $P(t)$ goes around U once in the counterclockwise direction. We may briefly summarize the behavior of $\cos t$ in the interval $[0, 2\pi]$ as follows.

Variation of t	Variation of $\cos t$
0 to $\pi/2$	1 to 0
$\pi/2$ to π	0 to -1
π to $3\pi/2$	-1 to 0
$3\pi/2$ to 2π	0 to 1

Using this information, together with the fact that the cosine function has period 2π, and plotting points as we did for the sine function, leads to the sketch in Figure 5.27. Note that the graph of $y = \cos t$ can be obtained by shifting the graph of $y = \sin t$ to the left a distance $\pi/2$.

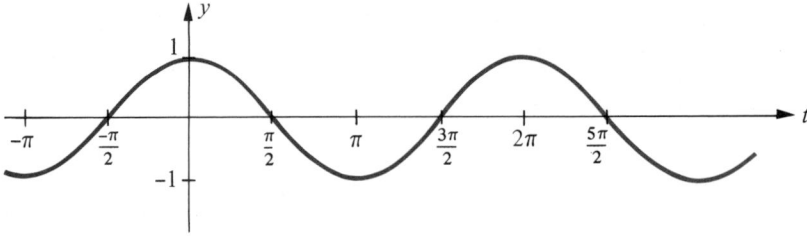

Figure 5.27. $y = \cos t$

Several values of the tangent function are given below.

t	$-\dfrac{\pi}{3}$	$-\dfrac{\pi}{4}$	$-\dfrac{\pi}{6}$	0	$\dfrac{\pi}{6}$	$\dfrac{\pi}{4}$	$\dfrac{\pi}{3}$
$\tan t$	$-\sqrt{3}$	-1	$\dfrac{-\sqrt{3}}{3}$	0	$\dfrac{\sqrt{3}}{3}$	1	$\sqrt{3}$

The corresponding points are plotted in Figure 5.28. The values of $\tan t$ near $t = \pi/2$ demand special consideration. As t increases through positive values toward $\pi/2$, the

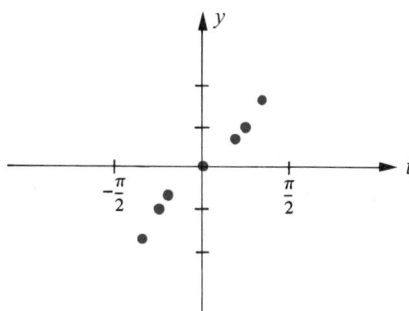

Figure 5.28

point $P(t) = (x, y)$ on the unit circle U that corresponds to t approaches the point $(0, 1)$. Since, by definition, $\tan t = y/x$, it follows that if y approaches 1 and x approaches 0 (through positive values), then $\tan t$ takes on large positive values. This can also be seen by referring to Table 3. First recall that $\pi/2 \approx \frac{1}{2}(3.1416) = 1.5708$. Next, we see from Table 3 that

$$\tan(1.5650) \approx 171.9$$
$$\tan(1.5679) \approx 343.8.$$

The following additional values were obtained using a hand-held calculator.

$$\tan(1.5700) \approx 1{,}255.8$$
$$\tan(1.5703) \approx 2{,}014.8$$
$$\tan(1.5706) \approx 5{,}093.5$$
$$\tan(1.5707) \approx 10{,}381.3$$
$$\tan(1.57079) \approx 158{,}057.9$$

Notice how rapidly $\tan t$ increases as t gets close to $\pi/2$. Indeed, $\tan t$ can be made arbitrarily large by choosing t sufficiently close to $\pi/2$. Using the terminology introduced in Section 3.8, we say that *tan t increases without bound as t approaches $\pi/2$ through values less than $\pi/2$*, written

$$\tan t \to \infty \quad \text{as} \quad t \to \frac{\pi^-}{2}.$$

Similarly, if t approaches $-\pi/2$ through values greater than $-\pi/2$, then tan t *decreases without bound*, written

$$\tan t \to -\infty \quad \text{as} \quad t \to -\frac{\pi}{2}^{+}.$$

This behavior in the open interval $(-\pi/2, \pi/2)$ is illustrated in Figure 5.29. The lines $t = \pi/2$ and $t = -\pi/2$ are vertical asymptotes for the graph. It is not difficult to show that the same pattern is repeated in the open intervals $(\pi/2, 3\pi/2)$, $(3\pi/2, 5\pi/2)$, and similar intervals of length π, as shown in Figure 5.29. Thus *the tangent function has period π*.

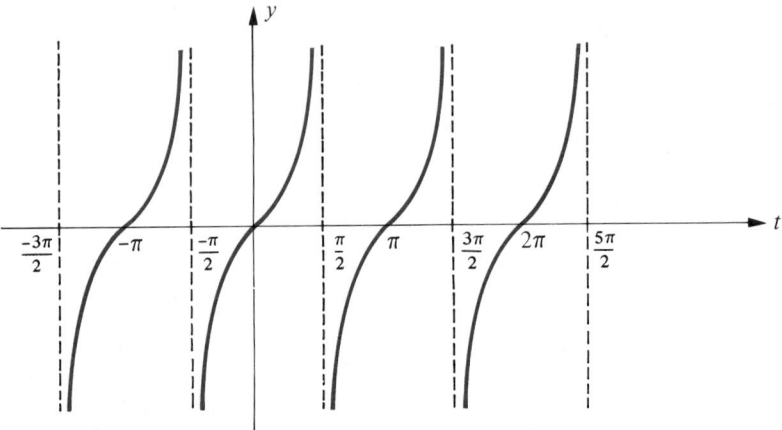

Figure 5.29. $y = \tan t$

The graphs of the sine and tangent functions are symmetric with respect to the origin, whereas the graph of the cosine function is symmetric with respect to the y-axis. These facts are consequences of the following rules, each of which holds for every real number t in the domain of the indicated function.

$$\begin{array}{l} \sin(-t) = -\sin t \\ \cos(-t) = \cos t \\ \tan(-t) = -\tan t \end{array}$$

To establish these formulas, let us recall from Section 5.1 that if t ranges from 0 to 2π, the point $P(t)$ traces the unit circle U once in the counterclockwise direction, whereas $P(-t)$ traces U once in the clockwise direction. Moreover, as illustrated in Figure 5.30, if $P(t)$ has coordinates (x, y) then $P(-t)$ has coordinates $(x, -y)$.

Applying the definition of the trigonometric functions (see Section 5.2) gives us

$$\sin(-t) = -y = -\sin t$$
$$\cos(-t) = \quad x = \cos t$$
$$\tan(-t) = \frac{-y}{x} = -\frac{y}{x} = -\tan t$$

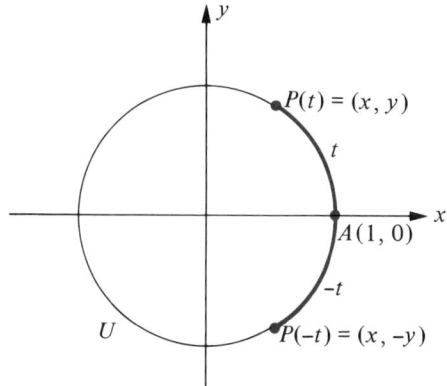

Figure 5.30

which is what we wished to prove. According to the tests for symmetry stated in Section 1.5, these identities imply that the graphs of the sine and tangent functions are symmetric with respect to the origin, whereas that of the cosine function is symmetric with respect to the y-axis.

The graphs of the remaining three trigonometric functions can now be easily obtained. For example, since $\csc t = 1/\sin t$, we may find an ordinate of a point on the graph of the cosecant function by taking the reciprocal of the corresponding ordinate of the sine graph. This is possible except for $t = n\pi$, where n is an integer, for in this case $\sin t = 0$ and hence $1/\sin t$ is undefined. As an aid to sketching the graph of the cosecant function it is convenient to sketch the graph of the sine function with dashes (see Figure 5.31) and then take reciprocals of ordinates to obtain points on the cosecant graph. Notice the manner in which the cosecant function increases or decreases without bound as t approaches $n\pi$, where n is an integer. The graph has vertical asymptotes as indicated in the figure.

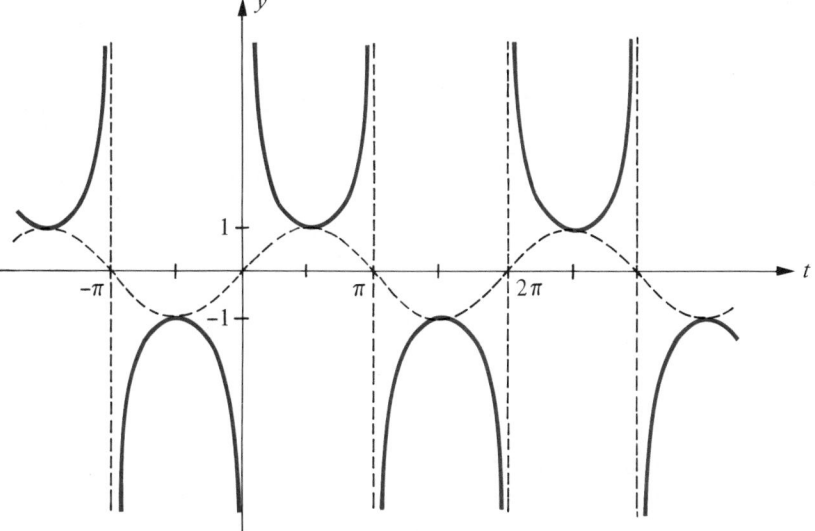

Figure 5.31. $y = \csc t$

Since $\sec t = 1/\cos t$ and $\cot t = 1/\tan t$, the graphs of the secant and cotangent functions may be obtained in similar fashion (see Figures 5.32 and 5.33). Their verifications are left as exercises.

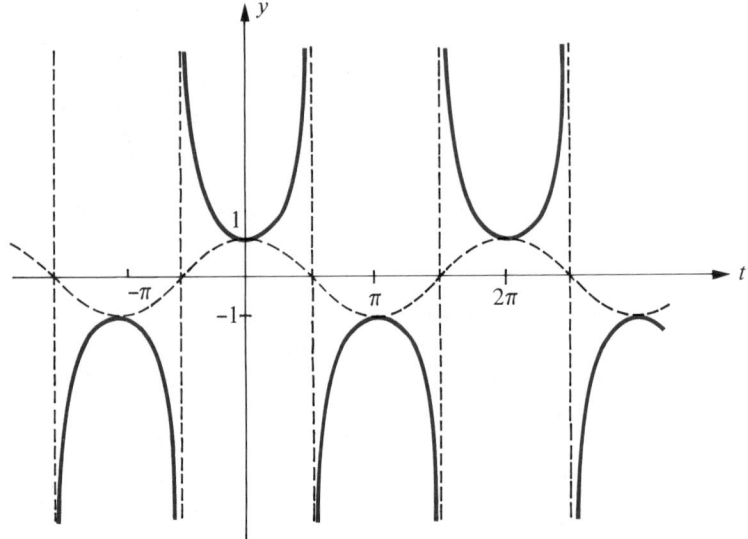

Figure 5.32. $y = \sec t$

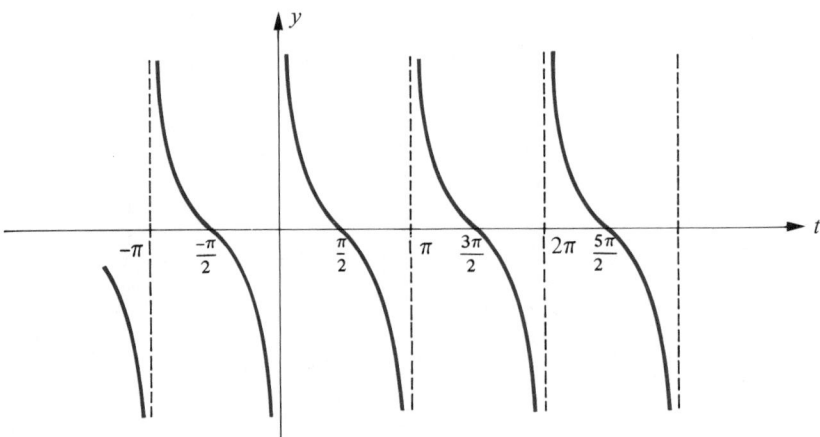

Figure 5.33. $y = \cot t$

Example   Sketch the graph of the function $f(t) = 2 \sin t$.

Solution   Although the graph could be obtained by plotting points, note that for each abscissa t_1 the ordinate $f(t_1)$ is always twice that of the corresponding ordinate on the sine graph. A simple graphical technique is to sketch the graph of

$y = \sin t$ with dashes and then double each ordinate to find points on the graph of $y = 2\sin t$ (see Figure 5.34).

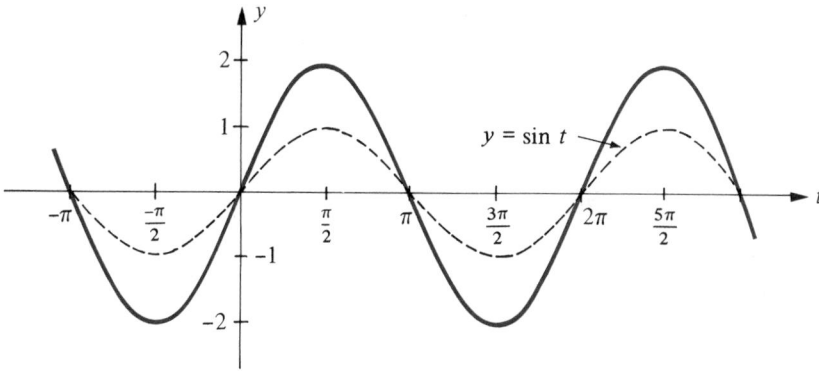

Figure 5.34. $y = 2\sin t$ ■

Graphs of the type given in the last example will be discussed more thoroughly in the next section.

EXERCISES 5.7

1 (a) Sketch the graph of the cosine function after plotting a sufficient number of points.

(b) Sketch the graph of the secant function by taking reciprocals in part (a).

(c) Describe the intervals between -2π and 2π in which the secant function is increasing.

(d) Describe the intervals between -2π and 2π in which the secant function is decreasing.

(e) With the aid of Table 3 or a hand-held calculator, discuss the behavior of $\sec t$ as t approaches $\pi/2$ through values less than $\pi/2$ and through values greater than $\pi/2$.

(f) Discuss the symmetry of the graph of the secant function.

2 (a) Sketch the graph of the cotangent function after plotting a sufficient number of points.

(b) In what intervals is the cotangent function increasing?

(c) In what intervals is the cotangent function decreasing?

(d) With the aid of Table 3 or a hand-held calculator, discuss the behavior of $\cot t$ as t approaches 0 through values greater than 0 and through values less than 0.

(e) Discuss the symmetry of the graph of the cotangent function.

3 Establish the following identities.

(a) $\csc(-t) = -\csc t$

(b) $\sec(-t) = \sec t$

(c) $\cot(-t) = -\cot t$

4 Prove that the graph of the cosecant function is symmetric with respect to the origin.

5 Practice sketching graphs of the sine, cosine, and tangent functions, taking different units of length on the horizontal and vertical axes. Continue this practice until you reach the stage at which if you were awakened from a sound sleep in the middle of the night and asked to sketch one of these graphs, you could do so in less than thirty seconds.

6 Repeat Exercise 5 for the cosecant, secant, and cotangent functions.

In Exercises 7–16 use the method illustrated in the Example of this section to sketch the graph of f.

7 $f(t) = 4 \sin t$

8 $f(t) = 3 \sin t$

9 $f(t) = \frac{1}{2} \sin t$

10 $f(t) = \frac{1}{4} \sin t$

11 $f(t) = 2 \cos t$

12 $f(t) = 4 \cos t$

13 $f(t) = \frac{1}{3} \cos t$

14 $f(t) = \frac{1}{2} \cos t$

15 $f(t) = -\sin t$

16 $f(t) = -\cos t$

CALCULATOR EXERCISES 5.7

If f is a function, the notation

$$f(t) \to L \quad \text{as} \quad t \to a^+$$

is sometimes used to express the fact that $f(t)$ gets very close to L as t gets close to a (through values greater than a). Use a hand-held calculator to support Exercises 1–6 by substituting the following values for t: $t = 0.1, t = 0.01, t = 0.001, t = 0.0001$.

1 $\dfrac{\sin t}{t} \to 1 \quad \text{as} \quad t \to 0^+$

2 $\dfrac{\tan t}{t} \to 1 \quad \text{as} \quad t \to 0^+$

3 $\dfrac{1 - \cos t}{t} \to 0 \quad \text{as} \quad t \to 0^+$

4 $\dfrac{\sin t}{1 + \cos t} \to 0 \quad \text{as} \quad t \to 0^+$

5 $t \cot t \to 1 \quad \text{as} \quad t \to 0^+$

6 $\dfrac{t + \tan t}{\sin t} \to 2 \quad \text{as} \quad t \to 0^+$

5.8 TRIGONOMETRIC GRAPHS

In this section we shall consider graphs of equations of the form

$$y = a \sin (bx + c)$$

where, a, b, c are real numbers. We shall also discuss graphs of similar equations that involve different trigonometric functions. Instead of using a ty-coordinate system as in Section 5.7, we shall now use the conventional xy-coordinate system. In this situation the letter x is used in place of t and hence is not to be regarded as the x used in the definition of the trigonometric functions. To sketch a graph we could begin by plotting many points; however, it is generally easier to use information about the graphs of the trigonometric functions discussed in the preceding section. Let us consider the special case in which $c = 0, b = 1$, and $a > 0$. Thus we wish to sketch the graph of the equation

$$y = a \sin x.$$

We may find the ordinate of a point on the graph by multiplying the corresponding ordinate on the graph of $y = \sin x$ by a. Thus if $y = 2 \sin x$, we multiply by 2; if $y = \frac{1}{2} \sin x$, we multiply by $\frac{1}{2}$; and so on. The graph of $y = 2 \sin x$ is sketched in Figure 5.34. The graph of $y = \frac{1}{2} \sin x$ is sketched in Figure 5.35 where for comparison we have indicated the graph of $y = \sin x$ with dashes.

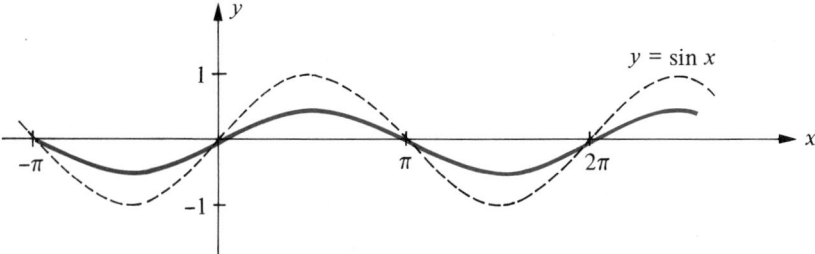

Figure 5.35. $y = \frac{1}{2}\sin x$

Example 1 Sketch the graph of the equation $y = 3\cos x$.

Solution The graph of the given equation may be obtained from the graph of $y = \cos x$ by multiplying ordinates of points by 3. We first sketch $y = \cos x$ with dashes and then we triple the ordinates. This gives us the sketch in Figure 5.36.

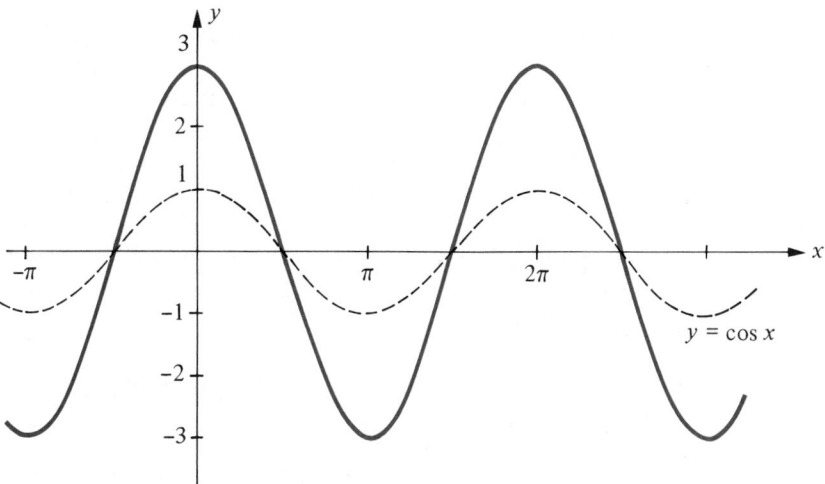

Figure 5.36. $y = 3\cos x$ ■

If $a < 0$, then the ordinates of points on the graph of $y = a\sin x$ are negatives of the corresponding ordinates of points on the graph of $y = |a|\sin x$, as illustrated in the next example.

Example 2 Sketch the graph of $y = -2\sin x$.

Solution As a guide we first sketch the graph of $y = \sin x$ with dashes and then we multiply each ordinate by -2. This gives us the sketch shown in Figure 5.37. The sketch may also be obtained by multiplying ordinates of the graph of $y = 2\sin x$ (see Figure 5.34) by -1. We sometimes refer to the graph of $y = -2\sin x$ as a *reflection through the x-axis* of the graph of $y = 2\sin x$.

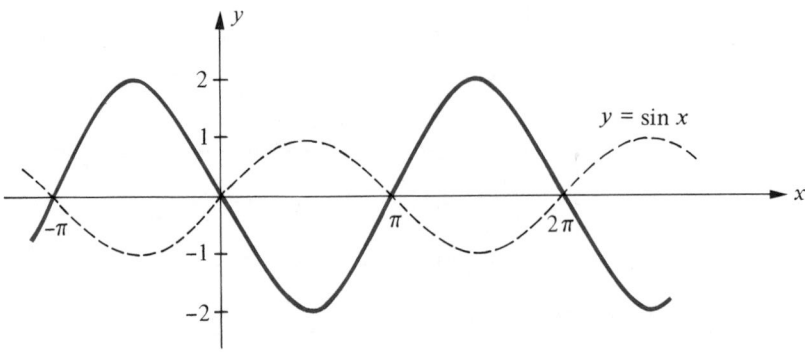

Figure 5.37. $y = -2 \sin x$

If $f(x) = a \sin (bx + c)$, then the **amplitude** of f (or of its graph) is defined as the largest ordinate of points on the graph. If $a > 0$, the largest ordinate occurs if $\sin (bx + c) = 1$, whereas if $a < 0$, we must take $\sin (bx + c) = -1$. In either case the amplitude is $|a|$. To obtain the amplitude from the graph of f we merely take the ordinate of the highest point on the graph. For example, if $f(x) = \frac{1}{2} \sin x$, then from Figure 5.34 we see that the amplitude is $\frac{1}{2}$. If $f(x) = -2 \sin x$, then from Figure 5.36 we see that the amplitude is 2. The same is true if f is given by $f(x) = a \cos (bx + c)$.

The next theorem provides some information about periods.

Theorem

If $f(x) = \sin bx$, where $b \neq 0$, then the period of f is $2\pi/|b|$.

Proof

If $b > 0$, then as bx varies from 0 to 2π, we obtain exactly one sine wave for the graph of f. The conclusion of the theorem follows from the fact that bx ranges from 0 to 2π if and only if x ranges from 0 to $2\pi/b$. A similar proof may be given if $b < 0$.

Example 3 Find the period and sketch the graph of f if

(a) $f(x) = \sin 2x$;   (b) $f(x) = \sin \frac{1}{2} x$.

Solutions   Both functions have the form given in the preceding theorem. Hence in part (a) the period is $2\pi/2 = \pi$, which means that there is exactly one sine wave of amplitude 1 corresponding to the interval $[0, \pi]$. The graph is sketched in Figure 5.38, where for convenience we have used different scales on the x- and y-axes. For part (b) the period is $2\pi/(1/2) = 4\pi$ and hence there is one sine wave of amplitude 1 corresponding to the interval $[0, 4\pi]$, as illustrated in Figure 5.39.

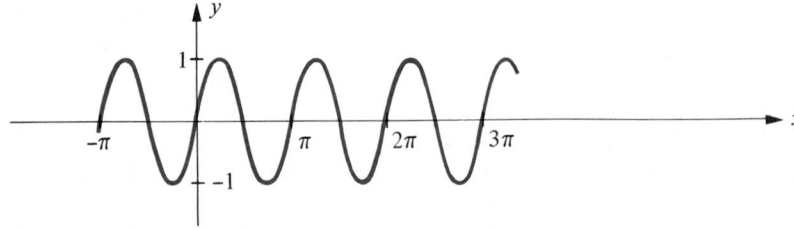

Figure 5.38. $y = \sin 2x$

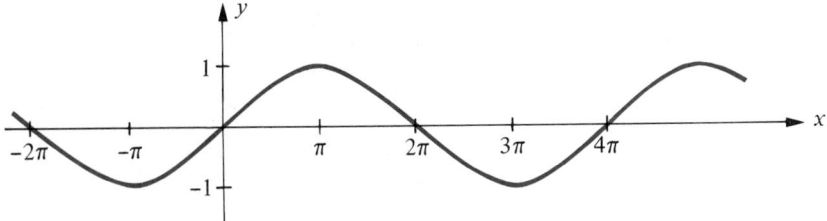

Figure 5.39. $y = \sin \frac{1}{2}x$ ∎

Given $f(x) = \sin bx$, then, roughly speaking, if b is large, $2\pi/b$ is small and the sine waves are close together. As a matter of fact, there are b sine waves in an interval of 2π units. However, if b is small, then $2\pi/b$ is large and the waves are shallow. For example, if $y = \sin (1/10)x$, then $1/10$ of a sine wave occurs in the interval $[0, 2\pi]$, and an interval 20π units long is required for one complete wave.

If $b < 0$, we can use the fact that $\sin (-t) = -\sin t$ to obtain the graph. To illustrate, the graph of $y = \sin (-2x)$ is the same as the graph of $y = -\sin 2x$.

By combining the above remarks we can arrive at a technique for sketching the graph of a function f defined by $f(x) = a \sin bx$. The graph has the basic sine wave pattern. However, the amplitude is $|a|$ and the period is $2\pi/|b|$. If $a < 0$ or $b < 0$, we make adjustments on the signs of ordinates, as discussed earlier.

Example 4   Sketch the graph of f if $f(x) = 3 \sin 2x$.

Solution   From the preceding discussion we see that the amplitude of f is 3 and the period is $2\pi/2 = \pi$. The graph is readily obtained by first sketching the graph of $y = \sin 2x$ with dashes and then multiplying the ordinates of each point by 3. This leads to the sketch in Figure 5.40.

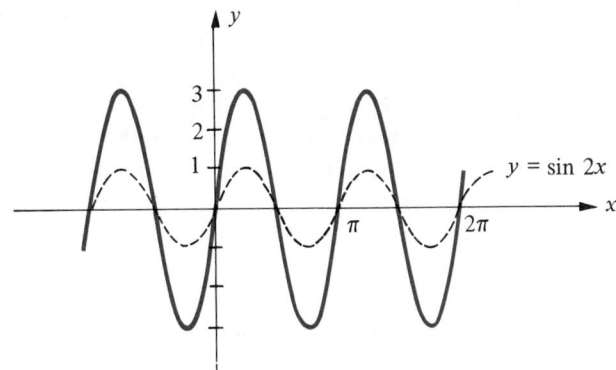

Figure 5.40. $y = 3 \sin 2x$ ∎

Example 5   Sketch the graph of f if $f(x) = 2 \sin (-3x)$.

Solution   Since $f(x) = -2 \sin 3x$, we see that the amplitude is 2 and the period is $2\pi/3$. Thus there is one sine wave in every interval of length $2\pi/3$. The minus sign indicates a reflection through the x-axis. If we consider the interval $[0, 2\pi/3]$ and sketch a sine wave of amplitude 2 (reflected through the x-axis), the shape of the graph is apparent. The configuration given in the interval $[0, 2\pi/3]$ is carried along periodically, as illustrated in Figure 5.41.

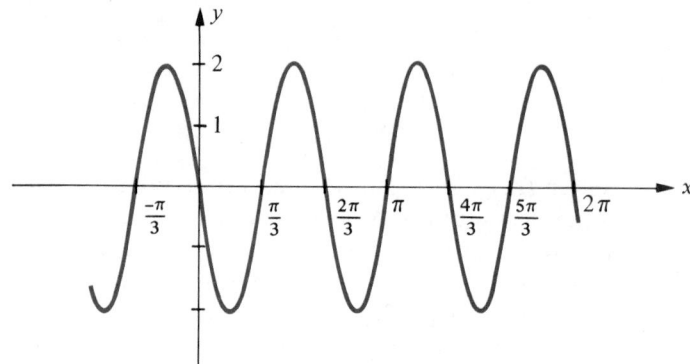

Figure 5.41. $y = -2\sin 3x$ ∎

Similar discussions can be given if f is defined by $f(x) = a\cos bx$ or by $f(x) = a\tan bx$. In the latter case the period is $\pi/|b|$ since the tangent function has period π. There is no largest ordinate for points on the graph of the tangent function and hence we do not refer to its amplitude; however, we may still use the process of multiplying tangent ordinates by a in order to obtain points on the graph of $y = a\tan bx$.

Let us conclude this section by considering the general situation

$$f(x) = a\sin(bx + c).$$

We have already observed that the amplitude is $|a|$. We also know that one sine wave for the graph of $y = \sin(bx + c)$ is obtained if $bx + c$ varies from 0 to 2π, that is, if bx ranges from $-c$ to $2\pi - c$. In turn, the latter variation is obtained by letting x range from $-c/b$ to $(2\pi - c)/b$. If $-c/b > 0$, this amounts graphically to *shifting* the graph of $y = a\sin bx$ to the right $-c/b$ units. If $-c/b < 0$, the shift is to the left. The number $-c/b$ is sometimes called the **phase shift** associated with the function. Similar remarks can be made for the other functions. It is unnecessary to remember a general formula for finding the phase shift. In any specific problem the interval which contains one complete sine wave can be found by solving the two equations

$$bx + c = 0 \quad \text{and} \quad bx + c = 2\pi$$

for x, as illustrated in the next example.

Example 6 Sketch the graph of $f(x) = 3\sin\left(2x - \frac{\pi}{2}\right)$.

Solution The equation is of the form discussed in this section, with $a = 3$, $b = 2$, and $c = -\pi/2$. It follows from our discussion that the graph has the sine wave pattern with amplitude 3 and period $2\pi/2 = \pi$. In order to obtain an interval containing exactly one sine wave we let $2x - (\pi/2)$ range from 0 to 2π. The endpoints of the interval can be found by solving the two equations

$$2x - \frac{\pi}{2} = 0 \quad \text{and} \quad 2x - \frac{\pi}{2} = 2\pi.$$

This gives us
$$x = \frac{\pi}{4} \quad \text{and} \quad x = \frac{5\pi}{4}.$$

Thus one sine wave of amplitude 3 will occur in the interval $[\pi/4, 5\pi/4]$. Sketching that wave and then repeating it to the right and left gives us the graph in Figure 5.42.

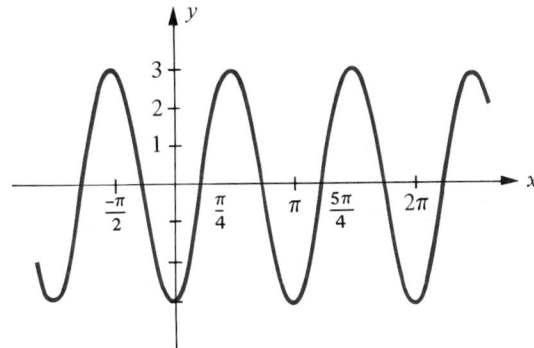

Figure 5.42. $\quad y = 3 \sin \left(2x - \dfrac{\pi}{2}\right)$

EXERCISES 5.8

1 Without plotting many points, sketch the graph and determine the amplitude and period of each function f defined as follows:

(a) $f(x) = 4 \sin x$

(b) $f(x) = \sin 4x$

(c) $f(x) = \frac{1}{4} \sin x$

(d) $f(x) = \sin (x/4)$

(e) $f(x) = 2 \sin (x/4)$

(f) $f(x) = \frac{1}{2} \sin 4x$

(g) $f(x) = -4 \sin x$

(h) $f(x) = \sin (-4x)$

2 Sketch the graphs of the functions involving the cosine which are analogous to those defined in Exercise 1.

3 Without plotting many points, sketch the graph of each function f defined as follows, determining the amplitude and period in each case:

(a) $f(x) = 3 \cos x$

(b) $f(x) = \cos 3x$

(c) $f(x) = \frac{1}{3} \cos x$

(d) $f(x) = \cos (x/3)$

(e) $f(x) = 2 \cos (x/3)$

(f) $f(x) = \frac{1}{3} \cos (2x)$

(g) $f(x) = -3\cos x$

(h) $f(x) = \cos(-3x)$

4 Sketch the graphs of the functions involving the sine which are analogous to those defined in Exercise 3.

Sketch the graphs of the equations in Exercises 5–44.

5 $y = \sin\left(x - \dfrac{\pi}{2}\right)$

6 $y = \cos\left(x + \dfrac{\pi}{4}\right)$

7 $y = 3\sin\left(x - \dfrac{\pi}{2}\right)$

8 $y = 4\cos\left(x + \dfrac{\pi}{4}\right)$

9 $y = \cos\left(x + \dfrac{\pi}{3}\right)$

10 $y = \sin\left(x - \dfrac{\pi}{3}\right)$

11 $y = 4\cos\left(x + \dfrac{\pi}{3}\right)$

12 $y = 5\sin\left(x - \dfrac{\pi}{3}\right)$

13 $y = \sin(3x + \pi)$

14 $y = \cos(3x - \pi)$

15 $y = -2\sin(3x + \pi)$

16 $y = 6\cos(3x - \pi)$

17 $y = 5\sin\left(3x - \dfrac{\pi}{2}\right)$

18 $y = -4\cos\left(2x + \dfrac{\pi}{3}\right)$

19 $y = 6\sin \pi x$

20 $y = 3\cos\frac{1}{2}\pi x$

21 $y = 2\cos\dfrac{\pi}{2}x$

22 $y = 4\sin 3\pi x$

23 $y = \frac{1}{2}\sin 2\pi x$

24 $y = \frac{1}{2}\cos\dfrac{\pi}{2}x$

25 $y = \frac{1}{2}\sec x$

26 $y = \frac{1}{4}\csc x$

27 $y = -2\csc x$

28 $y = -3\sec x$

29 $y = \sec 2x$

30 $y = \csc 3x$

31 $y = \csc\left(x - \dfrac{\pi}{4}\right)$

32 $y = \sec\left(x + \dfrac{\pi}{3}\right)$

33 $y = \tan(x/2)$

34 $y = \cot 2x$

35 $y = \tan\left(x + \dfrac{\pi}{2}\right)$

36 $y = \cot\left(x - \dfrac{\pi}{4}\right)$

37 $y = \frac{1}{2}\cot x$

38 $y = -2\tan x$

39 $y = \tan(-x)$

40 $y = -\cot x$

41 $y = 2\sin\left(x - \dfrac{\pi}{2}\right)$

42 $y = \cos\left(x - \dfrac{\pi}{2}\right)$

43 $y = \sin\left(x + \dfrac{\pi}{2}\right)$

44 $y = \cos\left(x + \dfrac{\pi}{2}\right)$

CALCULATOR EXERCISES 5.8

1–6 The graphs in Examples 1–6 of this section were obtained *qualitatively*, in the sense that we determined the general shape by referring to other graphs and not by plotting points. Verify them *quantitatively*, by calculating the coordinates of many points on the graphs.

5.9   ADDITIONAL GRAPHICAL TECHNIQUES

In mathematical applications it is common to encounter functions that are defined in terms of sums and products of expressions such as

$$f(x) = \sin 2x + \cos x \quad \text{or} \quad f(x) = 2^{-x}\sin x$$

When working with sums, the graphical technique called **addition of ordinates** described on the next page is useful. The method applies not only to trigonometric expressions but to arbitrary expressions as well. If f is a sum of two functions g and h having the same domain X, then

$$f(x) = g(x) + h(x)$$

for every x in X. A sketch of the graph of f may be obtained from the graphs of g and h as follows. We begin by sketching the graphs of the equations $y = g(x)$ and $y = h(x)$ on the same coordinate axes, as illustrated by the dashes in Figure 5.43.

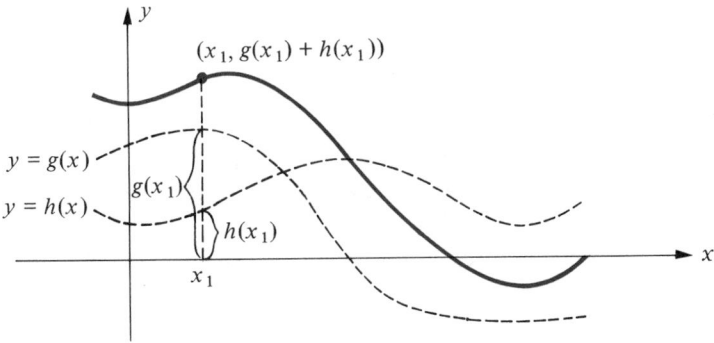

Figure 5.43. $y = g(x) + h(x)$

Since $f(x_1) = g(x_1) + h(x_1)$ for every x_1 in X, the ordinate of the point on the graph of $y = g(x) + h(x)$ with abscissa x_1 is the *sum* of the corresponding ordinates of points on the graphs of g and h. If we draw a vertical line at the point with coordinates $(x_1, 0)$, then the ordinates $g(x_1)$ and $h(x_1)$ may be added geometrically by means of a compass or ruler, as is illustrated in Figure 5.43. If either $g(x_1)$ or $h(x_1)$ is negative, then a *subtraction* of ordinates may be employed. By using this technique a sufficient number of times, we obtain a sketch of the graph of f, as illustrated in Figure 5.43.

Example 1   If $f(x) = \cos x + \sin x$, use the method of addition of ordinates to sketch the graph of f.

Solution   We begin by sketching (with dashes) the graphs of the equations $y = \cos x$ and $y = \sin x$. Next, for various numbers x_1, we add ordinates geometrically. After a sufficient number of ordinates are added and a pattern emerges, we draw a smooth curve through the points, as indicated by the sketch in Figure 5.44. As a check, it would be worthwhile to plot some points on the graph by substituting numbers for x. We shall leave such verifications to the reader. It can be seen from the graph that the function f is periodic with period 2π.

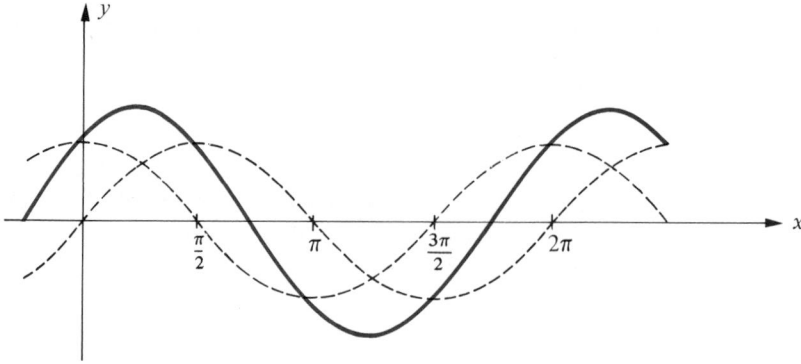

Figure 5.44. $y = \cos x + \sin x$ ■

Example 2   Sketch the graph of the equation $y = \cos x + \sin 2x$.

Solution   We sketch (with dashes) the graphs of the equations $y = \cos x$ and $y = \sin 2x$ on the same coordinate axes and use the method of addition of ordinates. The graph is illustrated in Figure 5.45. Evidently f is periodic with period 2π.

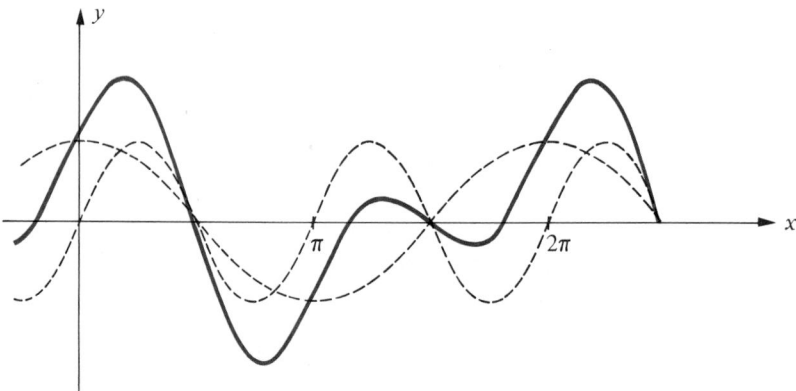

Figure 5.45. $y = \cos x + \sin 2x$ ■

The next example illustrates the fact that if f is the product of two functions g and h, then the graphs of g and h can often be used to supply information about the graph of f.

Example 3   Sketch the graph of f if $f(x) = 2^{-x} \sin x$.

Solution   The function f is the product of two functions g and h, where $g(x) = 2^{-x}$ and $h(x) = \sin x$. Using properties of absolute values, we see that

$|f(x)| = |2^{-x}| |\sin x|$, and since $|\sin x| \le 1$ and $2^{-x} > 0$, it follows that $|f(x)| \le 2^{-x}$ for all x. Consequently,

$$-2^{-x} \le f(x) \le 2^{-x}$$

for all x, which implies that the graph of f lies between the graphs of the equations $y = -2^{-x}$ and $y = 2^{-x}$. The graph of f will coincide with one of the latter graphs when $|\sin x| = 1$, that is, when $x = (\pi/2) + n\pi$, where n is an integer. Since $2^{-x} > 0$, the x-intercepts on the graph of f occur at $\sin x = 0$, that is, at $x = n\pi$. With this information we obtain the sketch shown in Figure 5.46.

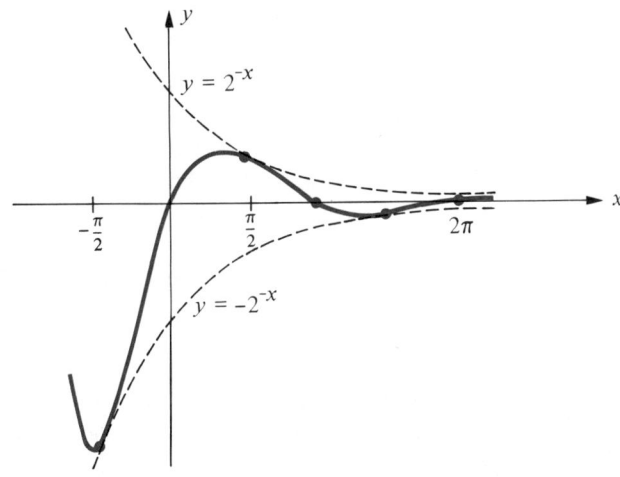

Figure 5.46. $y = 2^{-x} \sin x$ ∎

The graph of $y = 2^{-x} \sin x$ sketched in Figure 5.46 is called a **damped sine wave** and 2^{-x} is called the **damping factor**. By using different damping factors, we may obtain other variations of such compressed or expanded sine waves. The analysis of such graphs is important in physics and engineering.

EXERCISES 5.9

Use addition of ordinates to sketch the graphs of the equations in Exercises 1–20.

1 $y = \cos x + 3 \sin x$

2 $y = \sin x + 3 \cos x$

3 $y = 2 \cos x + 3 \sin x$

4 $y = 2 \sin x + 3 \cos x$

5 $y = \sin x + \cos 2x$

6 $y = 2 \cos x + \sin 2x$

7 $y = \cos x - \sin x$

8 $y = 2 \sin x - \cos x$

9 $y = 2 \cos x - \frac{1}{2} \sin 2x$

10 $y = 2 \cos x + \cos \frac{1}{2} x$

11 $y = 1 + \sin x$

12 $y = 2 + \tan x$

13 $y = \frac{1}{2}x + \sin x$

14 $y = 1 + \cos x$

15 $y = 2^x + \sin x$

16 $y = \csc x - 1$

17 $y = 2 + \sec x$

18 $y = x - \sin x$

19 $y = \cos x - 1$

20 $y = |x + 1| + \cos 2x$

In each of Exercises 21–30 sketch the graph of f.

21 $f(x) = 2^{-x} \cos x$ **22** $f(x) = 2^x \sin x$

23 $f(x) = |x| \sin x$ **24** $f(x) = (1 + x^2) \cos x$

25 $f(x) = \frac{1}{2} x^2 \sin x$ **26** $f(x) = \cot x \tan x$

27 $f(x) = |\sin x|$ **28** $f(x) = |\cos x| + 1$

29 $f(x) = \sin |x|$ **30** $f(x) = \cos |x + 1|$

CALCULATOR EXERCISES 5.9

Verify some of the graphs in this section quantitatively, by calculating the coordinates of many points.

5.10 APPLICATIONS INVOLVING RIGHT TRIANGLES

The early development of trigonometry was concerned with measurements of angles and finding lengths of sides of triangles. Solving problems of that type are no longer the most important applications; however, trigonometric techniques are still used to answer questions about triangles which arise in physical situations. We shall restrict our discussion in this section to right triangles. Triangles which do not contain a right angle will be considered in the next chapter.

In this section and in Chapter Six we shall often use the following notation. The vertices of a triangle will often be denoted by A, B, and C. The angles of the triangle at A, B, and C will be denoted by α, β, and γ, respectively, and the lengths of the sides opposite these angles by a, b, and c, respectively. The triangle itself will often be referred to as *triangle ABC*. To *solve* a triangle means to find all of its parts, that is, the lengths of the three sides and the measures of the three angles. If the triangle is a right triangle and if one of the acute angles and a side are known, or if two sides are given, then the formulas in Section 5.5 which express the trigonometric functions as ratios of sides of a triangle may be used to find the remaining parts.

Example 1   If, in triangle ABC, $\gamma = 90°$, $\alpha = 34°$, and $b = 10.5$, approximate the remaining parts of the triangle.

Solution   Since the sum of the angles is 180° it follows that $\beta = 56°$. Referring to Figure 5.47 and using the fact that the tangent equals opp/adj, we have

$$\tan 34° = \frac{a}{10.5} \quad \text{or} \quad a = (10.5) \tan 34°.$$

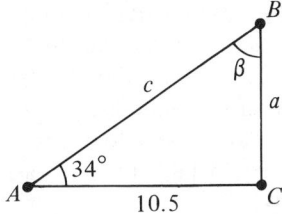

Figure 5.47

Substituting from Table 3 we obtain

$$a \approx (10.5)(0.6745) \approx 7.1.$$

Side c can be found using either the cosine or secant functions. Since the cosine equals adj/hyp, we may write

$$\cos 34° = \frac{10.5}{c} \quad \text{or} \quad c = \frac{10.5}{\cos 34°}.$$

Using Table 3 and dividing we obtain

$$c \approx \frac{10.5}{0.8290} \approx 12.7.$$

The division involved in finding c could have been avoided through the use of the secant, which equals hyp/adj. Thus

$$\sec 34° = \frac{c}{10.5} \quad \text{or} \quad c = (10.5)\sec 34°.$$

Referring to Table 3 and multiplying gives us

$$c \approx (10.5)(1.2062) = 12.6651 \approx 12.7. \qquad \blacksquare$$

As illustrated in Example 1, when working with triangles we shall round off answers. One reason for doing this is that in applications, the lengths of sides of triangles and measures of angles are usually found by mechanical devices, and hence are only approximations to the exact values. Consequently, a number such as 10.5 in Example 1 is assumed to have been rounded off to the nearest tenth. One cannot expect more accuracy in the calculated values for the remaining sides and, therefore, they should also be rounded off to the nearest tenth.

In some problems a large number of digits, such as 13,647.29, may be given for a number. If Table 3 is used, a number of this type should be rounded off to four significant figures and written as 13,650 before beginning any calculations. The situation here is similar to that discussed in Chapter Four with regard to the use of the table of logarithms. Since the values of the trigonometric functions given in Table 3 have been rounded off to four significant figures, we cannot expect more than four-figure accuracy in our computations.

As a final remark on approximations, answers should also be rounded off when Table 3 is used to find angles. In general, we shall use the following rules: if the sides of a triangle are known to four significant figures, then measures of angles calculated from Table 3 should be rounded off to the nearest minute; if the sides are known to three significant figures, then calculated measures of angles should be rounded off to the nearest multiple of ten minutes; and if the sides are known to only two significant figures, then calculations should be rounded off to the nearest degree. In order to justify these rules, we would have to make a much deeper analysis of problems involving approximate data.

Example 2   If, in triangle ABC, $\gamma = 90°$, $a = 12.3$, and $b = 31.6$, find the remaining parts.

Solution   Referring to the triangle illustrated in Figure 5.48 we have

$$\tan \alpha = \frac{12.3}{31.6} \approx 0.3892.$$

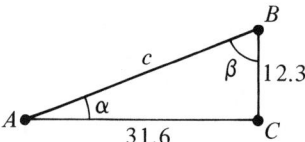

Figure 5.48

By the rule stated in the preceding paragraph, α should be rounded off to the nearest multiple of 10′. From Table 3, we see that $\alpha \approx 21°20'$. Consequently,

$$\beta \approx 90° - 21°20' = 68°40'.$$

Again referring to Figure 5.48,

$$\sec \alpha = \frac{c}{31.6} \quad \text{or} \quad c = (31.6)\sec \alpha$$

and hence $c \approx (31.6)\sec 21°20' \approx (31.6)(1.0736) \approx 33.9.$

Side c can also be found using the cosine. Referring to Figure 5.48 we see that

$$\cos \alpha = \frac{31.6}{c}$$

and hence $c = \dfrac{31.6}{\cos 21°20'} \approx \dfrac{31.6}{0.9315} \approx 33.9.$ ■

Right angles are useful in solving various types of applied problems. The following examples give two illustrations; others will be found in the exercises.

Example 3   From a point on level ground 135 feet from the foot of a tower, the angle of elevation of the top of the tower is $57°20'$. Find the height of the tower.

Solution   The **angle of elevation** is the angle which the line of sight makes with the horizontal. If we let d denote the height of the tower, then the given facts are represented by the triangle in Figure 5.49. Referring to the figure we see that

$$\tan 57°20' = \frac{d}{135} \quad \text{or} \quad d = (135)\tan 57°20'.$$

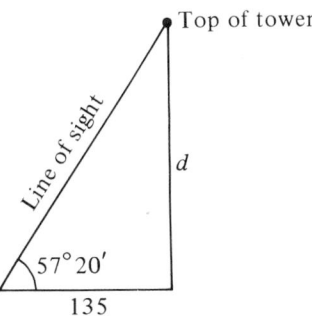

Figure 5.49

Substituting from Table 3 gives us

$$d = (135)(1.560) = 210.6 \approx 211.$$
∎

Example 4   From the top of a building which overlooks the ocean, a man sees a boat sailing directly toward him. If the man is 100 feet above sea level and if the angle of depression of the boat changes form 25° to 40° during the period of observation, find the approximate distance the boat travels during that time.

Solution   The **angle of depression** is the angle between the line of sight and the horizontal. Let A and B be the positions of the boat which correspond to the 25° and 40° angles, respectively. Suppose that the man is at point D and let C be the point 100 feet directly below him. Let d denote the distance the boat travels and let g denote the distance from B to C. This gives us the drawing in Figure 5.50, where α and β denote angles DAC and DBC, respectively. It follows that $\alpha = 25°$ and $\beta = 40°$. (Why?)

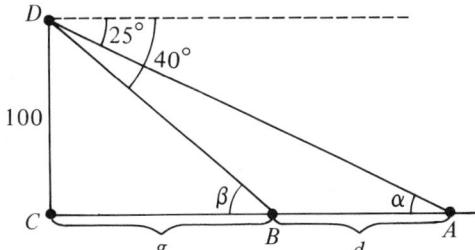

Figure 5.50

From triangle BCD we have

$$\cot \beta = \frac{g}{100} \quad \text{or} \quad g = 100 \cot \beta.$$

From triangle DAC we have

$$\cot \alpha = \frac{d + g}{100} \quad \text{or} \quad d + g = 100 \cot \alpha.$$

Consequently,
$$\begin{aligned}
d &= 100 \cot \alpha - g \\
&= 100 \cot \alpha - 100 \cot \beta \\
&= 100 \left(\cot \alpha - \cot \beta \right) \\
&= 100 \left(\cot 25° - \cot 40° \right) \\
&\approx 100(2.145 - 1.192) \\
&= 100(0.953) = 95.3
\end{aligned}$$

Hence $d \approx 95$ feet. ∎

In certain navigation and surveying problems the **direction**, or **bearing**, from a point P to a point Q is often specified by stating the acute angle which the half-line from P through Q varies to the east or west from the north-south line. Figure 5.51 illustrates four such lines. The north-south and east-west lines are labeled NS and WE, respectively. The bearing from P to Q_1 is 25° east of north and is denoted by N25°E. We also refer to the *direction* N25°E, meaning the direction from P to Q_1. The bearings from P to Q_2, Q_3, and Q_4 are represented in a similar manner in the figure.

In air navigation, directions and bearings are specified by measuring from the north in a clockwise direction. In this particular situation a positive measure is assigned to the angle instead of the negative measure to which we are accustomed for clockwise rotations. Thus, referring to (ii) of Figure 5.51, we see that the direction of PQ is 40°, whereas the direction of PR is 300°.

We shall use the above notations in Exercises 35–37.

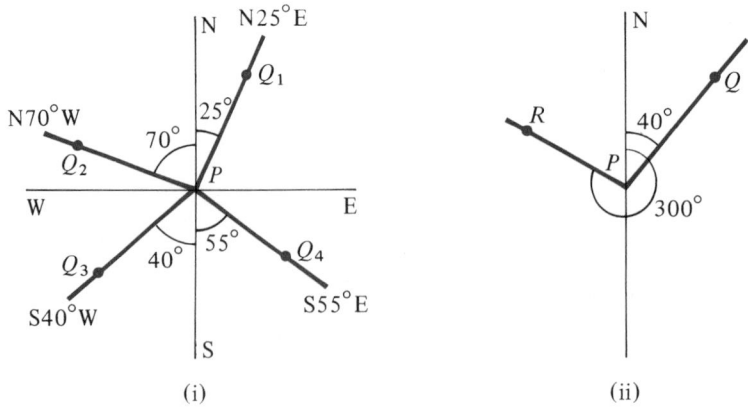

(i) (ii)

Figure 5.51

EXERCISES 5.10

Given the indicated parts of triangle ABC with $\gamma = 90°$, approximate the remaining parts of each triangle in Exercises 1–16.

1 $\alpha = 30°$, $b = 20$ **2** $\beta = 45°$, $b = 35$

3 $\beta = 52°$, $a = 15$ **4** $\alpha = 37°$, $b = 24$

5 $\alpha = 17°40'$, $a = 4.5$ **6** $\beta = 64°20'$, $a = 20.1$

7 $\beta = 71°51'$, $b = 240.0$ **8** $\alpha = 31°10'$, $a = 510$

9 $a = 25$, $b = 45$ **10** $a = 31$, $b = 9$

11 $c = 5.8$, $b = 2.1$ **12** $a = 0.42$, $c = 0.68$

13 $\alpha = 37°46'$, $b = 512.0$ **14** $\beta = 10°17'$, $b = 68.4$

15 $a = 614$, $c = 806$ **16** $c = 37.4$, $b = 21.6$

17 Approximate the angle of elevation of the sun if a boy 5 feet tall casts a shadow 4 feet long on level ground.

18 From a point 15 meters above level ground an observer measures the angle of depression of an object on the ground as $67°50'$. Approximately how far is the object from the point on the ground directly beneath the observer?

19 The string on a kite is taut and makes an angle of $54°20'$ with the horizontal. Find the approximate height of the kite above level ground if 85 meters of string are out and the end of the string is held 1.5 meters above the ground.

20 The side of a regular pentagon is 24 cm long. Approximate the radius of the circumscribed circle.

21 From a point P on level ground the angle of elevation of the top of a tower is $26°50'$. From a point 25 meters closer to the tower and on the same line with P and the base of the tower, the angle of elevation of the top is $53°30'$. Approximate the height of the tower.

22 A ladder 20 feet long leans against the side of a building. If the angle between the ladder and the building is $21°40'$, approximately how far is the bottom of the ladder from the building? If the distance from the bottom of the ladder to the building is increased by 3 feet, approximately how far does the top of the ladder move down the building?

23 As a weather balloon rises vertically its angle of elevation from a point P on the level ground 110 km from the point Q directly underneath the balloon changes from $19°20'$ to $31°50'$. Approximately how far does the balloon rise during this period?

24 From a point A, 8.2 meters above level ground, the angle of elevation of the top of a building is $31°20'$ and the angle of depression of the base of the building is $12°50'$. Approximate the height of the building.

25 In order to find the distance d between two points P and Q on opposite shores of a lake, a surveyor locates a point R which is 50 meters from P and such that RP is perpendicular to PQ. Next, using a transit, angle PRQ is found to measure $72°40'$. What is d?

26 A guy wire is attached to the top of a radio antenna and to a point on horizontal ground which is 40 meters from the base of the antenna. If the wire makes an angle of $58°20'$ with the ground, how long is the wire?

27 An octagon is inscribed in a circle of radius 12 cm. Find the perimeter of the octagon.

28 A builder wishes to construct a ramp 24 feet long which rises to a height of 4 feet above level ground. Approximate the angle that the ramp should make with the horizontal.

29 A rocket is fired at sea level and climbs at a constant angle of $75°$ through a distance of 10,000 km. Find its altitude.

30 A CB antenna is located on the top of a garage 16 feet tall. From a point on level ground which is 100 feet from a point directly below the antenna, the antenna subtends an angle of $12°$. What is the length of the antenna?

31 An airplane flying at an altitude of 10,000 feet passes directly over a fixed object on the ground. One minute later the angle of depression of the object is $42°$. What is the speed of the airplane?

32 A motorist is travelling along a level highway at a speed of 80 km per hour directly toward a distant mountain. She observes that between

1:00 P.M. and 1:10 P.M. the angle of elevation of the top of the mountain changes from 30° to 35°. Approximate the height of the mountain.

33 An airplane pilot wishes to make his approach to an airstrip at an angle of 10° with the horizontal. If he is flying at an altitude of 5,000 feet, how far from the airstrip should he begin his descent?

34 In order to measure the height h of a cloud cover, a spotlight is directed vertically upward from the ground. From a point on level ground which is d meters from the spotlight, the angle of elevation θ of the light image on the clouds is then measured. Find a formula which expresses h in terms of d and θ. As a special case, find h if $d = 1000$ meters and $\theta = 59°$.

35 A ship leaves port at 1:00 P.M. and sails in the direction N34°W at a rate of 24 miles per hour. Another ship leaves port at 1:30 P.M. and sails in the direction N56°E at a rate of 18 miles per hour. Approximately how far apart are the ships at 3:00 P.M.?

36 From an observation point A a forest ranger sights a fire in the direction S48°20′W. From a point B, 5 miles due west of A, another ranger sights the same fire in the direction S54°10′E. Approximate the distance of the fire from A.

37 An airplane flying at a speed of 360 miles per hour flies from a point A in the direction 137° for 30 minutes and then flies in the direction 227° for 45 minutes. Approximately how far is the airplane from A?

38 Generalize Exercise 19 to the case where the angle is α, the number of meters of string out is d, and the end of the string is held c meters above the ground. Express the height h of the kite in terms of α, d, and c.

39 Generalize Exercise 23 to the case where the distance from P to Q is d km and the angle of elevation changes from α to β.

40 Generalize Exercise 24 to the case where point A is d meters above ground and the angles of elevation and depression are α and β, respectively. Express the height h of the building in terms of d, α, and β.

CALCULATOR EXERCISES 5.10

Given the indicated parts of triangle ABC with $\gamma = 90°$, approximate the remaining parts of each triangle in Exercises 1–10. Round off answers to four significant figures.

1   $\alpha = 41.27°$, $a = 314.6$

2   $\beta = 24.96°$, $b = 209.3$

3   $\beta = 37.06°$, $a = 0.4613$

4   $\alpha = 17.69°$, $b = 1.307$

5   $\beta = 2.71°$, $b = 7149$

6   $\alpha = 84.07°$, $a = 0.1024$

7   $a = 46.87$, $b = 13.12$

8   $a = 6.948$, $b = 8.371$

9   $b = 2,462$, $c = 5,074$

10   $a = 88.12$, $c = 94.06$

5.11 REVIEW

Concepts

Define or discuss each of the following.

1 The point $P(t)$ on the unit circle U

2 The trigonometric functions

3 The Fundamental Identities

4 Domains and ranges of the trigonometric functions

5 Periodic function; period; periods of the trigonometric functions

6 Graphs of the trigonometric functions

7 Angles

8 Standard position of an angle

9 Initial and terminal sides

10 Coterminal angles

11 Positive and negative angles

12 Radian measure

13 Degree measure

14 Acute and obtuse angles

15 The relationship between radian measure and degree measure

16 Trigonometric functions of angles

17 Trigonometric functions as ratios

18 Reference numbers and angles

19 Finding values of trigonometric functions using reference numbers and angles

20 The graphs of $f(x) = a \sin (bx + c)$ and $f(x) = a \cos (bx + c)$

21 Addition of ordinates

22 Trigonometric solutions of right triangles

Exercises

In Exercises 1–3, $P(t)$ denotes the point on the unit circle U that corresponds to the real number t.

1 Find the rectangular coordinates of $P(7\pi)$, $P(-5\pi/2)$, $P(9\pi/2)$, $P(-3\pi/4)$, $P(18\pi)$, and $P(\pi/6)$.

2 If $P(t)$ has coordinates $(-3/5, -4/5)$, find the coordinates of $P(t + 3\pi)$, $P(t - \pi)$, $P(-t)$, and $P(2\pi - t)$.

3 Find the quadrant containing $P(t)$ if
   (a) $\sec t < 0$ and $\sin t > 0$.
   (b) $\cot t > 0$ and $\csc t < 0$.
   (c) $\cos t > 0$ and $\tan t < 0$.

4 Find the values of the remaining trigonometric functions if
   (a) $\sin t = -4/5$ and $\cos t = 3/5$.
   (b) $\csc t = \sqrt{13}/2$ and $\cot t = -3/2$.

5 Without the use of tables, find the values of the trigonometric functions corresponding to each of the following real numbers.
   (a) $9\pi/2$ (b) $-5\pi/4$ (c) 0
   (d) $11\pi/6$

6 Find the radian measures that correspond to the following degree measures: $330°$, $405°$, $-150°$, $240°$, $36°$.

7 Find the degree measures that correspond to the following radian measures: $9\pi/2$, $-2\pi/3$, $7\pi/4$, 5π, $\pi/5$.

8 Find the reference number t' if t equals: $5\pi/4$, $-5\pi/6$, $9\pi/8$.

9 Find the reference angle θ' if θ has measure: $245°$, $137°10'$, $892°$.

10 A central angle θ subtends an arc 20 cm long on a circle of radius 2 meters. What is the radian measure of θ?

11 Find the values of the six trigonometric functions of θ if θ is in standard position and

satisfies the stated condition.
(a) The point $(30, -40)$ is on the terminal side of θ.
(b) The terminal side of θ is in quadrant II and is parallel to the line $2x + 3y + 6 = 0$.
(c) $\theta = -90°$.

12 Find each of the following without the use of tables or calculators.
(a) $\cos 225°$ (b) $\tan 150°$

(c) $\sin(-\pi/6)$ (d) $\sec(4\pi/3)$

(e) $\cot(7\pi/4)$ (f) $\csc(300°)$

13 Use interpolation in Table 3 to approximate the following.
(a) $\cos 31°47'$ (b) $\tan 71°24'$
(c) $\csc 64°3'$

14 Use interpolation to approximate, to the nearest minute, the degree measure of all angles θ which are in the interval $[0°, 360°]$ if
(a) $\sin \theta = 0.8237$
(b) $\cos \theta = -0.7281$
(c) $\cot \theta = 1.238$

In each of Exercises 15–22, find the amplitude and period, and sketch the graph of f.

15 $f(x) = 5 \cos x$ 16 $f(x) = \frac{2}{3} \sin x$

17 $f(x) = -\tan x$ 18 $f(x) = 4 \sin 3x$

19 $f(x) = 3 \cos 4x$ 20 $f(x) = -4 \sin (x/3)$

21 $f(x) = 2 \sec (x/2)$ 22 $f(x) = \frac{1}{2} \csc (2x)$

Sketch the graph of the equation in each of Exercises 23–29.

23 $y = 2 \sin \left(x - \dfrac{2\pi}{3} \right)$ 24 $y = -4 \cos \left(x + \dfrac{\pi}{6} \right)$

25 $y = 5 \cos \left(2x + \dfrac{\pi}{2} \right)$ 26 $y = \tan \left(x - \dfrac{\pi}{4} \right)$

27 $y = 2 \sin x + \sin 2x$ 28 $y = 1 + x + \cos x$

29 $y = 2^{-x} \sin 2x$

30 Given the following parts of triangle ABC with $\gamma = 90°$, approximate the remaining parts.
(a) $\beta = 60°$, $b = 40$
(b) $\alpha = 54°40'$, $b = 220$
(c) $a = 62$, $b = 25$

6
ANALYTIC TRIGONOMETRY

In this chapter we shall examine various algebraic and geometric aspects of trigonometry. In Sections 6.1 through 6.5 the emphasis is on identities and equations. In Section 6.6 we consider inverse functions for the trigonometric functions. Methods for solving oblique triangles are explained in Sections 6.7 and 6.8. The chapter concludes with a discussion of vectors. In the course of our work we shall derive many important trigonometric identities and formulas. For convenient reference they are listed on the inside covers of the text.

6.1 THE FUNDAMENTAL IDENTITIES

The Fundamental Identities were introduced in Section 5.2. At that time we were primarily interested in their numerical significance. In this chapter they will be employed in a more general fashion. In particular, we shall see how the Fundamental Identities may be used to help simplify complicated expressions, derive other important formulas, and solve equations which involve trigonometric functions. Let us begin by restating them and working two numerical examples of the type considered in the previous chapter.

The Fundamental Identities

$$\csc t = \frac{1}{\sin t}, \qquad \sec t = \frac{1}{\cos t}, \qquad \cot t = \frac{1}{\tan t}$$

$$\tan t = \frac{\sin t}{\cos t}, \qquad \cot t = \frac{\cos t}{\sin t}$$

$$\sin^2 t + \cos^2 t = 1, \qquad 1 + \tan^2 t = \sec^2 t, \qquad 1 + \cot^2 t = \csc^2 t$$

Example 1 If $\sec \theta = 3$ and θ is a fourth quadrant angle, find $\tan \theta$.

Solution Using the identity $1 + \tan^2 \theta = \sec^2 \theta$ we obtain

$$\tan \theta = \pm \sqrt{\sec^2 \theta - 1}.$$

241

Since θ is in quadrant IV, $\tan\theta$ is negative, and hence

$$\tan\theta = -\sqrt{\sec^2\theta - 1} = -\sqrt{(3)^2 - 1} = -\sqrt{8} = -2\sqrt{2}. \qquad \blacksquare$$

Example 2   If $\csc t = 3/2$ and $\tan t < 0$ use fundamental identities to find the values of the remaining trigonometric functions of t.

Solution   Since $\csc t = 1/\sin t$, it follows that

$$\sin t = \frac{1}{\csc t} = \frac{1}{(3/2)} = 2/3.$$

Since $\csc t > 0$ and $\tan t < 0$, the angle (or point on the unit circle) corresponding to t is in quadrant II. Thus, $\cos t$ is negative and we may write

$$\cos t = -\sqrt{1 - \sin^2 t}$$
$$= -\sqrt{1 - (2/3)^2}$$
$$= -\sqrt{1 - (4/9)} = -\sqrt{5/9} = -\sqrt{5}/3.$$

Consequently, $$\sec t = \frac{1}{\cos t} = -\frac{3}{\sqrt{5}} = \frac{-3\sqrt{5}}{5}.$$

Next we see that

$$\tan t = \frac{\sin t}{\cos t} = \frac{(2/3)}{(-\sqrt{5}/3)} = -\frac{2}{\sqrt{5}} = \frac{-2\sqrt{5}}{5}$$

and $$\cot t = \frac{1}{\tan t} = -\frac{\sqrt{5}}{2}.$$

This gives us all of the functional values.

There are other ways of arriving at the values. For example, we could have started with

$$\cot^2 t = \csc^2 t - 1$$

and, since t is in quadrant II,

$$\cot t = -\sqrt{\csc^2 t - 1}$$
$$= -\sqrt{(3/2)^2 - 1}$$
$$= -\sqrt{(9/4) - 1}$$
$$= -\sqrt{5/4} = -\sqrt{5}/2.$$

Other identities could also have been employed. $\blacksquare$

The next example illustrates how fundamental identities may be used to express one of the trigonometric functions in terms of another. You are asked to provide similar illustrations in Exercises 33–38.

Example 3 Find a formula which expresses $\tan t$ in terms of $\cos t$.

Solution We may write

$$\tan t = \frac{\sin t}{\cos t} = \frac{\pm\sqrt{1 - \cos^2 t}}{\cos t}$$

where the sign depends on the particular value of t. ■

The following example illustrates how fundamental identities may be used to simplify trigonometric expressions. In the next section much more will be done along these lines.

Example 4 Use fundamental identities to simplify the expression

$$\sec t - \sin t \tan t.$$

Solution The reader should give reasons for each of the following steps:

$$\sec t - \sin t \tan t = \frac{1}{\cos t} - \sin t \left(\frac{\sin t}{\cos t}\right)$$

$$= \frac{1}{\cos t} - \frac{\sin^2 t}{\cos t}$$

$$= \frac{1 - \sin^2 t}{\cos t}$$

$$= \frac{\cos^2 t}{\cos t}$$

$$= \cos t. \qquad ■$$

Example 5 Use fundamental identities to transform

$$\frac{\sin^2 t}{1 + \cos t} \quad \text{into} \quad 1 - \cos t.$$

Solution We may proceed as follows:

$$\frac{\sin^2 t}{1 + \cos t} = \frac{1 - \cos^2 t}{1 + \cos t}$$

$$= \frac{(1 + \cos t)(1 - \cos t)}{1 + \cos t}$$

$$= 1 - \cos t. \qquad ■$$

EXERCISES 6.1

In each of Exercises 1–8 use fundamental identities to find the values of the remaining trigonometric functions of t.

1   $\cos t = 1/2$, $\sin t = -\sqrt{3}/2$

2   $\cos t = -\sqrt{3}/2$, $\csc t = 2$

3   $\sin t = \sqrt{2}/2$, $\sec t = -\sqrt{2}$

4   $\cos t = 0$, $\sin t = 1$

5   $\cot t = -\sqrt{3}$, $\csc t = -2$

6   $\sec t = \sqrt{2}$, $\csc t = -\sqrt{2}$

7   $\cos t = -1$, $\sin t = 0$

8   $\cot t = -1$, $\sec t = -\sqrt{2}$

In Exercises 9–16 use fundamental identities to find the values of all six trigonometric functions if the given conditions are true.

9   $\tan t = 3/4$ and $\sin t < 0$

10   $\cot t = -3/4$ and $\cos t > 0$

11   $\sec t = -13/5$ and $\tan t < 0$

12   $\cos t = -1/2$ and $\sin t > 0$

13   $\sin t = -2/3$ and $\sec t > 0$

14   $\sec t = -5$ and $\csc t < 0$

15   $\csc t = 8$ and $\cos t < 0$

16   $\cos t = 0$ and $\sin t = -1$

In Exercises 17–32 use fundamental identities to transform the first expression into the second.

17   $\cos t \csc t$, $\cot t$

18   $\tan t \csc t$, $\sec t$

19   $\tan x \cos x$, $\sin x$

20   $\cot x \sin x$, $\cos x$

21   $\dfrac{\sec \theta}{\tan \theta}$, $\csc \theta$

22   $\dfrac{\cot \theta}{\csc \theta}$, $\cos \theta$

23   $\dfrac{\cos \alpha}{\sec \alpha}$, $\cos^2 \alpha$

24   $\dfrac{\sec t}{\csc t}$, $\tan t$

25   $(1 + \cot^2 t)\tan^2 t$, $\sec^2 t$

26   $(1 - \sin^2 \theta)\sec^2 \theta$, 1

27   $\sin x(\csc x - \sin x)$, $\cos^2 x$

28   $(\sec t - \cos t)\cos t$, $\sin^2 t$

29   $\dfrac{\sin^2 \theta}{1 - \sin^2 \theta}$, $\tan^2 \theta$

30   $\dfrac{1 - \sin^2 \alpha}{1 - \cos^2 \alpha}$, $\cot^2 \alpha$

31   $\dfrac{1 + \tan^2 x}{\tan^2 x}$, $\csc^2 x$ 32 $\dfrac{\cos^2 t}{1 - \sin t}$, $1 + \sin t$

33   Express each trigonometric function in terms of $\sin t$.

34   Express each trigonometric function in terms of $\cos t$.

35   Express $\sin t$ in terms of $\tan t$.

36   Express $\cos t$ in terms of $\cot t$.

37   Express $\sec t$ in terms of $\csc t$.

38   Express $\csc t$ in terms of $\tan t$.

39   Prove that if $0 < t < \pi/2$, then

$$\log \csc t = -\log \sin t.$$

40   Prove that if θ is an acute angle, then

$$\log \tan \theta = \log \sin \theta - \log \cos \theta.$$

CALCULATOR EXERCISES 6.1

Use direct calculations to demonstrate that

$$\tan t = \frac{\sin t}{\cos t} \quad \text{and} \quad \sin^2 t + \cos^2 t = 1$$

for the values of t given in Exercises 1–8.

1 1.62°

2 37° 15′

3 $\pi/20$

4 137.48

5 sin 2

6 $\sqrt{\pi} + \log 9$

7 $e^{1.6}$

8 ln 13.42

9 If a calculator has keys for sin, cos, and tan, but not for csc, sec, and cot, how are values for the latter functions calculated?

10 Calculate $(\cos 5)^a$, where $a = \log(\tan(\sin 2))$.

6.2 TRIGONOMETRIC IDENTITIES

Any mathematical expression that contains symbols such as $\sin x$, $\cos \beta$, $\tan v$, etc., where the letters x, β, and v are variables, is referred to as a **trigonometric expression**. Trigonometric expressions may be very simple or very complicated. Some examples are

$$x + \sin x, \quad \frac{\cos(3y + 1)}{x^2 + \tan^2(z - y^2)}, \quad \frac{\sqrt{\theta} + 2^{\sin \theta}}{\sec(\cot \theta)}.$$

As usual, we assume that the domain of each variable is the set of real numbers (or angles) for which the expressions are meaningful. As indicated in Section 6.1 many trigonometric expressions can be simplified or changed in form by employing fundamental identities. The next example is another illustration of how this may be accomplished.

Example 1 Simplify the expression $(\sec \theta + \tan \theta)(1 - \sin \theta)$.

Solution The reader should supply reasons for the following steps.

$$(\sec \theta + \tan \theta)(1 - \sin \theta) = \left(\frac{1}{\cos \theta} + \frac{\sin \theta}{\cos \theta} \right)(1 - \sin \theta)$$

$$= \left(\frac{1 + \sin \theta}{\cos \theta} \right)(1 - \sin \theta)$$

$$= \frac{1 - \sin^2 \theta}{\cos \theta}$$

$$= \frac{\cos^2 \theta}{\cos \theta}$$

$$= \cos \theta \qquad \blacksquare$$

There are other ways to treat the expression in Example 1. We could begin by multiplying the two factors and then proceed to simplify and combine terms. The method we employed—of changing all expressions to expressions which involve only sines and cosines—is often worthwhile. However, that technique does not always lead to the shortest possible simplification.

A **trigonometric equation** is an equation which contains trigonometric expressions. When working with trigonometric equations, we may express solutions either in terms of real numbers or in terms of angles.

We encountered several trigonometric identities in our previous work. Some of them, such as the Fundamental Identities, are basic and should be memorized. Other identities are introduced merely to supply practice in manipulating trigonometric expressions and therefore should not be memorized. We shall take the latter point of view in this section. Thus the identities to be investigated are unimportant in their own right. The important thing is the manipulative practice that is gained. The ability to carry out trigonometric manipulations is essential for problems which are encountered in advanced courses in mathematics and science.

We shall often use the phrase "verify an identity" instead of "prove that an equation is an identity." When verifying an identity, we shall not use the definition of the trigonometric functions to return to the coordinates (x, y) of a point on the unit circle U. Although many of the identities could be proved in that way, it is not the type of practice we desire. Rather, we shall use fundamental identities and algebraic manipulations to change the form of trigonometric expressions in a manner similar to the solution of Example 1. The preferred method of showing that an equation is an identity is to transform one side into the other, as illustrated in the next three examples. The reader should supply reasons for all steps in the solutions.

Example 2   Verify the identity $\dfrac{\tan t + \cos t}{\sin t} = \sec t + \cot t$.

Solution   We shall transform the left side into the right side. Thus,

$$\frac{\tan t + \cos t}{\sin t} = \frac{\tan t}{\sin t} + \frac{\cos t}{\sin t}$$

$$= \frac{\left(\dfrac{\sin t}{\cos t}\right)}{\sin t} + \cot t$$

$$= \frac{1}{\cos t} + \cot t$$

$$= \sec t + \cot t. \qquad \blacksquare$$

Example 3   Verify the identity $\sec \alpha - \cos \alpha = \sin \alpha \tan \alpha$.

Solution

$$\sec \alpha - \cos \alpha = \frac{1}{\cos \alpha} - \cos \alpha$$

$$= \frac{1 - \cos^2 \alpha}{\cos \alpha}$$

$$= \frac{\sin^2 \alpha}{\cos \alpha} = \sin \alpha \left(\frac{\sin \alpha}{\cos \alpha} \right)$$

$$= \sin \alpha \tan \alpha.$$ ∎

Example 4 Verify the identity $\dfrac{\cos x}{1 - \sin x} = \dfrac{1 + \sin x}{\cos x}$.

Solution We begin by multiplying the numerator and denominator of the fraction on the left by $1 + \sin x$. Thus,

$$\frac{\cos x}{1 - \sin x} = \frac{\cos x}{1 - \sin x} \cdot \frac{1 + \sin x}{1 + \sin x}$$

$$= \frac{\cos x(1 + \sin x)}{1 - \sin^2 x}$$

$$= \frac{\cos x(1 + \sin x)}{\cos^2 x}$$

$$= \frac{1 + \sin x}{\cos x}.$$ ∎

In another technique for showing that an equation $p = q$ is an identity, we begin by transforming the left side p into another expression s, making sure that each step is *reversible* in the sense that it is possible to transform s back into p by reversing the procedure which has been used. In this case the equation $p = s$ is an identity. Next, as a *separate* exercise, it is shown that the right side q can also be transformed to the expression s by means of reversible steps and hence that $q = s$ is an identity. It then follows that $p = q$ is an identity. This method is illustrated in the next example.

Example 5 Verify the identity $(\tan \theta - \sec \theta)^2 = \dfrac{1 - \sin \theta}{1 + \sin \theta}$.

Solution We shall verify the identity by showing that each side of the equation can be transformed into the same expression. Starting with the left side we may write

$$(\tan \theta - \sec \theta)^2 = \tan^2 \theta - 2 \tan \theta \sec \theta + \sec^2 \theta$$

$$= \frac{\sin^2 \theta}{\cos^2 \theta} - \frac{2 \sin \theta}{\cos^2 \theta} + \frac{1}{\cos^2 \theta}$$

$$= \frac{\sin^2 \theta - 2 \sin \theta + 1}{\cos^2 \theta}.$$

The right side of the given equation may be changed by multiplying numerator and denominator by $1 - \sin\theta$. Thus

$$\frac{1 - \sin\theta}{1 + \sin\theta} = \frac{1 - \sin\theta}{1 + \sin\theta} \cdot \frac{1 - \sin\theta}{1 - \sin\theta}$$

$$= \frac{1 - 2\sin\theta + \sin^2\theta}{1 - \sin^2\theta}$$

$$= \frac{1 - 2\sin\theta + \sin^2\theta}{\cos^2\theta}$$

which is the same expression as that obtained for $(\tan\theta - \sec\theta)^2$. Since all steps are reversible, it follows that the given equation is an identity.   ∎

EXERCISES 6.2

Verify the following identities.

1   $\cos\theta\sec\theta = 1$

2   $\tan\alpha\cot\alpha = 1$

3   $\sin\theta\sec\theta = \tan\theta$

4   $\sin\alpha\cot\alpha = \cos\alpha$

5   $\dfrac{\csc x}{\sec x} = \cot x$

6   $\cot\beta\sec\beta = \csc\beta$

7   $(1 + \cos\alpha)(1 - \cos\alpha) = \sin^2\alpha$

8   $\cos^2 x(\sec^2 x - 1) = \sin^2 x$

9   $\cos^2 t - \sin^2 t = 2\cos^2 t - 1$

10   $(\tan\theta + \cot\theta)\tan\theta = \sec^2\theta$

11   $\dfrac{\sin t}{\csc t} + \dfrac{\cos t}{\sec t} = 1$

12   $1 - 2\sin^2 x = 2\cos^2 x - 1$

13   $(1 + \sin\alpha)(1 - \sin\alpha) = \dfrac{1}{\sec^2\alpha}$

14   $(1 - \sin^2 t)(1 + \tan^2 t) = 1$

15   $\sec\beta - \cos\beta = \tan\beta\sin\beta$

16   $\dfrac{\sin w + \cos w}{\cos w} = 1 + \tan w$

17   $\dfrac{\csc^2\theta}{1 + \tan^2\theta} = \cot^2\theta$

18   $\sin x + \cos x\cot x = \csc x$

19   $\sin t(\csc t - \sin t) = \cos^2 t$

20   $\cot t + \tan t = \csc t\sec t$

21   $\csc\theta - \sin\theta = \cot\theta\cos\theta$

22   $\cos\theta(\tan\theta + \cot\theta) = \csc\theta$

23   $\dfrac{\sec^2 u - 1}{\sec^2 u} = \sin^2 u$

24   $(\tan u + \cot u)(\cos u + \sin u) = \sec u + \csc u$

25   $(\cos^2 x - 1)(\tan^2 x + 1) = 1 - \sec^2 x$

26   $(\cot\alpha + \csc\alpha)(\tan\alpha - \sin\alpha) = \sec\alpha - \cos\alpha$

27   $\sec t\csc t + \cot t = \tan t + 2\cos t\csc t$

28   $\dfrac{1 + \cos^2 y}{\sin^2 y} = 2\csc^2 y - 1$

29   $\sec^2\theta\csc^2\theta = \sec^2\theta + \csc^2\theta$

30   $\dfrac{\sec x - \cos x}{\tan x} = \dfrac{\tan x}{\sec x}$

31   $\dfrac{1 + \cos t}{\sin t} + \dfrac{\sin t}{1 + \cos t} = 2\csc t$

32   $\tan^2\alpha - \sin^2\alpha = \tan^2\alpha\sin^2\alpha$

33   $\dfrac{1 + \tan^2 v}{\tan^2 v} = \csc^2 v$

34   $\dfrac{\sec\theta + \csc\theta}{\sec\theta - \csc\theta} = \dfrac{\sin\theta + \cos\theta}{\sin\theta - \cos\theta}$

35 $\dfrac{1 + \sin x}{1 - \sin x} - \dfrac{1 - \sin x}{1 + \sin x} = 4 \tan x \sec x$

36 $\dfrac{1}{1 - \cos \gamma} + \dfrac{1}{1 + \cos \gamma} = 2 \csc^2 \gamma$

37 $\dfrac{1 + \csc \beta}{\sec \beta} - \cot \beta = \cos \beta$

38 $\dfrac{\cos x \cot x}{\cot x - \cos x} = \dfrac{\cot x + \cos x}{\cos x \cot x}$

39 $(\sec u - \tan u)(\csc u + 1) = \cot u$

40 $\dfrac{\cot \theta - \tan \theta}{\sin \theta + \cos \theta} = \csc \theta - \sec \theta$

41 $\dfrac{\cot \alpha - 1}{1 - \tan \alpha} = \cot \alpha$

42 $\dfrac{1 + \sec \beta}{\tan \beta + \sin \beta} = \csc \beta$

43 $\csc^4 t - \cot^4 t = \cot^2 t + \csc^2 t$

44 $\cos^4 \theta + \sin^2 \theta = \sin^4 \theta + \cos^2 \theta$

45 $\dfrac{\cos \beta}{1 - \sin \beta} = \sec \beta + \tan \beta$

46 $\dfrac{1}{\csc y - \cot y} = \csc y + \cot y$

47 $\dfrac{\tan^2 x}{\sec x + 1} = \dfrac{1 - \cos x}{\cos x}$

48 $\dfrac{\cot x}{\csc x + 1} = \dfrac{\csc x - 1}{\cot x}$

49 $\dfrac{\cot u - 1}{\cot u + 1} = \dfrac{1 - \tan u}{1 + \tan u}$

50 $\dfrac{1 + \sec x}{\sin x + \tan x} = \csc x$

51 $\sin^4 r - \cos^4 r = \sin^2 r - \cos^2 r$

52 $\sin^4 \theta + 2 \sin^2 \theta \cos^2 \theta + \cos^4 \theta = 1$

53 $\tan^4 k - \sec^4 k = 1 - 2 \sec^2 k$

54 $\sec^4 u - \sec^2 u = \tan^4 u + \tan^2 u$

55 $(\sec t + \tan t)^2 = \dfrac{1 + \sin t}{1 - \sin t}$

56 $\sec^2 \gamma + \tan^2 \gamma = (1 - \sin^4 \gamma)(\sec^4 \gamma)$

57 $(\sin^2 \theta + \cos^2 \theta)^3 = 1$

58 $\dfrac{\sin t}{1 - \cos t} = \csc t + \cot t$

59 $\dfrac{1 + \csc \beta}{\cot \beta + \cos \beta} = \sec \beta$

60 $\dfrac{\sin z \tan z}{\tan z - \sin z} = \dfrac{\tan z + \sin z}{\sin z \tan z}$

61 $\left(\dfrac{\sin^2 x}{\tan^4 x}\right)^3 \left(\dfrac{\csc^3 x}{\cot^6 x}\right)^2 = 1$

62 $\dfrac{\cos^3 x - \sin^3 x}{\cos x - \sin x} = 1 + \sin x \cos x$

63 $\dfrac{\sin \theta + \cos \theta}{\tan^2 \theta - 1} = \dfrac{\cos^2 \theta}{\sin \theta - \cos \theta}$

64 $(\csc t - \cot t)^4 (\csc t + \cot t)^4 = 1$

65 $(a \cos t - b \sin t)^2 + (a \sin t + b \cos t)^2$
$= a^2 + b^2$

66 $\sin^6 v + \cos^6 v = 1 - 3 \sin^2 v \cos^2 v$

67 $\dfrac{\sin \alpha \cos \beta + \cos \alpha \sin \beta}{\cos \alpha \cos \beta - \sin \alpha \sin \beta} = \dfrac{\tan \alpha + \tan \beta}{1 - \tan \alpha \tan \beta}$

68 $\dfrac{\tan u - \tan v}{1 + \tan u \tan v} = \dfrac{\cot v - \cot u}{1 + \cot u \cot v}$

69 $\sqrt{\dfrac{1 - \cos t}{1 + \cos t}} = \dfrac{1 - \cos t}{|\sin t|}$

70 $\sqrt{\dfrac{1 - \sin \theta}{1 + \sin \theta}} = \dfrac{|\cos \theta|}{1 + \sin \theta}$

71 $\dfrac{\sin \alpha}{1 + \cos \alpha} + \dfrac{1 + \cos \alpha}{\sin \alpha} = 2 \csc \alpha$

72 $\dfrac{\csc x}{1 + \csc x} - \dfrac{\csc x}{1 - \csc x} = 2 \sec^2 x$

73 $\dfrac{1}{\tan \beta + \cot \beta} = \sin \beta \cos \beta$

74 $\dfrac{\cot y - \tan y}{\sin y \cos y} = \csc^2 y - \sec^2 y$

75 $\sec \theta + \csc \theta - \cos \theta - \sin \theta$
$= \sin \theta \tan \theta + \cos \theta \cot \theta$

76 $\sin^3 t + \cos^3 t = (1 - \sin t \cos t)(\sin t + \cos t)$

77 $(1 - \tan^2 \phi)^2 = \sec^4 \phi - 4 \tan^2 \phi$

78 $\cos^4 w + 1 - \sin^4 w = 2 \cos^2 w$

79 $\dfrac{\tan x}{1 - \cot x} + \dfrac{\cot x}{1 - \tan x} = 1 + \sec x \csc x$

80 $\dfrac{\cos \gamma}{1 - \tan \gamma} + \dfrac{\sin \gamma}{1 - \cot \gamma} = \cos \gamma + \sin \gamma$

81 $\sin(-t)\sec(-t) = -\tan t$

82 $\dfrac{\cot(-v)}{\csc(-v)} = \cos v$

83 $\log 10^{\tan t} = \tan t$

84 $10^{\log|\sin t|} = |\sin t|$

85 $\log \cot x = -\log \tan x$

86 $\log \sec \theta = -\log \cos \theta$

87 $-\log|\sec \theta - \tan \theta| = \log|\sec \theta + \tan \theta|$

88 $\log|\csc x - \cot x| = -\log|\csc x + \cot x|$

Show that the equations in Exercises 89–100 are not identities. (*Hint:* Find one number in the domain of t or θ for which the equation is false.)

89 $\cos t = \sqrt{1 - \sin^2 t}$

90 $\sqrt{\sin^2 t + \cos^2 t} = \sin t + \cos t$

91 $\sqrt{\sin^2 t} = \sin t$

92 $\sec t = \sqrt{\tan^2 t + 1}$

93 $(\sin \theta + \cos \theta)^2 = \sin^2 \theta + \cos^2 \theta$

94 $\log(1/\sin t) = 1/(\log \sin t)$

95 $\cos(-t) = -\cos t$ **96** $\sin(t + \pi) = \sin t$

97 $\cos(\sec t) = 1$ **98** $\cot(\tan \theta) = 1$

99 $\sin^2 t - 4\sin t - 5 = 0$

100 $3\cos^2 \theta + \cos \theta - 2 = 0$

6.3 TRIGONOMETRIC EQUATIONS

If a trigonometric equation is not an identity, then techniques similar to those used for algebraic equations may be employed to find the solutions. The main difference here is that we usually solve for $\sin x$, $\cos \theta$, and so on, and then find x and θ as illustrated below.

Example 1   Solve the equation $\sin \theta \tan \theta = \sin \theta$.

Solution   Each of the following is equivalent to the given equation:

$$\sin \theta \tan \theta - \sin \theta = 0$$
$$(\sin \theta)(\tan \theta - 1) = 0.$$

To find the solutions we set each factor on the left equal to zero, obtaining

$$\sin \theta = 0 \quad \text{and} \quad \tan \theta = 1.$$

The solutions of the equation $\sin \theta = 0$ consist of all multiples of π, that is, $\theta = n\pi$, where n is any integer.

Since the tangent function has period π, it is sufficient to find the solutions of the equation $\tan \theta = 1$ which are in the interval $[0, \pi)$, for once they are known, we may obtain all the others by adding multiples of π. The only solution of $\tan \theta = 1$

in $[0, \pi)$ is $\pi/4$, and hence every solution has the form

$$\theta = \frac{\pi}{4} + n\pi$$

where n is an integer.

It follows that the solutions of the given equation consist of all numbers of the form

$$n\pi \quad \text{and} \quad \frac{\pi}{4} + n\pi$$

where n is any integer. Some particular solutions are $0, \pm\pi, \pm 2\pi, \pm 3\pi, \pi/4, 5\pi/4, -3\pi/4,$ and $-7\pi/4$.   ∎

Note that in Example 1 it would have been incorrect to begin by dividing both sides by $\sin \theta$, for this manipulation would lose the solutions of $\sin \theta = 0$.

Example 2   Solve the equation $2 \sin^2 t - \cos t - 1 = 0$.

Solution   We first change the equation to an equation which involves only $\cos t$ and then we factor as follows:

$$2(1 - \cos^2 t) - \cos t - 1 = 0$$
$$-2 \cos^2 t - \cos t + 1 = 0$$
$$2 \cos^2 t + \cos t - 1 = 0$$
$$(2 \cos t - 1)(\cos t + 1) = 0.$$

As in Example 1, the solutions of the last equation are the solutions of

$$2 \cos t - 1 = 0 \quad \text{and} \quad \cos t + 1 = 0$$

or equivalently, $\cos t = \frac{1}{2}$ and $\cos t = -1$.

It is sufficient to find the solutions which are in the interval $[0, 2\pi)$, for once they are known, all solutions may be found by adding multiples of 2π.

If $\cos t = \frac{1}{2}$, then the reference number (or reference angle) is $\pi/3$ (or $60°$). Since $\cos t$ is positive, the point $P(t)$ on U (or the angle t) lies in quadrant I or quadrant IV. Hence in the interval $[0, 2\pi)$ we have

$$t = \frac{\pi}{3} \quad \text{or} \quad t = 2\pi - \frac{\pi}{3} = \frac{5\pi}{3}.$$

If $\cos t = -1$, then $t = \pi$.

It follows that the solutions of the given equation are

$$\frac{\pi}{3} + 2\pi n, \quad \frac{5\pi}{3} + 2\pi n, \quad \pi + 2\pi n$$

where n is any integer. If we wish to express the solutions in terms of degrees, we may write

$$60° + 360°n, \quad 300° + 360°n, \quad 180° + 360°n. \qquad \blacksquare$$

Example 3   Find the solutions of $4\sin^2 x \tan x - \tan x = 0$ which are in the interval $[0, 2\pi)$.

Solution   Factoring the left side we obtain

$$(\tan x)(4\sin^2 x - 1) = 0.$$

Setting each factor equal to zero gives us

$$\tan x = 0 \quad \text{and} \quad \sin^2 x = \tfrac{1}{4}$$

or equivalently,

$$\tan x = 0, \quad \sin x = \tfrac{1}{2}, \quad \text{and} \quad \sin x = -\tfrac{1}{2}.$$

The equation $\tan x = 0$ has solutions 0 and π in the interval $[0, 2\pi)$. The equation $\sin x = \tfrac{1}{2}$ has solutions $\pi/6$ and $5\pi/6$. The equation $\sin x = -\tfrac{1}{2}$ leads to the numbers between π and 2π which have reference number $\pi/6$. These are

$$\pi + \frac{\pi}{6} = \frac{7\pi}{6} \quad \text{and} \quad 2\pi - \frac{\pi}{6} = \frac{11\pi}{6}.$$

Hence the solutions of the given equation in the interval $[0, 2\pi)$ are

$$0, \pi, \frac{\pi}{6}, \frac{5\pi}{6}, \frac{7\pi}{6}, \frac{11\pi}{6}. \qquad \blacksquare$$

Example 4   Find the solutions of the equation $\csc^4 2u - 4 = 0$.

Solution   Factoring the left side we obtain

$$(\csc^2 2u - 2)(\csc^2 2u + 2) = 0.$$

Setting each factor equal to 0 leads to

$$\csc^2 2u = 2 \quad \text{and} \quad \csc^2 2u = -2.$$

The equation $\csc^2 2u = -2$ has no solutions. (Why?) The solutions of $\csc^2 2u = 2$ consist of the solutions of the two equations

$$\csc 2u = \sqrt{2} \quad \text{and} \quad \csc 2u = -\sqrt{2}.$$

If $\csc 2u = \sqrt{2}$, then

$$2u = \frac{\pi}{4} + 2\pi n \quad \text{and} \quad 2u = \frac{3\pi}{4} + 2\pi n$$

where n is any integer. Dividing both sides of the last two equations by 2 gives us the solutions

$$u = \frac{\pi}{8} + \pi n \quad \text{and} \quad u = \frac{3\pi}{8} + \pi n.$$

Similarly, from $\csc 2u = -\sqrt{2}$ we obtain

$$2u = \frac{5\pi}{4} + 2\pi n \quad \text{and} \quad 2u = \frac{7\pi}{4} + 2\pi n$$

where n is any integer. Dividing both sides of the last two equations by 2 gives us

$$u = \frac{5\pi}{8} + \pi n \quad \text{and} \quad u = \frac{7\pi}{8} + \pi n.$$

Collecting the above information, we see that all solutions are given by the formula

$$u = \frac{\pi}{8} + \frac{\pi}{4}n, \quad \text{where } n \text{ is any integer.} \qquad \blacksquare$$

Example 5 Approximate the solutions of the equation

$$5\sin\theta\tan\theta - 10\tan\theta + 3\sin\theta - 6 = 0$$

in the degree interval $[0°, 360°)$.

Solution The equation may be factored by grouping terms as follows:

$$5\tan\theta\,(\sin\theta - 2) + 3\,(\sin\theta - 2) = 0$$
$$(5\tan\theta + 3)(\sin\theta - 2) = 0.$$

Since the equation $\sin\theta = 2$ has no solutions (Why?), the solutions of the given equation are the same as those of

$$\tan\theta = -\tfrac{3}{5} = -0.6000.$$

Approximations to θ may be found by interpolating in Table 3. We first approximate the reference angle θ' as follows:

$$10' \left\{ d \begin{cases} \tan 30° 50' \approx 0.5969 \\ \tan\theta' \quad = 0.6000 \end{cases} 0.0031 \\ \tan 31° 00' \approx 0.6009 \end{cases} 0.0040$$

$$\frac{d}{10} = \frac{31}{40} \quad \text{or} \quad d = \frac{31}{4} \approx 8'$$

$$\theta' \approx 30° 58'.$$

Since θ lies in quadrant II or quadrant IV, this implies that

$$\theta \approx 180° - 30°58' = 149°2' \quad \text{or} \quad \theta \approx 360° - 30°58' = 329°2'. \qquad \blacksquare$$

A technique for solving trigonometric equations with the aid of a hand-held calculator will be discussed later in the chapter, after inverse trigonometric functions have been introduced.

EXERCISES 6.3

In each of Exercises 1–16 find all solutions of the given equation.

1 $2\cos t + 1 = 0$

2 $\cot \theta + 1 = 0$

3 $\tan^2 x = 1$

4 $4\cos \theta - 2 = 0$

5 $(\cos \theta - 1)(\sin \theta + 1) = 0$

6 $2\cos x = \sqrt{3}$

7 $\sec^2 \alpha - 4 = 0$

8 $3 - \tan^2 \beta = 0$

9 $\sqrt{3} + 2\sin \beta = 0$

10 $4\sin^2 x - 3 = 0$

11 $\cot^2 x - 3 = 0$

12 $(\sin t - 1)\cos t = 0$

13 $(2\sin \theta + 1)(2\cos \theta + 3) = 0$

14 $(2\sin u - 1)(\cos u - \sqrt{2}) = 0$

15 $\sin 2x(\csc 2x - 2) = 0$

16 $\tan \alpha + \tan^2 \alpha = 0$

In each of Exercises 17–36 find the solutions of the equation in the interval $[0, 2\pi)$ and also find the degree measure of each solution.

17 $2 - 8\cos^2 t = 0$

18 $\cot^2 \theta - \cot \theta = 0$

19 $2\sin^2 u = 1 - \sin u$

20 $2\cos^2 t + 3\cos t + 1 = 0$

21 $\tan^2 x \sin x = \sin x$

22 $\sec \beta \csc \beta = 2\csc \beta$

23 $2\cos^2 \gamma + \cos \gamma = 0$ 24 $\sin x - \cos x = 0$

25 $\sin^2 \theta + \sin \theta - 6 = 0$

26 $2\sin^2 u + \sin u - 6 = 0$

27 $1 - \sin t = \sqrt{3}\cos t$ 28 $\cos \theta - \sin \theta = 1$

29 $\cos \alpha + \sin \alpha = 1$ 30 $2\tan t - \sec^2 t = 0$

31 $\tan \theta + \sec \theta = 1$

32 $\cot \alpha + \tan \alpha = \csc \alpha \sec \alpha$

33 $2\sin^3 x + \sin^2 x - 2\sin x - 1 = 0$

34 $\sec^5 \theta = 4\sec \theta$

35 $2\tan t \csc t + 2\csc t + \tan t + 1 = 0$

36 $2\sin v \csc v - \csc v = 4\sin v - 2$

In each of Exercises 37–40 use Table 3 to approximate, to the nearest multiple of ten minutes, the solutions of the given equation in the interval $[0°, 360°)$.

37 $\sin^2 t - 4\sin t + 1 = 0$

38 $\tan^2 \theta + 3\tan \theta + 2 = 0$

39 $12\sin^2 u - 5\sin u - 2 = 0$

40 $5\cos^2 \alpha + 3\cos \alpha - 2 = 0$

6.4 THE ADDITION FORMULAS

Let t_1 and t_2 be any real numbers and let $P(t_1)$ and $P(t_2)$ be the corresponding points on the unit circle U. We shall denote these points in terms of rectangular coordinates by $P_1(x_1, y_1)$ and $P_2(x_2, y_2)$, respectively. Let us next consider the real number $t_1 - t_2$ and denote the rectangular coordinates of $P(t_1 - t_2)$ by $P_3(x_3, y_3)$. Applying the definition of the trigonometric functions we have

(*)
$$\cos t_1 = x_1, \quad \cos t_2 = x_2, \quad \cos (t_1 - t_2) = x_3,$$
$$\sin t_1 = y_1, \quad \sin t_2 = y_2, \quad \sin (t_1 - t_2) = y_3,$$

where the symbol (*) has been used for later reference to these formulas. Our goal is to obtain a formula for $\cos (t_1 - t_2)$ in terms of functional values of t_1 and t_2. For convenience let us assume that t_1 and t_2 are in the interval $[0, 2\pi]$ and $0 \le t_1 - t_2 < t_2$. In this case, $t_2 \le t_1$ and hence if A is the point $(1, 0)$ on U, then the length t_1 of $\overset{\frown}{AP_1}$ is greater than or equal to the length t_2 of $\overset{\frown}{AP_2}$. Also the length $t_1 - t_2$ of $\overset{\frown}{AP_3}$ is less than the length t_2 of $\overset{\frown}{AP_2}$. Figure 6.1 illustrates one arrangement of points P_1 and P_2 under these conditions.

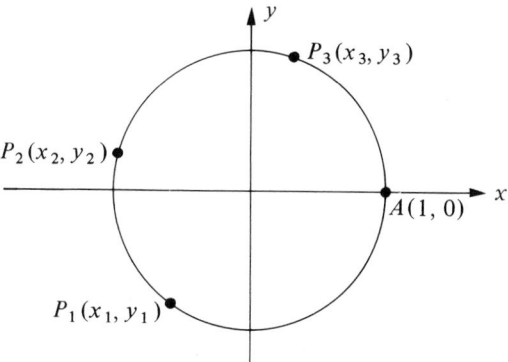

Figure 6.1

It is possible to extend our discussion to cover all values of t. Since the arc lengths of both $\overset{\frown}{P_2 P_1}$ and $\overset{\frown}{AP_3}$ equal $t_1 - t_2$, the line segments $P_2 P_1$ and AP_3 also have the same length; that is,

$$d(A, P_3) = d(P_1, P_2).$$

Using the Distance Formula, the last equation may be written

$$\sqrt{(x_3 - 1)^2 + (y_3 - 0)^2} = \sqrt{(x_2 - x_1)^2 + (y_2 - y_1)^2}.$$

Squaring both sides and expanding the terms underneath the radical gives us

$$x_3^2 - 2x_3 + 1 + y_3^2 = x_2^2 - 2x_1 x_2 + x_1^2 + y_2^2 - 2y_1 y_2 + y_1^2.$$

Since P_1, P_2, and P_3 lie on U and since an equation for U is $x^2 + y^2 = 1$, we may substitute 1 for each of $x_1^2 + y_1^2$, $x_2^2 + y_2^2$, and $x_3^2 + y_3^2$. Doing this and simplifying, we obtain

$$2 - 2x_3 = 2 - 2x_1x_2 - 2y_1y_2$$

which reduces to
$$x_3 = x_1x_2 + y_1y_2.$$

Substituting from the formulas stated in (*) at the beginning of this section gives us the desired formula, namely

$$\cos(t_1 - t_2) = \cos t_1 \cos t_2 + \sin t_1 \sin t_2.$$

In order to make the formula less cumbersome, we shall eliminate subscripts by changing the symbols for the variables from t_1 and t_2 to u and v, respectively. Our identity then takes on the form

$$\boxed{\cos(u - v) = \cos u \cos v + \sin u \sin v.}$$

Note that in most cases, $\cos(u - v) \neq \cos u - \cos v$.

Example 1　Find the exact value of $\cos 15°$.

Solution　　If we write $15° = 60° - 45°$, we may use the last formula with $u = 60°$ and $v = 45°$ as follows:

$$\cos 15° = \cos(60° - 45°)$$
$$= \cos 60° \cos 45° + \sin 60° \sin 45°$$
$$= \left(\frac{1}{2}\right)\left(\frac{\sqrt{2}}{2}\right) + \left(\frac{\sqrt{3}}{2}\right)\left(\frac{\sqrt{2}}{2}\right)$$
$$= \frac{\sqrt{2} + \sqrt{6}}{4}. \qquad \blacksquare$$

It is easy to obtain a formula for $\cos(u + v)$. We simply write $u + v = u - (-v)$ and employ the formula derived previously. Thus

$$\cos(u + v) = \cos[u - (-v)]$$
$$= \cos u \cos(-v) + \sin u \sin(-v).$$

It was shown in Section 5.7 that $\cos(-v) = \cos v$ and $\sin(-v) = -\sin v$ for all v, and hence

$$\boxed{\cos(u + v) = \cos u \cos v - \sin u \sin v.}$$

Example 2　Find $\cos(7\pi/12)$ by using $\pi/3$ and $\pi/4$.

Solution   Writing $7\pi/12 = (\pi/3) + (\pi/4)$ and using the formula for $\cos(u+v)$ gives
us

$$\cos\frac{7\pi}{12} = \cos\left(\frac{\pi}{3} + \frac{\pi}{4}\right)$$

$$= \cos\frac{\pi}{3}\cos\frac{\pi}{4} - \sin\frac{\pi}{3}\sin\frac{\pi}{4}$$

$$= \frac{1}{2}\frac{\sqrt{2}}{2} - \frac{\sqrt{3}}{2}\frac{\sqrt{2}}{2}$$

$$= \frac{\sqrt{2} - \sqrt{6}}{4}. \qquad \blacksquare$$

Similar identities are true for the sine function. Let us first establish the following identities, which are of some interest in themselves.

$$\boxed{\begin{array}{l} \cos\left(\dfrac{\pi}{2} - u\right) = \sin u \\[2mm] \sin\left(\dfrac{\pi}{2} - u\right) = \cos u \\[2mm] \tan\left(\dfrac{\pi}{2} - u\right) = \cot u \end{array}}$$

The first of these identities may be proved as follows:

$$\cos\left(\frac{\pi}{2} - u\right) = \cos\frac{\pi}{2}\cdot\cos u + \sin\frac{\pi}{2}\cdot\sin u$$

$$= 0\cdot\cos u + 1\cdot\sin u$$

$$= \sin u.$$

To obtain the second identity we substitute $(\pi/2) - v$ for u in the first identity, obtaining

$$\cos\left[\frac{\pi}{2} - \left(\frac{\pi}{2} - v\right)\right] = \sin\left(\frac{\pi}{2} - v\right)$$

or equivalently, $\cos v = \sin\left(\dfrac{\pi}{2} - v\right).$

The third identity may be established as follows:

$$\tan\left(\frac{\pi}{2} - u\right) = \frac{\sin\left(\dfrac{\pi}{2} - u\right)}{\cos\left(\dfrac{\pi}{2} - u\right)} = \frac{\cos u}{\sin u} = \cot u.$$

If θ denotes the degree measure of an angle, we may write

$$\sin(90° - \theta) = \cos\theta$$
$$\cos(90° - \theta) = \sin\theta$$
$$\tan(90° - \theta) = \cot\theta$$

If θ is acute, then θ and $90° - \theta$ are **complementary**, since their sum is $90°$. It is customary to refer to the sine and cosine functions as **cofunctions** of one another. Similarly, the tangent and cotangent functions are cofunctions, as are the secant and cosecant. Consequently, the last three formulas constitute a partial description of the fact that *any functional value of θ equals the cofunction of the complementary angle $90° - \theta$.*

The following identities may now be established.

$$\sin(u + v) = \sin u \cos v + \cos u \sin v$$
$$\sin(u - v) = \sin u \cos v - \cos u \sin v$$
$$\tan(u + v) = \frac{\tan u + \tan v}{1 - \tan u \tan v}$$
$$\tan(u - v) = \frac{\tan u - \tan v}{1 + \tan u \tan v}$$

We shall prove the first and third and leave the proofs of the remaining two as exercises. The reader should supply reasons for each of the following steps.

$$\sin(u + v) = \cos\left[\frac{\pi}{2} - (u + v)\right]$$
$$= \cos\left[\left(\frac{\pi}{2} - u\right) - v\right]$$
$$= \cos\left(\frac{\pi}{2} - u\right)\cos v + \sin\left(\frac{\pi}{2} - u\right)\sin v$$
$$= \sin u \cos v + \cos u \sin v.$$

To verify the formula for $\tan(u + v)$ we begin as follows:

$$\tan(u + v) = \frac{\sin(u + v)}{\cos(u + v)}$$
$$= \frac{\sin u \cos v + \cos u \sin v}{\cos u \cos v - \sin u \sin v}.$$

Next, dividing numerator and denominator by $\cos u \cos v$ (assuming, of course, that $\cos u \cos v \neq 0$), we obtain

$$\tan(u + v) = \frac{\left(\dfrac{\sin u}{\cos u}\right)\left(\dfrac{\cos v}{\cos v}\right) + \left(\dfrac{\cos u}{\cos u}\right)\left(\dfrac{\sin v}{\cos v}\right)}{\left(\dfrac{\cos u}{\cos u}\right)\left(\dfrac{\cos v}{\cos v}\right) - \left(\dfrac{\sin u}{\cos u}\right)\left(\dfrac{\sin v}{\cos v}\right)}$$
$$= \frac{\tan u + \tan v}{1 - \tan u \tan v}$$

The formulas derived in this section for functions of $u + v$ are known as **addition formulas**. Those for $u - v$ will be called **subtraction formulas**.

Example 3   Given $\sin \alpha = 4/5$, where α is an angle in quadrant I, and $\cos \beta = -12/13$, where β is in quadrant II, find $\sin (\alpha + \beta)$, $\tan (\alpha + \beta)$, and the quadrant containing $\alpha + \beta$.

Solution   It is convenient to represent α and β geometrically, as illustrated in Figure 6.2. There is no loss of generality in picturing α and β as positive angles between 0 and 2π as we have done in the figure. Since $\sin \alpha = 4/5$, the point $(3, 4)$ is on the terminal side of α. (Why?) Similarly, since $\cos \beta = -12/13$, we may choose the point $(-12, 5)$ on the terminal side of β. Referring to Figure 6.2 and using the Theorem on Trigonometric Functions as Ratios (see Section 5.5) we obtain

$$\cos \alpha = \tfrac{3}{5}, \quad \tan \alpha = \tfrac{4}{3}, \quad \sin \beta = \tfrac{5}{13}, \quad \text{and} \quad \tan \beta = -\tfrac{5}{12}.$$

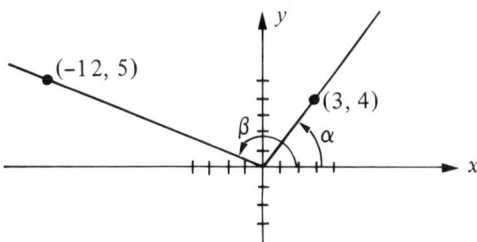

Figure 6.2

Using addition formulas gives us

$$\sin (\alpha + \beta) = \sin \alpha \cos \beta + \cos \alpha \sin \beta$$

$$= \left(\frac{4}{5}\right)\left(-\frac{12}{13}\right) + \left(\frac{3}{5}\right)\left(\frac{5}{13}\right)$$

$$= -\frac{33}{65}.$$

$$\tan (\alpha + \beta) = \frac{\tan \alpha + \tan \beta}{1 - \tan \alpha \tan \beta}$$

$$= \frac{\dfrac{4}{3} + \left(-\dfrac{5}{12}\right)}{1 - \left(\dfrac{4}{3}\right)\left(-\dfrac{5}{12}\right)}$$

$$= \frac{33}{56}.$$

Since $\sin (\alpha + \beta)$ is negative and $\tan (\alpha + \beta)$ is positive, it follows that $\alpha + \beta$ lies in quadrant III. ■

The addition formulas are very important in the study of calculus, where the type of simplification illustrated in the next example is required.

Example 4   If $f(x) = \sin x$ and $h \neq 0$, prove that

$$\frac{f(x+h) - f(x)}{h} = \sin x \left(\frac{\cos h - 1}{h}\right) + \cos x \left(\frac{\sin h}{h}\right).$$

Solution   Using the definition of f and the addition formula for the sine function we
obtain

$$\frac{f(x+h) - f(x)}{h} = \frac{\sin (x+h) - \sin x}{h}$$

$$= \frac{\sin x \cos h + \cos x \sin h - \sin x}{h}$$

$$= \frac{\sin x(\cos h - 1) + \cos x \sin h}{h}$$

$$= \sin x \left(\frac{\cos h - 1}{h}\right) + \cos x \left(\frac{\sin h}{h}\right). \qquad \blacksquare$$

Example 5   Prove that for every x,

$$a \cos Bx + b \sin Bx = A \cos (Bx - C)$$

where $A = \sqrt{a^2 + b^2}$ and $\tan C = b/a$.

Solution   Using the formula for $\cos (u - v)$ with $u = Bx$ and $v = C$ gives us

$$A \cos (Bx - C) = A (\cos Bx \cos C + \sin Bx \sin C).$$

We shall complete the proof by determining conditions on a and b such that

$$a \cos Bx + b \sin Bx = A \cos Bx \cos C + A \sin Bx \sin C.$$

The last formula is true for every x if and only if

$$a = A \cos C \quad \text{and} \quad b = A \sin C$$

(to verify this let $x = 0$ and $x = \pi/2$). Consequently,

$$a^2 + b^2 = A^2 \cos^2 C + A^2 \sin^2 C$$
$$= A^2 (\cos^2 C + \sin^2 C) = A^2$$

and we may choose

$$A = \sqrt{a^2 + b^2}.$$

Finally, we note that

$$\frac{b}{a} = \frac{A \sin C}{A \cos C} = \frac{\sin C}{\cos C} = \tan C.$$

This establishes the identity. $\qquad \blacksquare$

Example 6   The graph of $f(x) = \cos x + \sin x$ was sketched in Section 5.9 by adding ordinates (see Figure 5.44). Use the formula derived in Example 5 to find the amplitude, period, and phase shift of f.

Solution   Letting $a = 1$, $b = 1$, and $B = 1$ in Example 5 we obtain

$$A = \sqrt{a^2 + b^2} = \sqrt{1 + 1} = \sqrt{2} \quad \text{and} \quad \tan C = b/a = 1/1 = 1.$$

Since $\tan C = 1$ we may choose $C = \pi/4$. Substitution in the formula gives us

$$\cos x + \sin x = \sqrt{2} \cos\left(x - \frac{\pi}{4}\right).$$

It follows from our work in Section 5.8 that the amplitude is $\sqrt{2}$, the period is 2π, and the phase shift is $\pi/4$. The reader may check these facts by referring to Figure 5.44.   ■

EXERCISES 6.4

Write the expressions in Exercises 1–4 in terms of cofunctions of complementary angles.

1   (a) $\sin 46° 37'$ (b) $\cos 73° 12'$ (c) $\tan \pi/6$

2   (a) $\tan 24° 12'$ (b) $\sin 89° 41'$ (c) $\cos \pi/3$

3   (a) $\cos 7\pi/20$ (b) $\sin (1/4)$ (c) $\tan 1$

4   (a) $\sin \pi/12$ (b) $\cos (0.64)$ (c) $\tan \sqrt{2}$

Find the exact functional values in Exercises 5–10.

5   (a) $\cos \pi/4 + \cos \pi/3$ (b) $\cos 7\pi/12$

6   (a) $\sin 2\pi/3 + \sin \pi/4$ (b) $\sin 11\pi/12$

7   (a) $\tan 60° + \tan 225°$ (b) $\tan 285°$

8   (a) $\cos 135° - \cos 60°$ (b) $\cos 75°$

9   (a) $\sin 3\pi/4 - \sin \pi/6$ (b) $\sin 7\pi/12$

10   (a) $\tan 3\pi/4 - \tan \pi/6$ (b) $\tan 7\pi/12$

Express each of Exercises 11–16 in terms of one function of one angle.

11 $\cos 48° \cos 23° + \sin 48° \sin 23°$

12 $\cos 13° \cos 50° - \sin 13° \sin 50°$

13 $\cos 10° \sin 5° - \sin 10° \cos 5°$

14 $\sin 57° \cos 4° + \cos 57° \sin 4°$

15 $\cos 3 \sin (-2) - \cos 2 \sin 3$

16 $\sin (-5) \cos 2 + \cos 5 \sin (-2)$

17   If α and β are acute angles such that $\cos \alpha = 4/5$ and $\tan \beta = 8/15$, find $\cos (\alpha + \beta)$, $\sin (\alpha + \beta)$, and the quadrant containing $\alpha + \beta$.

18   If $\sin \alpha = -4/5$ and $\sec \beta = 5/3$, where α is a third-quadrant angle and β is a first-quadrant angle, find $\sin (\alpha + \beta)$, $\tan (\alpha + \beta)$, and the quadrant containing $\alpha + \beta$.

19   If $\tan \alpha = -7/24$ and $\cot \beta = 3/4$, where α is in the second quadrant and β is in the third quadrant, find $\sin (\alpha + \beta)$, $\cos (\alpha + \beta)$, $\tan (\alpha + \beta)$, $\sin (\alpha - \beta)$, $\cos (\alpha - \beta)$, and $\tan (\alpha - \beta)$.

20   If $P(t_1)$ and $P(t_2)$ are in quadrant III, and if $\cos t_1 = -2/5$ and $\cos t_2 = -3/5$, find $\sin (t_1 - t_2)$, $\cos (t_1 - t_2)$, and the quadrant containing $P(t_1 - t_2)$.

Verify the identities in Exercises 21–40.

21 $\sin\left(x + \dfrac{\pi}{2}\right) = \cos x$

22 $\cos\left(x + \dfrac{\pi}{2}\right) = -\sin x$

23 $\cos\left(\theta + \dfrac{3\pi}{2}\right) = \sin\theta$

24 $\sin\left(\alpha - \dfrac{3\pi}{2}\right) = \cos\alpha$

25 $\sin\left(\theta + \dfrac{\pi}{4}\right) = \left(\dfrac{\sqrt{2}}{2}\right)(\sin\theta + \cos\theta)$

26 $\cos\left(\theta + \dfrac{\pi}{4}\right) = \left(\dfrac{\sqrt{2}}{2}\right)(\cos\theta - \sin\theta)$

27 $\tan\left(u + \dfrac{\pi}{4}\right) = \dfrac{1 + \tan u}{1 - \tan u}$

28 $\tan\left(x - \dfrac{\pi}{4}\right) = \dfrac{\tan x - 1}{\tan x + 1}$

29 $\tan\left(u + \dfrac{\pi}{2}\right) = -\cot u$

30 $\cot\left(t - \dfrac{\pi}{3}\right) = \dfrac{\sqrt{3}\tan t + 1}{\tan t - \sqrt{3}}$

31 $\sin(u + v)\cdot\sin(u - v) = \sin^2 u - \sin^2 v$

32 $\cos(u + v)\cdot\cos(u - v) = \cos^2 u - \sin^2 v$

33 $\cos(u + v) + \cos(u - v) = 2\cos u\cos v$

34 $\sin(u + v) + \sin(u - v) = 2\sin u\cos v$

35 $\sin 2u = 2\sin u\cos u$ (Hint: $2u = u + u$)

36 $\cos 2u = \cos^2 u - \sin^2 u$

37 $\dfrac{\sin(u + v)}{\cos(u - v)} = \dfrac{\tan u + \tan v}{1 + \tan u\tan v}$

38 $\dfrac{\cos(u + v)}{\cos(u - v)} = \dfrac{1 - \tan u\tan v}{1 + \tan u\tan v}$

39 $\dfrac{\sin(u + v)}{\sin(u - v)} = \dfrac{\tan u + \tan v}{\tan u - \tan v}$

40 $\tan u + \tan v = \dfrac{\sin(u + v)}{\cos u\cos v}$

41 Express $\sin(u + v + w)$ in terms of functions of u, v, and w. (Hint: Write $\sin(u + v + w) = \sin[(u + v) + w]$ and use an addition formula.)

42 Express $\tan(u + v + w)$ in terms of functions of u, v, and w.

43 Derive the formula

$$\cot(u + v) = \frac{\cot u\cot v - 1}{\cot u + \cot v}.$$

44 If α and β are complementary angles, prove that

$$\sin^2\alpha + \sin^2\beta = 1.$$

45 Derive the subtraction formulas for the sine and tangent functions.

46 Prove each of the following:
(a) $\sec(\pi/2 - u) = \csc u$
(b) $\csc(\pi/2 - u) = \sec u$
(c) $\cot(\pi/2 - u) = \tan u$

47 If $f(x) = \cos x$, prove that

$$\frac{f(x + h) - f(x)}{h}$$
$$= \cos x\left(\frac{\cos h - 1}{h}\right) - \sin x\left(\frac{\sin h}{h}\right).$$

48 If $f(x) = \tan x$, prove that

$$\frac{f(x + h) - f(x)}{h}$$
$$= \sec^2 x\left(\frac{\sin h}{h}\right)\frac{1}{\cos h - \sin h\tan x}.$$

In Exercises 49 and 50 find the solutions of the given equation which are in the interval $[0, 2\pi)$ and also find the degree measure of each solution.

49 $\sin 4t\cos t = \sin t\cos 4t$

50 $\cos 5t\cos 3t = 2^{-1} + \sin(-5t)\sin 3t$

In each of Exercises 51–54 use the formula given in Example 5 to express f in terms of the cosine function and determine the amplitude, period, and phase shift.

51 $f(x) = \sqrt{3}\cos 2x + \sin 2x$

52 $f(x) = \cos 4x + \sqrt{3}\sin 4x$

53 $f(x) = 2\cos 3x - 2\sin 3x$

54 $f(x) = -5\cos 10x + 5\sin 10x$

CALCULATOR EXERCISES 6.4

1 Demonstrate, by direct calculations, that the addition formulas for the sine, cosine, and tangent functions are true for the following values of u and v: $u = 1.46$, $v = 8.27$.

2 If u and v are in the interval $[0, \pi]$ such that $\cos u = -0.4163$ and $\sin v = 0.7216$, find $\cos (u + v)$ and $\sin (u + v)$.

6.5 MULTIPLE-ANGLE FORMULAS

In the preceding section we were interested primarily in trigonometric identities which involve values of $u \pm v$. In this section we shall obtain formulas for values of nu, where n represents a certain integer or rational number. The formulas are referred to as **multiple-angle formulas**. In particular, the following identities are called the **double-angle formulas**, because of the expression $2u$ which appears.

Double-Angle Formulas

$$\sin 2u = 2 \sin u \cos u$$
$$\cos 2u = \cos^2 u - \sin^2 u = 1 - 2 \sin^2 u = 2 \cos^2 u - 1$$
$$\tan 2u = \frac{2 \tan u}{1 - \tan^2 u}$$

These identities may be proved by letting $u = v$ in the appropriate addition formulas. If we use the formula for $\sin (u + v)$, then

$$\sin 2u = \sin (u + u)$$
$$= \sin u \cos u + \cos u \sin u$$
$$= 2 \sin u \cos u.$$

Similarly, using the formula for $\cos (u + v)$ gives us

$$\cos 2u = \cos (u + u)$$
$$= \cos u \cos u - \sin u \sin u$$
$$= \cos^2 u - \sin^2 u.$$

The other two forms for $\cos 2u$ may be obtained by using the fundamental identity $\sin^2 u + \cos^2 u = 1$. Thus,

$$\cos 2u = \cos^2 u - \sin^2 u$$
$$= (1 - \sin^2 u) - \sin^2 u$$
$$= 1 - 2 \sin^2 u.$$

Similarly, if we substitute for $\sin^2 u$ instead of $\cos^2 u$, we obtain

$$\cos 2u = \cos^2 u - (1 - \cos^2 u)$$
$$= 2 \cos^2 u - 1.$$

The formula for $\tan 2u$ may be obtained by taking $u = v$ in the formula for $\tan(u + v)$.

Example 1   Find $\sin 2\alpha$ and $\cos 2\alpha$ if $\sin \alpha = \frac{4}{5}$ and α is in quadrant I.

Solution   As in Example 3 of the preceding section, $\cos \alpha = 3/5$. Substitution in double-angle formulas gives us

$$\sin 2\alpha = 2 \sin \alpha \cos \alpha = 2\left(\frac{4}{5}\right)\left(\frac{3}{5}\right) = \frac{24}{25}$$

$$\cos 2\alpha = \cos^2 \alpha - \sin^2 \alpha = \frac{9}{25} - \frac{16}{25} = -\frac{7}{25}. \qquad \blacksquare$$

Example 2   Express $\cos 3\theta$ in terms of $\cos \theta$.

Solution
$$\begin{aligned}
\cos 3\theta &= \cos(2\theta + \theta) \\
&= \cos 2\theta \cos \theta - \sin 2\theta \sin \theta \\
&= (2\cos^2 \theta - 1)\cos \theta - (2 \sin \theta \cos \theta) \sin \theta \\
&= 2\cos^3 \theta - \cos \theta - 2 \cos \theta \sin^2 \theta \\
&= 2\cos^3 \theta - \cos \theta - 2 \cos \theta(1 - \cos^2 \theta) \\
&= 2\cos^3 \theta - \cos \theta - 2 \cos \theta + 2\cos^3 \theta \\
&= 4\cos^3 \theta - 3 \cos \theta. \qquad \blacksquare
\end{aligned}$$

The next three identities are useful for simplifying certain expressions involving powers of trigonometric functions.

$$\sin^2 u = \frac{1 - \cos 2u}{2}; \qquad \cos^2 u = \frac{1 + \cos 2u}{2}; \qquad \tan^2 u = \frac{1 - \cos 2u}{1 + \cos 2u}$$

The first and second of these identities may be verified by solving the equations

$$\cos 2u = 1 - 2 \sin^2 u \quad \text{and} \quad \cos 2u = 2 \cos^2 u - 1$$

for $\sin^2 u$ and $\cos^2 u$, respectively. The third identity may be obtained from the first two by using the fact that $\tan^2 u = \sin^2 u / \cos^2 u$.

Example 3   Verify the identity $\sin^2 x \cos^2 x = \frac{1}{8}(1 - \cos 4x)$.

Solution   Using the preceding identities for $\sin^2 u$ and $\cos^2 u$ with $u = x$, we obtain

$$\sin^2 x \cos^2 x = \left(\frac{1 - \cos 2x}{2}\right)\left(\frac{1 + \cos 2x}{2}\right)$$

$$= \frac{1}{4}(1 - \cos^2 2x)$$

$$= \frac{1}{4}\sin^2 2x.$$

Finally, using the formula $\sin^2 u = (1 - \cos 2u)/2$ with $u = 2x$, gives us

$$\sin^2 x \cos^2 x = \frac{1}{4}\left(\frac{1 - \cos 4x}{2}\right)$$

$$= \frac{1}{8}(1 - \cos 4x).$$

Another method of proof is to use the fact that $\sin 2x = 2 \sin x \cos x$ and hence that

$$\sin x \cos x = \tfrac{1}{2} \sin 2x.$$

Squaring both sides, we have

$$\sin^2 x \cos^2 x = \tfrac{1}{4} \sin^2 2x.$$

The remainder of the solution is the same as that given above. ∎

Example 4 Express $\cos^4 t$ in terms of values of cos with exponent 1.

Solution The reader should supply reasons for the following steps.

$$\cos^4 t = (\cos^2 t)^2$$

$$= \left(\frac{1 + \cos 2t}{2}\right)^2$$

$$= \frac{1}{4}(1 + 2\cos 2t + \cos^2 2t)$$

$$= \frac{1}{4}\left(1 + 2\cos 2t + \frac{1 + \cos 4t}{2}\right)$$

$$= \frac{3}{8} + \frac{1}{2}\cos 2t + \frac{1}{8}\cos 4t. \quad ∎$$

Substituting $v/2$ for u in the last three formulas for $\sin^2 u$, $\cos^2 u$, and $\tan^2 u$ gives us

$$\sin^2 \frac{v}{2} = \frac{1 - \cos v}{2}$$

$$\cos^2 \frac{v}{2} = \frac{1 + \cos v}{2}$$

$$\tan^2 \frac{v}{2} = \frac{1 - \cos v}{1 + \cos v}.$$

If we take the square root of both sides of the latter equations and use the fact that $\sqrt{a^2} = |a|$ for every real number a, the following identities result. They are called *half-*

angle formulas because of the expression $v/2$ which appears.

Half-Angle Formulas

$$\left|\sin\frac{v}{2}\right| = \sqrt{\frac{1 - \cos v}{2}}$$

$$\left|\cos\frac{v}{2}\right| = \sqrt{\frac{1 + \cos v}{2}}$$

$$\left|\tan\frac{v}{2}\right| = \sqrt{\frac{1 - \cos v}{1 + \cos v}}$$

The absolute value signs may be eliminated if more information is known about $v/2$. For example, if the point $P(v/2)$ on the unit circle U is in either quadrant I or II (or, equivalently, if the angle determined by $v/2$ lies in one of these quadrants), then $\sin v/2$ is positive and we may write

$$\sin\frac{v}{2} = \sqrt{\frac{1 - \cos v}{2}}.$$

However, if $v/2$ leads to a point (or angle) in either quadrant III or IV, then

$$\sin\frac{v}{2} = -\sqrt{\frac{1 - \cos v}{2}}.$$

Similar remarks are true for the other formulas.

An alternate form for $\tan v/2$ can be obtained. Multiplying numerator and denominator of the radicand in the third half-angle formula by $1 - \cos v$ gives us

$$\left|\tan\frac{v}{2}\right| = \sqrt{\frac{1 - \cos v}{1 + \cos v} \cdot \frac{1 - \cos v}{1 - \cos v}}$$

$$= \sqrt{\frac{(1 - \cos v)^2}{\sin^2 v}}$$

$$= \frac{1 - \cos v}{|\sin v|}.$$

The absolute value sign is unnecessary in the numerator since $1 - \cos v$ is never negative. It can be shown that $\tan v/2$ and $\sin v$ always have the same sign. For example, if $0 < v < \pi$, then $0 < v/2 < \pi/2$, and hence both $\sin v$ and $\tan v/2$ are positive. If $\pi < v < 2\pi$, then $\pi/2 < v/2 < \pi$, and hence both $\sin v$ and $\tan v/2$ are negative. It is possible to generalize these remarks to all values of v for which the expressions $\tan v/2$ and $(1 - \cos v)/|\sin v|$ have meaning. This gives us the first of the next two identities. The second identity for $\tan(v/2)$ may be obtained by multiplying numerator and denominator of the radicand in the third half-angle formula by $1 + \cos v$.

$$\tan\frac{v}{2} = \frac{1 - \cos v}{\sin v}, \qquad \tan\frac{v}{2} = \frac{\sin v}{1 + \cos v}$$

Example 5   Find the exact values of sin 22.5° and cos 22.5°.

Solution   Using the formula for sin $(v/2)$ and the fact that 22.5 is in quadrant I, we have

$$\sin 22.5° = \sin \frac{45°}{2} = \sqrt{\frac{1 - \cos 45°}{2}}$$

$$= \sqrt{\frac{1 - \sqrt{2}/2}{2}}$$

$$= \frac{\sqrt{2 - \sqrt{2}}}{2}$$

$$\cos 22.5° = \sqrt{\frac{1 + \cos 45°}{2}}$$

$$= \sqrt{\frac{1 + \sqrt{2}/2}{2}}$$

$$= \frac{\sqrt{2 + \sqrt{2}}}{2}. \qquad\blacksquare$$

Example 6   If $\tan \alpha = -4/3$, where α is in quadrant IV, find $\tan \alpha/2$.

Solution   If we choose the point $(3, -4)$ on the terminal side of α as illustrated in Figure 6.3, then $\sin \alpha = -4/5$ and $\cos \alpha = 3/5$. (Why?) Applying a half-angle formula, we have

$$\tan \frac{\alpha}{2} = \frac{1 - \cos \alpha}{\sin \alpha} = \frac{1 - \frac{3}{5}}{-\frac{4}{5}} = -\frac{1}{2}.$$

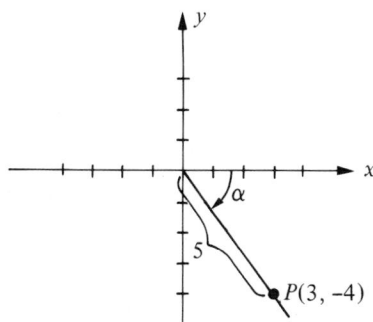

Figure 6.3 $\qquad\qquad\qquad\qquad\qquad\qquad\qquad\qquad\blacksquare$

Example 7   Find the solutions of the equation $\cos 2x + \cos x = 0$ which are in the interval $[0, 2\pi)$. Express the solutions both in radian measure and degree measure.

Solution   We first use a double-angle formula to write the equation in terms of cos x and then solve by factoring as follows:

$$\cos 2x + \cos x = 0$$
$$(2\cos^2 x - 1) + \cos x = 0$$
$$2\cos^2 x + \cos x - 1 = 0$$
$$(2\cos x - 1)(\cos x + 1) = 0.$$

Setting each factor equal to zero we obtain

$$\cos x = \frac{1}{2} \quad \text{and} \quad \cos x = -1.$$

The solutions of the last two equations (and hence of the given equation) in $[0, 2\pi)$ are

$$\pi/3, \quad 5\pi/3, \quad \text{and} \quad 3\pi/2.$$

The corresponding degree measures are

$$60°, \quad 300°, \quad \text{and} \quad 270°. \qquad \blacksquare$$

EXERCISES 6.5

In Exercises 1–4 find the exact values of sin 2θ, cos 2θ, and tan 2θ subject to the given conditions.

1 $\cos \theta = \frac{3}{5}$ and θ acute

2 $\cot \theta = \frac{4}{3}$ and $180° < \theta < 270°$

3 $\sec \theta = -3$ and $90° < \theta < 180°$

4 $\sin \theta = -\frac{4}{5}$ and $270° < \theta < 360°$

In Exercises 5–8 find the exact values of sin $\theta/2$, cos $\theta/2$, and tan $\theta/2$ subject to the given conditions.

5 $\sec \theta = \frac{5}{4}$ and θ acute

6 $\csc \theta = -\frac{5}{3}$ and $-90° < \theta < 0°$

7 $\tan \theta = 1$ and $-180° < \theta < -90°$

8 $\sec \theta = -4$ and $180° < \theta < 270°$

In Exercises 9 and 10, use half-angle formulas to find exact values for the given numbers.

9 (a) $\cos 67° \, 30'$ (b) $\sin 15°$
(c) $\tan 3\pi/8$

10 (a) $\cos 165°$ (b) $\sin 157° \, 30'$
(c) $\tan \pi/8$

Verify the identities stated in Exercises 11–28.

11 $\sin 10\theta = 2\sin 5\theta \cos 5\theta$

12 $\cos^2 3x - \sin^2 3x = \cos 6x$

13 $4\sin \frac{x}{2}\cos \frac{x}{2} = 2\sin x$

14 $\dfrac{\sin^2 2\alpha}{\sin^2 \alpha} = 4 - 4\sin^2 \alpha$

15 $(\sin t + \cos t)^2 = 1 + \sin 2t$

16 $\csc 2u = \frac{1}{2}\sec u \csc u$

17 $\sin 3u = \sin u(3 - 4\sin^2 u)$

18 $\sin 4t = 4\cos t \sin t(1 - 2\sin^2 t)$

19 $\cos 4\theta = 8\cos^4 \theta - 8\cos^2 \theta + 1$

20 $\cos 6t = 32\cos^6 t - 48\cos^4 t + 18\cos^2 t - 1$

21 $\sin^4 t = \frac{3}{8} - \frac{1}{2}\cos 2t + \frac{1}{8}\cos 4t$

22 $\cos^4 x - \sin^4 x = \cos 2x$

23 $\sec 2\theta = \dfrac{\sec^2 \theta}{2 - \sec^2 \theta}$

24 $\cot 2u = \dfrac{\cot^2 u - 1}{2\cot u}$

25 $2\sin^2 2t + \cos 4t = 1$

26 $\tan\theta + \cot\theta = 2\csc 2\theta$

27 $\tan 3u = \dfrac{(3 - \tan^2 u)\tan u}{1 - 3\tan^2 u}$

28 $\dfrac{1 + \sin 2v + \cos 2v}{1 + \sin 2v - \cos 2v} = \cot v$

Write the expressions in Exercises 29 and 30 in terms of values of cos with exponent 1.

29 $\cos^4 (\theta/2)$ **30** $\sin^4 2x$

In Exercises 31–38 find all solutions of the given equations in the interval $[0, 2\pi)$. Express the solutions both in radian measure and degree measure.

31 $\sin 2t + \sin t = 0$ **32** $\cos t - \sin 2t = 0$

33 $\cos u + \cos 2u = 0$ **34** $\cos 2\theta - \tan\theta = 1$

35 $\tan 2x = \tan x$ **36** $\tan 2t - 2\cos t = 0$

37 $\sin\frac{1}{2}u + \cos u = 1$ **38** $2 - \cos^2 x$
$= 4\sin^2 \frac{1}{2}x.$

39 If $a > 0$, $b > 0$, and $0 < u < \pi/2$, prove that

$$a\sin u + b\cos u = \sqrt{a^2 + b^2}\,\sin(u + v)$$

where $0 < v < \pi/2$, $\sin v = b/\sqrt{a^2 + b^2}$, and $\cos v = a/\sqrt{a^2 + b^2}$.

40 Use Exercise 39 to express $8\sin u + 15\cos u$ in the form $c\sin(u + v)$.

CALCULATOR EXERCISES 6.5

1 Approximate $\sin 2\theta$ if θ is a fourth-quadrant angle and $\sin\theta = -0.4637$.

2 Approximate $\cos(\theta/2)$ if θ is a third-quadrant angle and $\tan\theta = 3.8105$.

In each of Exercises 3–6 demonstrate, by direct calculations, that the indicated formula is true for the stated value of x.

3 $\sin 2x = 2\sin x \cos x$; $x = 3.624$

4 $\cos 2x = \cos^2 x - \sin^2 x$; $x = \sqrt{2\pi} + e^{\sqrt{7}}$

5 $\tan 2x = 2\tan x/(1 - \tan^2 x)$; $x = \log(13.6)^5$

6 $|\sin(x/2)| = \sqrt{(1 - \cos x)/2}$; $x = \tan 0.1634$

6.6 SUM AND PRODUCT FORMULAS

The following identities may be used to change the form of certain trigonometric expressions.

Product Formulas

$$\sin (u + v) + \sin (u - v) = 2 \sin u \cos v$$
$$\sin (u + v) - \sin (u - v) = 2 \cos u \sin v$$
$$\cos (u + v) + \cos (u - v) = 2 \cos u \cos v$$
$$\cos (u - v) - \cos (u + v) = 2 \sin u \sin v$$

Each of these identities is a consequence of our work in Section 6.4. For example, to verify the first product formula, we merely add the left- and right-hand sides of the identities we obtained for $\sin (u + v)$ and $\sin (u - v)$. The remaining product formulas are obtained in similar fashion.

By starting with the expression on the *right* side of each product formula, we can express certain products as sums, as illustrated in the next example.

Example 1   Express each of the following as a sum.

(a) $\sin 4\theta \cos 3\theta$ (b) $\sin 3x \sin x$

Solution

(a)   Using the first product formula with $u = 4\theta$ and $v = 3\theta$ gives us

$$2 \sin 4\theta \cos 3\theta = \sin (4\theta + 3\theta) + \sin (4\theta - 3\theta)$$

or $\sin 4\theta \cos 3\theta = \frac{1}{2} \sin 7\theta + \frac{1}{2} \sin \theta.$

This relationship could also have been obtained by using the second product formula.

(b)   Using the fourth product formula with $u = 3x$ and $v = x$ gives us

$$2 \sin 3x \sin x = \cos (3x - x) - \cos (3x + x)$$

or $\sin 3x \sin x = \frac{1}{2} \cos 2x - \frac{1}{2} \cos 4x.$ ∎

The Product Formulas may also be employed to express a sum as a product. In order to obtain a form which can be applied more easily, we shall change the notation as follows. If we let

$$u + v = a \quad \text{and} \quad u - v = b$$

then $(u + v) + (u - v) = a + b$, which simplifies to

$$u = \frac{a + b}{2}.$$

Similarly, from $(u + v) - (u - v) = a - b$ we obtain

$$v = \frac{a - b}{2}.$$

If we now substitute for $u + v$ and $u - v$ on the left sides of the Product Formulas, and for u and v on the right sides, we obtain the following.

Factoring Formulas

$$\sin a + \sin b = 2 \sin \frac{a + b}{2} \cos \frac{a - b}{2}$$

$$\sin a - \sin b = 2 \cos \frac{a + b}{2} \sin \frac{a - b}{2}$$

$$\cos a + \cos b = 2 \cos \frac{a + b}{2} \cos \frac{a - b}{2}$$

$$\cos b - \cos a = 2 \sin \frac{a + b}{2} \sin \frac{a - b}{2}$$

Example 2   Express $\sin 5x - \sin 3x$ as a product.

Solution   Using the second factoring formula with $a = 5x$ and $b = 3x$,

$$\sin 5x - \sin 3x = 2 \cos \frac{5x + 3x}{2} \sin \frac{5x - 3x}{2}$$

$$= 2 \cos 4x \sin x. \qquad \blacksquare$$

Example 3   Verify the identity $\dfrac{\sin 3t + \sin 5t}{\cos 3t - \cos 5t} = \cot t.$

Solution   Using the first and fourth factoring formulas,

$$\frac{\sin 3t + \sin 5t}{\cos 3t - \cos 5t} = \frac{2 \sin \dfrac{3t + 5t}{2} \cos \dfrac{3t - 5t}{2}}{2 \sin \dfrac{5t + 3t}{2} \sin \dfrac{5t - 3t}{2}}$$

$$= \frac{2 \sin 4t \cos (-t)}{2 \sin 4t \sin t}$$

$$= \frac{\cos (-t)}{\sin t}$$

$$= \frac{\cos t}{\sin t}$$

$$= \cot t \qquad \blacksquare$$

Example 4   Find the solutions of the equation $\cos t - \sin 2t - \cos 3t = 0.$

Solution   Using the fourth factoring formula,

$$\cos t - \cos 3t = 2 \sin \frac{3t + t}{2} \sin \frac{3t - t}{2}$$

$$= 2 \sin 2t \sin t.$$

Hence the given equation is equivalent to

$$2 \sin 2t \sin t - \sin 2t = 0$$

or to

$$\sin 2t (2 \sin t - 1) = 0.$$

Thus the solutions of the given equation are the solutions of

$$\sin 2t = 0 \quad \text{and} \quad \sin 2t = \frac{1}{2}.$$

The first of these equations has solutions

$$2t = n\pi \quad \text{or} \quad t = \frac{\pi}{2} n$$

where n is any integer. The second equation has solutions

$$t = \frac{\pi}{6} + 2n\pi \quad \text{and} \quad t = \frac{5\pi}{6} + 2n\pi$$

where n is any integer.   ∎

The addition formulas may also be employed to derive the so-called **reduction formulas**. The latter formulas can be used to write expressions such as

$$\sin\left(\theta + n \cdot \frac{\pi}{2}\right) \quad \text{and} \quad \cos\left(\theta + n \cdot \frac{\pi}{2}\right)$$

where n is any integer, in terms of only $\sin \theta$ or $\cos \theta$. Similar formulas are true for the other trigonometric functions. Instead of deriving general reduction formulas we shall illustrate several special cases in the next example.

Example 5   Express $\sin(\theta - 3\pi/2)$ and $\cos(\theta + \pi)$ in terms of a function of θ.

Solution   We may proceed as follows.

$$\sin\left(\theta - \frac{3\pi}{2}\right) = \sin\theta \cos\frac{3\pi}{2} - \cos\theta \sin\frac{3\pi}{2}$$

$$= \sin\theta \cdot (0) - \cos\theta \cdot (-1)$$

$$= \cos\theta$$

$$\cos(\theta + \pi) = \cos\theta \cos\pi - \sin\theta \sin\pi$$

$$= \cos\theta \cdot (-1) - \sin\theta \cdot (0)$$

$$= -\cos\theta.$$ ∎

We have derived many important identities in this chapter. In order to make them readily available for reference, they are listed on the inside of the covers of the text.

EXERCISES 6.6

In each of Exercises 1–8 express the product as a sum or difference.

1 $2 \sin 9\theta \cos 3\theta$ **2** $2 \cos 5\theta \sin 5\theta$

3 $\sin 7t \sin 3t$ **4** $\sin (-4x) \cos 8x$

5 $\cos 6u \cos (-4u)$ **6** $\sin 4t \sin 6t$

7 $3 \cos x \sin 2x$ **8** $5 \cos u \sin 5u$

In each of Exercises 9–16 write the expression as a product.

9 $\sin 6\theta + \sin 2\theta$ **10** $\sin 4\theta - \sin 8\theta$

11 $\cos 5x - \cos 3x$ **12** $\cos 5t + \cos 6t$

13 $\sin 3t - \sin 7t$ **14** $\cos \theta - \cos 5\theta$

15 $\cos x + \cos 2x$ **16** $\sin 8t + \sin 2t$

Verify the identities in Exercises 17–24.

17 $\dfrac{\sin 4t + \sin 6t}{\cos 4t - \cos 6t} = \cot t$

18 $\dfrac{\sin \theta + \sin 3\theta}{\cos \theta + \cos 3\theta} = \tan 2\theta$

19 $\dfrac{\sin u + \sin v}{\cos u + \cos v} = \tan \dfrac{u + v}{2}$

20 $\dfrac{\sin u - \sin v}{\cos u - \cos v} = -\cot \dfrac{u + v}{2}$

21 $\dfrac{\sin u - \sin v}{\sin u + \sin v} = \dfrac{\tan \frac{1}{2}(u - v)}{\tan \frac{1}{2}(u + v)}$

22 $\dfrac{\cos u - \cos v}{\cos u + \cos v} = \tan \frac{1}{2}(u + v) \tan \frac{1}{2}(u - v)$

23 $\sin 2x + \sin 4x + \sin 6x = 4 \cos x \cos 2x \sin 3x$

24 $\dfrac{\cos t + \cos 4t + \cos 7t}{\sin t + \sin 4t + \sin 7t} = \cot 4t$

25 Express $(\sin ax)(\cos bx)$ as a sum.

26 Express $(\cos mu)(\cos nu)$ as a sum.

In Exercises 27–30 find the solutions of the given equations.

27 $\sin 5t + \sin 3t = 0$ **28** $\sin t + \sin 3t = \sin 2t$

29 $\cos x = \cos 3x$ **30** $\cos 4x - \cos 3x = 0$

In Exercises 31–36 verify the given reduction formulas by using addition formulas.

31 $\sin (\theta + \pi) = -\sin \theta$

32 $\sin \left(\theta + \dfrac{3\pi}{2}\right) = -\cos \theta$

33 $\cos \left(\theta - \dfrac{5\pi}{2}\right) = \sin \theta$

34 $\cos (\theta - 3\pi) = -\cos \theta$

35 $\tan (\pi - \theta) = -\tan \theta$

$\left(Hint: \tan (\pi - \theta) = \dfrac{\sin (\pi - \theta)}{\cos (\pi - \theta)}\right)$

36 $\tan (\theta + \pi/2) = -\cot \theta$

CALCULATOR EXERCISES 6.6

1 Demonstrate, by direct calculations, that the first product formula is true if $u = 4.631$ and $v = -8.059$.

2 Demonstrate that the first factoring formula is true if $a = 10^{\sqrt{148}}$ and $b = \log 735$.

6.7 THE INVERSE TRIGONOMETRIC FUNCTIONS

Inverse functions were discussed in Section 2.4. At that time it was pointed out that if f is a one-to-one function with domain X and range Y, then, as illustrated in (i) of Figure 6.4, for each number u in Y there is exactly one number v in X such that $f(v) = u$. In this case, as in (ii) of Figure 6.4, we can define a function f^{-1} from Y to X by letting $f^{-1}(u) = v$. The function f^{-1} is called the *inverse function* of f. In words, we say that f^{-1} *reverses* the correspondence from X to Y given by f; that is,

(A)
$$\boxed{f^{-1}(u) = v \quad \text{if and only if} \quad f(v) = u}$$

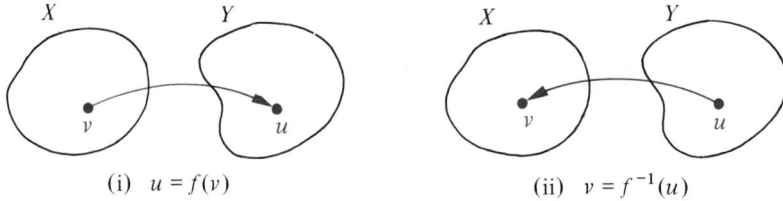

(i) $u = f(v)$ (ii) $v = f^{-1}(u)$

Figure 6.4

for all u in Y and v in X. In the remainder of this section we shall use the relationship in (A) to obtain the *inverse trigonometric functions*. Recall, from Chapter Two, that the inverse function f^{-1} is characterized by the identities

(B)
$$\boxed{\begin{aligned} f^{-1}(f(v)) &= v \quad \text{for every } v \text{ in } X \\ f(f^{-1}(u)) &= u \quad \text{for every } u \text{ in } Y. \end{aligned}}$$

Since the trigonometric functions are not one-to-one, they do not have inverses. However, by restricting the domains, it is possible to obtain functions which behave in the same way as the trigonometric functions (over the smaller domains) and which do possess inverse functions. Given a trigonometric function we shall choose a subset S of the domain throughout which the function is either increasing or decreasing, since this will guarantee the existence of an inverse. Of course, we must also choose S in such a way that the function takes on all of its values.

Let us first consider the sine function, whose domain is $\mathbb{R}$ and whose range is the set of real numbers in the interval $[-1, 1]$. The sine function is not one-to-one since, for example, numbers such as $\pi/6$, $5\pi/6$, and $-7\pi/6$ give rise to the same functional value, $\frac{1}{2}$. If we restrict x to the interval $[-\pi/2, \pi/2]$, then as x varies from $-\pi/2$ to $\pi/2$, $\sin x$ takes on each value between -1 and 1 once and only once. Note that in this interval the sine function is increasing, and hence the new function obtained by restricting the domain to $[-\pi/2, \pi/2]$ has an inverse function. Applying the relationship in (A) leads to the following definition.

Definition

> The **inverse sine function**, denoted by $\sin^{-1}$, is defined by
>
> $$\sin^{-1} u = v \quad \text{if and only if} \quad \sin v = u$$
>
> where $-1 \leq u \leq 1$ and $-\pi/2 \leq v \leq \pi/2$.

It is also customary to refer to $\sin^{-1}$ as the **arcsine function** and to use the notation arcsin u in place of $\sin^{-1} u$. The expression arcsin u is used because if $v = $ arcsin u, then $\sin v = u$, that is, v is a number (or an *arc*length) whose sine is u. Since both notations $\sin^{-1}$ and arcsin are commonly used in mathematics and its applications, we shall employ both of them in our work. Note that by definition we have

$$-\frac{\pi}{2} \leq \sin^{-1} u \leq \frac{\pi}{2}$$

or equivalently,

$$-\frac{\pi}{2} \leq \text{arcsin } u \leq \frac{\pi}{2}.$$

Example 1 Sketch the graph of $y = \sin^{-1} x$.

Solution   By definition, the graph of $y = \sin^{-1} x$ is the same as the graph of $\sin y = x$, except that the variables are restricted as follows:

$$-\frac{\pi}{2} \leq y \leq \frac{\pi}{2} \quad \text{and} \quad -1 \leq x \leq 1.$$

Several coordinates of points on the graph are given in the following table.

y	$-\dfrac{\pi}{2}$	$-\dfrac{\pi}{3}$	$-\dfrac{\pi}{6}$	0	$\dfrac{\pi}{6}$	$\dfrac{\pi}{3}$	$\dfrac{\pi}{2}$
x	-1	$-\dfrac{\sqrt{3}}{2}$	$-\dfrac{1}{2}$	0	$\dfrac{1}{2}$	$\dfrac{\sqrt{3}}{2}$	1

The graph is sketched in Figure 6.5.

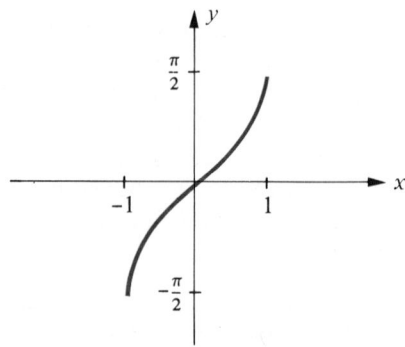

Figure 6.5. $y = \arcsin x = \sin^{-1} x$

Using the relationships stated in (B) gives us the following important identities:

$$
\begin{array}{ll}
\sin^{-1}(\sin v) = v & \text{if} \quad -\dfrac{\pi}{2} \le v \le \dfrac{\pi}{2} \\[2mm]
\sin(\sin^{-1} u) = u & \text{if} \quad -1 \le u \le 1.
\end{array}
$$

These may also be written in the form

$$
\arcsin(\sin v) = v \quad \text{and} \quad \sin(\arcsin u) = u
$$

provided that v and u are suitably restricted.   ∎

Example 2   Find $\sin^{-1}(\sqrt{2}/2)$ and $\arcsin(-1/2)$.

Solution   If $v = \sin^{-1}(\sqrt{2}/2)$, then $\sin v = \sqrt{2}/2$, and consequently, $v = \pi/4$. Note that it is essential to choose v in the interval $[-\pi/2, \pi/2]$. A number such as $3\pi/4$ is incorrect, even though $\sin(3\pi/4) = \sqrt{2}/2$.

In like manner, if $v = \arcsin(-1/2)$, then $\sin v = -1/2$. Since, by definition, we must choose v in the interval $[-\pi/2, \pi/2]$, it follows that $v = -\pi/6$.   ∎

Example 3   Find $\sin^{-1}(\tan 3\pi/4)$.

Solution   If we let

$$
v = \sin^{-1}\left(\tan \frac{3\pi}{4}\right) = \sin^{-1}(-1)
$$

then, by definition, $\sin v = -1$.

Consequently, $v = -\pi/2$.   ∎

The other trigonometric functions may also be used to introduce inverse functions. The procedure is first to determine a convenient subset of the domain so that a one-to-one function is obtained and then to apply the formulas in (A).

If the domain of the cosine function is restricted to the interval $[0, \pi]$ and if v increases from 0 to π, then $\cos v$ decreases and takes on each value between -1 and 1 once and only once. This leads to the following.

Definition

> The **inverse cosine function**, denoted by $\cos^{-1}$, is defined by
>
> $$\cos^{-1} u = v \quad \text{if and only if} \quad \cos v = u$$
>
> where $-1 \le u \le 1$ and $0 \le v \le \pi$.

The inverse cosine function is also referred to as the **arccosine function** and the notation arccos u is used interchangeably with $\cos^{-1} u$.

Applying the formulas in (B) we see that if $0 \le v \le \pi$ and $-1 \le u \le 1$, then

> $$\cos(\cos^{-1} u) = \cos(\arccos u) = u$$
> $$\cos^{-1}(\cos v) = \arccos(\cos v) = v.$$

By definition, the graph of $y = \cos^{-1} x$ is the same as the graph of $\cos y = x$, where $0 \le y \le \pi$. It is left as an exercise to verify that the graph has the appearance indicated in Figure 6.6.

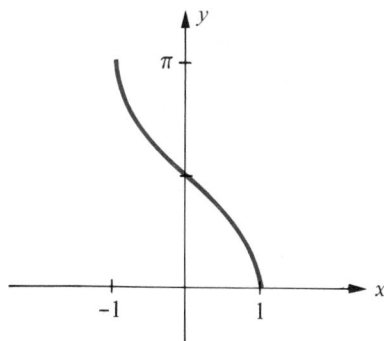

Figure 6.6. $y = \arccos x = \cos^{-1} x$

Example 4

(a) Find $\cos^{-1}(-\sqrt{3}/2)$.

(b) Approximate $\cos^{-1}(-0.7951)$.

Solution

(a) If we let $v = \cos^{-1}(-\sqrt{3}/2)$, then $\cos v = -\sqrt{3}/2$. The reference number for v is $\pi/6$. (Why?) By definition, v must be chosen in the interval $[0, \pi]$, and hence $v = \pi - (\pi/6) = 5\pi/6$. Thus,

$$\cos^{-1}(-\sqrt{3}/2) = 5\pi/6.$$

(b) If $v = \cos^{-1}(-0.7951)$, then $\cos v = -0.7951$. If v' is the reference number for v, then $\cos v' = 0.7951$. From Table 3 we see that $v' \approx 0.6516$. According to the definition of the inverse cosine function, v must be chosen between 0 and π. Hence,

$$v = \pi - v' \approx 3.1416 - 0.6516.$$

That is, $\cos^{-1}(-0.7951) \approx 2.4900.$ ■

If we restrict the domain of the tangent function to the open interval $(-\pi/2, \pi/2)$, then a one-to-one correspondence is obtained. We may, therefore, adopt the following definition.

Definition

> The **inverse tangent function** or **arctangent function**, denoted by $\tan^{-1}$ or arctan, is defined by
>
> $$\tan^{-1} u = \arctan u = v \quad \text{if and only if} \quad \tan v = u$$
>
> where u is any real number and $-\pi/2 < v < \pi/2$.

Note that the domain of the arctangent function is all of $\mathbb{R}$ and the range is the open interval $(-\pi/2, \pi/2)$. In Exercise 42 you are asked to verify the graph of the inverse tangent function sketched in Figure 6.7.

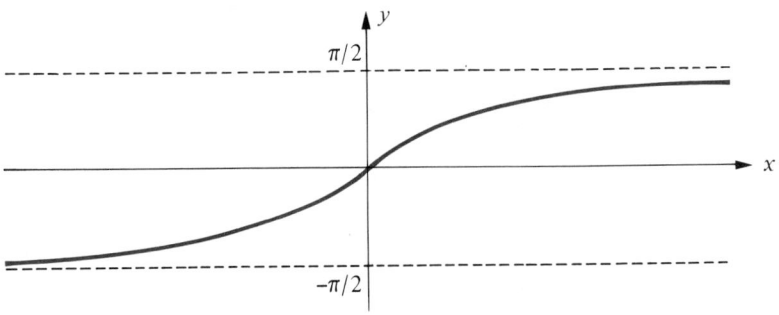

Figure 6.7. $y = \tan^{-1} x = \arctan x$

An analogous procedure may be used for the remaining trigonometric functions. The functions we have defined are generally referred to as the **inverse trigonometric functions**.

Example 5   Without using tables or calculators find $\sec(\arctan 2/3)$.

Solution   If $v = \arctan 2/3$, then $\tan v = 2/3$. Since $\sec^2 v = 1 + \tan^2 v$ and $0 < v < \pi/2$, we may write

$$\sec v = \sqrt{1 + \tan^2 v} = \sqrt{1 + \left(\frac{2}{3}\right)^2}$$

$$= \sqrt{1 + \frac{4}{9}} = \sqrt{\frac{13}{9}} = \frac{\sqrt{13}}{3}.$$

Hence, $\sec(\arctan \frac{2}{3}) = \sec v = \sqrt{13}/3.$ ■

The next two examples illustrate some of the manipulations that can be carried out with the inverse trigonometric functions.

Example 6   Evaluate $\sin(\arctan \frac{1}{2} - \arccos \frac{4}{5})$.

Solution   If we let $v = \arctan 1/2$ and $w = \arccos 4/5$, then $\tan v = 1/2$ and $\cos w = 4/5$. We wish to find $\sin(v - w)$. Since v and w are in the interval $[0, \pi/2]$, they may be considered as the radian measure of positive acute angles, and other functional values of v and w may be found by referring to the right triangles in Figure 6.8. This gives us $\sin v = 1/\sqrt{5}$, $\cos v = 2/\sqrt{5}$, and $\sin w = 3/5$. Consequently,

$$\sin(v - w) = \sin v \cos w - \cos v \sin w$$

$$= \frac{1}{\sqrt{5}}\frac{4}{5} - \frac{2}{\sqrt{5}}\frac{3}{5}$$

$$= \frac{-2}{5\sqrt{5}} = \frac{-2\sqrt{5}}{25}.$$

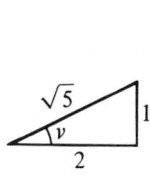
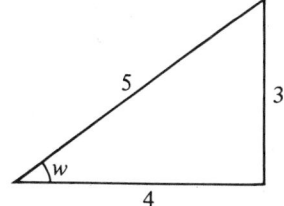

Figure 6.8   ■

Example 7   Write $\cos(\sin^{-1} u)$ as an algebraic expression in u.

Solution   Let $v = \sin^{-1} u$, so that $\sin v = u$. We wish to find $\cos v$. Since $-\pi/2 \leq v \leq \pi/2$, it follows that $\cos v \geq 0$, and hence

$$\cos v = \sqrt{1 - \sin^2 v} = \sqrt{1 - u^2}.$$

Consequently, $\cos{(\sin^{-1} u)} = \sqrt{1 - u^2}$.   ∎

Most of the trigonometric equations considered in Section 6.3 had solutions which were rational multiples of π, such as $\pi/3$, $3\pi/4$, π, and so on. If solutions are not of that type we can sometimes use inverse functions to express them in exact form, as illustrated in the next example.

Example 8   Find the solutions of the equation

$$5 \sin^2 t + 3 \sin t - 1 = 0$$

which are in the interval $[-\pi/2, \pi/2]$.

Solution   The given equation may be regarded as a quadratic equation in $\sin t$. Applying the quadratic formula gives us

$$\sin t = \frac{-3 \pm \sqrt{9 + 20}}{10} = \frac{-3 \pm \sqrt{29}}{10}.$$

Hence, by the definition of the inverse sine function, with $v = t$ and $u = (-3 \pm \sqrt{29})/10$, we obtain the solutions

$$t = \sin^{-1} \frac{1}{10}(-3 + \sqrt{29})$$

and

$$t = \sin^{-1} \frac{1}{10}(-3 - \sqrt{29}).$$ ∎

If approximations to the solutions in Example 8 are desired, one may refer to Table 3. If a hand-held calculator is available the task is much simpler. To illustrate, using a typical calculator we first calculate

$$\frac{1}{10}(-3 + \sqrt{29}) \approx 0.2385.$$

If we next press the inverse sine key the result is

$$t \approx \sin^{-1}(0.2385) \approx 0.2408.$$

In like manner,

$$t = \sin^{-1} \frac{1}{10}(-3 - \sqrt{29}) \approx \sin^{-1}(-0.8385) \approx -0.9946.$$

EXERCISES 6.7

Find the numbers in Exercises 1–10 without the use of tables or calculators.

1 (a) $\sin^{-1}(1/2)$ (b) $\cos^{-1}(1/2)$

2 (a) $\sin^{-1}\sqrt{3}/2$ (b) $\sin^{-1}(-\sqrt{3}/2)$

3 (a) $\arcsin 0$ (b) $\arccos 0$

4 (a) $\cos^{-1}\sqrt{2}/2$ (b) $\cos^{-1}(-\sqrt{2}/2)$

5 (a) $\arcsin 1$ (b) $\arccos 1$

6 (a) $\sin^{-1}(-1)$ (b) $\cos^{-1}(-1)$

7 (a) $\tan^{-1}\sqrt{3}$ (b) $\arctan(-\sqrt{3})$

8 (a) $\tan^{-1}(-1)$ (b) $\arccos(-1/2)$

9 (a) $\arcsin(-\sqrt{2}/2)$ (b) $\tan^{-1}(-\sqrt{3}/3)$

10 (a) $\tan^{-1}0$ (b) $\arctan 1$

In Exercises 11–14 use Table 3 to approximate the given numbers.

11 $\sin^{-1}(-0.6494)$ **12** $\cos^{-1}(-0.9112)$

13 $\arctan(2.1775)$ **14** $\arcsin(0.8004)$

Determine the numbers in Exercises 15–26 without the use of tables or calculators.

15 $\sin[\cos^{-1}\frac{1}{2}]$ **16** $\cos[\sin^{-1}0]$

17 $\sin[\arccos\frac{4}{5}]$ **18** $\tan[\tan^{-1}5]$

19 $\arcsin(\sin 5\pi/4)$

20 $\cos^{-1}(\cos 3\pi/4)$

21 $\cos[\sin^{-1}\frac{4}{5} + \tan^{-1}\frac{3}{4}]$

22 $\sin[\arcsin\frac{1}{2} + \arccos 0]$

23 $\tan[\arctan\frac{4}{3} + \arccos\frac{8}{17}]$

24 $\cos[2\sin^{-1}\frac{15}{17}]$

25 $\sin[2\arccos(-\frac{3}{5})]$

26 $\cos[\frac{1}{2}\tan^{-1}\frac{8}{15}]$

In each of Exercises 27–30 rewrite the given expression as an algebraic expression in u.

27 $\sin(\tan^{-1}u)$ **28** $\tan(\arccos u)$

29 $\cos(\frac{1}{2}\arccos u)$ **30** $\cos(2\tan^{-1}u)$

Verify the identities in Exercises 31–38.

31 $\sin^{-1}u + \cos^{-1}u = \pi/2$

32 $\arctan u + \arctan 1/u = \pi/2, u > 0$

33 $\arcsin\dfrac{2u}{1+u^2} = 2\arctan u, |u| \le 1$

34 $2\cos^{-1}u = \cos^{-1}(2u^2 - 1)$, where $0 \le u \le 1$

35 $\arcsin(-x) = -\arcsin x$

36 $\arccos(-x) = \pi - \arccos x$

37 $\sin^{-1}u = \tan^{-1}\dfrac{u}{\sqrt{1-u^2}}$

38 $\tan^{-1}u + \tan^{-1}v = \tan^{-1}\dfrac{u+v}{1-uv}$

39 Define $\cot^{-1}$ by restricting the domain of cot to the interval $(0, \pi)$.

40 Define $\sec^{-1}$ by restricting the domain of sec to $[0, \pi/2) \cup [\pi, 3\pi/2)$.

41 Verify the sketch in Figure 6.6.

42 Verify the sketch in Figure 6.7.

Sketch the graphs of the equations in Exercises 43–52.

43 $y = \sin^{-1}2x$ **44** $y = \cos^{-1}(x/2)$

45 $y = \frac{1}{2}\sin^{-1}x$ **46** $y = 2\cos^{-1}x$

47 $y = 2\tan^{-1}x$ **48** $y = \tan^{-1}2x$

49 $y = 2 + \tan^{-1}x$ **50** $y = \sin^{-1}(x+1)$

51 $y = \sin(\arccos x)$ **52** $y = \sin(\sin^{-1}x)$

Prove that the equations in Exercises 53 and 54 are not identities.

53 $\tan^{-1}x = \dfrac{1}{\tan x}$

54 $(\arcsin x)^2 + (\arccos x)^2 = 1$

In each of Exercises 55–60, use inverse trigonometric functions to state the solutions of the given equation in the given interval.

55 $2\tan^2 t + 9\tan t + 3 = 0; (-\pi/2, \pi/2)$

56 $3 \sin^2 t + 7 \sin t + 3 = 0;\ [-\pi/2, \pi/2]$

57 $15 \cos^4 x - 14 \cos^2 x + 3 = 0;\ [0, \pi]$

58 $3 \tan^4 \theta - 19 \tan^2 \theta + 2 = 0;\ (-\pi/2, \pi/2)$

59 $6 \sin^3 \theta + 18 \sin^2 \theta - 5 \sin \theta - 15 = 0;$ $(-\pi/2, \pi/2)$

60 $6 \sin 2x - 8 \cos x + 9 \sin x - 6 = 0;$ $(-\pi/2, \pi/2)$

CALCULATOR EXERCISES 6.7

1–6 Approximate the solutions of the equations in Exercises 55–60 to four decimal places.

7 Show, by actual calculations, that $\sin (\arcsin x) = x$ for the following values of x:

   (a) 0.4631   (b) $\log 3.64$ (c) $1/\sqrt{\pi}$

8 Show that $\tan^{-1} (\tan x) = x$ for the following values of x:

   (a) 74.85   (b) $\log (\pi^2 + 94.7)$

   (c) $10^{6.39}$

Exercises 9 and 10 employ the notation introduced in the Calculator Exercises of Section 5.7. Use a hand-held calculator to support the given statements by substituting the following values for t: 0.1, 0.01, 0.001, 0.0001.

9 $\dfrac{\arcsin 2x}{\arcsin x} \to 2$ as $x \to 0^+$

10 $\csc x \tan^{-1} x \to 1$ as $x \to 0^+$

6.8 THE LAW OF SINES

A triangle that does not contain a right angle is referred to as an **oblique triangle**. Since it is always possible to divide such triangles into two right triangles, methods developed in Chapter Five may be used for solving them; however, sometimes it is cumbersome to proceed in that manner. In this, and the next section, formulas are obtained which aid in simplifying solutions of oblique triangles.

If two angles and a side of a triangle are known, or if two sides and an angle opposite one of them are known, then the remaining parts of the triangle may be found by means of the formula given in this section. We shall use the letters $A, B, C, a, b, c, \alpha, \beta,$ and γ for parts of triangles as they were used in Section 5.10. Given triangle ABC, let us place angle α in standard position on a rectangular coordinate system so that B is on the positive x-axis. The case where α is obtuse is illustrated in Figure 6.9. The type of argument we shall give may also be used if α is acute.

Consider the line through C parallel to the y-axis and intersecting the x-axis at point D. Suppose that $d(C, D) = h$, so that the ordinate of C is h. It follows that

$$\sin \alpha = \frac{h}{b} \quad \text{or} \quad h = b \sin \alpha.$$

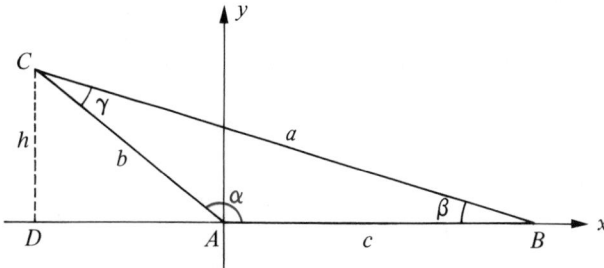

Figure 6.9

Referring to triangle BDC we see that

$$\sin \beta = \frac{h}{a} \quad \text{or} \quad h = a \sin \beta.$$

Consequently, $\qquad\qquad\qquad\qquad a \sin \beta = b \sin \alpha$

which may be written $\qquad\qquad \dfrac{a}{\sin \alpha} = \dfrac{b}{\sin \beta}.$

If α is taken in standard position, so that C is on the positive x-axis, then by the same reasoning,

$$\frac{a}{\sin \alpha} = \frac{c}{\sin y}.$$

The last two equalities give us the following result.

**The Law
of Sines**

> If ABC is any oblique triangle labeled in the usual manner, then
>
> $$\frac{a}{\sin \alpha} = \frac{b}{\sin \beta} = \frac{c}{\sin \gamma}.$$

The next example illustrates a method of applying the Law of Sines to the case in which two angles and a side of a triangle are known. We shall use the rules for rounding off answers discussed in Section 5.10.

Example 1 Given triangle ABC with $\alpha = 48°20'$, $\gamma = 57°30'$, and $b = 47.3$, approximate the remaining parts.

Solution The triangle is represented in Figure 6.10. Since the sum of the angles is 180°,

$$\beta = 180° - (57°30' + 48°20') = 74°10'.$$

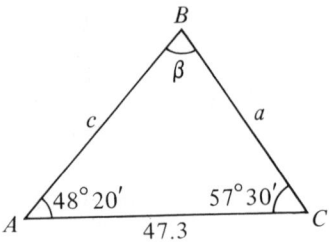

Figure 6.10

Applying the Law of Sines,

$$a = \frac{b \sin \alpha}{\sin \beta} = \frac{(47.3) \sin 48°20'}{\sin 74°10'}.$$

Consulting Table 3, $a \approx \dfrac{(47.3)(0.7470)}{(0.9621)} \approx 36.7.$

Similarly, $c \approx \dfrac{b \sin \gamma}{\sin \beta} = \dfrac{(47.3) \sin 57°30'}{\sin 74°10'}$

$$\approx \frac{(47.3)(0.8434)}{(0.9621)} \approx 41.5. \qquad \blacksquare$$

Data such as that given in Example 1 always gives us a unique triangle *ABC*. However, if two sides and an angle opposite one of them are given, a unique triangle is not always determined. To illustrate, suppose that two numbers *a* and *b* are to be lengths of sides of a triangle *ABC*. In addition, suppose that there is given an angle α which is to be opposite the side of length *a*. Let us consider the case in which α is acute. Place α in standard position on a rectangular coordinate system and consider the line segment *AC* of length *b* on the terminal side of α, as shown in Figure 6.11. The third vertex *B* should be somewhere on the *x*-axis. Since the length *a* of the side opposite α is given, *B* may be found by striking off a circular arc of length *a* with center at *C*. There are four possible outcomes for the construction, as illustrated in Figure 6.12, where the coordinate axes have been deleted.

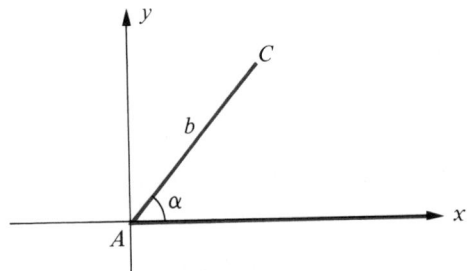

Figure 6.11

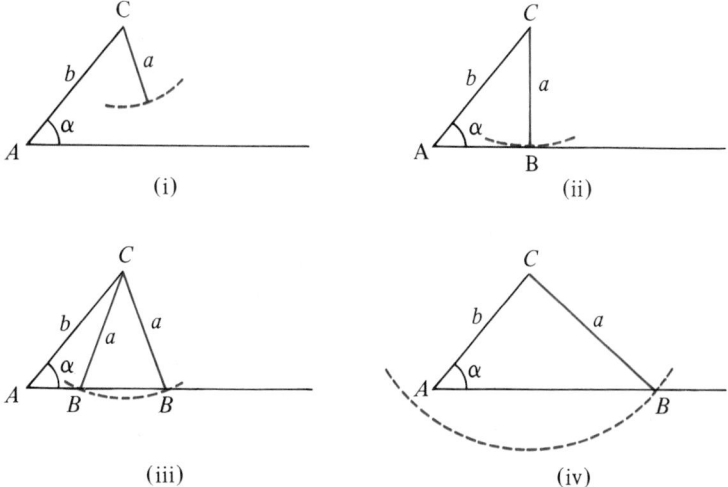

Figure 6.12

The four possibilities may be listed as follows:

(i)   The arc does not intersect the x-axis and no triangle is formed.

(ii)   The arc is tangent to the x-axis and a right triangle is formed.

(iii)   The arc intersects the positive x-axis in two distinct points and two triangles are formed.

(iv)   The arc intersects both the positive and nonpositive parts of the x-axis and one triangle is formed.

Since the distance from C to the x-axis is $b \sin \alpha$ (Why?), we see that (i) occurs if $a < b \sin \alpha$, (ii) occurs if $a = b \sin \alpha$, (iii) occurs if $b \sin \alpha < a < b$, and (iv) occurs if $a \geq b$. It is unnecessary to memorize these facts, since in any specific problem the case that occurs will become evident when the solution is attempted. For example, in solving the equation

$$\frac{a}{\sin \alpha} = \frac{b}{\sin \beta}$$

suppose that we obtain $|\sin \beta| > 1$. This will indicate that no triangle exists. If the equation $\sin \beta = 1$ is obtained, then $\beta = 90°$ and hence case (ii) occurs. On the other hand, if $|\sin \beta| < 1$, then there are two possible choices for the angle β. By checking both possibilities, it will become apparent if (iii) or (iv) occurs.

If the measure of α is greater than $90°$, then, as in Figure 6.13, a triangle exists if and only if $a > b$. Needless to say, our discussion is independent of the symbols we have used; that is, we might be given b, c, β, or a, c, γ, and so on.

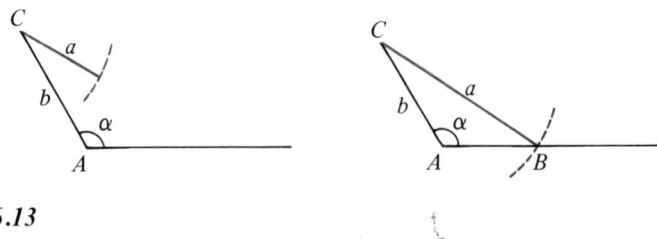

Figure 6.13

Since different possibilities may arise, the case in which two sides and an angle opposite one of them are given is sometimes called the **ambiguous case**.

Example 2   Solve triangle ABC if $\alpha = 67°$, $c = 125$, and $a = 100$.

Solution   Using $\sin \gamma = \dfrac{c \sin \alpha}{a}$, we obtain

$$\sin \gamma = \frac{(125) \sin 67°}{100}$$

$$\approx \frac{(125)(0.9205)}{100}$$

$$\approx 1.1506.$$

Since $\sin \gamma > 1$, there is no triangle with the given parts.   ■

Example 3   Approximate the remaining parts of triangle ABC if $a = 12.4$, $b = 8.7$, and $\beta = 36°40'$.

Solution   Using $\sin \alpha = \dfrac{a \sin \beta}{b}$ we have

$$\sin \alpha = \frac{(12.4) \sin 36°40'}{8.7}$$

$$\approx \frac{(12.4)(0.5972)}{8.7} \approx 0.8512.$$

There are two possible angles α between $0°$ and $180°$ such that $\sin \alpha \approx 0.8512$. If we let α' denote the reference angle for α, then from Table 3 we obtain

$$\alpha' \approx 58°20'.$$

Consequently, the two possibilities for α are

$$\alpha_1 \approx 58°20' \quad \text{and} \quad \alpha_2 \approx 121°40'.$$

If we let γ_1 and γ_2 denote the third angle of the triangle corresponding to the angles α_1 and α_2, respectively, then

$$\gamma_1 \approx 180° - (36°40' + 58°20') = 85°$$

and

$$\gamma_2 \approx 180° - (36°40' + 121°40') = 21°40'.$$

Thus there are two possible triangles which have the given parts. They are the triangles A_1BC and A_2BC shown in Figure 6.14.

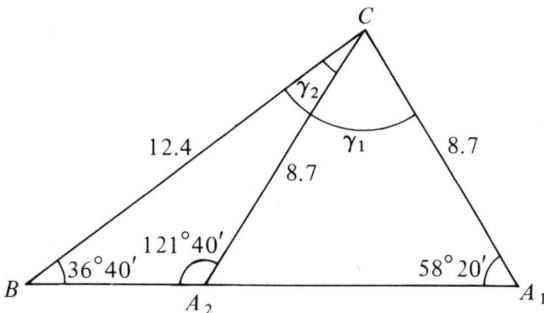

Figure 6.14

If c_1 is the side opposite γ_1 in triangle A_1BC, then

$$c_1 = \frac{a\sin\gamma_1}{\sin\alpha_1}$$

$$c_1 \approx \frac{(12.4)\sin 85°}{\sin 58°20'}$$

$$\approx \frac{(12.4)(0.9962)}{0.8511} \approx 14.5.$$

Hence the solution for triangle A_1BC is $\alpha_1 \approx 58°20'$, $\gamma_1 \approx 85°$, and $c_1 \approx 14.5$.

Similarly, if c_2 is the side opposite angle γ_2 in triangle A_2BC, we have

$$c_2 = \frac{a\sin\gamma_2}{\sin\alpha_2}$$

or

$$c_2 \approx \frac{(12.4)\sin 21°40'}{\sin 121°40'} \approx \frac{(12.4)(0.3692)}{0.8511} \approx 5.4.$$

Consequently, the solution for triangle A_2BC is $\alpha_2 \approx 121°40'$, $\gamma_2 \approx 21°40'$, and $c_2 \approx 5.4$. ∎

EXERCISES 6.8

In each of Exercises 1–12 approximate the remaining parts of triangle ABC.

1 $\alpha = 41°$, $\gamma = 77°$, $a = 10.5$

2 $\beta = 20°$, $\gamma = 31°$, $b = 210$

3 $\alpha = 27°40'$, $\beta = 52°10'$, $a = 32.4$

4 $\alpha = 42°10'$, $\gamma = 61°20'$, $b = 19.7$

5 $\beta = 50°50'$, $\gamma = 70°30'$, $c = 537$

6 $\alpha = 7°10'$, $\beta = 11°40'$, $a = 2.19$

7 $\alpha = 65°10'$. $a = 21.3$, $b = 18.9$

8 $\beta = 30°$, $b = 17.9$, $a = 35.8$

9 $\gamma = 53°20'$, $a = 140$, $c = 115$

10 $\alpha = 27°30'$, $c = 52.8$, $a = 28.1$

11 $\beta = 113°10'$. $b = 248$, $c = 195$

12 $\gamma = 81°$, $c = 11$, $b = 12$

13 It is desired to find the distance between two points A and B lying on opposite banks of a river. A line segment AC of length 240 yards is laid off and the measures of angles BAC and ACB are found to be $63°20'$ and $54°10'$, respectively. Approximate the distance from A to B.

14 In order to determine the distance between two points A and B, a surveyor chooses a point C which is 375 yards from A and 530 yards from B. If angle BAC has measure $49°30'$, approximate the required distance.

15 When the angle of elevation of the sun is $64°$, a telegraph pole that is tilted at an angle of $12°$ directly away from the sun casts a shadow 34 feet long on level ground. Approximate the length of the pole.

16 A straight road makes an angle of $15°$ with the horizontal. When the angle of elevation of the sun is $57°$, a vertical pole at the side of the road casts a shadow 75 feet long directly down the road. Approximate the length of the pole.

17 The angles of elevation of a balloon from two points A and B on level ground are $24°10'$ and $47°40'$, respectively. If A and B are 8.4 miles apart and the balloon is between A and B in the same vertical plane, approximate the height of the balloon above the ground.

18 An airport A is 480 miles due east of airport B. A pilot flew in the direction $235°$ from A to C and then in the direction $320°$ from C to B. Approximate the total distance he flew.

19 A forest ranger at an observation point A sights a fire in the direction N27°10′E. Another ranger at an observation point B, 6 miles due east of A, sights the same fire at N52°40′W. Approximately how far is the fire from each of the observation points?

20 A surveyor notes that the direction from point A to point B is S63°W and the direction from A to C is S38°W. If the distance from A to B is 239 yards and the distance from B to C is 374 yards, approximate the distance and direction from A to C.

21 A straight road makes an angle of $22°$ with the horizontal. From a certain point P on the road the angle of elevation of an airplane is $57°$. At the same instant, from another point 100 meters farther up the road, the angle of elevation is $63°$. How far is the airplane from P?

22 A point P on level ground is 3 km due north of a point Q. If a person jogged in the direction N25°E from Q to a point R, and then from R to P in the direction S40°W, approximate the total distance jogged.

CALCULATOR EXERCISES 6.8

In the following, find the remaining parts of triangle *ABC*.

1 $\beta = 25.6°$, $\gamma = 34.7°$, $b = 184.8$

2 $\alpha = 6.24°$, $\beta = 14.08°$, $a = 4.56$

3 $\alpha = 103.45°$, $\gamma = 27.19°$, $b = 38.84$

4 $\gamma = 47.74°$, $a = 131.08$, $c = 97.84$

5 $\beta = 121.624°$, $b = 0.283$, $c = 0.178$

6 $\alpha = 32.32°$, $c = 574.3$, $a = 263.6$

6.9 THE LAW OF COSINES

If two sides and the included angle or the three sides of a triangle are given, then the Law of Sines cannot be applied directly. We may, however, use the following result.

The Law of Cosines

If *ABC* is any triangle labeled in the usual manner, then

$$a^2 = b^2 + c^2 - 2bc \cos \alpha$$
$$b^2 = a^2 + c^2 - 2ac \cos \beta$$
$$c^2 = a^2 + b^2 - 2ab \cos \gamma.$$

Instead of memorizing each of the formulas given in this law, it is more convenient to remember the following statement, which takes all of them into account.

The square of the length of any side of a triangle equals the sum of the squares of the lengths of the other two sides minus twice the product of the lengths of the other two sides and the cosine of the angle between them.

We shall use the distance formula to establish the Law of Cosines. We again place α in standard position on a rectangular coordinate system, as illustrated in Figure 6.15.

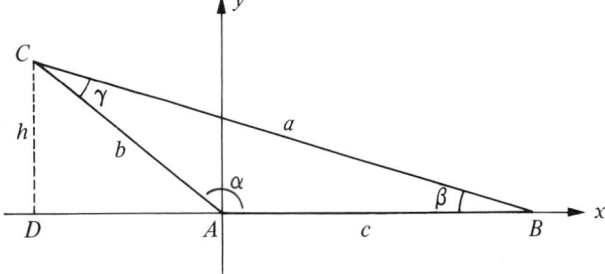

Figure 6.15

Although α is pictured as an obtuse angle, our development is also true if α is acute. Since the segment *AB* has length *c*, the coordinates of *B* are $(c, 0)$. It follows from the Theorem on Trigonometric Functions as Ratios (see Section 5.5), that the coordinates of C are $(b \cos \alpha, b \sin \alpha)$. We also have $a = d(B, C)$ and hence $a^2 = [d(B, C)]^2$. Using the Distance Formula leads to the following equations:

$$a^2 = (b \cos \alpha - c)^2 + (b \sin \alpha - 0)^2$$

$$= b^2 \cos^2 \alpha - 2bc \cos \alpha + c^2 + b^2 \sin^2 \alpha$$

$$= b^2 (\cos^2 \alpha + \sin^2 \alpha) + c^2 - 2bc \cos \alpha$$

$$= b^2 + c^2 - 2bc \cos \alpha.$$

This gives us the first formula stated in the Law of Cosines. The second and third formulas may be obtained by placing β and γ, respectively, in standard position on a rectangular coordinate system and using a similar procedure.

Example 1   Approximate the remaining parts of triangle ABC if $a = 5.0$, $c = 8.0$, and $\beta = 77°10'$.

Solution   Using the Law of Cosines

$$b^2 = (5.0)^2 + (8.0)^2 - 2(5.0)(8.0) \cos 77°10'$$

$$\approx 25 + 64 - (80)(0.2221)$$

$$\approx 71.2.$$

Consequently, $b \approx \sqrt{71.2} \approx 8.44$. We next use the Law of Sines to find γ. Thus

$$\sin \gamma \approx \frac{(8.0) \sin 77°10'}{8.44} \approx \frac{(8.0)(0.9750)}{8.44} \approx 0.9242.$$

Consulting Table 3 we see that $\gamma \approx 67°30'$. Hence

$$\alpha \approx 180° - (77°10' + 67°30') = 35°20'. \qquad \blacksquare$$

Example 2   Given sides $a = 90$, $b = 70$, and $c = 40$ of triangle ABC, approximate angles α, β, and γ.

Solution   According to the first formula stated in the Law of Cosines,

$$\cos \alpha = \frac{b^2 + c^2 - a^2}{2bc}$$

$$= \frac{4900 + 1600 - 8100}{5600}$$

$$\approx -0.2857.$$

From Table 3 we see that the reference angle for α is approximately $73°20'$. Hence $\alpha \approx 180° - 73°20' = 106°40'$.

Similarly, from the second formula stated in the Law of Cosines,

$$\cos \beta = \frac{a^2 + c^2 - b^2}{2ac}$$

$$= \frac{8100 + 1600 - 4900}{7200}$$

$$\approx 0.6667$$

and hence $\beta \approx 48°10'$. Finally,

$$\gamma \approx 180° - (106°40' + 48°10') = 25°10'. \qquad ■$$

EXERCISES 6.9

In Exercises 1–10 approximate the remaining parts of triangle ABC.

1 $\alpha = 60°$, $b = 20$, $c = 30$

2 $\gamma = 45°$, $b = 10$, $a = 15$

3 $\beta = 150°$, $a = 150$, $c = 30$

4 $\beta = 73°50'$, $c = 14$, $a = 87$

5 $\gamma = 115°10'$, $a = 1.1$, $b = 2.1$

6 $\alpha = 23°40'$, $c = 4.3$, $b = 70$

7 $a = 2$, $b = 3$, $c = 4$

8 $a = 10$, $b = 15$, $c = 12$

9 $a = 25$, $b = 80$, $c = 60$

10 $a = 20$, $b = 20$, $c = 10$

11 A parallelogram has sides of length 30 inches and 70 inches. If one of the angles has measure 65°, approximate the length of each diagonal.

12 The angle at one corner of a triangular plot of ground has measure 73°40'. If the sides that meet at this corner are 175 feet and 150 feet long, approximate the length of the third side.

13 A vertical pole 40 feet tall stands on a hillside that makes an angle of 17° with the horizontal. What is the minimal length of rope that will reach from the top of the pole to a point directly down the hill 72 feet from the base of the pole?

14 To find the distance between two points A and B, a surveyor chooses a point C which is 420 yards from A and 540 yards from B. If angle ACB has measure 63°10', approximate the distance.

15 Two automobiles leave from the same point and travel along straight highways which differ in direction by 84°. If their speeds are 60 miles per hour and 45 miles per hour, respectively, approximately how far apart will they be at the end of 20 minutes?

16 A triangular plot of land has sides of length 420 feet, 350 feet, and 180 feet. Find the smallest angle between the sides.

17 A ship leaves P at 1:00 P.M. and travels S35°E at the rate of 24 miles per hour. Another ship leaves P at 1:30 P.M. and travels S20°W at 18 miles per hour. Approximately how far apart are the ships at 3:00 P.M.?

18 An airplane flies 165 miles from point A in the direction 130° and then travels in the direction 245° for 80 miles. Approximately how far is the airplane from A?

19 A jogger, running at a constant speed of one mile every eight minutes, runs in the direction S40°E for 20 minutes and then in the direction N20°E for the next 16 minutes. How far is the jogger from the starting point?

20 Two points P and Q on level ground are on opposite sides of a building. In order to find the distance d between the points, a surveyor chooses a point R which is 300 feet from P and 438 feet from Q, and then determines that angle PRQ has measure $37°40'$. Find d.

21 A motor boat traveled along a triangular course of sides 2 km, 4 km, and 3 km, respectively. If the first side was traversed in the direction N20°W and the second in the southwest direction, in what direction was the third side traversed?

22 If a rhombus has sides of length 100 cm and if the angle at one of the vertices is $70°$, find the lengths of the diagonals.

23 Show that for every triangle ABC,
(a) $a^2 + b^2 + c^2 =$
$\quad 2(bc \cos \alpha + ac \cos \beta + ab \cos \gamma).$
(b) $\dfrac{\cos \alpha}{a} + \dfrac{\cos \beta}{b} + \dfrac{\cos \gamma}{c} = \dfrac{a^2 + b^2 + c^2}{2abc}.$

24 Show that if ABC is a right triangle, then one part of the Law of Cosines reduces to the Pythagorean Theorem.

CALCULATOR EXERCISES 6.9

Find the remaining parts of triangle ABC in Exercises 1–6

1 $\alpha = 48.3°,\quad b = 24.7,\quad c = 52.8$

2 $\beta = 137.8°,\quad a = 178.2,\quad c = 431.4$

3 $\gamma = 6.85°,\quad a = 0.846,\quad b = 0.364$

4 $a = 435,\quad b = 482,\quad c = 78$

5 $a = 5,150,\quad b = 1814,\quad c = 3,429$

6 $a = 0.64,\quad b = 0.27,\quad c = 0.49$

6.10 VECTORS

In previous work we assigned directions to certain lines, such as the x- and y-axes. In similar fashion, a **directed line segment** is a line segment to which a direction has been assigned. Another name for a directed line segment is a **vector**. If a vector extends from a point A (called the **initial point**) to a point B (called the **terminal point**), it is customary to place an arrowhead at B and use $\overrightarrow{AB}$ to represent the vector (see Figure 6.16). The length of the directed line segment is called the **magnitude** of the vector $\overrightarrow{AB}$ and is

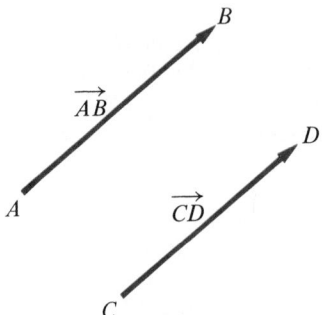

Figure 6.16

denoted by $|\overrightarrow{AB}|$. The vectors $\overrightarrow{AB}$ and $\overrightarrow{CD}$ are considered **equal**, and we write $\overrightarrow{AB} = \overrightarrow{CD}$ if and only if they have the same magnitude and direction, as illustrated in Figure 6.16. Consequently, vectors may be translated from one position to another, provided neither the magnitude nor direction is changed. Vectors of this type are often referred to as **free vectors**.

Many physical concepts may be represented by vectors. To illustrate, suppose an airplane is descending at a constant rate of 100 mph and the line of flight makes an angle of 20° with the horizontal. Both of these facts are represented by the vector in (i) of Figure 6.17, where it is assumed that units have been chosen so that the magnitude is 100. As shown in the figure, we will use a bold face letter such as **v** to denote a vector whose endpoints are not specified. The vector **v** in this illustration is called a **velocity vector**. The magnitude $|\mathbf{v}|$ of **v** is the speed of the airplane.

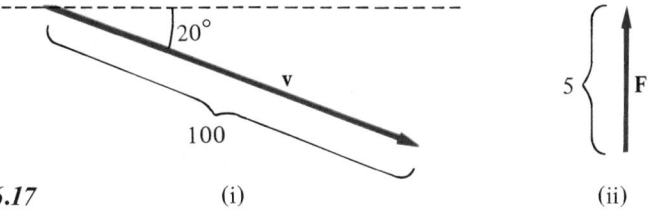

Figure 6.17   (i)   (ii)

As a second illustration, suppose a person pulls directly upward on an object with a force of 5 kg, as would be the case in lifting a 5 kg weight. We may indicate this fact by the vector **F** in (ii) of Figure 6.17. A vector which represents a pull or push of some type is called a **force vector**.

Another use for vectors is to let $\overrightarrow{AB}$ represent the path of a point (or some physical particle) as it moves along the line segment from A to B. We then refer to $\overrightarrow{AB}$ as a **displacement** of the point (or particle). As illustrated in Figure 6.18, a displacement $\overrightarrow{AB}$ followed by a displacement $\overrightarrow{BC}$ will lead to the same point as the single displacement $\overrightarrow{AC}$. The latter vector is called the **sum** of the first two, and we write

$$\overrightarrow{AB} = \overrightarrow{AB} + \overrightarrow{BC}.$$

Since we are working with free vectors, any two vectors can be added by placing the initial point of one on the terminal point of the other and then proceeding as in Figure 6.18.

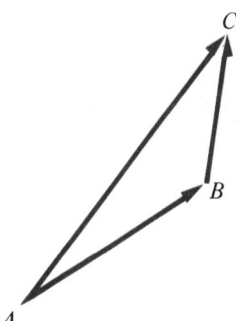

Figure 6.18 A

Another way to find the sum of two vectors is to consider vectors which are equal to the given ones and have the same initial point, as illustrated by $\vec{PQ}$ and $\vec{PR}$ in Figure 6.19. If we construct the parallelogram $RPQS$ with adjacent sides $\vec{PR}$ and $\vec{PQ}$, then since $\vec{PR} = \vec{QS}$, it follows that

$$\boxed{\vec{PS} = \vec{PQ} + \vec{PR}.}$$

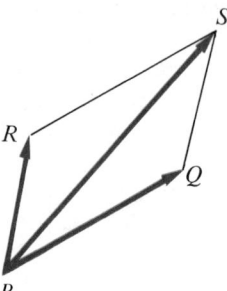

Figure 6.19

If $\vec{PQ}$ and $\vec{PR}$ are two forces acting at P, then it can be shown experimentally that $\vec{PS}$ is the **resultant** force, that is, the single force that produces the same effect as the two combined forces.

If c is a real number and $\vec{AB}$ is a vector, then the product $c\vec{AB}$ is defined as a vector whose magnitude is $|c|$ times the magnitude of $\vec{AB}$ and whose direction is the same as $\vec{AB}$ if $c > 0$, and opposite that of $\vec{AB}$ if $c < 0$. Geometric illustrations are given in Figure 6.20. We refer to $c\vec{AB}$ as a **scalar multiple** of $\vec{AB}$.

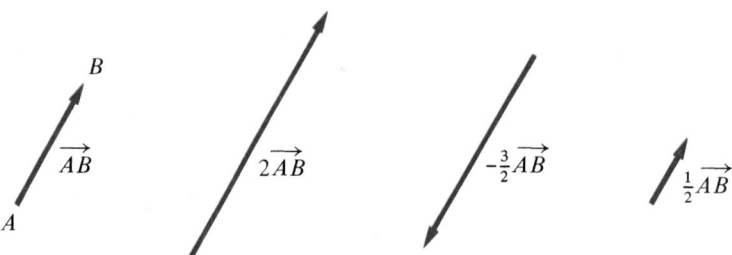

Figure 6.20

Example 1   Two forces $\vec{PQ}$ and $\vec{PR}$ of magnitudes 7 and 5, respectively, act at a point P. If $\vec{PQ}$ is directed due east, and if $\vec{PR}$ has the direction N30°E, find the magnitude and direction of the resultant $\vec{PS}$.

Solution   The given vectors and the resultant $\vec{PS}$ (obtained using a parallelogram) are sketched in Figure 6.21. It follows that angle PQS has measure 120° and the length of the line segment QS is 5. (Why?) Applying the Law of Cosines to triangle PQS gives us

$$|\vec{PS}|^2 = 7^2 + 5^2 - 2(7)(5)\cos 120°$$
$$= 49 + 25 - 70(-1/2)$$
$$= 49 + 25 + 35 = 109$$

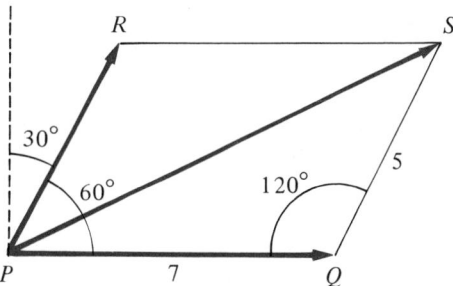

Figure 6.21

and hence
$$|\overrightarrow{PS}| = \sqrt{109}.$$

The Law of Sines may be used to find the direction of $\overrightarrow{PS}$. If we denote angle SPQ by θ, then

$$\frac{\sin\theta}{5} = \frac{\sin 120°}{\sqrt{109}}$$

and
$$\sin\theta = \frac{5\sin 120°}{\sqrt{109}} = \frac{5(\sqrt{3}/2)}{\sqrt{109}} \approx 0.4148.$$

Referring to Table 3 gives us $\theta \approx 24°30'$ and hence the direction of $\overrightarrow{PS}$ is approximately N65°30′E. ∎

Let us next introduce a coordinate plane and assume that all vectors under discussion are in that plane. Since the position of a vector may be changed, provided the magnitude and direction are not altered, we may place the initial point of each vector at the origin. The terminal point of a typical vector $\overrightarrow{OP}$ may then be assigned coordinates (a, b), as shown in Figure 6.22. Conversely, every ordered pair (a, b) determines the vector $\overrightarrow{OP}$, where P has coordinates (a, b). We thus obtain a one-to-one correspondence between vectors and ordered pairs. This allows us to regard a vector in a plane as an ordered pair of real numbers instead of a directed line segment. To avoid confusion with the notation for open intervals or points, we shall use the symbol $\langle a, b \rangle$ for an ordered pair which represents a vector. Moreover, we shall refer to $\langle a, b \rangle$ as a vector and denote it by a boldface letter. The numbers a and b are called the **components** of the vector $\langle a, b \rangle$. The magnitude of $\langle a, b \rangle$ is, by definition, the distance from the origin to the point $P(a, b)$. This may also be stated as follows.

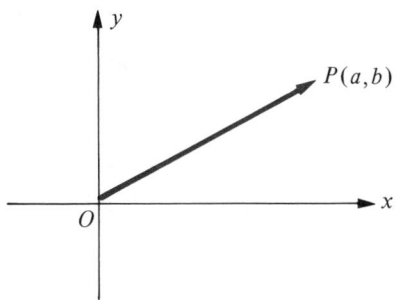

Figure 6.22

Definition

> The magnitude $|\mathbf{v}|$ of the vector $\mathbf{v} = \langle a, b \rangle$ is given by $|\mathbf{v}| = \sqrt{a^2 + b^2}$.

Example 2   Sketch the vectors corresponding to each of the following, and find the magnitude and the smallest positive angle θ from the positive x-axis to each vector.

(a) $\mathbf{a} = \langle -3, -3 \rangle$ (b) $\mathbf{b} = \langle 0, -2 \rangle$ (c) $\mathbf{c} = \langle 4/5, 3/5 \rangle$

Solutions   The vectors are sketched in Figure 6.23. Applying the definition of magnitude, and our knowledge of trigonometry, we obtain:

(a) $|\mathbf{a}| = \sqrt{9 + 9} = 3\sqrt{2}; \quad \theta = 5\pi/4$

(b) $|\mathbf{b}| = \sqrt{0 + 4} = 2; \quad \theta = 3\pi/2$

(c) $|\mathbf{c}| = \sqrt{(16/25) + (9/25)} = 1; \quad \theta = \arctan 3/4.$

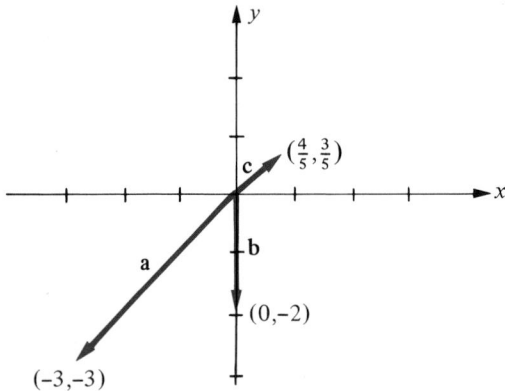

Figure 6.23 $(-3,-3)$ ■

If, as in Figure 6.24, we consider vectors $\overrightarrow{OQ}$ and $\overrightarrow{OR}$ corresponding to $\langle a, b \rangle$ and $\langle c, d \rangle$, respectively, and if we let $\overrightarrow{OS}$ be the vector corresponding to $\langle a + c, b + d \rangle$, then it can be shown that O, Q, S, and R are vertices of a parallelogram. It follows that

$$\overrightarrow{OQ} + \overrightarrow{OR} = \overrightarrow{OS}.$$

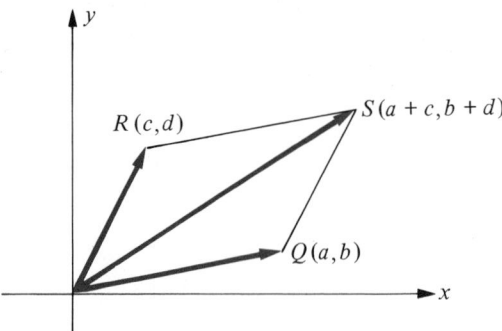

Figure 6.24

Expressing this fact in terms of ordered pairs gives us the following rule for addition of vectors.

> $$\langle a, b \rangle + \langle c, d \rangle = \langle a + c, b + d \rangle$$

Although we shall not prove it, the rule corresponding to a scalar multiple $k\overrightarrow{OP}$ of a vector is as follows.

$$k\langle a,b\rangle = \langle ka,kb\rangle$$

Example 3   If $\mathbf{a} = \langle 2,1\rangle$, find $3\mathbf{a}$ and $-2\mathbf{a}$, and represent all three vectors geometrically.

Solution   If $\mathbf{a} = \langle 2,1\rangle$, then, by the preceding rule for scalar multiples, $3\mathbf{a} = \langle 6,3\rangle$ and $-2\mathbf{a} = \langle -4,-2\rangle$. The geometric representations are shown in Figure 6.25.

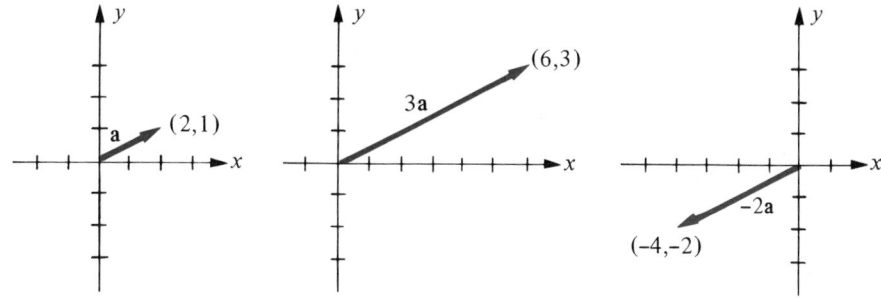

Figure 6.25

Example 4   If $\mathbf{a} = \langle 3,-2\rangle$ and $\mathbf{b} = \langle -6,7\rangle$, find $\mathbf{a}+\mathbf{b}$, $4\mathbf{a}$, and $2\mathbf{a}+3\mathbf{b}$.

Solution   Using the rules for addition and scalar multiples of vectors,

$$\mathbf{a}+\mathbf{b} = \langle 3,-2\rangle + \langle -6,7\rangle = \langle -3,5\rangle$$

$$4\mathbf{a} = 4\langle 3,-2\rangle = \langle 12,-8\rangle$$

$$2\mathbf{a}+3\mathbf{b} = \langle 6,-4\rangle + \langle -18,21\rangle = \langle -12,17\rangle.$$

By definition, the **zero vector 0** corresponds to $\langle 0,0\rangle$. Also, if $\mathbf{a} = \langle a,b\rangle$, then we define $-\mathbf{a} = \langle -a,-b\rangle$. Using these definitions we may establish the following properties, where $\mathbf{a}$, $\mathbf{b}$, and $\mathbf{c}$ denote arbitrary vectors.

**Properties
of Addition
of Vectors**

$$\begin{aligned}
\mathbf{a}+\mathbf{b} &= \mathbf{b}+\mathbf{a} \\
\mathbf{a}+(\mathbf{b}+\mathbf{c}) &= (\mathbf{a}+\mathbf{b})+\mathbf{c} \\
\mathbf{a}+\mathbf{0} &= \mathbf{a} \\
\mathbf{a}+(-\mathbf{a}) &= \mathbf{0}
\end{aligned}$$

The proof of each property follows readily from the rule for addition of vectors and properties of real numbers. For example, if $\mathbf{a} = \langle a_1,a_2\rangle$ and $\mathbf{b} = \langle b_1,b_2\rangle$, then, since

$$a_1 + b_1 = b_1 + a_1 \text{ and } a_2 + b_2 = b_2 + a_2,$$

$$\begin{aligned} \mathbf{a} + \mathbf{b} &= \langle a_1 + b_1, a_2 + b_2 \rangle \\ &= \langle b_1 + a_1, b_2 + a_2 \rangle \\ &= \mathbf{b} + \mathbf{a}. \end{aligned}$$

The remainder of the proof is left as an exercise (see Exercise 15). The reader should also give geometric interpretations for each property.

The operation of **subtraction** of vectors, denoted by "$-$", is defined as follows.

Definition

$$\boxed{\mathbf{a} - \mathbf{b} = \mathbf{a} + (-\mathbf{b})}$$

If we use the ordered pair notation for $\mathbf{a}$ and $\mathbf{b}$, then since $-\mathbf{b} = \langle -b_1, -b_2 \rangle$,

$$\mathbf{a} - \mathbf{b} = \langle a_1, a_2 \rangle + \langle -b_1, -b_2 \rangle$$

and hence,

$$\boxed{\mathbf{a} - \mathbf{b} = \langle a_1 - b_1, a_2 - b_2 \rangle.}$$

Thus, to find $\mathbf{a} - \mathbf{b}$, we merely subtract the components of $\mathbf{b}$ from the corresponding components of $\mathbf{a}$.

Example 5   If $\mathbf{a} = \langle 2, -4 \rangle$ and $\mathbf{b} = \langle 6, 7 \rangle$, find $3\mathbf{a} - 5\mathbf{b}$.

Solution   We proceed as follows:

$$3\mathbf{a} - 5\mathbf{b} = \langle 6, -12 \rangle - \langle 30, 35 \rangle = \langle -24, -47 \rangle. \qquad \blacksquare$$

The following properties can be proved for any vectors $\mathbf{a}$, $\mathbf{b}$ and real numbers c, d.

**Properties
of Scalar
Multiples
of Vectors**

$$\begin{aligned} c(\mathbf{a} + \mathbf{b}) &= c\mathbf{a} + c\mathbf{b} \\ (c + d)\mathbf{a} &= c\mathbf{a} + d\mathbf{a} \\ (cd)\mathbf{a} &= c(d\mathbf{a}) = d(c\mathbf{a}) \\ 1\mathbf{a} &= \mathbf{a} \\ 0\mathbf{a} &= \mathbf{0} = c\mathbf{0} \end{aligned}$$

We shall prove the first property and leave the remaining proofs as an exercise (see Exercise 16). Letting $\mathbf{a} = \langle a_1, a_2 \rangle$ and $\mathbf{b} = \langle b_1, b_2 \rangle$, we have

$$\begin{aligned} c(\mathbf{a} + \mathbf{b}) &= c\langle a_1 + b_1, a_2 + b_2 \rangle \\ &= \langle ca_1 + cb_1, ca_2 + cb_2 \rangle \\ &= \langle ca_1, ca_2 \rangle + \langle cb_1, cb_2 \rangle \\ &= c\mathbf{a} + c\mathbf{b}. \end{aligned}$$

The special vectors **i** and **j** are defined as follows:

$$\mathbf{i} = \langle 1, 0 \rangle, \quad \mathbf{j} = \langle 0, 1 \rangle.$$

The vectors **i** and **j** can be used to obtain an alternate way of denoting vectors. Specifically, if $\mathbf{a} = \langle a_1, a_2 \rangle$, then we may write

$$\mathbf{a} = \langle a_1, 0 \rangle + \langle 0, a_2 \rangle$$
$$= a_1 \langle 1, 0 \rangle + a_2 \langle 0, 1 \rangle$$

that is,

$$\mathbf{a} = \langle a_1, a_2 \rangle = a_1 \mathbf{i} + a_2 \mathbf{j}.$$

The vector sum on the right in the last formula is called a **linear combination** of **i** and **j**. If this notation is employed, then previous rules for addition, subtraction, and multiplication by a scalar may be written as follows, where $\mathbf{b} = \langle b_1, b_2 \rangle = b_1 \mathbf{i} + b_2 \mathbf{j}$:

$$(a_1 \mathbf{i} + a_2 \mathbf{j}) + (b_1 \mathbf{i} + b_2 \mathbf{j}) = (a_1 + b_1) \mathbf{i} + (a_2 + b_2) \mathbf{j}$$
$$(a_1 \mathbf{i} + a_2 \mathbf{j}) - (b_1 \mathbf{i} + b_2 \mathbf{j}) = (a_1 - b_1) \mathbf{i} + (a_2 - b_2) \mathbf{j}$$
$$c(a_1 \mathbf{i} + a_2 \mathbf{j}) = (ca_1) \mathbf{i} + (ca_2) \mathbf{j}.$$

These formulas show that linear combinations of **i** and **j** may be regarded as ordinary algebraic sums.

Example 6   If $\mathbf{a} = 5\mathbf{i} + \mathbf{j}$ and $\mathbf{b} = 4\mathbf{i} - 7\mathbf{j}$, express $3\mathbf{a} - 2\mathbf{b}$ as a linear combination of **i** and **j**.

Solution
$$3\mathbf{a} - 2\mathbf{b} = 3(5\mathbf{i} + \mathbf{j}) - 2(4\mathbf{i} - 7\mathbf{j})$$
$$= (15\mathbf{i} + 3\mathbf{j}) - (8\mathbf{i} - 14\mathbf{j})$$
$$= 7\mathbf{i} + 17\mathbf{j} \qquad \blacksquare$$

A **unit vector** is a vector of magnitude 1. The vectors **i** and **j** are unit vectors, as is the vector **c** in Example 2 of this section.

The formula $\mathbf{a} = a_1 \mathbf{i} + a_2 \mathbf{j}$ for the vector $\mathbf{a} = \langle a_1, a_2 \rangle$ has an interesting geometric interpretation. Vectors corresponding to **i**, **j**, and **a** are illustrated in (i) of Figure 6.26.

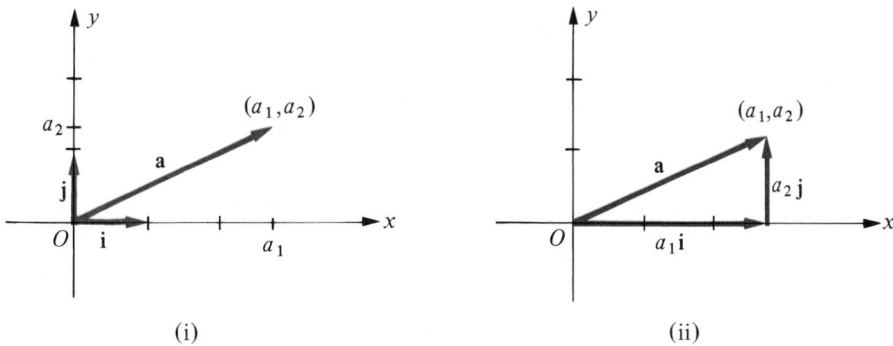

(i)   (ii)

Figure 6.26

Since **i** and **j** are unit vectors, $a_1\mathbf{i}$ and $a_2\mathbf{j}$ may be represented by horizontal and vertical vectors of magnitudes $|a_1|$ and $|a_2|$, respectively, as illustrated in (ii) of Figure 6.26. The vector **a** may be regarded as the sum of these vectors. For this reason a_1 is called the **horizontal component** and a_2 the **vertical component** of the vector **a**.

EXERCISES 6.10

In each of Exercises 1–10 find $\mathbf{a} + \mathbf{b}$, $\mathbf{a} - \mathbf{b}$, $4\mathbf{a} + 5\mathbf{b}$, and $4\mathbf{a} - 5\mathbf{b}$.

1 $\mathbf{a} = \langle 2, -3 \rangle$, $\mathbf{b} = \langle 1, 4 \rangle$

2 $\mathbf{a} = \langle -2, 6 \rangle$, $\mathbf{b} = \langle 2, 3 \rangle$

3 $\mathbf{a} = -\langle 7, -2 \rangle$, $\mathbf{b} = 4\langle -2, 1 \rangle$

4 $\mathbf{a} = 2\langle 5, -4 \rangle$, $\mathbf{b} = -\langle 6, 0 \rangle$

5 $\mathbf{a} = \mathbf{i} + 2\mathbf{j}$, $\mathbf{b} = 3\mathbf{i} - 5\mathbf{j}$

6 $\mathbf{a} = -3\mathbf{i} + \mathbf{j}$, $\mathbf{b} = -3\mathbf{i} + \mathbf{j}$

7 $\mathbf{a} = -(4\mathbf{i} - \mathbf{j})$, $\mathbf{b} = 2(\mathbf{i} - 3\mathbf{j})$

8 $\mathbf{a} = 8\mathbf{j}$, $\mathbf{b} = (-3)(-2\mathbf{i} + \mathbf{j})$

9 $\mathbf{a} = 2\mathbf{j}$, $\mathbf{b} = -3\mathbf{i}$

10 $\mathbf{a} = 0$, $\mathbf{b} = \mathbf{i} + \mathbf{j}$

In Exercises 11–14 sketch vectors corresponding to **a**, **b**, $\mathbf{a} + \mathbf{b}$, $2\mathbf{a}$, and $-3\mathbf{b}$.

11 $\mathbf{a} = 3\mathbf{i} + 2\mathbf{j}$, $\mathbf{b} = -\mathbf{i} + 5\mathbf{j}$

12 $\mathbf{a} = -5\mathbf{i} + 2\mathbf{j}$, $\mathbf{b} = \mathbf{i} - 3\mathbf{j}$

13 $\mathbf{a} = \langle -4, 6 \rangle$, $\mathbf{b} = \langle -2, 3 \rangle$

14 $\mathbf{a} = \langle 2, 0 \rangle$, $\mathbf{b} = \langle -2, 0 \rangle$

15 Prove the second, third, and fourth properties of addition of vectors.

16 Prove the second through the fifth properties of scalar multiples of vectors.

Prove each of the properties in Exercises 17–24, where $\mathbf{a} = \langle a_1, a_2 \rangle$, $\mathbf{b} = \langle b_1, b_2 \rangle$, and c is any real number.

17 $(-1)\mathbf{a} = -\mathbf{a}$

18 $(-c)\mathbf{a} = -c\mathbf{a}$

19 $-(\mathbf{a} + \mathbf{b}) = -\mathbf{a} - \mathbf{b}$

20 $c(\mathbf{a} - \mathbf{b}) = c\mathbf{a} - c\mathbf{b}$

21 If $\mathbf{a} + \mathbf{b} = \mathbf{0}$, then $\mathbf{b} = -\mathbf{a}$.

22 If $\mathbf{a} + \mathbf{b} = \mathbf{a}$, then $\mathbf{b} = \mathbf{0}$.

23 If $c\mathbf{a} = \mathbf{0}$ and $c \neq 0$, then $\mathbf{a} = \mathbf{0}$.

24 If $c\mathbf{a} = \mathbf{0}$ and $\mathbf{a} \neq \mathbf{0}$, then $c = 0$.

25 If $\mathbf{v} = \langle a, b \rangle$, prove each of the following.
  (a) The magnitude of $2\mathbf{v}$ is twice the magnitude of **v**.
  (b) The magnitude of $\frac{1}{2}\mathbf{v}$ is one-half the magnitude of **v**.
  (c) The magnitude of $-2\mathbf{v}$ is twice the magnitude of **v**.
  (d) If k is any real number, then the magnitude of $k\mathbf{v}$ is $|k|$ times the magnitude of **v**.

26 If $\mathbf{v} = \langle a, b \rangle$ and $\mathbf{w} = \langle c, d \rangle$, give a geometric interpretation for $\mathbf{v} - \mathbf{w}$.

If $\mathbf{v} = a_1\mathbf{i} + a_2\mathbf{j}$ and $\mathbf{u} = b_1\mathbf{i} + b_2\mathbf{j}$, the **inner product** $\mathbf{v} \cdot \mathbf{u}$ is defined by $\mathbf{v} \cdot \mathbf{u} = a_1b_1 + a_2b_2$. Prove the properties in Exercises 27–30.

27 $\mathbf{v} \cdot \mathbf{u} = \mathbf{u} \cdot \mathbf{v}$

28 $\mathbf{v} \cdot \mathbf{v} = |\mathbf{v}|^2$

29 $(c\mathbf{v}) \cdot \mathbf{u} = c(\mathbf{v} \cdot \mathbf{u})$, for every real number c

30 $\mathbf{v} \cdot (\mathbf{u} + \mathbf{w}) = \mathbf{v} \cdot \mathbf{u} + \mathbf{v} \cdot \mathbf{w}$, where $\mathbf{w} = c_1\mathbf{i} + c_2\mathbf{j}$

In Exercises 31–38, find the magnitude of **a** and the smallest positive angle θ from the positive x-axis to the vector $\overrightarrow{OP}$ corresponding to **a**.

31 $\mathbf{a} = \langle 3, -3 \rangle$ **32** $\mathbf{a} = \langle -2, -2\sqrt{3} \rangle$

33 $\mathbf{a} = \langle -5, 0 \rangle$ **34** $\mathbf{a} = \langle 0, 10 \rangle$

35 $\mathbf{a} = -4\mathbf{i} + 5\mathbf{j}$ **36** $\mathbf{a} = 10\mathbf{i} - 10\mathbf{j}$

37 $\mathbf{a} = -18\mathbf{j}$ **38** $\mathbf{a} = 2\mathbf{i} - 3\mathbf{j}$

39 Two forces of magnitudes 9 and 12 are acting at a point P. If the respective directions are N75°W and S5°E, find the magnitude and direction of the resultant.

40 Rework Exercise 39 if the magnitudes and directions are 20, S17°W, and 30, N82°E.

6.11 REVIEW

Concepts

Define or discuss each of the following:

1 The Fundamental Identities

2 Verifying identities

3 Trigonometric equations

4 The addition formulas

5 The double-angle formulas

6 The half-angle formulas

7 The product formulas

8 The factoring formulas

9 Reduction formulas

10 The inverse trigonometric functions

11 The Law of Sines

12 The Law of Cosines

13 Vector

14 Magnitude of a vector

15 Addition of vectors

16 Scalar multiple of a vector

17 Vectors as ordered pairs

18 Components of a vector

19 Subtraction of vectors

20 Unit vector

Exercises

Verify the identities in Exercise 1–16.

1 $(\cot^2 x + 1)(1 - \cos^2 x) = 1$

2 $\cos \theta + \sin \theta \tan \theta = \sec \theta$

3 $\dfrac{(\sec^2 \theta - 1) \cot \theta}{\tan \theta \sin \theta + \cos \theta} = \sin \theta$

4 $(\tan x + \cot x)^2 = \sec^2 x \csc^2 x$

5 $\dfrac{1}{1 + \sin t} = (\sec t - \tan t) \sec t$

6 $\dfrac{\sin (\alpha - \beta)}{\cos (\alpha + \beta)} = \dfrac{\tan \alpha - \tan \beta}{1 - \tan \alpha \tan \beta}$

7 $\dfrac{2 \cot u}{\csc^2 u - 2} = \tan 2u$

8 $\cos^2 \dfrac{v}{2} = \dfrac{1 + \sec v}{2 \sec v}$

9 $\dfrac{\tan^3 \phi - \cot^3 \phi}{\tan^2 \phi + \csc^2 \phi} = \tan \phi - \cot \phi$

10 $\dfrac{\sin u + \sin v}{\csc u + \csc v} = \dfrac{1 - \sin u \sin v}{-1 + \csc u \csc v}$

11 $\cos \left(x - \dfrac{5\pi}{2} \right) = \sin x$

12 $\tan \left(x + \dfrac{3\pi}{4} \right) = \dfrac{\tan x - 1}{\tan x + 1}$

13 $\frac{1}{4} \sin 4\beta = \sin \beta \cos^3 \beta - \cos \beta \sin^3 \beta$

14 $\tan \frac{1}{2} \theta = \csc \theta - \cot \theta$

15 $\sin 8\theta =$
$8 \sin \theta \cos \theta (1 - 2 \sin^2 \theta)(1 - 8 \sin^2 \theta \cos^2 \theta)$

16 $\arcsin \dfrac{2u}{1 + u^2} = 2 \arctan u,$ where $|u| \le 1$

In each of Exercises 17–28 find the solutions of the given equation which are in the interval $[0, 2\pi)$, and also find the degree measure of each solution.

17 $2 \cos^3 \theta - \cos \theta = 0$

18 $2 \cos \alpha + \tan \alpha = \sec \alpha$

19 $\sin \theta = \tan \theta$

20 $\csc^5 \theta - 4 \csc \theta = 0$

21 $2 \cos^3 t + \cos^2 t - 2 \cos t - 1 = 0$

22 $\cos x \cot^2 x = \cos x$

23 $\sin \beta + 2 \cos^2 \beta = 1$

24 $\cos 2x + 3 \cos x + 2 = 0$

25 $2 \sec u \sin u + 2 = 4 \sin u + \sec u$

26 $\sin 2u = \sin u$

27 $2 \cos^2 \frac{1}{2}\theta - 3 \cos \theta = 0$

28 $\sec 2x \csc 2x = 2 \csc 2x$

In Exercises 29–32 find the exact values without the use of tables or calculators.

29 $\cos 75°$ **30** $\tan 285°$

31 $\sin 195°$ **32** $\csc \pi/8$

If θ and ϕ are acute angles such that $\csc \theta = 5/3$ and $\cos \phi = 8/17$, find the numbers in Exercises 33–41 without the use of tables or calculators.

33 $\sin (\theta + \phi)$ **34** $\cos (\theta + \phi)$

35 $\tan (\theta - \phi)$ **36** $\sin (\phi - \theta)$

37 $\sin 2\phi$ **38** $\cos 2\phi$

39 $\tan 2\theta$ **40** $\sin \theta/2$

41 $\tan \theta/2$

42   Express $\cos (\alpha + \beta + \gamma)$ in terms of functions of α, β, and γ.

43   Express each of the following products as a sum or difference.
(a) $\sin 7t \sin 4t$
(b) $\cos (u/4) \cos (-u/6)$
(c) $6 \cos 5x \sin 3x$

44   Express each of the following as a product.
(a) $\sin 8u + \sin 2u$
(b) $\cos 3\theta - \cos 8\theta$
(c) $\sin (t/4) - \sin (t/5)$

Find each of the numbers in Exercises 45–53 without the use of tables or calculators.

45 $\cos^{-1} \left(\dfrac{-\sqrt{3}}{2} \right)$

46 $\sin^{-1} \left(\dfrac{\sqrt{2}}{2} \right)$

47 $\arccos (-1)$

48 $\arctan \left(\dfrac{-\sqrt{3}}{3} \right)$

49 $\sin \arccos \left(\dfrac{-\sqrt{3}}{2} \right)$

50 $\cos \left[\sin^{-1} \dfrac{15}{17} - \sin^{-1} \dfrac{8}{17} \right]$

51 $\cos [2 \sin^{-1} \frac{4}{5}]$

52 $\sin [\sin^{-1} \frac{2}{3}]$

53 $\cos^{-1}(\sin 0)$

Sketch the graphs of the equations in Exercises 54–56.

54 $y = \cos^{-1} 3x$

55 $y = 4 \sin^{-1} x$

56 $y = 1 - \sin^{-1} x$

Without using tables or calculators, find the remaining parts of triangle ABC in each of Exercises 57–60.

57 $\alpha = 60°$, $\beta = 45°$, $b = 100$

58 $\gamma = 30°$, $a = 2\sqrt{3}$, $c = 2$

59 $\alpha = 60°$, $b = 6$, $c = 7$

60 $a = 2$, $b = 3$, $c = 4$

61   If $\mathbf{a} = \langle -4, 5 \rangle$ and $\mathbf{b} = \langle 2, -8 \rangle$, sketch vectors corresponding to $\mathbf{a} + \mathbf{b}$, $\mathbf{a} - \mathbf{b}$, $2\mathbf{a}$, and $-\frac{1}{2}\mathbf{b}$.

62   If $\mathbf{a} = 2\mathbf{i} + 5\mathbf{j}$ and $\mathbf{b} = 4\mathbf{i} - \mathbf{j}$, find the vectors or numbers corresponding to:
(a) $4\mathbf{a} + \mathbf{b}$ (b) $2\mathbf{a} - 3\mathbf{b}$
(c) $|\mathbf{a} - \mathbf{b}|$ (d) $|\mathbf{a}| - |\mathbf{b}|$.

7

COMPLEX NUMBERS

Although real numbers are adequate for many mathematical and scientific problems, there is a serious defect in the system when it comes to solving some equations. Indeed, since the square of a real number cannot be negative, an equation such as $x^2 = -5$ has no solutions if x is restricted to $\mathbb{R}$. For many applications we need a mathematical system which contains the real numbers and which has the additional property that equations such as $x^2 = -5$ do have solutions. Fortunately, it is possible to construct such a system: the system of complex numbers discussed in this chapter.

7.1 DEFINITION OF COMPLEX NUMBERS

Let us consider the problem of inventing a new mathematical system $\mathbb{C}$ which contains the real number system $\mathbb{R}$ and which has certain other properties. If $\mathbb{C}$ is to be used to find solutions of equations, then it must possess *operations*, that is, rules which may be applied to every pair of its elements to obtain another element. As a matter of fact, we would like to define operations called *addition* and *multiplication* in such a way that if we restrict the elements to the subset $\mathbb{R}$, then the operations behave in the same way as addition and multiplication of real numbers. Since we are extending the notions of addition and multiplication to the set $\mathbb{C}$, we shall continue to use the symbols $+$ and $\cdot$ for those operations.

In order to gain some insight into the construction of $\mathbb{C}$, let us begin by taking an intuitive approach. We first note that if we want equations of the form $x^2 = -k$, where k is a positive real number, to have solutions, then in particular when $k = 1$ it is necessary for $\mathbb{C}$ to contain some element i such that $i^2 = -1$. If b is in $\mathbb{R}$, then b is also in $\mathbb{C}$, and since $\mathbb{C}$ is to be closed relative to multiplication, bi must be in $\mathbb{C}$. Moreover, if a is in $\mathbb{R}$ and if $\mathbb{C}$ is to be closed relative to addition, then $a + bi$ is in $\mathbb{C}$. Thus, $\mathbb{C}$ contains elements of the form $a + bi, c + di$, and so on, where $a, b, c,$ and d are real numbers and $i^2 = -1$. If we want the Commutative, Associative, and Distributive Properties to be valid, then these elements must add as follows:

$$(a + bi) + (c + di) = (a + c) + (bi + di)$$

or

(A)
$$(a + bi) + (c + di) = (a + c) + (b + d)i$$

Similarly, if $i^2 = -1$, the following manipulations should be valid:

$$
\begin{aligned}
(a + bi)(c + di) &= (a + bi)c + (a + bi)(di) \\
&= ac + (bi)c + a(di) + (bi)(di) \\
&= ac + (bc)i + (ad)i + (bd)(i^2) \\
&= ac + (bd)(-1) + (ad)i + (bc)i \\
&= (ac - bd) + (ad + bc)i.
\end{aligned}
$$

To summarize, the following rule for multiplication must hold in $\mathbb{C}$:

(M)
$$(a + bi)(c + di) = (ac - bd) + (ad + bc)i.$$

The   preceding discussion indicates the manner in which elements behave if a system of the required type is to exist. Moreover, our discussion provides a key to the actual construction of $\mathbb{C}$. Thus, we begin by *defining* a **complex number** as any symbol of the form $a + bi$, where a and b are real numbers. The real number a is called the **real part** of the complex number and bi is called the **imaginary part**. At the outset the letter i is given no specific meaning and the $+$ sign which appears in $a + bi$ is not to be interpreted as the symbol for addition, but only as part of the notation for a complex number. As above, $\mathbb{C}$ will denote the set of all complex numbers. Two complex numbers $a + bi$ and $c + di$ are said to be **equal**, and we write

$$a + bi = c + di \quad \text{if and only if} \quad a = c \text{ and } b = d.$$

Next we *define* addition and multiplication of complex numbers by means of formulas (A) and (M). It should be observed that the $+$ sign is used in three different ways in (A). First, by our previous remarks, it is part of the symbol for a complex number. Second, it is used to denote addition of the complex numbers $a + bi$ and $c + di$. Third, it is the addition sign for real numbers, as in the expressions $a + c$ and $b + d$. The need for remembering this threefold use of $+$ will disappear after we agree on the notational conventions which follow.

Let us consider the subset $\mathbb{R}'$ of $\mathbb{C}$ consisting of all complex numbers of the form $a + 0i$, where a is a real number. By associating a with $a + 0i$, we obtain a one-to-one correspondence between the sets $\mathbb{R}$ and $\mathbb{R}'$. Applying (A) and (M) to the elements of $\mathbb{R}'$ (by letting $b = d = 0$), we obtain

$$
\begin{aligned}
(a + 0i) + (c + 0i) &= (a + c) + 0i \\
(a + 0i)(c + 0i) &= ac + 0i.
\end{aligned}
$$

This shows that in order to add (or multiply) two elements of $\mathbb{R}'$, we merely add (or multiply) the real parts, *disregarding* the imaginary parts. Hence, as far as properties of addition and multiplication are concerned, the only difference between $\mathbb{R}$ and $\mathbb{R}'$ is the notation for the elements. Accordingly, we shall use the symbol a in place of $a + 0i$. For example, an element such as 3 of $\mathbb{R}$ (or $\mathbb{C}$) is considered the same as the element $3 + 0i$ of $\mathbb{C}$. It is also convenient to abbreviate the complex number $0 + bi$ by the symbol bi. Applying (A) gives us

$$(a + 0i) + (0 + bi) = (a + 0) + (0 + b)i = a + bi.$$

This indicates that $a + bi$ may be thought of as the sum of two complex numbers a and bi (that is, $a + 0i$ and $0 + bi$). With these agreements on notation, all the $+$ signs in (A) may be regarded as addition of complex numbers.

Example 1 Express each of the following in the form $a + bi$, where a and b are real numbers.
  (a) $(3 + 4i) + (2 + 5i)$ (b) $(3 + 4i)(2 + 5i)$ (c) $(3 + 4i)^2$

Solutions
  (a) Applying (A),

$$(3 + 4i) + (2 + 5i) = (3 + 2) + (4 + 5)i = 5 + 9i.$$

  (b) Using (M),

$$(3 + 4i)(2 + 5i) = (3 \cdot 2 - 4 \cdot 5) + (3 \cdot 5 + 4 \cdot 2)i = -14 + 23i.$$

  (c) Exponents are defined in $\mathbb{C}$ exactly as they are in $\mathbb{R}$. Thus,

$$
\begin{aligned}
(3 + 4i)^2 &= (3 + 4i)(3 + 4i) \\
&= (3 \cdot 3 - 4 \cdot 4) + (3 \cdot 4 + 4 \cdot 3)i \\
&= -7 + 24i.
\end{aligned}
$$ ■

It is not difficult to show that addition and multiplication of complex numbers are both commutative and associative. The distributive property is also true. The identity element relative to addition is 0 (or, equivalently, $0 + 0i$), since

$$
\begin{aligned}
(a + bi) + 0 &= (a + bi) + (0 + 0i) \\
&= (a + 0) + (b + 0)i \\
&= a + bi.
\end{aligned}
$$

As usual, we refer to 0 as **zero** or the **zero element**. It follows from (M) that the *product* of zero and any complex number is zero. We may also use (M) to prove that 1 (that is, $1 + 0i$) is the identity element relative to multiplication.

If $(-a) + (-bi)$ is added to $a + bi$, we obtain 0. This implies that $(-a) + (-b)i$ is the additive inverse of $a + bi$, that is,

$$\boxed{-(a + bi) = (-a) + (-b)i.}$$

We shall postpone the discussion of multiplicative inverses until the next section. **Subtraction** of complex numbers is defined using additive inverses as follows:

$$(a + bi) - (c + di) = (a + bi) + [-(c + di)].$$

Since $-(c + di) = (-c) + (-d)i$, it follows that

$$(a + bi) - (c + di) = (a - c) + (b - d)i.$$

The special case with $b = c = 0$ gives us

$$(a + 0i) - (0 + di) = (a - 0) + (0 - d)i$$

which may be written in the form

$$a - di = a + (-d)i.$$

The preceding formula is useful when the real number associated with the imaginary part of a complex number is negative.

If c, d, and k are real numbers, then by (M) and our agreement on notation,

$$k(c + di) = (k + 0i)(c + di) = (kc - 0d) + (kd + 0c)i$$

that is

$$k(c + di) = kc + (kd)i.$$

One illustration of this formula is

$$3(5 + 2i) = 15 + 6i.$$

The special case with $k = -1$ gives us

$$(-1)(c + di) = (-c) + (-d)i = -(c + di).$$

Hence, as in $\mathbb{R}$, the additive inverse of a complex number may be found by multiplying it by -1.

The complex number $0 + 1i$ (or, equivalently, $1i$) will be denoted by i. Observe that

$$b(0 + 1i) = (b \cdot 0) + (b \cdot 1)i = 0 + bi = bi.$$

Thus the symbol bi which has been used throughout this section may be regarded as the *product* of b and i.

Finally, using (M) with $a = c = 0$ and $b = d = 1$, we obtain

$$i^2 = (0 + 1i)(0 + 1i)$$
$$= (0 \cdot 0 - 1 \cdot 1) + (0 \cdot 1 + 1 \cdot 0)i$$
$$= -1 + 0i.$$

This gives us the following important rule:

$$i^2 = -1.$$

If we collect all of the formulas and remarks made in this section it becomes evident that, when working with complex numbers, *we may treat all symbols just as though they represented real numbers with exactly one exception: wherever the symbol i^2 appears it* *may be replaced by* -1. Consequently, manipulations can be carried out without referring to (A) or (M), which is what we had in mind from the very beginning of our discussion! We shall use this technique in the solution of the next example. If, as in Example 2, we are asked to write an expression in the form $a + bi$, we shall also accept the form $a - di$ since we have seen that it equals $a + (-d)i$.

Example 2 Write each of the following in the form $a + bi$.

(a) $4(2 + 5i) - (3 - 4i)$ (b) $(4 - 3i)(2 + i)$ (c) $i(3 - 2i)^2$

(d) i^{51}

Solutions

(a) $4(2 + 5i) - (3 - 4i) = 8 + 20i - 3 + 4i = 5 + 24i$

(b) $(4 - 3i)(2 + i) = 8 - 6i + 4i - 3i^2 = 11 - 2i$

(c) $i(3 - 2i)^2 = i(9 - 12i + 4i^2) = i(5 - 12i) = 5i - 12i^2 = 12 + 5i$

(d) Taking successive powers of i, we obtain $i^1 = i, i^2 = -1, i^3 = -i, i^4 = 1$, and then the cycle starts over: $i^5 = i, \ i^6 = i^2 = -1$, etc. In particular, $i^{51} = i^{48}i^3 = (i^4)^{12}i^3 = (1)^{12}i^3 = i^3 = -i$. ∎

Finally, it should be pointed out that there is another way to define complex numbers. Observe that each symbol $a + bi$ determines a unique ordered pair (a, b) of real numbers. Conversely, every ordered pair (a, b) can be used to obtain a symbol $a + bi$. In this way we obtain a one-to-one correspondence between the symbols $a + bi$ and ordered pairs (a, b). The correspondence suggests using ordered pairs of real numbers to define the system $\mathbb{C}$. Formulas (A) and (M) can then be used to motivate definitions for addition and multiplication. Specifically, we define $\mathbb{C}$ as the set of all ordered pairs of real numbers subject to the following two laws:

$$(a, b) + (c, d) = (a + c, b + d)$$
$$(a, b)(c, d) = (ac - bd, ad + bc).$$

Notice the manner in which (A) and (M) were used to help formulate the definition. We merely replaced symbols such as $a + bi$ by (a, b) and translated the rules accordingly. By a suitable change in notation we can obtain the $a + bi$ form for complex numbers introduced in this section.

EXERCISES 7.1

In each of Exercises 1–36 write the expression in the form $a + bi$.

1 $(3 + 2i) + (-5 + 4i)$

2 $(8 - 5i) + (2 - 3i)$

3 $(-4 + 5i) + (2 - i)$

4 $(5 + 7i) + (-8 - 4i)$

5 $(16 + 10i) - (9 + 15i)$

6 $(2 - 6i) - (7 + 2i)$

7 $-(-2 + 7i) + (-6 + 6i)$

8 $-(5 - 3i) - (-3 - 4i)$

29 $i(3 - 2i)(5 + i)$

30 $(1 + i)^4$

9 $7 - (3 - 7i)$

10 $-9 + (5 + 9i)$

11 $5i - (6 + 2i)$

12 $(10 + 7i) - 12i$

31 $\left(-\dfrac{1}{2} - \dfrac{\sqrt{3}}{2}i\right)^3$

32 $\left(-\dfrac{1}{2} + \dfrac{\sqrt{3}}{2}i\right)^3$

13 $(4 + 3i)(-1 + 2i)$

14 $(3 - 6i)(2 + i)$

33 i^{42}

34 i^{23}

15 $(-7 + i)(-3 + i)$

16 $(5 + 2i)(5 - 2i)$

35 i^{157}

36 $(-i)^{50}$

17 $(3 + 4i)(3 - 4i)$

18 $7i(13 + 8i)$

In Exercises 37–40 solve for x and y, if x and y are real.

19 $-9i(4 - 8i)$

20 $(6i)(-2i)$

37 $5x + 6i = -8 + 2yi$

21 $4(8 - 11i)$

22 $-3(-6 + 12i)$

38 $7 - 4yi = 9x + 3i$

23 $(-3i)(5i)$

24 $(1 - i)(1 + i)$

39 $i(2x - 4y) = 4x + 2 + 3yi$

25 $(\sqrt{7} + \sqrt{3}i)(\sqrt{7} - \sqrt{3}i)$

40 $(2x + y) + (3x - 4y)i = (x - 2) + (2y - 5)i$

26 $(-7 + 3i)^2$

27 $(3 + 2i)^2$

28 $4i(2 + 5i)^2$

7.2   CONJUGATES AND INVERSES

The complex number $a - bi$ is called the **conjugate** of the complex number $a + bi$. Since

$$\boxed{(a + bi)(a - bi) = a^2 + b^2}$$

we see that the product of a complex number and its conjugate is a real number. If $a^2 + b^2 \neq 0$, then multiplying both sides of the last equation by $1/(a^2 + b^2)$ and rearranging terms on the left-hand side gives us

$$\left(\frac{1}{a^2 + b^2}\right)(a - bi)(a + bi) = 1.$$

Hence, if $a + bi \neq 0$, then the complex number $a + bi$ has a multiplicative inverse denoted by $(a + bi)^{-1}$, or $1/(a + bi)$, where

$$\boxed{\frac{1}{a + bi} = \left(\frac{1}{a^2 + b^2}\right)(a - bi).}$$

If $c + di \neq 0$, we define the **quotient**

$$\frac{a + bi}{c + di}$$

to be the product of $a + bi$ and $1/(c + di)$. We can write this quotient in the form $u + vi$, where u and v are real numbers, by multiplying numerator and denominator by the conjugate $c - di$ of the denominator as follows:

$$\frac{a + bi}{c + di} = \frac{a + bi}{c + di} \cdot \frac{c - di}{c - di}$$

$$= \frac{(ac + bd) + (bc - ad)i}{c^2 + d^2}$$

$$= \left(\frac{ac + bd}{c^2 + d^2}\right) + \left(\frac{bc - ad}{c^2 + d^2}\right) i.$$

The above technique may also be used to find the multiplicative inverse of $(a + bi)$. Specifically, we multiply numerator and denominator of $1/(a + bi)$ by $a - bi$ as follows:

$$\frac{1}{a + bi} = \frac{1}{a + bi} \cdot \frac{a - bi}{a - bi}$$

$$= \frac{a - bi}{a^2 + b^2} = \frac{1}{a^2 + b^2}(a - bi).$$

Example 1 Express each of the following in the form $a + bi$.

(a) $\dfrac{1}{9 + 2i}$ (b) $\dfrac{7 - i}{3 - 5i}$

Solutions

(a)
$$\frac{1}{9 + 2i} = \frac{1}{9 + 2i} \cdot \frac{9 - 2i}{9 - 2i} = \frac{9 - 2i}{81 + 4} = \frac{9}{85} - \frac{2}{85}i$$

(b)
$$\frac{7 - i}{3 - 5i} = \frac{7 - i}{3 - 5i} \cdot \frac{3 + 5i}{3 + 5i}$$

$$= \frac{21 - 3i + 35i - 5i^2}{9 - 25i^2}$$

$$= \frac{26 + 32i}{34} = \frac{13}{17} + \frac{16}{17}i$$ ∎

Conjugates of complex numbers have several interesting and useful properties. To simplify the notation, if $z = a + bi$ is a complex number, then its conjugate will be denoted by $\bar{z}$, that is, $\bar{z} = a - bi$.

Theorem on Conjugates

If z and w are complex numbers, then

(i) $\overline{z + w} = \bar{z} + \bar{w}$

(ii) $\overline{z \cdot w} = \bar{z} \cdot \bar{w}$

(iii) $\overline{z^n} = \bar{z}^n$, for every positive integer n

(iv) $\bar{z} = z$ if and only if z is real.

Proof

Let $z = a + bi$ and $w = c + di$, where a, b, c, and d are real numbers. Since $z + w = (a + c) + (b + d)i$, we have, by the definition of conjugate and properties of addition of complex numbers.

$$\overline{z + w} = (a + c) - (b + d)i$$
$$= (a - bi) + (c - di)$$
$$= \overline{z} + \overline{w}.$$

That proves (i).

Multiplication gives us $z \cdot w = (ac - bd) + (ad + bc)i$ and hence the conjugate is

$$\overline{z \cdot w} = (ac - bd) - (ad + bc)i = (a - bi)(c - di) = \overline{z} \cdot \overline{w},$$

which proves (ii).

If we set $w = z$ in (ii), then $\overline{z \cdot z} = \overline{z} \cdot \overline{z}$, that is, $\overline{z^2} = \overline{z}^2$. We may then write $\overline{z^3} = \overline{z^2 \cdot z} = \overline{z^2} \cdot \overline{z} = \overline{z}^2 \cdot \overline{z} = \overline{z}^3$. Continuing in this manner, it appears that $\overline{z^n} = \overline{z}^n$ for all positive integers n. A complete proof of (iii) requires the method of mathematical induction discussed in Chapter Nine.

Finally, let us prove (iv). If $z = a + bi$ is real, then $b = 0$ and $\overline{z} = a - 0i = a + 0i = z$. Conversely, if $\overline{z} = z$, then $a - bi = a + bi$ and $-b = b$. It follows that $b = 0$; that is, z is real.

It is not difficult to extend (i) and (ii) of the preceding theorem to more than two complex numbers. For example, if z, w, and u are complex numbers, then applying (i) twice we have

$$\overline{(z + w) + u} = \overline{z + w} + \overline{u} = \overline{z} + \overline{w} + \overline{u}.$$

The analogous result holds for more than three complex numbers. This fact may be stated: "The conjugate of a sum of complex numbers equals the sum of the conjugates." A similar result is true for products.

Real numbers may be represented geometrically by means of points on a coordinate line. We can also obtain geometric representations for complex numbers by using points in a coordinate plane. Specifically, each complex number $a + bi$ determines a unique ordered pair (a, b). The corresponding point $P(a, b)$ in a coordinate plane is called the **geometric representation of** $a + bi$. To emphasize that we are assigning complex numbers to points in a plane, the point $P(a, b)$ will be labeled $a + bi$. A coordinate plane with a complex number assigned to each point is referred to as the **complex plane** instead of the xy-plane. Also, according to this scheme, the x-axis is called the **real axis** and the y-axis the **imaginary axis**. In Figure 7.1 we have indicated the geometric representations of several complex numbers. Note that to obtain the point corresponding to the conjugate $a - bi$ of any complex number $a + bi$ we simply reflect through the real axis.

In Chapter One we defined the concept of the absolute value $|a|$ of a real number a and noted that geometrically, a is the distance between the origin and the point on a coordinate line which corresponds to a. It is natural, therefore, to interpret the absolute

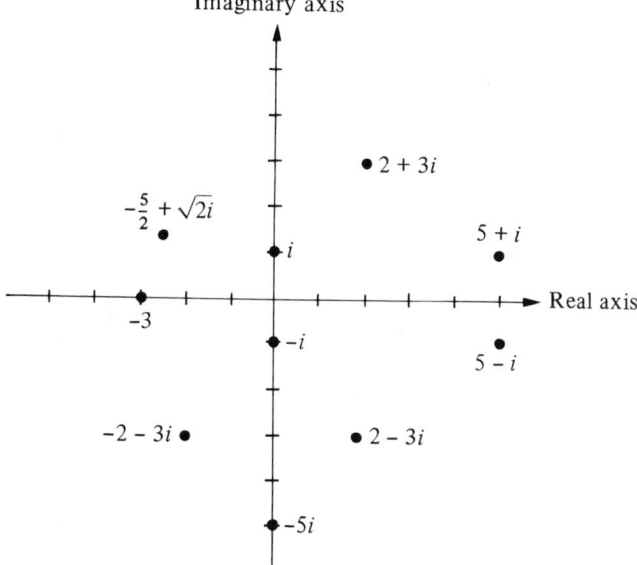

Figure 7.1

value $a + bi$ of a complex number as the distance $\sqrt{a^2 + b^2}$ between the origin of a complex plane and the point (a, b) corresponding to $a + bi$. This motivates the following definition.

Definition

> The **absolute value** of a complex number $a + bi$ is denoted by $|a + bi|$ and is defined to be the nonnegative real number $\sqrt{a^2 + b^2}$.

Example 2 Find (a) $|2 - 6i|$; (b) $|3i|$.

Solutions Using the definition of absolute value we obtain

(a) $|2 - 6i| = \sqrt{4 + 36} = \sqrt{40} = 2\sqrt{10}$;

(b) $|3i| = \sqrt{0 + 9} = 3$. ∎

It is worth noting that the points which correspond to all of the complex numbers having a fixed absolute value lie on a circle with center at the origin in the complex plane. For example, the points corresponding to the complex numbers z with $|z| = 1$ lie on a unit circle.

EXERCISES 7.2

In each of Exercises 1–28 express the given number in the form $a + bi$.

1 $\dfrac{1}{3 + 2i}$

2 $\dfrac{1}{5 + 8i}$

3 $\dfrac{7}{5 - 6i}$

4 $\dfrac{-3}{2 - 5i}$

5 $\dfrac{4 - 3i}{2 + 4i}$

6 $\dfrac{4 + 3i}{-1 + 2i}$

7 $\dfrac{6 + 4i}{1 - 5i}$

8 $\dfrac{7 - 6i}{-5 - i}$

9 $\dfrac{21 - 7i}{i}$

10 $\dfrac{10 + 9i}{-3i}$

11 $\dfrac{2 - 3i}{1 + i} + \dfrac{7 + 4i}{3 + 5i}$

12 $\dfrac{6 - 2i}{3 + i} - \dfrac{3 - 7i}{i}$

13 $8 + \dfrac{4 - i}{1 + 2i}$

14 $\dfrac{1}{10 - i} + 5i$

15 $\dfrac{1}{(1 + i)^3}$

16 $\dfrac{(1 - i)^3}{1 + i}$

17 $\dfrac{4 - i^2}{i - 2}$

18 $\dfrac{4i^2 - 25}{5 + 2i}$

19 $\left(\dfrac{1}{5i}\right)^3$

20 $\dfrac{1}{(3 + 2i)^2}$

21 $|3 - 4i|$

22 $|5 + 8i|$

23 $|-6 - 7i|$

24 $|1 - i|$

25 $|8i|$

26 $|i^7|$

27 $|i^{500}|$

28 $|-15i|$

Represent the complex numbers in Exercises 29–38 geometrically.

29 $4 + 2i$

30 $-5 + 3i$

31 $3 - 5i$

32 $-2 - 6i$

33 $-(3 - 6i)$

34 $(1 + 2i)^2$

35 $2i(2 + 3i)$

36 $(-3i)(2 - i)$

37 $(1 + i)^2$

38 $4(-1 + 2i)$

In each of Exercises 39–42 find all complex numbers z that satisfy the given equation, and express z in the form $a + bi$.

39 $5z + 3i = 2iz + 4$

40 $2iz - 6 = 9i + z$

41 $(z - 2i)^2 = (z + 3i)^2$

42 $z(z + 4i) = (z + i)(z - 3i)$

If $z = a + bi$ and $w = c + di$, verify the identities in Exercises 43–50.

43 $\overline{\overline{z}} = z$

44 $\overline{z - w} = \overline{z} - \overline{w}$

45 $|z| = \sqrt{z\overline{z}}$

46 $|-z| = |z|$

47 $|\overline{z}| = |z|$

48 $|z| = 0$ if and only if $z = 0$

49 $|zw| = |z||w|$

50 $|z/w| = |z|/|w|, \quad w \neq 0$

7.3 COMPLEX ZEROS OF POLYNOMIALS

It is easy to see that if p is any positive real number, then the equation $x^2 = -p$ has solutions in $\mathbb{C}$. As a matter of fact, one solution is $\sqrt{p}\,i$, since

$$(\sqrt{p}\,i)^2 = (\sqrt{p})^2 i^2 = p(-1) = -p.$$

Similarly, $-\sqrt{p}\,i$ is also a solution. Moreover, they are the only solutions, for if a complex number z is a solution, then $z^2 + p = 0$ and hence

$$(z + \sqrt{p}\,i)(z - \sqrt{p}\,i) = 0.$$

This implies that either $z = -\sqrt{p}\,i$ or $z = \sqrt{p}\,i$.

The next definition is motivated by the fact that $(\sqrt{p}\,i)^2 = -p$.

Definition

> If p is a positive real number, then the **principal square root** of $-p$ is denoted by $\sqrt{-p}$ and is defined to be the complex number $\sqrt{p}\,i$.

As illustrations, we have

$$\sqrt{-9} = \sqrt{9}i = 3i, \quad \sqrt{-5} = \sqrt{5}i, \quad \sqrt{-1} = \sqrt{1}i = i.$$

Care must be taken in using the radical sign if the radicand is negative. For example, the formula $\sqrt{a}\sqrt{b} = \sqrt{ab}$ which holds for positive real numbers is not true when both a and b are negative. To illustrate,

$$\sqrt{-3}\sqrt{-3} = (\sqrt{3}i)(\sqrt{3}i) = (\sqrt{3})^2 i^2 = 3(-1) = -3$$

whereas,

$$\sqrt{(-3)(-3)} = \sqrt{9} = 3.$$

Hence,

$$\sqrt{-3}\sqrt{-3} \neq \sqrt{(-3)(-3)}.$$

However, if only *one* of a or b is negative, then it can be shown that $\sqrt{a}\sqrt{b} = \sqrt{ab}$. In general, we shall not apply laws of radicals if radicands are negative. Instead, we shall change the form of radicals before performing any operations, as illustrated in the next example.

Example 1   Express $(5 - \sqrt{-3})(-1 + \sqrt{-4})$ in the form $a + bi$.

Solution $(5 - \sqrt{-3})(-1 + \sqrt{-4}) = (5 - \sqrt{3}i)(-1 + 2i)$

$$= -5 - 2\sqrt{3}i^2 + 10i + \sqrt{3}i$$
$$= (-5 + 2\sqrt{3}) + (10 + \sqrt{3})i \qquad \blacksquare$$

In Section 3.2 we proved that if a, b and c are real numbers such that $b^2 - 4ac \geq 0$ and $a \neq 0$, then the solutions of the quadratic equation $ax^2 + bx + c = 0$ are

$$\frac{-b + \sqrt{b^2 - 4ac}}{2a} \quad \text{and} \quad \frac{-b - \sqrt{b^2 - 4ac}}{2a}.$$

We may now extend this fact to the case where $b^2 - 4ac < 0$. Indeed, the same manipulations used to obtain the quadratic formula, together with the developments in this chapter, show that if $b^2 - 4ac < 0$, then the solutions of $ax^2 + bx + c = 0$ are the two *complex* numbers given above. Notice that the solutions are conjugates of one another. In terms of quadratic functions, given $f(x) = ax^2 + bx + c$, where $a \neq 0$, if $b^2 - 4ac = 0$, then f has one real zero; if $b^2 - 4ac > 0$, then f has two distinct real zeros; and if $b^2 - 4ac < 0$, then f has two distinct conjugate complex zeros.

Example 2   Find the solutions of the equation $5x^2 + 2x + 1 = 0$.

Solution   By the quadratic formula we have

$$x = \frac{-2 \pm \sqrt{4 - 20}}{10} = \frac{-2 \pm \sqrt{-16}}{10} = \frac{-2 \pm 4i}{10}.$$

Dividing numerator and denominator by 2, we see that the solutions of the equation are $-\frac{1}{5} + (\frac{2}{5})i$ and $-\frac{1}{5} - (\frac{2}{5})i$. ■

Example 3   Find the roots of the equation $x^3 - 1 = 0$.

Solution   The given equation may be written as

$$(x - 1)(x^2 + x + 1) = 0.$$

Setting each factor equal to zero and solving the resulting equations, we obtain the solutions

$$1, \quad \frac{-1 \pm \sqrt{1 - 4}}{2}$$

which may be written as

$$1, \quad -\frac{1}{2} + \frac{\sqrt{3}}{2}i, \quad -\frac{1}{2} - \frac{\sqrt{3}}{2}i. \qquad ■$$

The three roots of $x^3 - 1 = 0$ are called the **cube roots of unity.** It can be shown that if n is any positive integer, then the equation $x^n - 1 = 0$ has n distinct complex roots. They are called the **nth roots of unity.**

Quadratic equations with complex coefficients may also be considered. The solutions are again given by the quadratic formula. Since in this case $b^2 - 4ac$ may be complex, solving such an equation may involve finding the square root of a complex number. A technique for determining roots of complex numbers is discussed in Section 7.5.

In certain areas of advanced mathematics it is necessary to consider polynomials $f(x)$ of the form

$$f(x) = k_n x^n + k_{n-1} x^{n-1} + \cdots + k_1 x + k_0$$

where each coefficient k_i is a complex number. We may define the **degree, addition, multiplication,** and **factors** of such polynomials exactly as we did in Chapter Three, where all the coefficients were real. The Division Algorithm, the Remainder Theorem, and the Factor Theorem can be extended to polynomials with complex coefficients. There is a crucial difference, however, when it comes to the *existence* of zeros. Specifically, we know that there are polynomials with real coefficients, such as $x^2 + 1$, which have no real zeros. The next theorem states that, if *complex* solutions are allowed, then every polynomial with real or complex coefficients has a complex zero. Because of its importance, the theorem is given a title, as follows.

Fundamental Theorem of Algebra

> If a polynomial $f(x)$ has complex coefficients and has degree greater than 0, then $f(x)$ has at least one complex zero.

The usual proof of this theorem requires results from the field of mathematics called *functions of a complex variable*. In turn, a prerequisite for studying the latter field is a strong background in calculus. The first proof of the Fundamental Theorem of Algebra was given by the German mathematician Carl Friedrich Gauss (1777–1855), who is considered by many to be the greatest mathematician of all time.

As a special case, if all the coefficients of $f(x)$ are real, then $f(x)$ has at least one complex zero. We should also remark that if $a + bi$ is a complex zero of a polynomial, it may happen that $b = 0$, in which case we refer to the number as a **real zero**. If the Fundamental Theorem is combined with the Factor Theorem, the following useful corollary is obtained.

Corollary

> Every polynomial that has complex coefficients and degree greater than 0 has a factor of the form $x - c$, where c is a complex number.

The last result enables us, at least in theory, to express any nth degree polynomial $f(x)$ as a product of polynomials of degree 1. We begin by applying the corollary, obtaining

$$f(x) = (x - c_1)f_1(x)$$

where c_1 is a complex number and $f_1(x)$ is a polynomial of degree $n - 1$. If $n - 1 > 0$, we may apply the corollary again, obtaining

$$f_1(x) = (x - c_2)f_2(x)$$

where c_2 is a complex number and $f_2(x)$ is a polynomial of degree $n - 2$. Hence,

$$f(x) = (x - c_1)(x - c_2)f_2(x).$$

Continuing this process, after n steps we arrive at a polynomial $f_n(x)$ of degree 0. Thus, $f_n(x) = k$ for some nonzero complex number k and we may write

$$f(x) = k(x - c_1)(x - c_2)\cdots(x - c_n)$$

where $c_1, c_2, \ldots, c_n$ are complex numbers and each c_i is a zero of $f(x)$. Evidently, the leading coefficient of the polynomial on the right in the last equation is k. Since two polynomials are equal if and only if corresponding coefficients are equal, it follows that k is the leading coefficient of $f(x)$. We have proved the following theorem.

Theorem

> If $f(x)$ is a polynomial of degree $n > 0$, with complex coefficients, then there exist n complex numbers $c_1, c_2, \ldots, c_n$ such that
>
> $$f(x) = k(x - c_1)(x - c_2)\cdots(x - c_n)$$
>
> where k is the leading coefficient of $f(x)$.

Of course, not all the c_i in the last theorem need be different. If a factor $x - c$ occurs m times in the factorization, then as in Section 3.6, c is called a **zero of multiplicity** m of

$f(x)$. If a zero of multiplicity m is counted as m zeros, then the theorem tells us that a polynomial $f(x)$ of degree $n > 0$ has *at least* n zeros (not necessarily all different). Combining this with the fact that $f(x)$ has at *most* n zeros, we obtain the next result.

Theorem

> If a polynomial $f(x)$ has complex coefficients and degree $n > 0$, and if a zero of multiplicity m is counted m times, then $f(x)$ has precisely n complex zeros.

Example 4   Express $f(x) = x^5 - 4x^4 + 13x^3$ as a product of linear factors and list the five zeros of $f(x)$.

Solution   We begin by writing

$$f(x) = x^3(x^2 - 4x + 13).$$

By the quadratic formula, the zeros of the polynomial $x^2 - 4x + 13$ are given by

$$\frac{4 \pm \sqrt{16 - 52}}{2} = \frac{4 \pm \sqrt{-36}}{2} = \frac{4 \pm 6i}{2} = 2 \pm 3i.$$

Hence, by the Factor Theorem, $x^2 - 4x + 13$ has factors $x - (2 + 3i)$ and $x - (2 - 3i)$ and we obtain the desired factorization

$$f(x) = x \cdot x \cdot x \cdot (x - 2 - 3i)(x - 2 + 3i).$$

Since $x - 0$ occurs as a factor three times, the number 0 is a zero of multiplicity three, and the five zeros of $f(x)$ are 0, 0, 0, $2 + 3i$, and $2 - 3i$. ∎

We shall conclude this section by returning to polynomials with real coefficients. An interesting fact about such polynomials is illustrated in the preceding example, where the two complex zeros of $x^5 - 4x^4 + 13x^3$ are conjugates of one another. That is no accident, since the following general result is true.

Theorem

> If $f(x)$ is a polynomial of degree $n > 0$, with real coefficients, and if z is a complex zero of $f(x)$, then the conjugate $\bar{z}$ is also a zero of $f(x)$.

Proof   We may write

$$f(x) = a_n x^n + a_{n-1} x^{n-1} + \cdots + a_1 x + a_0$$

where the a_i are real numbers and $a_n \neq 0$. If $f(z) = 0$, then

$$a_n z^n + a_{n-1} z^{n-1} + \cdots + a_1 z + a_0 = 0.$$

If two complex numbers are equal, then so are their conjugates. Consequently, the conjugates of each side of the latter equation are equal, that is,

$$\overline{a_n z^n + a_{n-1} z^{n-1} + \cdots + a_1 z + a_0} = \bar{0} = 0.$$

(The fact that $\overline{0} = 0$ follows from (iv) of the Theorem on Conjugates.) As pointed out in the preceding section, the conjugate of a sum of complex numbers equals the sum of the conjugates, and therefore,

$$\overline{a_n z^n} + \overline{a_{n-1} z^{n-1}} + \cdots + \overline{a_1 z} + \overline{a_0} = 0.$$

Since, for each i,

$$\overline{a_i z^i} = \overline{a_i} \cdot \overline{z^i} = \overline{a_i} \cdot \overline{z}^i = a_i \overline{z}^i,$$

we obtain

$$a_n \overline{z}^n + a_{n-1} \overline{z}^{n-1} + \cdots + a_1 \overline{z} + a_0 = 0.$$

The last equation states that $f(\overline{z}) = 0$, which completes the proof.

Example 5 Find a polynomial $f(x)$ of degree 4 that has real coefficients and zeros $2 + i$ and $-3i$.

Solution By the preceding theorem, $f(x)$ must also have zeros $2 - i$ and $3i$, and hence, by the Factor Theorem, $f(x)$ has factors $x - (2 + i), x - (2 - i), x - (-3i)$, and $x - (3i)$. Multiplying those factors gives us a polynomial of the required type. Thus,

$$\begin{aligned} f(x) &= [x - (2 + i)][x - (2 - i)](x - 3i)(x + 3i) \\ &= (x^2 - 4x + 5)(x^2 + 9) \\ &= x^4 - 4x^3 + 14x^2 - 36x + 45. \end{aligned}$$ ∎

If a polynomial with real coefficients is factored as on page 315, some of the factors $x - c_i$ may have a complex coefficient c_i. However, it is possible to obtain a factorization into polynomials with real coefficients, as stated in the next theorem.

Theorem

> Every polynomial with real coefficients and positive degree n can be expressed as a product of linear and quadratic polynomials with real coefficients, where the quadratic factors have no real zeros.

Proof Since $f(x)$ has precisely n complex zeros $c_1, c_2, \ldots, c_n$, we may write

$$f(x) = k(x - c_1)(x - c_2) \cdots (x - c_n)$$

where k is the leading coefficient of $f(x)$. Of course, some of the c_i may be real; that is, their imaginary parts may be zero. In such cases we obtain the linear factors referred to in the statement of the theorem. If a zero c_i is not real, then by the preceding theorem the conjugate $\overline{c_i}$ is also a zero of $f(x)$ and hence must be one of the numbers in the set $c_1, c_2, \ldots, c_n$. This implies that both $x - c_i$ and $x - \overline{c_i}$ appear in the factorization of $f(x)$. If those factors are multiplied, we obtain

$$(x - c_i)(x - \overline{c_i}) = x^2 - (c_i + \overline{c_i})x + c_i \overline{c_i}$$

which has *real* coefficients, since $c_i + \overline{c_i}$ and $c_i\overline{c_i}$ are real numbers (see Exercise 45). Thus, the complex zeros of $f(x)$ and their conjugates give rise to quadratic polynomials which are irreducible over $\mathbb{R}$. This completes the proof.

Example 6   Express $x^4 - 2x^2 - 3$ (a) as a product of linear polynomials and   (b) as a product of linear and quadratic polynomials with real coefficients which are irreducible over $\mathbb{R}$.

Solutions

(a) The zeros of the given polynomial can be found by solving the equation $x^4 - 2x^2 - 3 = 0$, which may be regarded as quadratic in x^2. Solving for x^2 by means of the quadratic formula, we obtain

$$x^2 = \frac{2 \pm \sqrt{4 + 12}}{2} = \frac{2 \pm 4}{2}$$

or $x^2 = 3$ and $x^2 = -1$. Hence the zeros are $\sqrt{3}, -\sqrt{3}, i,$ and $-i$, and so we obtain the factorization

$$x^4 - 2x^3 - 3 = (x - \sqrt{3})(x + \sqrt{3})(x - i)(x + i).$$

(b) Multiplying the last two factors in the preceding factorization gives us

$$x^4 - 2x^2 - 3 = (x - \sqrt{3})(x + \sqrt{3})(x^2 + 1)$$

which is of the form stated in the preceding theorem.
The solution of this example could also have been obtained by factoring the original expression without first finding the zeros. Thus,

$$x^4 - 2x^2 - 3 = (x^2 - 3)(x^2 + 1)$$
$$= (x + \sqrt{3})(x - \sqrt{3})(x + i)(x - i). \qquad \blacksquare$$

EXERCISES 7.3

In each of Exercises 1–12 express the given number in the form $a + bi$.

1 $\sqrt{-5}\sqrt{-5}$ **2** $(-\sqrt{5})(\sqrt{-5})$

3 $(3 + \sqrt{-16}) - (6 - \sqrt{-9})$

4 $(4 - \sqrt{-4}) + (3 - \sqrt{-25})$

5 $(7 + \sqrt{-5})(7 - \sqrt{-5})$

6 $(\sqrt{-9} + 3)(\sqrt{-16} - 5)$

7 $\sqrt{-2}(\sqrt{2} + \sqrt{-2})$

8 $\sqrt{-3}(3 - \sqrt{-12})$ **9** $(\sqrt{-10})^3$

10 $\sqrt{(-10)^3}$ **11** $\dfrac{5 + \sqrt{-4}}{1 + \sqrt{-9}}$

12 $\dfrac{1}{6 + \sqrt{-3}}$

Find the solutions of the equations in Exercises 13–26.

13 $x^2 - 3x + 10 = 0$ **14** $x^2 - 5x + 20 = 0$

15 $x^2 + 2x + 5 = 0$ **16** $x^2 + 3x + 6 = 0$

17 $4x^2 + x + 3 = 0$ **18** $-3x^2 + x - 5 = 0$

19 $x^3 - 125 = 0$ **20** $x^3 + 27 = 0$

21 $x^6 - 64 = 0$ **22** $x^4 = 81$

23 $4x^4 + 25x^2 + 36 = 0$

24 $27x^4 + 21x^2 + 4 = 0$

25 $x^3 + 3x^2 + 4x = 0$

26 $8x^3 - 12x^2 + 2x - 3 = 0$

In each of Exercises 27–38 find a quadratic equation with the given roots.

27 $4 + i, \quad 4 - i$

28 $1 + 3i, \quad 1 - 3i$

29 $-3 + 2i, \quad -3 - 2i$

30 $\frac{1}{2} + \frac{1}{3}i, \quad \frac{1}{2} - \frac{1}{3}i$

31 $-5i, \quad 5i$

32 $\sqrt{2}i, \quad -\sqrt{2}i$

33 $10, \quad i$ **34** $6, \quad -6i$

35 $i, \quad 1/i$ **36** $i, \quad i^3$

37 $2i, \quad 3i$ **38** $1 + i, \quad 2 + 2i$

39 Show that 3 is a zero of multiplicity two of the polynomial $f(x) = x^4 - 6x^3 + 13x^2 - 24x + 36$, and express $f(x)$ as a product of linear factors.

40 Show that 2 is a zero of multiplicity two of the polynomial $f(x) = x^4 - 4x^3 + 13x^2 - 36x + 36$, and express $f(x)$ as a product of linear factors.

41 Find a polynomial that has real coefficients, degree 2, and root $2 - 3i$.

42 Find a polynomial that has real coefficients, degree 3, and roots $4i$ and -3.

43 Find a polynomial that has real coefficients, degree 4, and roots $2 + i$ and $1 + 2i$.

44 Find a polynomial that has real coefficients, degree 4, and roots i and $3 - \sqrt{2}i$.

45 Prove that if z is any complex number, then $z + \bar{z}$ is a real number. Is $z - \bar{z}$ necessarily a real number?

46 Prove that the sum and product of the roots of the equation $ax^2 + bx + c = 0$ are $-b/a$ and c/a, respectively.

7.4 TRIGONOMETRIC FORM FOR COMPLEX NUMBERS

The geometric representation of a complex number as a point in a coordinate plane leads to a method of representation that involves trigonometric functions. Let us consider a nonzero complex number $z = a + bi$ and its geometric representation $P(a, b)$. Let θ be any angle in standard position whose terminal side lies on the segment OP and let $r = |z|$, that is, $r = d(O, P) = \sqrt{a^2 + b^2}$. An illustration of this situation is shown in Figure 7.2.

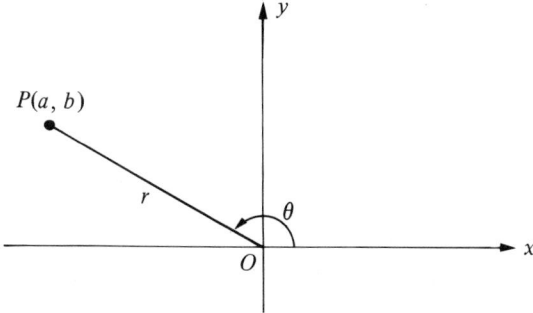

Figure 7.2. $z = a + bi = r(\cos \theta + i \sin \theta)$

Since $a = r\cos\theta$, $b = r\sin\theta$, and $z = a + bi$, we obtain

$$z = (r\cos\theta) + (r\sin\theta)i$$

which may be written in the form

$$z = r(\cos\theta + i\sin\theta).$$

It is important to remember that the number r is the absolute value of z. If $z = 0$, then $r = 0$, and hence the preceding form may be used to represent the complex number 0 where θ is *any* angle. We call $r(\cos\theta + i\sin\theta)$ the **trigonometric form** or **polar form** for the complex number $z = a + bi$. The trigonometric form for z is not unique since there are an unlimited number of different choices for the angle θ. When the trigonometric form is used, the absolute value r of z is often referred to as the **modulus** of z and the angle θ associated with z is called the **argument** (or **amplitude**) of z.

Example 1   Express each of the following complex numbers in trigonometric form.

(a) $-4 + 4i$ (b) $2\sqrt{3} - 2i$ (c) $2 + 7i$

Solutions   The three complex numbers are represented geometrically in Figure 7.3. Using the trigonometric form gives us the following.

(a) $$-4 + 4i = 4\sqrt{2}\left[\cos\frac{3\pi}{4} + i\sin\frac{3\pi}{4}\right]$$

(b) $$2\sqrt{3} - 2i = 4\left[\cos\left(-\frac{\pi}{6}\right) + i\sin\left(-\frac{\pi}{6}\right)\right]$$

(c) $$2 + 7i = \sqrt{53}\left[\cos\left(\arctan\frac{7}{2}\right) + i\sin\left(\arctan\frac{7}{2}\right)\right]$$

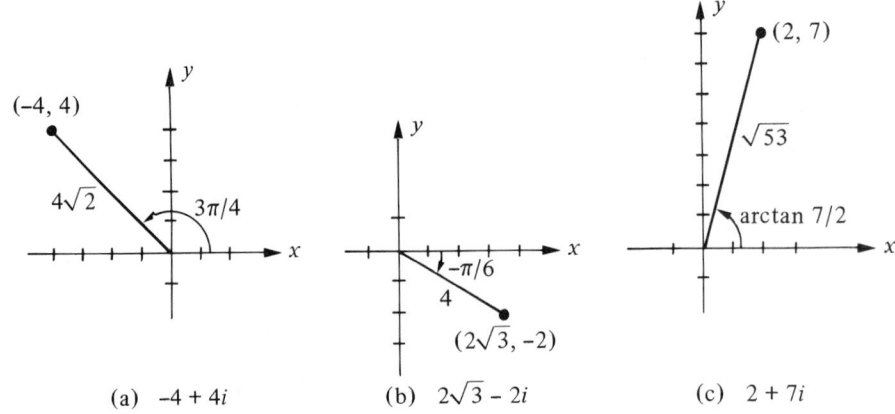

(a) $-4 + 4i$ (b) $2\sqrt{3} - 2i$ (c) $2 + 7i$

Figure 7.3

If complex numbers are expressed in trigonometric form, then multiplications and divisions may be carried out in the simple manner indicated by the next theorem.

Theorem

> If the trigonometric forms for two complex numbers z_1 and z_2 are given by
>
> $$z_1 = r_1(\cos\theta_1 + i\sin\theta_1) \quad \text{and} \quad z_2 = r_2(\cos\theta_2 + i\sin\theta_2)$$
>
> then
>
> (i) $z_1 z_2 = r_1 r_2[\cos(\theta_1 + \theta_2) + i\sin(\theta_1 + \theta_2)]$
>
> (ii) $\dfrac{z_1}{z_2} = \dfrac{r_1}{r_2}[\cos(\theta_1 - \theta_2) + i\sin(\theta_1 - \theta_2)], \quad z_2 \neq 0.$

Proof

We shall prove (i) and leave (ii) as an exercise. Thus,

$$z_1 z_2 = r_1(\cos\theta_1 + i\sin\theta_1) \cdot r_2(\cos\theta_2 + i\sin\theta_2)$$
$$= r_1 r_2\{[\cos\theta_1 \cos\theta_2 - \sin\theta_1 \sin\theta_2] + i[\sin\theta_1 \cos\theta_2 + \cos\theta_1 \sin\theta_2]\}.$$

Applying the addition formulas for $\cos(\theta_1 + \theta_2)$ and $\sin(\theta_1 + \theta_2)$ gives us (i).

Part (i) of the preceding theorem states that the modulus of a product of two complex numbers is the product of their moduli, and the argument is the sum of their arguments. An analogous statement can be made for (ii).

Example 2 Use trigonometric forms to find $z_1 z_2$ and z_1/z_2 if $z_1 = 2\sqrt{3} - 2i$ and $z_2 = -1 + \sqrt{3}i$. Check by using the methods of Sections 1 and 2.

Solution From Example 1 we have

$$z_1 = 2\sqrt{3} - 2i = 4\left[\cos\left(-\frac{\pi}{6}\right) + i\sin\left(-\frac{\pi}{6}\right)\right].$$

Similarly, the reader may verify that

$$z_2 = -1 + \sqrt{3}i = 2\left[\cos\frac{2\pi}{3} + i\sin\frac{2\pi}{3}\right].$$

Applying (i) of the preceding theorem, we have

$$z_1 z_2 = 8\left[\cos\left(-\frac{\pi}{6} + \frac{2\pi}{3}\right) + i\sin\left(-\frac{\pi}{6} + \frac{2\pi}{3}\right)\right]$$
$$= 8\left[\cos\frac{\pi}{2} + i\sin\frac{\pi}{2}\right]$$
$$= 8i.$$

As a check, using the methods of Section 1, we have

$$z_1 z_2 = (2\sqrt{3} - 2i)(-1 + \sqrt{3}i)$$
$$= (-2\sqrt{3} + 2\sqrt{3}) + i(2 + 6)$$
$$= 8i$$

which is in agreement with our previous answer.

Applying (ii) of the preceding theorem we obtain

$$\frac{z_1}{z_2} = 2\left[\cos\left(-\frac{\pi}{6} - \frac{2\pi}{3}\right) + i\sin\left(-\frac{\pi}{6} - \frac{2\pi}{3}\right)\right]$$
$$= 2\left[\cos\left(-\frac{5\pi}{6}\right) + i\sin\left(-\frac{5\pi}{6}\right)\right]$$
$$= 2\left[-\frac{\sqrt{3}}{2} + i\left(-\frac{1}{2}\right)\right]$$
$$= -\sqrt{3} - i.$$

Using the methods of Section 2 gives us

$$\frac{z_1}{z_2} = \frac{2\sqrt{3} - 2i}{-1 + \sqrt{3}i} \cdot \frac{-1 - \sqrt{3}i}{-1 - \sqrt{3}i}$$
$$= \frac{(-2\sqrt{3} - 2\sqrt{3}) + i(2 - 6)}{4}$$
$$= -\sqrt{3} - i. \qquad\blacksquare$$

EXERCISES 7.4

Change the complex numbers in Exercises 1–20 to trigonometric form.

of this section. Check by using the methods of Section 2.

1 $1 - i$

2 $\sqrt{3} + i$

3 $-4\sqrt{3} + 4i$

4 $-2 - 2i$

5 $-20i$

6 15

7 $-5(1 + \sqrt{3}i)$

8 $-6i$

9 -7

10 $2i(1 - \sqrt{3}i)$

11 $2 + i$

12 $4 - 3i$

13 $-4 - 4i$

14 $-10 + 10i$

15 $6i$

16 -5

17 $-4\sqrt{3} + 4i$

18 $-5 - 5\sqrt{3}i$

19 12

20 0

In Exercises 21–28 find $z_1 z_2$ and z_1/z_2 by changing to trigonometric form and then using the last theorem

21 $z_1 = -1 + i, \quad z_2 = 1 + i$

22 $z_1 = \sqrt{3} - i, \quad z_2 = -\sqrt{3} - i$

23 $z_1 = -2 - 2\sqrt{3}i, \quad z_2 = 5i$

24 $z_1 = -5 + 5i, \quad z_2 = -3i$

25 $z_1 = -10, \quad z_2 = -4$

26 $z_1 = 2i, \quad z_2 = -3i$

27 $z_1 = 4, \quad z_2 = 2 - i$

28 $z_1 = -3, \quad z_2 = 5 + 2i$

29 Extend (i) of the last theorem in this section to the case of three complex numbers. Generalize to n complex numbers.

30 Prove (ii) of the last theorem in this section.

7.5 DE MOIVRE'S THEOREM AND nTH ROOTS OF COMPLEX NUMBERS

If z is a complex number and n is a positive integer, then a complex number w is called an **nth root** of z if $w^n = z$. In this section we shall show that every nonzero complex number has n distinct nth roots. Since $\mathbb{R}$ is in $\mathbb{C}$, it will also follow that every nonzero real number has n distinct nth roots. If a is a positive real number and $n = 2$, then we already know that the roots are $\sqrt{a}$ and $-\sqrt{a}$.

If, in the last theorem of Section 7.4, we let both z_1 and z_2 equal $z = r(\cos \theta + i \sin \theta)$, we obtain

$$z^2 = r^2(\cos 2\theta + i \sin 2\theta).$$

If we now apply the same theorem to z and z^2, then

$$z^2 \cdot z = (r^2 \cdot r)[\cos(2\theta + \theta) + i \sin (2\theta + \theta)]$$

or
$$z^3 = r^3(\cos 3\theta + i \sin 3\theta).$$

Next, applying the theorem to z^3 and z produces

$$z^4 = r^4(\cos 4\theta + i \sin 4\theta).$$

In general, we have the following result, which is named after the French mathematician Abraham De Moivre (1667–1754).

De Moivre's Theorem

> For every integer n,
> $$[r(\cos \theta + i \sin \theta)]^n = r^n[\cos n\theta + i \sin n\theta].$$

A complete proof of the theorem may be given with the aid of the method of mathematical induction discussed in Chapter Nine.

Example 1   Find $(1 + i)^{20}$.

Solution   Introducing the trigonometric form,

$$1 + i = \sqrt{2}\left(\cos \frac{\pi}{4} + i \sin \frac{\pi}{4}\right).$$

Next, by De Moivre's Theorem,

$$(1 + i)^{20} = (2^{1/2})^{20}\left[\cos 20\left(\frac{\pi}{4}\right) + i \sin 20\left(\frac{\pi}{4}\right)\right]$$

$$= 2^{10}(\cos 5\pi + i \sin 5\pi)$$

$$= -1024. \qquad \blacksquare$$

If a nonzero complex number z has an nth root w, then $w^n = z$. If the trigonometric forms for w and z are

$$w = s(\cos \alpha + i \sin \alpha) \quad \text{and} \quad z = r(\cos \theta + i \sin \theta),$$

then, applying De Moivre's Theorem to $w^n = z$, we get

$$s^n[\cos n\alpha + i \sin n\alpha] = r(\cos \theta + i \sin \theta).$$

If two complex numbers are equal, then so are their absolute values. Consequently, $s^n = r$, and since s and r are nonnegative, $s = \sqrt[n]{r}$. Substituting s^n for r in the last displayed equation and dividing both sides by s^n, we obtain

$$\cos n\alpha + i \sin n\alpha = \cos \theta + i \sin \theta.$$

It follows that

$$\cos n\alpha = \cos \theta \quad \text{and} \quad \sin n\alpha = \sin \theta.$$

Since both the sine and cosine functions have period 2π, the last two equations are true if and only if $n\alpha$ and θ differ by a multiple of 2π. Thus, for some integer k,

$$n\alpha = \theta + 2\pi k$$

and hence

$$\alpha = \frac{\theta + 2\pi k}{n}.$$

Substituting in the trigonometric form for w, we obtain the formula

$$w = \sqrt[n]{r}\left[\cos\left(\frac{\theta + 2\pi k}{n}\right) + i \sin\left(\frac{\theta + 2\pi k}{n}\right)\right].$$

If we substitute $k = 0, 1, \ldots, n - 1$ successively, there result n distinct values for w and hence n distinct nth roots of z. No other value of k will produce a new nth root. For example, if $k = n$, we obtain the angle $(\theta + 2\pi n)/n$, or $\theta/n + 2\pi$, which gives us the same nth root as $k = 0$. Similarly, $k = n + 1$ yields the same nth root as $k = 1$, and so on. The same is true for negative values of k. Our discussion also shows that the numbers given by the formula are the only possible nth roots of z. We have proved the following theorem.

Theorem on nth Roots

If $z = r(\cos \theta + i \sin \theta)$ is any nonzero complex number and if n is any positive integer, then z has precisely n distinct nth roots. Moreover, the roots are given by

$$\sqrt[n]{r}\left[\cos\left(\frac{\theta + 2\pi k}{n}\right) + i \sin\left(\frac{\theta + 2\pi k}{n}\right)\right]$$

where $k = 0, 1, \ldots, n - 1$.

Note that the *n*th roots of *z* all have modulus $\sqrt[n]{r}$ and hence they lie on a circle of radius $\sqrt[n]{r}$ with center at *O*. Moreover, they are equispaced on this circle since the difference in the arguments of successive *n*th roots is $2\pi/n$.

It is sometimes convenient to use degree measure for θ. In this event the formula in the preceding theorem becomes

$$\sqrt[n]{r}\left[\cos\left(\frac{\theta + k \cdot 360°}{n}\right) + i\sin\left(\frac{\theta + k \cdot 360°}{n}\right)\right]$$

where $k = 0, 1, \ldots, n - 1$.

Example 2   Find the four fourth roots of $-8(1 + i\sqrt{3})$.

Solution   The geometric representation of the given number is shown in Figure 7.4.

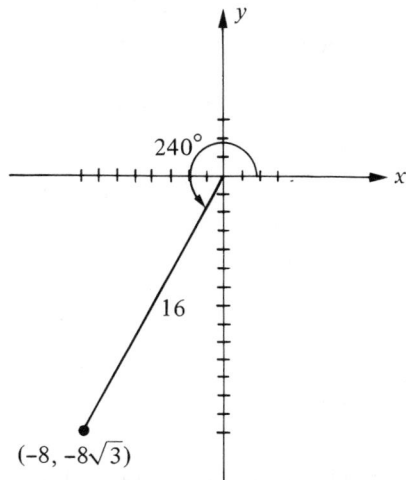

Figure 7.4

Introducing trigonometric form, we have

$$-8(1 + i\sqrt{3}) = 16[\cos 240° + i\sin 240°].$$

By the preceding theorem, with $n = 4$, the fourth roots are given by

$$2\left[\cos\left(\frac{240° + k \cdot 360°}{4}\right) + i\sin\left(\frac{240° + k \cdot 360°}{4}\right)\right]$$

where $k = 0, 1, 2,$ and 3. The formula may be rewritten as

$$2[\cos(60° + k \cdot 90°) + i\sin(60° + k \cdot 90°)].$$

Substituting 0, 1, 2, and 3 for k yields the following fourth roots:

$$2(\cos 60° \ + i \sin 60°) \ = 1 + \sqrt{3}i$$
$$2(\cos 150° + i \sin 150°) = -\sqrt{3} + i$$
$$2(\cos 240° + i \sin 240°) = -1 - \sqrt{3}i$$
$$2(\cos 330° + i \sin 330°) = \sqrt{3} - i.$$
■

Example 3   Find the six sixth roots of -1.

Solution   Writing $-1 = 1\,(\cos \pi + i \sin \pi)$ and using the Theorem on nth Roots with $n = 6$, we find that the sixth roots of -1 are given by

$$\cos\left(\frac{\pi + 2\pi k}{6}\right) + i \sin\left(\frac{\pi + 2\pi k}{6}\right)$$

or

$$\cos\left(\frac{\pi}{6} + \frac{\pi}{3}k\right) + i \sin\left(\frac{\pi}{6} + \frac{\pi}{3}k\right).$$

Substituting 0, 1, 2, 3, 4, and 5 for k gives us the sixth roots of -1, namely,

$$\cos \ \ \pi/6 + i \sin \ \ \pi/6 = \sqrt{3}/2 + (1/2)i$$
$$\cos \ \ \pi/2 + i \sin \ \ \pi/2 = i$$
$$\cos \ 5\pi/6 + i \sin \ 5\pi/6 = -\sqrt{3}/2 + (1/2)i$$
$$\cos \ 7\pi/6 + i \sin \ 7\pi/6 = -\sqrt{3}/2 - (1/2)i$$
$$\cos \ 3\pi/2 + i \sin \ 3\pi/2 = -i$$
$$\cos 11\pi/6 + i \sin 11\pi/6 = \sqrt{3}/2 - (1/2)i.$$
■

The special case in which $z = 1$ is of particular interest. The n distinct nth roots of 1 are called the **nth roots of unity**.

Example 4   Find the three third roots of unity.

Solution   Writing $1 = \cos 0 + i \sin 0$ and using the Theorem on nth Roots with $n = 3$, we obtain the three roots

$$\cos \frac{2\pi k}{3} + i \sin \frac{2\pi k}{3}$$

where $k = 0$, 1, and 2. Substituting for k gives us

$$\cos 0 + \ \ i \sin 0 = 1$$

$$\cos \frac{2\pi}{3} + i \sin \frac{2\pi}{3} = -\frac{1}{2} + \left(\frac{\sqrt{3}}{2}\right)i$$

$$\cos \frac{4\pi}{3} + i \sin \frac{4\pi}{3} = -\frac{1}{2} - \left(\frac{\sqrt{3}}{2}\right)i.$$
■

The reader should compare this solution with Example 3 of Section 7.3.

EXERCISES 7.5

In Exercises 1–12 use De Moivre's Theorem to express the given numbers in the form $a + bi$, where a and b are real numbers.

1 $(3 + 3i)^5$

2 $(1 + i)^{12}$

3 $(1 - i)^{10}$

4 $(-1 + i)^8$

5 $(1 - \sqrt{3}i)^3$

6 $(1 - \sqrt{3}i)^5$

7 $\left(-\dfrac{\sqrt{2}}{2} + \dfrac{\sqrt{2}}{2}i\right)^{15}$

8 $\left(\dfrac{\sqrt{2}}{2} + \dfrac{\sqrt{2}}{2}i\right)^{25}$

9 $\left(-\dfrac{\sqrt{3}}{2} - \dfrac{1}{2}i\right)^{20}$

10 $\left(-\dfrac{\sqrt{3}}{2} - \dfrac{1}{2}i\right)^{50}$

11 $(\sqrt{3} + i)^7$

12 $(-2 - 2i)^{10}$

13 Find the two square roots of $\sqrt{3} + i$.

14 Find the two square roots of $-9i$.

15 Find the four fourth roots of $-1 - \sqrt{3}i$.

16 Find the four fourth roots of $-8 + 8\sqrt{3}i$.

17 Find the three cube roots of $-27i$.

18 Find the three cube roots of $64i$.

19 Find the six sixth roots of unity.

20 Find the eight eighth roots of unity.

21 Find the five fifth roots of $1 + i$.

22 Find the five fifth roots of $-\sqrt{3} - i$.

Find the solutions of the equations in Exercises 23–30.

23 $x^4 - 16 = 0$

24 $x^6 - 64 = 0$

25 $x^6 + 64 = 0$

26 $x^5 + 1 = 0$

27 $x^3 + 8i = 0$

28 $x^3 - 64i = 0$

29 $x^5 - 243 = 0$

30 $x^4 + 81 = 0$

7.6 REVIEW

Concepts

Define or discuss the following.

1 The system of complex numbers

2 The conjugate of a complex number

3 The absolute value of a complex number

4 The nth roots of unity

5 Geometric representation of a complex number

6 The Fundamental Theorem of Algebra

7 Trigonometric form for complex numbers

8 De Moivre's Theorem

Exercises

Express each number in Exercises 1–20 in the form $a + bi$ where a and b are real numbers.

1 $(7 + 5i) + (-8 + 3i)$

2 $(-4 + 7i) - (3 - 3i)$

3 $(4 + 2i)(-5 + 4i)$

4 $(-8 + 2i)(-7 + 5i)$

5 $8i(7 - 5i)$

6 $(-5 + 6i)(-5 - 6i)$

7 $(9 + 4i)(9 - 4i)$

8 $(3 + 8i)^2$

9 $i(6 + i)(2 - 3i)$

10 $(6 - 3i)/(2 + 7i)$

11 $(4 + 6i)/(5 - i)$

12 $(20 - 8i)/4i$

13 $1/(10 - 8i)$

14 $3/(-2i)$

15 $|6 - 10i|$

16 $|-11 + 10i|$

17 $-5(3 + i) - (4 + 2i)$

18 $(2 + i)^4$

19 $(1 + i)^{-1} - (1 - i)^{-1}$

20 i^{99}

Find the solutions of the equations in Exercises 21–24.

21 $5x^2 - 2x + 3 = 0$

22 $3x^2 + x + 6 = 0$

23 $6x^4 + 29x^2 + 28 = 0$

24 $x^4 + x^2 + 1 = 0$

In each of Exercises 25–28 find a quadratic equation with the given roots.

25 $3 - 5i, \quad 3 + 5i$ **26** $1 - i, i$

27 $-4 - i, \quad -4 + i$ **28** $10i, -10i$

Express each of the polynomials in Exercises 29–32 as a product of polynomials which are (a) irreducible over $\mathbb{R}$, (b) irreducible over $\mathbb{C}$.

29 $81 - 16x^4$ **30** $x^3 + 125$

31 $3x^4 - 2x^3 + 12x^2 - 8x$

32 $x^4 - 64$

33 Find a polynomial that has real coefficients, degree 4, and roots $2i$ and $\sqrt{3}i$.

34 Find a polynomial that has real coefficients, degree 3, and roots $1 + 2i$ and -2.

Change the complex numbers in Exercises 35–40 to trigonometric form.

35 $-10 + 10i$ **36** $2 - 2\sqrt{3}i$

37 -17 **38** $-12i$

39 $-5\sqrt{3} - 5i$ **40** $4 + 5i$

In each of Exercises 41–44 use De Moivre's Theorem to write the given number in the form $a + bi$, where a and b are real numbers.

41 $(-\sqrt{3} + i)^9$ **42** $\left(\dfrac{\sqrt{2}}{2} - \dfrac{\sqrt{2}}{2}i \right)^{30}$

43 $(3 - 3i)^5$ **44** $(2 + 2\sqrt{3}i)^{10}$

45 Find the three cube roots of -27.

46 Find the solutions of the equation $x^5 - 32 = 0$.

SYSTEMS OF EQUATIONS AND INEQUALITIES

In certain types of mathematical problems it is necessary to work simultaneously with more than one equation in several variables. We then refer to the given equations as a system of equations. It is usually desirable to find the solutions which are common to all equations in the system. In this chapter we shall investigate special types of systems of equations and develop methods for finding their common solutions. Of particular importance are the matrix techniques introduced for systems of linear equations. We shall also touch briefly on systems of inequalities and on linear programming.

8.1 SYSTEMS OF EQUATIONS

An ordered pair (a, b) is a **solution** of an equation in two variables x and y if a true statement is obtained when a and b are substituted for x and y, respectively. Two equations in x and y are **equivalent** if they have exactly the same solutions. For example, the equation

$$x^2 - 4y - 2 = 3x + 2y + 6$$

is equivalent to $\qquad x^2 - 6y - 3x - 8 = 0.$

As we know, the **graph** of an equation in x and y consists of all points in a coordinate plane which correspond to solutions.

By a **system** of two equations in x and y we mean any two equations in those variables. An ordered pair (a, b) is called a **solution of the system** if (a, b) is a solution of both equations. It follows that the points which correspond to the solutions are precisely the points at which the graphs of the two equations intersect.

As a concrete example, consider the system

$$x^2 - y = 0, \quad y - 2x - 3 = 0.$$

The following table exhibits some solutions of the equation $x^2 - y = 0$.

x	-2	-1	0	1	2	3	4
y	4	1	0	1	4	9	16

The graph is the parabola sketched in Figure 8.1. A similar table for $y - 2x - 3 = 0$ is

x	-2	-1	0	1	2	3	4
y	-1	1	3	5	7	9	11

The graph is the line in Figure 8.1. Note that the pairs $(3, 9)$ and $(-1, 1)$ are solutions of both equations and hence are solutions of the system. We shall see later that they are the *only* solutions. As pointed out before, the two pairs which are solutions of the system represent points of intersection of the two graphs.

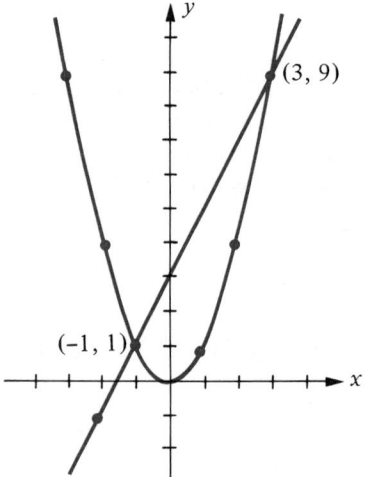

Figure 8.1. $\begin{cases} x^2 - y = 0 \\ y - 2x - 3 = 0 \end{cases}$

The graphical analysis we have used is obviously of little value if the solutions of the system involve rational numbers such as 37/849, or irrational numbers, or if the equations are more complicated than those we have considered. Let us illustrate how the solutions of the preceding system can be found without reference to graphs. We shall begin by solving the first equation for y in terms of x, obtaining $y = x^2$. It follows that if an ordered pair (x, y) is a solution of the system, then it must be of the form (x, x^2). In particular, (x, x^2) must be a solution of the equation $y - 2x - 3 = 0$, that is,

$$x^2 - 2x - 3 = 0.$$

Factoring gives us $\qquad (x - 3)(x + 1) = 0$

and hence either $x = 3$ or $x = -1$. The corresponding values for y (obtained from $y = x^2$) are 9 and 1, respectively. Thus the only possible solutions of the system are the ordered pairs $(3, 9)$ and $(-1, 1)$. That these are, indeed, solutions can be seen by checking each equation of the system.

To recapitulate, in order to find the solutions algebraically, we began by solving one equation of the system for y in terms of x, and then we substituted for y in the other equation, obtaining an equation in one variable, x. The solutions of the latter equation were the only possible x-values for the solutions of the system. The corresponding y-values were found by means of the equation which expressed y in terms of x. This algebraic technique is called the *method of substitution*. Sometimes it is more convenient to begin by solving one equation for x in terms of y, and then substituting for x in the other equation. In general the steps used in this process may be listed as follows.

The Method of Substitution

(i) Solve one of the equations for one variable in terms of the other variable.

(ii) Substitute the expression obtained in step (i) in the other equation, obtaining an equation in one variable.

(iii) Find the solutions of the equation obtained in step (ii).

(iv) Use the solutions from step (iii) together with the expression obtained in step (i) to find the solutions of the system.

In the next example we find the solutions of the system considered at the beginning of this section by first solving one equation for x in terms of y.

Example 1 Find the solutions of the system

$$x^2 - y = 0, \quad y - 2x - 3 = 0.$$

Solution We may solve the second equation for x in terms of y as follows:

$$2x = y - 3, \quad \text{or} \quad x = \frac{y - 3}{2}.$$

Substituting for x in the first equation gives us

$$\left(\frac{y - 3}{2}\right)^2 - y = 0$$

$$\frac{y^2 - 6y + 9}{4} - y = 0$$

$$y^2 - 6y + 9 - 4y = 0$$

$$y^2 - 10y + 9 = 0.$$

Factoring we obtain $\qquad (y - 9)(y - 1) = 0$

which has solutions $y = 9$ and $y = 1$. If we now substitute for y in the equation $x = (y - 3)/2$ we get the corresponding values $x = 3$ and $x = -1$. Hence, as before, the solutions of the system are $(3, 9)$ and $(-1, 1)$. ■

Example 2   Find the solutions of the system $x^2 + y^2 = 25$, $x^2 + y = 19$. Sketch the graph of each equation, showing the points of intersection.

Solution   Solving the second equation for y we obtain $y = 19 - x^2$. Substituting for y in the first equation leads to the following chain of equivalent equations:

$$x^2 + (19 - x^2)^2 = 25$$

$$x^4 - 37x^2 + 336 = 0$$

$$(x^2 - 16)(x^2 - 21) = 0.$$

The solutions of the last equation are 4, -4, $\sqrt{21}$, and $-\sqrt{21}$. The corresponding y values are found by substituting for x in the equation $y = 19 - x^2$. Substitution of 4 or -4 for x gives us $y = 3$, whereas substitution of $\sqrt{21}$ or $-\sqrt{21}$ gives us $y = -2$. Hence the only possible solutions of the system are

$$(4, 3), \quad (-4, 3), \quad (\sqrt{21}, -2), \quad \text{and} \quad (-\sqrt{21}, -2).$$

It can be seen by direct substitution in each of the given equations that all four pairs are solutions. The graph of $x^2 + y^2 = 25$ is a circle of radius 5 with center at the origin, and the graph of $y = 19 - x^2$ is a parabola with a vertical axis. The graphs and their points of intersection are illustrated in Figure 8.2.

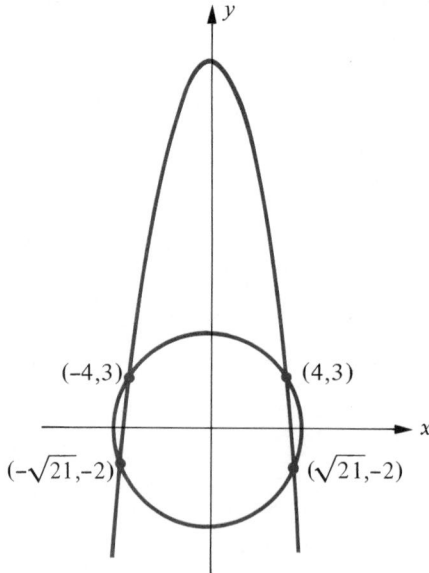

Figure 8.2. $\begin{cases} x^2 + y^2 = 25 \\ x^2 + y = 19 \end{cases}$

Example 3   Find the solutions of the system $y - \log(x + 3) = 1, 2 - y = \log x$.

Solution   From the second equation we obtain $y = 2 - \log x$. Substituting in the first equation and simplifying, we obtain

$$\log x + \log (x + 3) = 1.$$

This leads to the following chain of equivalent equations (supply the reasons):

$$\log x(x + 3) = 1$$
$$x(x + 3) = 10$$
$$x^2 + 3x - 10 = 0$$
$$(x + 5)(x - 2) = 0.$$

Hence the only possible solutions are $x = -5$ and $x = 2$. If we substitute -5 for x in the equation $y = 2 - \log x$, we arrive at the logarithm of a negative number. Since $\log x$ is defined only if $x > 0$, this value of x is extraneous. If we substitute 2 for x we get $y = 2 - \log 2$. Checking, we see that the system has one solution $(2, 2 - \log 2)$. ∎

We can also consider equations in three variables x, y, and z, such as

$$x^2 y + x \log z + 3^y = 4z^3.$$

We say that the equation has a **solution** (a, b, c) if substitution of a, b, and c for x, y and z, respectively, produces a true statement. We refer to (a, b, c) as an **ordered triple** of real numbers. Equivalent equations are defined as before, A system of equations in three variables and the corresponding solutions are defined as in the two-variable case. In like manner, we can consider systems of *any* number of equations in *any* number of variables.

The method of substitution can be extended to these more complicated systems. For example, given three equations in three variables, suppose that it is possible to solve one of the equations for one variable in terms of the remaining two variables. By substituting that expression in each of the other equations, we obtain a system of two equations in two variables. The solutions of the latter system can then be used to find the solutions of the original system, as illustrated in the following example.

Example 4   Find the solutions of the system $x - y + z = 2$, $xyz = 0$, $2y + z = 1$.

Solution   Solving the third equation for z gives us

$$z = 1 - 2y.$$

Substituting for z in the first two equations of the system we obtain

$$\begin{cases} x - y + (1 - 2y) = 2 \\ xy(1 - 2y) = 0 \end{cases}$$

or equivalently,

$$\begin{cases} x - 3y - 1 = 0 \\ xy(1 - 2y) = 0. \end{cases}$$

We now find the solutions of the latter system. Solving the first equation for x in terms of y gives us

$$x = 3y + 1.$$

Substituting $3y + 1$ for x in the second equation $xy(1 - 2y) = 0$, we obtain

$$(3y + 1)y(1 - 2y) = 0$$

which has, for its solutions, the numbers $-\frac{1}{3}, 0, \frac{1}{2}$. These are the only possible y-values for the solutions of the system. To obtain the corresponding x-values, we use the equation $x = 3y + 1$, obtaining $x = 0$, 1, and $\frac{5}{2}$, respectively. Finally, returning to $z = 1 - 2y$, we obtain the z-values $\frac{5}{3}$, 1, and 0. It follows that the solutions of the original system consist of the ordered triples $(0, -\frac{1}{3}, \frac{5}{3})$, $(1, 0, 1)$, and $(\frac{5}{2}, \frac{1}{2}, 0)$. ∎

EXERCISES 8.1

In Exercises 1–34 use the method of substitution to find the solutions of each system of equations.

1 $\begin{cases} y = x^2 - 4 \\ y = 2x - 1 \end{cases}$

2 $\begin{cases} y = x^2 + 1 \\ x + y = 3 \end{cases}$

3 $\begin{cases} y^2 = 1 - x \\ x + 2y = 1 \end{cases}$

4 $\begin{cases} y^2 = x \\ x + 2y + 3 = 0 \end{cases}$

5 $\begin{cases} 2y = x^2 \\ y = 4x^3 \end{cases}$

6 $\begin{cases} x - y^3 = 1 \\ 2x = 9y^2 + 2 \end{cases}$

7 $\begin{cases} x + 2y = -1 \\ 2x - 3y = 12 \end{cases}$

8 $\begin{cases} 3x - 4y + 20 = 0 \\ 3x + 2y + 8 = 0 \end{cases}$

9 $\begin{cases} 2x - 3y = 1 \\ -6x + 9y = 4 \end{cases}$

10 $\begin{cases} 4x - 5y = 2 \\ 8x - 10y = -5 \end{cases}$

11 $\begin{cases} x + 3y = 5 \\ x^2 + y^2 = 25 \end{cases}$

12 $\begin{cases} 3x - 4y = 25 \\ x^2 + y^2 = 25 \end{cases}$

13 $\begin{cases} x^2 + y^2 = 8 \\ y - x = 4 \end{cases}$

14 $\begin{cases} x^2 + y^2 = 25 \\ 3x + 4y = -25 \end{cases}$

15 $\begin{cases} x^2 + y^2 = 9 \\ y - 3x = 2 \end{cases}$

16 $\begin{cases} x^2 + y^2 = 16 \\ y + 2x = -1 \end{cases}$

17 $\begin{cases} x^2 + y^2 = 16 \\ 2y - x = 4 \end{cases}$

18 $\begin{cases} x^2 + y^2 = 1 \\ y + 2x = -3 \end{cases}$

19 $\begin{cases} (x - 1)^2 + (y + 2)^2 = 10 \\ x + y = 1 \end{cases}$

20 $\begin{cases} xy = 2 \\ 3x - y + 5 = 0 \end{cases}$

21 $\begin{cases} y = 20/x^2 \\ y = 9 - x^2 \end{cases}$

22 $\begin{cases} x = y^2 - 4y + 5 \\ x - y = 1 \end{cases}$

23 $\begin{cases} y^2 - 4x^2 = 4 \\ 9y^2 + 16x^2 = 140 \end{cases}$

24 $\begin{cases} 25y^2 - 16x^2 = 400 \\ 9y^2 - 4x^2 = 36 \end{cases}$

25 $\begin{cases} x^2 - y^2 = 4 \\ x^2 + y^2 = 12 \end{cases}$

26 $\begin{cases} 6x^3 - y^3 = 1 \\ 3x^3 + 4y^3 = 5 \end{cases}$

27 $\begin{cases} x + 2y - z = -1 \\ 2x - y + z = 9 \\ x + 3y + 3z = 6 \end{cases}$

28 $\begin{cases} 2x - 3y - z^2 = 0 \\ x - y - z^2 = -1 \\ x^2 - xy = 0 \end{cases}$

29 $\begin{cases} y = 5^x + 4 \\ y = 5^{2x} - 2 \end{cases}$

30 $\begin{cases} x + 3(2^y) = 2^{2y} \\ x - 2^{y+1} = 6 \end{cases}$

31 $\begin{cases} \log(3r + 1) = s + 5 \\ \log(r - 2) = s + 4 \end{cases}$

32 $\begin{cases} \log u = v - 3 \\ 5 = v + \log(u - 2) \end{cases}$

33 $\begin{cases} \log(3x + 5) - \log y = 1 \\ 5x - 2y = 6 \end{cases}$

34 $\begin{cases} y + 2 = \log(2x + 5) \\ y - 1 = \log x \end{cases}$

Solve Exercises 35–40 by introducing several variables and using a suitable system of equations.

35 Find two positive integers whose difference is 4 and whose squares differ by 88. (*Hint:* Let x and y denote the two integers.)

36 Find two real numbers whose difference and product both equal 4.

37 The perimeter of a rectangle is 40 inches and its area is 96 square inches. Find the length and width.

38 Generalize Exercise 37 to the case where the perimeter is P and area is A. Express the length

and width in terms of P and A. What restrictions are necessary?

39 Find three numbers whose sum and product are 20 and 60 respectively, and such that one of the numbers equals the sum of the other two.

40 Find the values of b for which the system

$$\begin{cases} x^2 + y^2 = 4 \\ y = x + b \end{cases}$$

has solutions consisting of (a) one real number, (b) two real numbers, (c) no real numbers. Interpret the three cases geometrically.

8.2 SYSTEMS OF LINEAR EQUATIONS IN TWO VARIABLES

An equation of the form $ax + by + c = 0$ (or, equivalently, $ax + by = -c$) is called a *linear equation* in x and y. Similarly, a linear equation in three variables x, y, and z is an equation of the form $ax + by + cz = d$, where the coefficients are real numbers. Linear equations in any number of variables may be defined in like manner. If a large number of variables occurs, a subscript notation is usually employed. If we let $x_1, x_2, \ldots, x_n$ denote variables, where n is any positive integer, then an expression of the form

$$a_1 x_1 + a_2 x_2 + \cdots + a_n x_n = a$$

where $a_1, a_2, \ldots, a_n$ and a are real numbers, is called a **linear equation in n variables** with real coefficients.

In modern applications of mathematics, perhaps the most common systems of equations are those in which all the equations are linear. In this section we shall only consider systems of two linear equations in two variables. Systems involving more than two variables are discussed in Section 8.3.

The method used to solve a system of equations in several variables is similar to that used for one equation in one variable, in the sense that the given system is replaced by a chain of equivalent systems until a stage is reached from which the solutions are easily obtained. Some general rules which allow us to transform a system of equations into an equivalent system are given in the next theorem. Since we prefer not to specify the number or types of variables, we shall use the notation $p = 0$ and $q = 0$ for typical equations in a system. The reader should bear in mind that the symbols p and q represent expressions in the variables under consideration. For example, if

$p = 3x - 2y + 5z - 1$ and $q = 6x + y - 7z - 4$, then the equations $p = 0$ and $q = 0$ have the form

$$3x - 2y + 5z - 1 = 0, \quad 6x + y - 7z - 4 = 0.$$

Transformations that Lead to Equivalent Systems of Equations

> The following transformations do not change the solutions of a system of equations.
>
> (i)   Interchanging the position of any two equations.
> (ii) Multiplying both sides of an equation in the system by a nonzero real number.
> (iii) Replacing an equation $q = 0$ of the system by $kp + q = 0$, where $p = 0$ is any other equation in the system and k is any real number.

Proof

It is easy to show that (i) and (ii) do not change the solutions of the system and therefore we shall omit the proofs.

To prove (iii), we first note that a solution of the original system is a solution of both equations $p = 0$ and $q = 0$. Accordingly, each of the expressions p and q equals zero if the variables are replaced by appropriate real numbers, and hence the expression $kp + q$ will also equal zero. Since none of the other equations has been changed, this shows that any solution of the original system is also a solution of the transformed system obtained by replacing the equation $q = 0$ by $kp + q = 0$.

Conversely, given a solution of the transformed system, both of the expressions $kp + q$ and p equal zero if the variables are replaced by appropriate numbers. This implies, however, that $(kp + q) - kp$ equals 0. Since $(kp + q) - kp = q$, we see that q must also equal zero when the substitution is made. Thus a solution of the transformed system is also a solution of the original system. This completes the proof.

For convenience, we shall describe the process of using (iii) of the preceding theorem by the phrase "add to one equation of the system k times any other equation of the system." Of course, to *add* two equations means to add corresponding sides. To *multiply* an equation by k means to multiply both sides of the equation by k. The process may also be applied if 0 is not on one side of each equation, as illustrated in the next example.

Example 1   Find the solutions of the system

$$\begin{cases} x + 3y = -1 \\ 2x - y = 5. \end{cases}$$

Solution   By (ii) of the last theorem we may multiply the second equation by 3. This gives us the equivalent system

$$\begin{cases} x + 3y = -1 \\ 6x - 3y = 15. \end{cases}$$

Next, by (iii) of the theorem, we may add to the second equation 1 times the first. This gives us

$$\begin{cases} x + 3y = -1 \\ 7x \qquad = 14. \end{cases}$$

We see from the last equation that the only possible value for x is 2. The corresponding value for y may be found by substituting for x in the first equation. This gives us the following:

$$2 + 3y = -1, \quad 3y = -3, \quad y = -1.$$

Consequently, the given system has one solution, $(2, -1)$.

There are other methods of solution. For example, we could begin by multiplying the first equation by -2, obtaining

$$\begin{cases} -2x - 6y = 2 \\ 2x - y = 5. \end{cases}$$

If we next add the first equation to the second, we get

$$\begin{cases} -2x - 6y = 2 \\ -7y = 7. \end{cases}$$

The last equation implies that $y = -1$. Substitution for y in the first equation gives us

$$-2x - 6(-1) = 2. \quad -2x = -4, \quad x = 2.$$

Again we see that the solution is $(2, -1)$.

The graphs of the two equations, showing the point of intersection $(2, -1)$, are sketched in Figure 8.3.

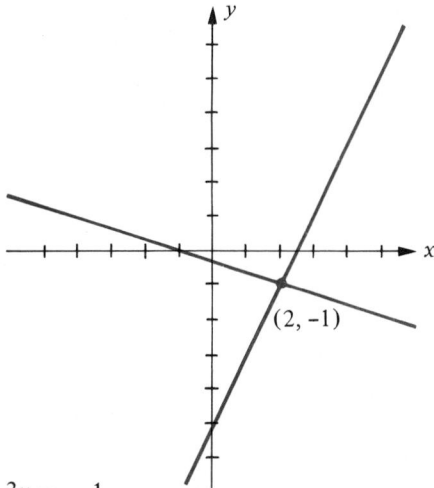

Figure 8.3. $\begin{cases} x + 3y = -1 \\ 2x - y = 5. \end{cases}$

The technique used in Example 1 is called the **method of elimination**, since it involves eliminating a variable from one of the equations. The method of elimination usually leads to solutions in fewer steps than the method of substitution discussed in Section 8.1.

Example 2   Find the solutions of the system

$$\begin{cases} 3x + y = 6 \\ 6x + 2y = 12. \end{cases}$$

Solution   Multiplying the second equation by $\frac{1}{2}$ gives us

$$\begin{cases} 3x + y = 6 \\ 3x + y = 6. \end{cases}$$

Consequently, (a, b) is a solution if and only if $3a + b = 6$, that is, $b = 6 - 3a$. It follows that the solutions consist of all ordered pairs of the form $(a, 6 - 3a)$, where a is a real number. If we wish to find particular solutions we may substitute various values for a. For example, a few solutions are $(0, 6)$, $(1, 3)$, $(3, -3)$, $(-2, 12)$, $(\sqrt{2}, 6 - 3\sqrt{2})$. ∎

Example 3   Find the solutions of the system

$$\begin{cases} 3x + y = 6 \\ 6x + 2y = 20. \end{cases}$$

Solution   If we add to the second equation -2 times the first equation we obtain the equivalent system

$$\begin{cases} 3x + y = 6 \\ 0 = 8. \end{cases}$$

Since the last equation is never true, the system has no solutions. Of course, this means that the graphs of the given equations do not intersect (see Figure 8.4).

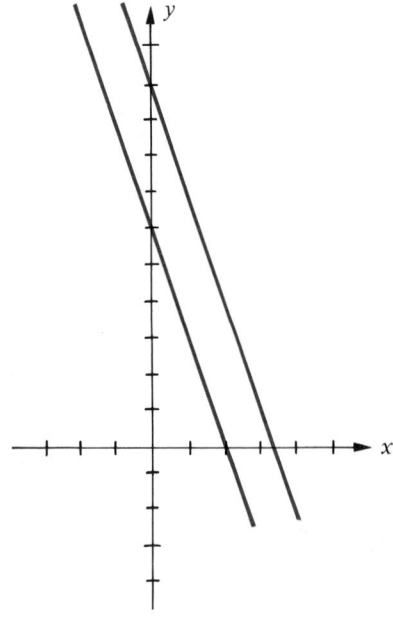

Figure 8.4. $\begin{cases} 3x + y = 6 \\ 6x + 2y = 20 \end{cases}$ ∎

Every system of two linear equations in two variables x and y with real coefficients has the form

$$\begin{cases} a_1 x + a_2 y = a \\ b_1 x + b_2 y = b \end{cases}$$

where the coefficients are real numbers. We know from our work in Chapter Two that the graph of each equation in the system is a straight line. It follows that precisely *one* of the following three possibilities must occur:

(i)   The lines are identical.

(ii)   The lines are parallel.

(iii)   The lines intersect in one and only one point.

Since the pairs which are solutions of the system exhibit the coordinates of the points of intersection of the graphs of the two equations, we may interpret statements (i)–(iii) in the following way.

> For each system of two linear equations in two variables, one and only one of the following statements is true:
>
> (i)   The system has an infinite number of solutions.
>
> (ii)   The system has no solutions.
>
> (iii)   The system has exactly one solution.

If (i) occurs, then every solution of one equation of the system is also a solution of the other, and we say that the equations are **dependent**.

If (ii) occurs, there is no solution and the system is said to be **inconsistent**.

If (iii) occurs, there is a unique solution and the system is called **consistent**.

In practice, there should be little difficulty in determining which of the three cases occurs. In case (i) we have a solution similar to that for Example 2, where one of the equations can be transformed into the other. In case (ii) the lack of a solution is indicated by an absurdity such as the statement $0 = 8$, which appeared in Example 3. The case of the unique solution (iii) will be apparent when suitable transformations are applied to the system, as illustrated in Example 1.

EXERCISES 8.2

Find the solutions of the systems in Exercises 1–20.

1 $\begin{cases} 2x + 3y = 2 \\ x - 2y = 8 \end{cases}$

2 $\begin{cases} 4x + 5y = 13 \\ 3x + y = -4 \end{cases}$

5 $\begin{cases} 3r + 4s = 3 \\ r - 2s = -4 \end{cases}$

6 $\begin{cases} 9u + 2v = 0 \\ 3u - 5v = 17 \end{cases}$

3 $\begin{cases} 2x + 5y = 16 \\ 3x - 7y = 24 \end{cases}$

4 $\begin{cases} 7x - 8y = 9 \\ 4x + 3y = -10 \end{cases}$

7 $\begin{cases} 5x - 6y = 4 \\ 3x + 7y = 8 \end{cases}$

8 $\begin{cases} 2x + 8y = 7 \\ 3x - 5y = 4 \end{cases}$

9 $\begin{cases} \frac{1}{3}c + \frac{1}{2}d = 5 \\ c - \frac{2}{3}d = -1 \end{cases}$ **10** $\begin{cases} \frac{1}{2}t - \frac{1}{3}v = \frac{3}{2} \\ \frac{2}{3}t + \frac{1}{4}v = \frac{5}{12} \end{cases}$

11 $\begin{cases} \sqrt{3}x - \sqrt{2}y = 2\sqrt{3} \\ 2\sqrt{2}x + \sqrt{3}y = \sqrt{2} \end{cases}$

12 $\begin{cases} 0.11x - 0.03y = 0.25 \\ 0.12x + 0.05y = 0.70 \end{cases}$

13 $\begin{cases} 2x - 3y = 5 \\ -6x + 9y = 12 \end{cases}$

14 $\begin{cases} 3p - q = 7 \\ -12p + 4q = 3 \end{cases}$ **15** $\begin{cases} 3m - 4n = 2 \\ -6m + 8n = -4 \end{cases}$

16 $\begin{cases} x - 5y = 2 \\ 3x - 15y = 6 \end{cases}$ **17** $\begin{cases} 2y - 5x = 0 \\ 3y + 4x = 0 \end{cases}$

18 $\begin{cases} 3x + 7y = 9 \\ y = 5 \end{cases}$

19 $\begin{cases} \dfrac{2}{x} + \dfrac{3}{y} = -2 \\ \dfrac{4}{x} - \dfrac{5}{y} = 1 \end{cases}$ (*Hint:* Let $x' = 1/x$ and $y' = 1/y$.)

20 $\begin{cases} \dfrac{3}{x-1} + \dfrac{4}{y+2} = 2 \\ \dfrac{6}{x-1} - \dfrac{7}{y+2} = -3 \end{cases}$

21 Determine a and b so that the straight line with equation $ax + by = 10$ passes through the points $P_1(-3, 4)$ and $P_2(3, 1)$.

22 Generalize Exercise 21 to the case where the graph of $ax + by = 10$ passes through the points $P_1(x_1, y_1)$ and $P_2(x_2, y_2)$. (Express a and b in terms of x_1, y_1, x_2, and y_2.) Is it necessary to restrict P_1 or P_2 in any way?

Solve each of Exercises 23–30 by employing a system of linear equations.

23 The price of admission for a certain event was $2.25 for adults and $1.50 for children. If 450 tickets were sold for a total of $777.75, how many of each kind were purchased?

24 A chemist wishes to obtain 80 cc of a 30% acid solution by mixing a 50% solution and a 25% solution. How many cc of each solution should be used?

25 The sum of the digits of a certain two-digit number is 14. If the digits are reversed, the number is increased by 18. Find the number.

26 A man rows a boat 500 feet upstream against a constant current in 10 minutes. He then rows downstream (with the same current), covering 300 feet in 5 minutes. Find the speed of the current and the equivalent rate at which he can row in still water.

27 A collection of dimes and quarters amounts to $7.00. If there are seven more dimes than quarters, find the number of quarters.

28 A man has $10,000 invested in two funds which pay simple interest rates of 8% and 7%, respectively. If he receives yearly interest of $772, how much is invested in each fund?

29 A man receives income from two investments at simple interest rates of $6\frac{1}{2}\%$ and 8%, respectively. He has twice as much invested at $6\frac{1}{2}\%$ as at 8%. If his annual income from the two investments is $698.25, find how much is invested at each rate.

30 Two water pipes running at the same time can fill a swimming pool in 4 hours. If both pipes run for 2 hours and the first is then shut off, it takes 3 more hours for the second to fill the pool. How long does it take each pipe to fill the pool by itself?

8.3 SYSTEMS OF LINEAR EQUATIONS IN MORE THAN TWO VARIABLES

For systems of equations containing more than two variables we can use either the method of substitution explained in Section 8.1 or the techniques developed in Section

8.2. Let us illustrate this by solving a system in two different ways, first by the method of substitution and second by the method of elimination. You will see that the method of elimination is the shorter and more straightforward technique for finding solutions.

Example 1   Find the solutions of the system

$$\begin{cases} x - 2y + 3z = 4 \\ 2x + y - 4z = 3 \\ -3x + 4y - z = -2. \end{cases}$$

Solution 1   Using the method of substitution we solve the first equation for x, obtaining

$$x = 4 + 2y - 3z.$$

Substituting for x in the second and third equations gives us

$$\begin{cases} 2(4 + 2y - 3z) + y - 4z = 3 \\ -3(4 + 2y - 3z) + 4y - z = -2 \end{cases}$$

which simplifies to

$$\begin{cases} y - 2z = -1 \\ y - 4z = -5. \end{cases}$$

If we solve the first equation for y, obtaining $y = 2z - 1$, and then substitute for y in the second equation, we find that

$$(2z - 1) - 4z = -5 \quad \text{or} \quad -2z = -4.$$

The last equation has the solution $z = 2$. To find y we return to $y = 2z - 1$, obtaining $y = 2(2) - 1 = 3$. The corresponding value of x is found by substituting for y and z in the equation $x = 4 + 2y - 3z$. This gives us $x = 4 + 2(3) - 3(2) = 4$. Hence the given system has only one solution, the ordered triple $(4, 3, 2)$. We shall leave it to the reader to show that this triple is a solution of *each* of the given equations. Of course, other approaches to the problem could be used, such as solving the third equation for z and substituting in the first and second equations, obtaining two equations in x and y.

Solution 2   We begin by eliminating x from the second and third equations. Adding to the second equation -2 times the first equation gives us the equivalent system

$$\begin{cases} x - 2y + 3z = 4 \\ 5y - 10z = -5 \\ -3x + 4y - z = -2. \end{cases}$$

Next, adding to the third equation 3 times the first equation, we obtain the equivalent system

$$\begin{cases} x - 2y + 3z = 4 \\ 5y - 10z = -5 \\ -2y + 8z = 10. \end{cases}$$

To simplify computations we multiply the second equation by $\frac{1}{3}$, obtaining

$$\begin{cases} x - 2y + 3z = \quad 4 \\ \quad\quad y - 2z = -1 \\ \quad -2y + 8z = \quad 10. \end{cases}$$

We now eliminate y from the third equation by adding to it 2 times the second equation. This gives us the system

$$\begin{cases} x - 2y + 3z = \quad 4 \\ \quad\quad y - 2z = -1 \\ \quad\quad\quad 4z = \quad 8. \end{cases}$$

The solutions of the last system are easy to obtain. From the third equation it is clear that the only possible z-value is 2. Substituting 2 for z in the second equation, $y - 2z = -1$, we get $y - 2(2) = -1$, or $y = 3$. Finally, the x-value is found by substituting for y and z in the first equation. This gives us $x - 2(3) + 3(2) = 4$ and hence $x = 4$. Thus there is one solution, $(4, 3, 2)$. ■

The general form for a system of three linear equations in three variables with real coefficients is

$$\begin{cases} a_1 x + a_2 y + a_3 z = a \\ b_1 x + b_2 y + b_3 z = b \\ c_1 x + c_2 y + c_3 z = c \end{cases}$$

where the coefficients are real numbers. It can be shown that the preceding system has either a *unique solution*, an *infinite number of solutions*, or *no solutions*. As with the case of two equations in two variables, the terminology used to describe these cases is *consistent*, *dependent*, or *inconsistent*, respectively.

Similar remarks and methods of solution hold for systems of four linear equations in four variables or, for that matter, systems of n linear equations in n variables, where n is any positive integer. Using the method of substitution we can always solve one of the equations for one variable in terms of the remaining variables. By substituting in each of the remaining equations we obtain a system containing one less equation in one less variable. We keep repeating this process until we obtain a form from which the solution may be easily found. Of course, the method of elimination can also be used.

Sometimes it is necessary to work with systems in which the number of equations is not the same as the number of variables. Such systems can be attacked by using methods similar to those we have discussed.

Example 2   Find the solutions of the system

$$\begin{cases} 2x + 3y + 4z = 1 \\ 3x + 4y + 5z = 3. \end{cases}$$

Solution We can eliminate x from the second equation by adding to it $-3/2$ times the first equation, obtaining the equivalent system

$$\begin{cases} 2x + 3y + 4z = 1 \\ \quad\ -\tfrac{1}{2}y - \ z = \tfrac{3}{2}. \end{cases}$$

To clear of fractions we multiply the second equation by -2, obtaining

$$\begin{cases} 2x + 3y + 4z = 1 \\ \quad\quad\ y + 2z = -3. \end{cases}$$

There are an infinite number of solutions (b, c) for the second equation. Indeed, if we write $y = -3 - 2z$, then by substituting any number c for z, the corresponding value b of y is given by $b = -3 - 2c$. For each such solution (b,c) of $y + 2z = -3$ there corresponds a number a such that the triple (a, b, c) is a solution of the first equation, that is,

$$2a + 3b + 4c = 1.$$

Solving the latter equation for a we obtain

$$a = \tfrac{1}{2}(1 - 3b - 4c) = \tfrac{1}{2} - \tfrac{3}{2}b - 2c.$$

Since $b = -3 - 2c$, we have

$$a = \tfrac{1}{2} - \tfrac{3}{2}(-3 - 2c) - 2c$$

which reduces to $a = 5 + c.$

Consequently, the solutions of the system consist of all ordered triples of the form

$$(5 + c, -3 - 2c, c)$$

where c is any real number. These solutions may be checked by substituting $5 + c$ for x, $-3 - 2c$ for y, and c for z in the two given equations. We can obtain any number of solutions for the system by substituting specific real numbers for c. For example, if $c = 0$, we obtain $(5, -3, 0)$; if $c = 2$, we have $(7, -7, 2)$; if $c = -1/2$, we get $(9/2, -2, -1/2)$; and so on. ∎

Another method of solution for Example 2 is to rewrite the original system as

$$\begin{cases} 2x + 3y = 1 - 4z \\ 3x + 4y = 3 - 5z. \end{cases}$$

If a number c is substituted for z, we obtain a system of two equations in two variables x and y, and the solutions can be found as in Section 8.2. For a different approach we could solve for two *other* variables in terms of the third. This method can be extended. For example, given two equations in four variables, say x, y, z, and w, it may be possible to solve for two of the variables, say x and y, in terms of the remaining variables z and w.

If we then substitute numbers c and d for z and w, respectively, and calculate the corresponding values for x and y, we obtain the solutions.

The general system of three linear equations in x, y, z discussed prior to Example 2 is called **homogeneous** if $a = b = c = 0$. The same terminology is used for more than three equations. A system of homogeneous equations always has the **trivial solution** obtained by substituting zero for all the variables. Nontrivial solutions sometimes exist. The procedure for finding solutions is the same as that used for nonhomogeneous systems.

Example 3 Find the solutions of the system

$$\begin{cases} x - y + 4z = 0 \\ 2x + y - z = 0 \\ -x - y + 2z = 0. \end{cases}$$

Solution   Adding to the second equation -2 times the first equation and then adding the first equation to the third gives us the equivalent system

$$\begin{cases} x - y + 4z = 0 \\ 3y - 9z = 0 \\ -2y + 6z = 0. \end{cases}$$

Multiplying the second equation by $\frac{1}{3}$ and the third equation by $-\frac{1}{2}$ leads to the same equation, $y - 3z = 0$. Hence the given system is equivalent to

$$\begin{cases} x - y + 4z = 0 \\ y - 3z = 0. \end{cases}$$

The solutions may now be found as in the previous example. If we substitute any value c for z, then the second equation gives us $y = 3c$. Substituting c for z and $3c$ for y in the first equation, we obtain $x - 3c + 4c = 0$, or $x = -c$. Thus the solutions consist of all ordered triples of the form $(-c, 3c, c)$, where c is any real number.   ■

Example 4 Find the solutions of the system

$$\begin{cases} x + y + z = 0 \\ x - y + z = 0 \\ x - y - z = 0. \end{cases}$$

Solution   By the usual process we obtain the equivalent system

$$\begin{cases} x + y + z = 0 \\ -2y = 0 \\ -2y - 2z = 0. \end{cases}$$

Multiplying the second and third equations by $-\frac{1}{2}$ gives us

$$\begin{cases} x + y + z = 0 \\ y = 0 \\ y + z = 0. \end{cases}$$

From the second equation of this system we see that the only possible y-value is 0. Substituting 0 for y in the last equation we obtain $0 + z = 0$, or $z = 0$. Finally, substituting 0 for both y and z in the first equation we see that x also must equal 0. Hence the only solution for the system is the trivial one, $(0, 0, 0)$.   ∎

The next example is an illustration of an applied problem which can be solved by means of a system of three linear equations.

Example 5   A merchant wishes to mix two grades of peanuts costing $1.50 and $2.50 per pound, respectively, with cashews costing $4.00 per pound, to obtain 130 pounds of a mixture costing $3.00 per pound. If the merchant also wants the amount of cheaper grade peanuts to be twice that of the better grade, how many pounds of each variety should be mixed?

Solution   Let us introduce three variables as follows:

$$x = \text{pounds of peanuts at \$1.50 per pound}$$
$$y = \text{pounds of peanuts at \$2.50 per pound}$$
$$z = \text{pounds of cashews at \$4.00 per pound.}$$

We refer to the statement of the problem and write the following system of equations

$$\begin{cases} x + y + z = 130 \\ 1.50x + 2.50y + 4.00z = 3.00(130) \\ x = 2y. \end{cases}$$

We shall use a combination of the methods of substitution and elimination to find the solution. Let us begin by substituting $2y$ for x in the first two equations, obtaining

$$\begin{cases} 2y + y + z = 130 \\ 1.5(2y) + 2.5y + 4z = 3(130) \end{cases}$$

or equivalently,

$$\begin{cases} 3y + z = 130 \\ 5.5y + 4z = 390. \end{cases}$$

We next eliminate z from the second equation by adding to it -4 times the first equation. This gives us

$$\begin{cases} 3y + z = 130 \\ -6.5y = -130. \end{cases}$$

It follows that

$$y = \frac{-130}{-6.5} = 20$$

$$x = 2y = 2(20) = 40$$

$$z = 130 - x - y = 130 - 40 - 20 = 70.$$

Thus the merchant should use 40 pounds of the \$1.50 peanuts, 20 pounds of the \$2.50 peanuts, and 70 pounds of cashews. The check is left to the reader.   ∎

EXERCISES 8.3

Solve the systems in Exercises 1–26.

1. $\begin{cases} x - 2y - 3z = -1 \\ 2x + y + z = 6 \\ x + 3y - 2z = 13 \end{cases}$

2. $\begin{cases} x + 3y - z = -3 \\ 3x - y + 2z = 1 \\ 2x - y + z = -1 \end{cases}$

3. $\begin{cases} 5x + 2y - z = -7 \\ x - 2y + 2z = 0 \\ 3y + z = 17 \end{cases}$

4. $\begin{cases} 4x - y + 3z = 6 \\ -8x + 3y - 5z = -6 \\ 5x - 4y = -9 \end{cases}$

5. $\begin{cases} 2x + 6y - 4z = 1 \\ x + 3y - 2z = 4 \\ 2x + y - 3z = -7 \end{cases}$

6. $\begin{cases} 2x - y = 5 \\ 5y + 3z = -2 \\ x - 7z = 3 \end{cases}$

7. $\begin{cases} 2x - 3y + 2z = -3 \\ -3x + 2y + z = 1 \\ 4x + y - 3z = 4 \end{cases}$

8. $\begin{cases} 2x - 3y + z = 2 \\ 3x + 2y - z = -5 \\ 5x - 2y + z = 0 \end{cases}$

9. $\begin{cases} x + 3y + z = 0 \\ x + y - z = 0 \\ x - 2y - 4z = 0 \end{cases}$

10. $\begin{cases} 2x - y + z = 0 \\ x - y - 2z = 0 \\ 2x - 3y + z = 0 \end{cases}$

11. $\begin{cases} 2x + y + z = 0 \\ x - 2y - 2z = 0 \\ x + y + z = 0 \end{cases}$

12. $\begin{cases} x + y - 2z = 0 \\ x - y - 4z = 0 \\ z + y = 0 \end{cases}$

13. $\begin{cases} 3x - 2y + 5z = 7 \\ x + 4y - z = -2 \end{cases}$

14. $\begin{cases} 2x - y + 4z = 8 \\ -3x + y - 2z = 5 \end{cases}$

15. $\begin{cases} 4x - 2y + z = 5 \\ 3x + y - 4z = 0 \end{cases}$

16. $\begin{cases} 5x + 2y - z = 10 \\ y + z = -3 \end{cases}$

17. $\begin{cases} x + 2y - z - 3w = 2 \\ 3x + y - 2z - w = 6 \\ x + y + 3z - 2w = -3 \\ 4x - 3y - z - 2w = -8 \end{cases}$

18. $\begin{cases} x - 2y - 5z + w = -1 \\ 2x - y + z + w = 1 \\ 3x - 2y - 4z - 2w = 1 \\ x + y + 3z - 2w = 2 \end{cases}$

19. $\begin{cases} 5x + 2z = 1 \\ y - 3z = 2 \\ 2x + y = 3 \end{cases}$

20. $\begin{cases} 2x - 5y = 4 \\ 3y + 2z = -3 \\ 7x - 3z = 1 \end{cases}$

21. $\begin{cases} 4x - 3y = 1 \\ 2x + y = -7 \\ -x + y = -1 \end{cases}$

22. $\begin{cases} 2x + 3y = -2 \\ x + y = 1 \\ x - 2y = 13 \end{cases}$

23. $\begin{cases} 2x + 3y = 5 \\ x - 3y = 4 \\ x + y = -2 \end{cases}$

24. $\begin{cases} 4x - y = 2 \\ 2x + 2y = 1 \\ 4x - 5y = 3 \end{cases}$

25 $\begin{cases} \dfrac{3}{x} - \dfrac{3}{y} + \dfrac{1}{z} = -1 \\ \dfrac{1}{x} + \dfrac{4}{y} + \dfrac{1}{z} = 5 \\ \dfrac{2}{x} + \dfrac{1}{y} - \dfrac{4}{z} = 0 \end{cases}$ 26 $\begin{cases} \dfrac{1}{x} + \dfrac{1}{y} - \dfrac{1}{z} = 4 \\ \dfrac{2}{x} - \dfrac{1}{y} + \dfrac{1}{z} = 2 \\ \dfrac{4}{x} + \dfrac{2}{y} + \dfrac{3}{z} = -3 \end{cases}$

27 A collection of nickels, dimes, and quarters amounts to $8.00. If there are 72 coins in all and there are twice as many nickels as there are quarters, find the number of dimes.

28 A three-digit number is equal to 31 times the sum of its digits. If the three digits are reversed, the resulting number exceeds the given number by 99. The tens digit is three less than the sum of the hundreds and units digits. Find the number.

29 A chemist has three solutions containing a certain acid. The first contains 10% acid, the second 30%, and the third 50%. He wishes to use all three solutions to obtain a mixture of 50 liters containing 32% acid, using twice as much of the 50% solution as the 30% solution. How many liters of each solution should be used?

30 A swimming pool can be filled by three pipes A, B, and C. Pipe A can fill the pool by itself in 8 hours. If pipes A and C are used together, the pool can be filled in 6 hours. If B and C are used together, it takes 10 hours. How long does it take to fill the pool if all three pipes are used?

31 Find an equation of the circle which passes through the three points $P_1(2, 1)$, $P_2(-1, -4)$ and $P_3(3, 0)$. (*Hint:* An equation of the circle has the form $x^2 + y^2 + ax + by + c = 0$.)

32 Determine a, b, and c such that the graph of the equation

$$y = ax^2 + bx + c$$

passes through the points $P_1(3, -1)$, $P_2(1, -7)$, and $P_3(-2, 14)$.

8.4 MATRIX SOLUTIONS OF SYSTEMS OF LINEAR EQUATIONS

In Example 1 of Section 8.3 we used two different methods to solve the system

$$\begin{cases} x - 2y + 3z = 4 \\ 2x + y - 4z = 3 \\ -3x + 4y - z = -2. \end{cases}$$

If we analyze the second method (the method of elimination), we see that the symbols used for the variables really have little effect on the solution of the problem. The *coefficients* of the variables are the important things to consider. Thus if different symbols such as r, s, and t are used for the variables in that example, we obtain the system

$$\begin{cases} r - 2s + 3t = 4 \\ 2r + s - 4t = 3 \\ -3r + 4s - t = -2. \end{cases}$$

The method of elimination could then proceed exactly as before. Since this is true, it is possible to use a process which reduces the amount of labor involved. Specifically, we introduce a notation for keeping track of the coefficients in such a way that the variables do not have to be written down. Referring to the preceding system and checking that corresponding variables appear underneath one another, and terms not involving variables are to the right of the equals sign, we list the numbers which are involved in the equations in the following manner:

$$\begin{bmatrix} 1 & -2 & 3 & 4 \\ 2 & 1 & -4 & 3 \\ -3 & 4 & -1 & -2 \end{bmatrix}$$

If some variable had not appeared in one of the equations, we would have used a zero in the appropriate position. An array of numbers of this type is called a **matrix**. The **rows** of the matrix are the numbers which appear next to one another *horizontally*. Thus the first row is 1 -2 3 4, the second row is 2 1 -4 3, and the third row is -3 4 -1 -2. The **columns** of the matrix are the sets of numbers which form a *vertical* pattern. For example, the second column consists of the numbers -2, 1, 4 (in that order); the fourth column consists of 4, 3, -2; and so on.

Before discussing a method of solving a system of equations by means of a matrix, let us introduce a general definition of matrix. It will be convenient to use a **double subscript notation** for the symbols which represent the numbers in the matrix. Specifically, the symbol a_{ij} will denote the number which appears in row i and column j. We call i the **row subscript** and j the **column subscript** of a_{ij}.

Definition

If m and n are positive integers, then an $m \times n$ **matrix** over the real number system is an array of the form

$$\begin{bmatrix} a_{11} & a_{12} & a_{13} & \cdots & a_{1n} \\ a_{21} & a_{22} & a_{23} & \cdots & a_{2n} \\ a_{31} & a_{32} & a_{33} & \cdots & a_{3n} \\ \cdot & \cdot & \cdot & & \cdot \\ \cdot & \cdot & \cdot & & \cdot \\ \cdot & \cdot & \cdot & & \cdot \\ a_{m1} & a_{m2} & a_{m3} & \cdots & a_{mn} \end{bmatrix}$$

where each a_{ij} is a real number.

The symbol $m \times n$ in the definition is read "m by n." It is also possible to consider matrices where the elements a_{ij} belong to other mathematical systems; however, we shall not do so in this book. The rows and columns of a matrix are defined as before. Thus the matrix in the definition has m rows and n columns. The reader should carefully examine the notation for a matrix, observing, for example, that the symbol a_{23} appears in row 2 and column 3, whereas a_{32} appears in row 3 and column 2. Each a_{ij} is called an **element of the matrix**. The elements a_{11}, a_{22}, a_{33},... are called the **main diagonal elements**. If $m \neq n$, the matrix is sometimes referred to as a **rectangular matrix**. If $m = n$, we refer to the matrix as a **square matrix of order** n.

It is possible to develop a comprehensive theory for matrices. In succeeding sections we shall define *equality* of matrices, the *sum* of two matrices, the *product* of two matrices, and the *inverse* of a matrix. These ideas are very important and useful in mathematics and applications. In particular, matrix techniques are well suited for the types of problems which are solved by computers. Our main purpose in introducing matrices at this time is for use as a tool in solving systems of linear equations.

Let us return to the system of equations considered at the beginning of this section. The matrix we formed at that time is called the **matrix of the system**. In certain cases we may wish to consider only the *coefficients* of the variables of a system of linear equations. The corresponding matrix is called the **coefficient matrix**. The coefficient matrix for the system considered earlier is

$$\begin{bmatrix} 1 & -2 & 3 \\ 2 & 1 & -4 \\ -3 & 4 & -1 \end{bmatrix}.$$

In order to distinguish the matrix of the system from the coefficient matrix, the former is sometimes called the **augmented matrix**. After forming the matrix of the system, we work with the rows of the matrix *just as though they were equations*. The only items missing are the symbols for the variables, the addition signs used between terms, and the equals signs. We simply keep in mind that the numbers in the first column are the coefficients of the first variable, the numbers in the second column are the coefficients of the second variable, and so on. Our rules for transforming a matrix are formulated in such a way that they always produce a matrix of an equivalent system of equations. As a matter of fact, we have the following theorem for matrices, which is a perfect analogue of the Theorem on Transformations of Systems stated in Section 8.2.

Matrix Row Transformation Theorem

Given a matrix of a system of linear equations, each of the following transformations results in a matrix of an equivalent system of linear equations:

(i)   Interchanging any two rows.

(ii) Multiplying all of the elements in a row by the same nonzero real number k.

(iii) Adding to the elements in a row k times the corresponding elements of any other row, where k is any real number.

For convenience we shall refer to (ii) as the process of "multiplying a row by the number k" and (iii) will be described by saying "add k times any other row to a row." Rules (i)–(iii) of the preceding theorem are called the **elementary row transformations** of a matrix.

Let us find the solution of the system

$$\begin{cases} r - 2s + 3t = 4 \\ 2r + s - 4t = 3 \\ -3r + 4s - t = -2 \end{cases}$$

by means of the preceding theorem. The reader should compare our work with the second solution of Example 1 of Section 8.3. We shall use arrows to indicate that the form of the matrices has been changed by means of elementary row transformations, and we shall justify each transformation by stating the reason for each step. Thus

$$\begin{bmatrix} 1 & -2 & 3 & 4 \\ 2 & 1 & -4 & 3 \\ -3 & 4 & -1 & -2 \end{bmatrix} \rightarrow \begin{bmatrix} 1 & -2 & 3 & 4 \\ 0 & 5 & -10 & -5 \\ -3 & 4 & -1 & -2 \end{bmatrix}$$ Add to the second row -2 times the first row.

$$\rightarrow \begin{bmatrix} 1 & -2 & 3 & 4 \\ 0 & 5 & -10 & -5 \\ 0 & -2 & 8 & 10 \end{bmatrix}$$ Add to the third row 3 times the first row.

$$\rightarrow \begin{bmatrix} 1 & -2 & 3 & 4 \\ 0 & 1 & -2 & -1 \\ 0 & -2 & 8 & 10 \end{bmatrix}$$ Multiply row 2 by $\frac{1}{5}$.

$$\rightarrow \begin{bmatrix} 1 & -2 & 3 & 4 \\ 0 & 1 & -2 & -1 \\ 0 & 0 & 4 & 8 \end{bmatrix}$$ Add to the third row 2 times the second row.

The last matrix has zeros everywhere below the main diagonal and is said to be in **echelon form**. We now use that matrix to return to the system of equations

$$\begin{cases} r - 2s + 3t = & 4 \\ s - 2t = & -1 \\ 4t = & 8 \end{cases}$$

which is equivalent to the given system. The solutions may now be found as was done in the previous section.

The method used above is completely general. Beginning with *any* system of linear equations, we can use elementary row transformations of type (iii) many times, and also interchange rows if necessary, to introduce zeros everywhere in the first column after the first row; that is, we can obtain a matrix of an equivalent system which has the form

$$\begin{bmatrix} a_{11} & a_{12} & a_{13} & \cdots \\ 0 & b_{22} & b_{23} & \cdots \\ 0 & c_{32} & c_{33} & \cdots \\ 0 & \cdot & \cdot & \cdots \\ \cdot & \cdot & \cdot & \cdots \\ \cdot & \cdot & \cdot & \cdots \\ \cdot & \cdot & \cdot & \cdots \\ 0 & \cdot & \cdot & \cdots \end{bmatrix}$$

where the dots indicate that possibly more numbers appear. If the latter matrix has a nonzero element in the second column and after the first row, then again using elementary row transformations of type (iii), we may transform the matrix into the form

$$\begin{bmatrix} a_{11} & a_{12} & a_{13} & \cdots \\ 0 & k_{22} & k_{23} & \cdots \\ 0 & 0 & l_{33} & \cdots \\ \cdot & \cdot & \cdot & \cdots \\ \cdot & \cdot & \cdot & \cdots \\ \cdot & \cdot & \cdot & \cdots \\ 0 & 0 & \cdot & \cdots \end{bmatrix}$$

where only zeros appear below k_{22} in the second column. This process is continued until we obtain the matrix of a system which is in echelon form. That will be true when all elements below the main diagonal are zero. The final matrix is then used to obtain a system of equations which is equivalent to the original system. The solutions are then found as in previous sections. Let us illustrate the technique by solving a system of four linear equations.

Example   Find the solutions of the system

$$\begin{cases} x & - 2z + 2w = & 1 \\ -2x + 3y + 4z & = -1 \\ & y + z - w = & 0 \\ 3x + y - 2z - w = & 3. \end{cases}$$

Solution   Notice that we have arranged the equations so that the same variables appear in vertical columns. We shall begin with the matrix of the system and proceed to the solution.

$$\begin{bmatrix} 1 & 0 & -2 & 2 & 1 \\ -2 & 3 & 4 & 0 & -1 \\ 0 & 1 & 1 & -1 & 0 \\ 3 & 1 & -2 & -1 & 3 \end{bmatrix} \rightarrow \begin{bmatrix} 1 & 0 & -2 & 2 & 1 \\ 0 & 3 & 0 & 4 & 1 \\ 0 & 1 & 1 & -1 & 0 \\ 0 & 1 & 4 & -7 & 0 \end{bmatrix}$$

Add to the second row 2 times the first row and then add to the fourth row -3 times the first row.

$$\rightarrow \begin{bmatrix} 1 & 0 & -2 & 2 & 1 \\ 0 & 1 & 1 & -1 & 0 \\ 0 & 3 & 0 & 4 & 1 \\ 0 & 1 & 4 & -7 & 0 \end{bmatrix}$$

Interchange rows 2 and 3.

$$\rightarrow \begin{bmatrix} 1 & 0 & -2 & 2 & 1 \\ 0 & 1 & 1 & -1 & 0 \\ 0 & 0 & -3 & 7 & 1 \\ 0 & 0 & 3 & -6 & 0 \end{bmatrix}$$

Add to the third row -3 times the second row and then add to the fourth row -1 times the second row.

$$\rightarrow \begin{bmatrix} 1 & 0 & -2 & 2 & 1 \\ 0 & 1 & 1 & -1 & 0 \\ 0 & 0 & -3 & 7 & 1 \\ 0 & 0 & 0 & 1 & 1 \end{bmatrix}$$

Add row 3 to row 4.

The final matrix corresponds to the system of equations

$$\begin{cases} x & - 2z + 2w = 1 \\ y + z - w = 0 \\ - 3z + 7w = 1 \\ w = 1. \end{cases}$$

From the last equation we see that $w = 1$. Substituting in the third equation, we get $-3z + 7(1) = 1$, or $z = 2$. Using the second equation, we get $y + 2 - 1 = 0$, or $y = -1$. Finally, from the first equation, we have $x - 2(2) + 2(1) = 1$, or $x = 3$. Hence the system has only one solution, $x = 3$, $y = -1$, $z = 2$, and $w = 1$. ∎

EXERCISES 8.4

1–20 Use matrices to solve Exercises 1–20 of Section 8.3.

21–28 Use matrices to solve Exercises 1–8 of Section 8.2.

Solve Exercises 29 and 30 by means of matrices.

29 $\begin{cases} 2x - y - 2z + 2s - 5t = 2 \\ x + 3y - 2z + s - 2t = -5 \\ -x + 4y + 2z - 3s + 8t = -4 \\ 3x - 2y - 4z + s - 3t = -3 \\ 4x - 6y + z - 2s + t = 10 \end{cases}$

30 $\begin{cases} 3x + 2y + z + 3u + v + w = 1 \\ 2x + y - 2z + 3u - v + 4w = 6 \\ 6x + 3y + 4z - u + 2v + w = -6 \\ x + y + z + u - v - w = 8 \\ -2x - 2y + z - 3u + 2v - 3w = -10 \\ x - 3y + 2z + u + 3v + w = -1 \end{cases}$

8.5 THE ALGEBRA OF MATRICES

It is possible to develop a comprehensive theory for matrices which has many mathematical and scientific applications. In this section we shall discuss several basic algebraic properties of matrices which serve as the starting point for such a theory. Our work will be restricted to matrices whose elements are real numbers.

In order to conserve space it is sometimes convenient to denote an $m \times n$ matrix A of the type displayed in the definition given in Section 8.4 by the symbol (a_{ij}). If (b_{ij}) denotes another $m \times n$ matrix B, then we say that A and B are **equal** and we write

$$\boxed{A = B \quad \text{if and only if} \quad a_{ij} = b_{ij}}$$

for every i and j. For example,

$$\begin{bmatrix} 1 & 0 & 5 \\ \sqrt[3]{8} & 3^2 & -2 \end{bmatrix} = \begin{bmatrix} (-1)^2 & 0 & \sqrt{25} \\ 2 & 9 & -2 \end{bmatrix}.$$

If $A = (a_{ij})$ and $B = (b_{ij})$ are $m \times n$ matrices, then their **sum** $A + B$ is defined as the $m \times n$ matrix $C = (c_{ij})$, where $c_{ij} = a_{ij} + b_{ij}$ for all i and j. Thus, to add two matrices we add the elements which appear in corresponding positions in each matrix. Note that two matrices can be added only if they have the same number of rows and the same number of columns. Using the parentheses notation, we have

$$(a_{ij}) + (b_{ij}) = (a_{ij} + b_{ij}).$$

Although we have used the symbol $+$ in two different ways, there is little chance for confusion, since whenever $+$ appears between symbols for matrices it refers to matrix addition, and when $+$ is used between real numbers it denotes their sum. An example of the sum of two 3×2 matrices is

$$\begin{bmatrix} 4 & -5 \\ 0 & 4 \\ -6 & 1 \end{bmatrix} + \begin{bmatrix} 3 & 2 \\ 7 & -4 \\ -2 & 1 \end{bmatrix} = \begin{bmatrix} 7 & -3 \\ 7 & 0 \\ -8 & 2 \end{bmatrix}.$$

It is not difficult to prove that addition of matrices is both commutative and associative, that is,

$$A + B = B + A, \quad A + (B + C) = (A + B) + C$$

for $m \times n$ matrices A, B, and C.

The **$m \times n$ zero matrix**, denoted by O, is the matrix with m rows and n columns in which every element is 0. It is an identity element relative to addition, since

$$A + O = A$$

for every $m \times n$ matrix A. For example,

$$\begin{bmatrix} a_{11} & a_{12} \\ a_{21} & a_{22} \\ a_{31} & a_{32} \end{bmatrix} + \begin{bmatrix} 0 & 0 \\ 0 & 0 \\ 0 & 0 \end{bmatrix} = \begin{bmatrix} a_{11} & a_{12} \\ a_{21} & a_{22} \\ a_{31} & a_{32} \end{bmatrix}.$$

The **additive inverse** $-A$ of the matrix $A = (a_{ij})$ is, by definition, the matrix $(-a_{ij})$ obtained by changing the sign of each element of A. For example,

$$-\begin{bmatrix} 2 & -3 & 4 \\ -1 & 0 & 5 \end{bmatrix} = \begin{bmatrix} -2 & 3 & -4 \\ 1 & 0 & -5 \end{bmatrix}.$$

It follows that for every $m \times n$ matrix A,

$$A + (-A) = O.$$

Subtraction of two $m \times n$ matrices is defined by

$$A - B = A + (-B).$$

Using the parentheses notation for matrices, this implies that

$$(a_{ij}) - (b_{ij}) = (a_{ij}) + (-b_{ij}) = (a_{ij} - b_{ij}).$$

Thus, to subtract two matrices, we merely subtract the elements in the same positions.

The **product** of a real number c and an $m \times n$ matrix $A = (a_{ij})$ is defined by

$$cA = (ca_{ij}).$$

Thus, to find cA, we multiply each element of A by c. For example,

$$3 \begin{bmatrix} 4 & -1 \\ 2 & 3 \end{bmatrix} = \begin{bmatrix} 12 & -3 \\ 6 & 9 \end{bmatrix}.$$

The following results may be established, where A and B are $m \times n$ matrices, and c and d are any real numbers.

$$c(A + B) = cA + cB$$
$$(c + d)A = cA + dA$$
$$(cd)A = c(dA).$$

The following definition of the product of two matrices may appear unusual to the beginning student; however, there are many applications which justify the form of the definition. In order to define the product AB of two matrices A and B, *the number of columns of A must be the same as the number of rows of B.* Suppose that $A = (a_{ij})$ is $m \times n$ and $B = (b_{ij})$ is $n \times p$. To determine the element c_{ij} of the product, we single out row i of A and column j of B as follows:

$$\begin{bmatrix} a_{11} & a_{12} & \cdots & a_{1n} \\ & & & \\ & & & \\ \boxed{a_{i1}} & a_{i2} & \cdots & a_{in} \\ & & & \\ & & & \\ a_{m1} & a_{m2} & \cdots & a_{mn} \end{bmatrix} \begin{bmatrix} b_{11} & \cdots & b_{1j} & \cdots & b_{1p} \\ b_{21} & \cdots & b_{2j} & \cdots & b_{2p} \\ & & & & \\ & & & & \\ b_{n1} & \cdots & b_{nj} & \cdots & b_{np} \end{bmatrix}.$$

Next we multiply pairs of elements and then add them, according to the following formula:

$$c_{ij} = a_{i1}b_{1j} + a_{i2}b_{2j} + \cdots + a_{in}n_{nj}.$$

For example, the element c_{11} in the first row and the first column of AB is given by

$$c_{11} = a_{11}b_{11} + a_{12}b_{21} + \cdots + b_{1n}b_{n1}.$$

The element c_{12} in the first row and second column of AB is

$$c_{12} = a_{11}b_{12} + a_{12}b_{22} + \cdots + a_{1n}b_{n2}.$$

By definition, the product AB has the same number of rows as A and the same number of columns as B. In particular, if A is $m \times n$ and B is $n \times p$, then AB is $m \times p$. This is illustrated by the following product of a 2×3 matrix and a 3×4 matrix.

$$\begin{bmatrix} 1 & 2 & -3 \\ 4 & 0 & -2 \end{bmatrix} \begin{bmatrix} 5 & -4 & 2 & 0 \\ -1 & 6 & 3 & 1 \\ 7 & 0 & 4 & 8 \end{bmatrix} = \begin{bmatrix} -18 & 8 & -4 & -22 \\ 6 & -16 & 0 & -16 \end{bmatrix}.$$

Here are some typical computations of the elements c_{ij} in the product:

$$c_{11} = (1)(5) + (2)(-1) + (-3)(7) = 5 - 2 - 21 = -18$$
$$c_{13} = (1)(2) + (2)(3) + (-3)(4) = 2 + 6 - 12 = -4$$
$$c_{23} = (4)(2) + (0)(3) + (-2)(4) = 8 + 0 - 8 = 0$$
$$c_{24} = (4)(0) + (0)(1) + (-2)(8) = 0 + 0 - 16 = -16.$$

The reader should check the remaining elements.

The product operation for matrices is not commutative. Indeed, if A is 2×3 and B is 3×4, then AB may be found, but BA is undefined since the number of columns of B is different from the number of rows of A. Even if AB and BA are both defined, it is often true that these products are different. This is illustrated in the next example, along with the fact that the product of two nonzero matrices may equal a zero matrix.

Example   If $A = \begin{bmatrix} 2 & 2 \\ -1 & -1 \end{bmatrix}$ and $B = \begin{bmatrix} 1 & 2 \\ 1 & 2 \end{bmatrix}$, show that $AB \neq BA$.

Solution   Using the definition of product we obtain

$$AB = \begin{bmatrix} 4 & 8 \\ -2 & -4 \end{bmatrix} \quad \text{and} \quad BA = \begin{bmatrix} 0 & 0 \\ 0 & 0 \end{bmatrix}.$$

Hence $AB \neq BA$. Note that BA is a zero matrix.   ■

It is possible to show that matrix multiplication is associative in the sense that

$$\boxed{A(BC) = (AB)C}$$

provided that the indicated products are defined. That will be the case when A is $m \times n$, B is $n \times p$, and C is $p \times q$. The Distributive Properties also hold if the matrices involved have the proper number of rows and columns. Thus if A_1 and A_2 are $m \times n$ matrices and if B_1 and B_2 are $n \times p$ matrices, then

$$\boxed{A_1(B_1 + B_2) = A_1 B_1 + A_1 B_2}$$

and

$$\boxed{(A_1 + A_2)B_1 = A_1 B_1 + A_2 B_1.}$$

As a special case, if all matrices are square, of order n, then the Associative and Distributive Properties are true. We shall not give proofs for these theorems.

EXERCISES 8.5

In Exercises 1–8 find $A + B$, $A - B$, $2A$, and $-3B$.

1 $A = \begin{bmatrix} 5 & -2 \\ 1 & 3 \end{bmatrix}$, $B = \begin{bmatrix} 4 & 1 \\ -3 & 2 \end{bmatrix}$

2 $A = \begin{bmatrix} 3 & 0 \\ -1 & 2 \end{bmatrix}$, $B = \begin{bmatrix} 3 & -4 \\ 1 & 1 \end{bmatrix}$

3 $A = \begin{bmatrix} 6 & -1 \\ 2 & 0 \\ -3 & 4 \end{bmatrix}$, $B = \begin{bmatrix} 3 & 1 \\ -1 & 5 \\ 6 & 0 \end{bmatrix}$

4 $A = \begin{bmatrix} 0 & -2 & 7 \\ 5 & 4 & -3 \end{bmatrix}$, $B = \begin{bmatrix} 8 & 4 & 0 \\ 0 & 1 & 4 \end{bmatrix}$

5 $A = \begin{bmatrix} 4 & -3 & 2 \end{bmatrix}$, $B = \begin{bmatrix} 7 & 0 & -5 \end{bmatrix}$

6 $A = \begin{bmatrix} 7 \\ -16 \end{bmatrix}$, $B = \begin{bmatrix} -11 \\ 9 \end{bmatrix}$

7 $A = \begin{bmatrix} 0 & 4 & 0 & 3 \\ 1 & 2 & 0 & -5 \end{bmatrix}$,

$B = \begin{bmatrix} -3 & 0 & 1 & 3 \\ 2 & 0 & 7 & -2 \end{bmatrix}$

8 $A = \begin{bmatrix} -7 \end{bmatrix}$, $B = \begin{bmatrix} 9 \end{bmatrix}$

In Exercises 9–18 find AB and BA.

9 $A = \begin{bmatrix} 2 & 6 \\ 3 & -4 \end{bmatrix}$, $B = \begin{bmatrix} 5 & -2 \\ 1 & 7 \end{bmatrix}$

10 $A = \begin{bmatrix} 4 & -2 \\ -2 & 1 \end{bmatrix}$, $B = \begin{bmatrix} 2 & 1 \\ 4 & 2 \end{bmatrix}$

11 $A = \begin{bmatrix} 3 & 0 & -1 \\ 0 & 4 & 2 \\ 5 & -3 & 1 \end{bmatrix}$, $B = \begin{bmatrix} 1 & -5 & 0 \\ 4 & 1 & -2 \\ 0 & -1 & 3 \end{bmatrix}$

12 $A = \begin{bmatrix} 5 & 0 & 0 \\ 0 & -3 & 0 \\ 0 & 0 & 2 \end{bmatrix}$, $B = \begin{bmatrix} 3 & 0 & 0 \\ 0 & 4 & 0 \\ 0 & 0 & -2 \end{bmatrix}$

13 $A = \begin{bmatrix} 4 & -3 & 1 \\ -5 & 2 & 2 \end{bmatrix}$, $B = \begin{bmatrix} 2 & 1 \\ 0 & 1 \\ -4 & 7 \end{bmatrix}$

14 $A = \begin{bmatrix} 2 & 1 & -1 & 0 \\ 3 & -2 & 0 & 5 \\ -2 & 1 & 4 & 2 \end{bmatrix}$,

$B = \begin{bmatrix} 5 & -3 & 1 \\ 1 & 2 & 0 \\ -1 & 0 & 4 \\ 0 & -2 & 3 \end{bmatrix}$

15 $A = \begin{bmatrix} 1 & 2 & 3 \\ 4 & 5 & 6 \\ 7 & 8 & 9 \end{bmatrix}$, $B = \begin{bmatrix} 1 & 0 & 0 \\ 0 & 1 & 0 \\ 0 & 0 & 1 \end{bmatrix}$

16 $A = \begin{bmatrix} 1 & 2 & 3 \\ 2 & 3 & 1 \\ 3 & 1 & 2 \end{bmatrix}$, $B = \begin{bmatrix} 2 & 0 & 0 \\ 0 & 2 & 0 \\ 0 & 0 & 2 \end{bmatrix}$

17 $A = \begin{bmatrix} -3 & 7 & 2 \end{bmatrix}$, $B = \begin{bmatrix} 1 \\ 4 \\ -5 \end{bmatrix}$

18 $A = \begin{bmatrix} 4 & 8 \end{bmatrix}$, $B = \begin{bmatrix} -3 \\ 2 \end{bmatrix}$

In Exercises 19–22 find AB.

19 $A = \begin{bmatrix} 4 & -2 \\ 0 & 3 \\ -7 & 5 \end{bmatrix}$, $B = \begin{bmatrix} 3 \\ 4 \end{bmatrix}$

20 $A = \begin{bmatrix} 4 \\ -3 \\ 2 \end{bmatrix}$, $B = \begin{bmatrix} 5 & 1 \end{bmatrix}$

21 $A = \begin{bmatrix} 2 & 1 & 0 & -3 \\ -7 & 0 & -2 & 4 \end{bmatrix}$,

$B = \begin{bmatrix} 4 & -2 & 0 \\ 1 & 1 & -2 \\ 0 & 0 & 5 \\ -3 & -1 & 0 \end{bmatrix}$

22 $A = \begin{bmatrix} 1 & 2 & -3 \\ 4 & -5 & 6 \end{bmatrix}$,

$B = \begin{bmatrix} 1 & -1 & 0 & 2 \\ -2 & 3 & 1 & 0 \\ 0 & 4 & 0 & -3 \end{bmatrix}$

23 If $A = \begin{bmatrix} 1 & 2 \\ 0 & -3 \end{bmatrix}$, and $B = \begin{bmatrix} 2 & -1 \\ 3 & 1 \end{bmatrix}$, show that $(A + B)(A - B) \neq A^2 - B^2$, where $A^2 = AA$ and $B^2 = BB$.

24 If A and B are the matrices of Exercise 23, show that $(A + B)(A + B) \neq A^2 + 2AB + B^2$.

25 If A and B are the matrices of Exercise 23 and $C = \begin{bmatrix} 3 & 1 \\ -2 & 0 \end{bmatrix}$, prove that $A(B + C) = AB + AC$.

26 If A, B, and C are the matrices of Exercise 25 prove that $A(BC) = (AB)C$.

Prove the identities given in Exercises 27–30, where
$$A = \begin{bmatrix} a_{11} & a_{12} \\ a_{21} & a_{22} \end{bmatrix}, B = \begin{bmatrix} b_{11} & b_{12} \\ b_{21} & b_{22} \end{bmatrix}, C = \begin{bmatrix} c_{11} & c_{12} \\ c_{21} & c_{22} \end{bmatrix},$$
and c, d are real numbers.

27 $c(A + B) = cA + cB$

28 $(c + d)A = cA + dA$

29 $A(B + C) = AB + AC$

30 $A(BC) = (AB)C$

8.6 INVERSES OF MATRICES

Throughout this section we shall concentrate on square matrices. The symbol I_n will be used to denote the square matrix of order n which has 1 in each position on the main diagonal and 0 elsewhere. For example,

$$I_2 = \begin{bmatrix} 1 & 0 \\ 0 & 1 \end{bmatrix}, \quad I_3 = \begin{bmatrix} 1 & 0 & 0 \\ 0 & 1 & 0 \\ 0 & 0 & 1 \end{bmatrix}$$

and so on. It can be shown that if A is any square matrix of order n, then

$$AI_n = A = I_nA.$$

For that reason I_n is called an **identity matrix of order n.** To illustrate, if $A = (a_{ij})$ is of order 2, then a direct calculation shows that

$$\begin{bmatrix} a_{11} & a_{12} \\ a_{21} & a_{22} \end{bmatrix}\begin{bmatrix} 1 & 0 \\ 0 & 1 \end{bmatrix} = \begin{bmatrix} a_{11} & a_{12} \\ a_{21} & a_{22} \end{bmatrix} = \begin{bmatrix} 1 & 0 \\ 0 & 1 \end{bmatrix}\begin{bmatrix} a_{11} & a_{12} \\ a_{21} & a_{22} \end{bmatrix}.$$

Some, but not all, $n \times n$ matrices A have an **inverse** in the sense that there is a matrix B such that $AB = I_n = BA$. If A has an inverse we denote it by A^{-1} and write

$$AA^{-1} = I_n = A^{-1}A.$$

The symbol A^{-1} is read "A inverse." In matrix theory it is *not* acceptable to use the symbol $1/A$ in place of A^{-1}.

Let us describe a technique for finding the inverse of a square matrix A, whenever it exists. We shall not attempt to justify the following procedure, since that would require concepts which are beyond the scope of this text. Given the $n \times n$ matrix $A = (a_{ij})$, we begin by forming the $n \times (2n)$ matrix:

$$\begin{bmatrix} a_{11} & a_{12} & \cdots & a_{1n} & 1 & 0 & 0 & \cdots & 0 \\ a_{21} & a_{22} & \cdots & a_{2n} & 0 & 1 & 0 & \cdots & 0 \\ \vdots & \vdots & \vdots\vdots\vdots & \vdots & \vdots & \vdots & \vdots & \vdots\vdots\vdots & \vdots \\ a_{n1} & a_{n2} & \cdots & a_{nn} & 0 & 0 & 0 & \cdots & 1 \end{bmatrix}$$

where the $n \times n$ identity matrix I_n appears "to the right" of the matrix A, as indicated. We next apply a succession of elementary row transformations until we arrive at a matrix of the form

$$\begin{bmatrix} 1 & 0 & 0 & \cdots & 0 & b_{11} & b_{12} & \cdots & b_{1n} \\ 0 & 1 & 0 & \cdots & 0 & b_{21} & b_{22} & \cdots & b_{2n} \\ \vdots & \vdots & \vdots & \vdots\vdots\vdots & \vdots & \vdots & \vdots & \vdots\vdots\vdots & \vdots \\ 0 & 0 & 0 & \cdots & 1 & b_{n1} & b_{n2} & \cdots & b_{nn} \end{bmatrix}$$

where the identity matrix I_n appears "to the left" of the $n \times n$ matrix (b_{ij}). It can be shown that (b_{ij}) is the desired inverse A^{-1}.

Example 1   Find A^{-1} if $A = \begin{bmatrix} 3 & 5 \\ 1 & 4 \end{bmatrix}$.

Solution   We begin with the matrix

$$\begin{bmatrix} 3 & 5 & 1 & 0 \\ 1 & 4 & 0 & 1 \end{bmatrix}.$$

We next perform elementary row transformations until the identity matrix I_2 appears on the left, as follows:

$$\begin{bmatrix} 3 & 5 & 1 & 0 \\ 1 & 4 & 0 & 1 \end{bmatrix} \rightarrow \begin{bmatrix} 1 & 4 & 0 & 1 \\ 3 & 5 & 1 & 0 \end{bmatrix} \quad \text{Interchange rows 1 and 2.}$$

$$\rightarrow \begin{bmatrix} 1 & 4 & 0 & 1 \\ 0 & -7 & 1 & -3 \end{bmatrix} \quad \text{Add to the second row } -3 \text{ times the first row.}$$

$$\rightarrow \begin{bmatrix} 1 & 4 & 0 & 1 \\ 0 & 1 & -\frac{1}{7} & \frac{3}{7} \end{bmatrix} \quad \text{Multiply the second row by } -1/7.$$

$$\rightarrow \begin{bmatrix} 1 & 0 & \frac{4}{7} & -\frac{5}{7} \\ 0 & 1 & -\frac{1}{7} & \frac{3}{7} \end{bmatrix} \quad \text{Add to the first row } -4 \text{ times the second row.}$$

According to the previous discussion,

$$A^{-1} = \begin{bmatrix} \frac{4}{7} & -\frac{5}{7} \\ -\frac{1}{7} & \frac{3}{7} \end{bmatrix}.$$

To check our work the reader should verify that

$$\begin{bmatrix} 3 & 5 \\ 1 & 4 \end{bmatrix}\begin{bmatrix} \frac{4}{7} & -\frac{5}{7} \\ -\frac{1}{7} & \frac{3}{7} \end{bmatrix} = \begin{bmatrix} 1 & 0 \\ 0 & 1 \end{bmatrix} = \begin{bmatrix} \frac{4}{7} & -\frac{5}{7} \\ -\frac{1}{7} & \frac{3}{7} \end{bmatrix}\begin{bmatrix} 3 & 5 \\ 1 & 4 \end{bmatrix}.$$

■

Example 2 Find A^{-1} if $A = \begin{bmatrix} -1 & 3 & 1 \\ 2 & 5 & 0 \\ 3 & 1 & -2 \end{bmatrix}$.

Solution As in Example 1, we begin with

$$\begin{bmatrix} -1 & 3 & 1 & 1 & 0 & 0 \\ 2 & 5 & 0 & 0 & 1 & 0 \\ 3 & 1 & -2 & 0 & 0 & 1 \end{bmatrix}$$

and apply elementary row operations until I_3 appears on the left side of the matrix. Thus

$$\begin{bmatrix} -1 & 3 & 1 & 1 & 0 & 0 \\ 2 & 5 & 0 & 0 & 1 & 0 \\ 3 & 1 & -2 & 0 & 0 & 1 \end{bmatrix}$$

$\rightarrow \begin{bmatrix} 1 & -3 & -1 & -1 & 0 & 0 \\ 2 & 5 & 0 & 0 & 1 & 0 \\ 3 & 1 & -2 & 0 & 0 & 1 \end{bmatrix}$ Multiply the first row by -1.

$\rightarrow \begin{bmatrix} 1 & -3 & -1 & -1 & 0 & 0 \\ 0 & 11 & 2 & 2 & 1 & 0 \\ 0 & 10 & 1 & 3 & 0 & 1 \end{bmatrix}$ Add to the second row -2 times the first row and then add to the third row -3 times the first row.

$\rightarrow \begin{bmatrix} 1 & -3 & -1 & -1 & 0 & 0 \\ 0 & 1 & 1 & -1 & 1 & -1 \\ 0 & 10 & 1 & 3 & 0 & 1 \end{bmatrix}$ Add to the second row -1 times the third row.

$\rightarrow \begin{bmatrix} 1 & 0 & 2 & -4 & 3 & -3 \\ 0 & 1 & 1 & -1 & 1 & -1 \\ 0 & 0 & -9 & 13 & -10 & 11 \end{bmatrix}$ Add to the first row 3 times the second row and then add to the third row -10 times the second row.

$\rightarrow \begin{bmatrix} 1 & 0 & 2 & -4 & 3 & -3 \\ 0 & 1 & 1 & -1 & 1 & -1 \\ 0 & 0 & 1 & -\frac{13}{9} & \frac{10}{9} & -\frac{11}{9} \end{bmatrix}$ Multiply the third row by $-1/9$.

$\rightarrow \begin{bmatrix} 1 & 0 & 0 & -\frac{10}{9} & \frac{7}{9} & -\frac{5}{9} \\ 0 & 1 & 0 & \frac{4}{9} & -\frac{1}{9} & \frac{2}{9} \\ 0 & 0 & 1 & -\frac{13}{9} & \frac{10}{9} & -\frac{11}{9} \end{bmatrix}$ Add to the first row -2 times the third row and then add to the second row -1 times the third row.

Consequently,

$$A^{-1} = \begin{bmatrix} -\frac{10}{9} & \frac{7}{9} & -\frac{5}{9} \\ \frac{4}{9} & -\frac{1}{9} & \frac{2}{9} \\ -\frac{13}{9} & \frac{10}{9} & -\frac{11}{9} \end{bmatrix} = \frac{1}{9}\begin{bmatrix} -10 & 7 & -5 \\ 4 & -1 & 2 \\ -13 & 10 & -11 \end{bmatrix}.$$

The reader should check the fact that

$$AA^{-1} = I_3 = A^{-1}A. \qquad \blacksquare$$

There are many uses for inverses of matrices. One application concerns solutions of systems of linear equations. To illustrate, let us consider the case of two linear equations in two unknowns:

$$\begin{cases} a_{11}x + a_{12}y = k_1 \\ a_{21}x + a_{22}y = k_2. \end{cases}$$

We may express the system by means of matrices as follows:

$$\begin{bmatrix} a_{11}x + a_{12}y \\ a_{21}x + a_{22}y \end{bmatrix} = \begin{bmatrix} k_1 \\ k_2 \end{bmatrix}.$$

If we let

$$A = \begin{bmatrix} a_{11} & a_{12} \\ a_{21} & a_{22} \end{bmatrix}, \quad X = \begin{bmatrix} x \\ y \end{bmatrix}, \quad \text{and} \quad B = \begin{bmatrix} k_1 \\ k_2 \end{bmatrix},$$

then we may write $\qquad AX = B.$

If A^{-1} exists, then multiplying both sides of this *matrix equation* by A^{-1} gives us $A^{-1}AX = A^{-1}B$. Since $A^{-1}A = I_n$ and $I_nX = X$, this leads to

$$X = A^{-1}B$$

from which the solution (x, y) may be found. The above technique may be extended to systems of n linear equations in n unknowns.

Example 3   Solve the following system of equations:

$$\begin{cases} -x + 3y + z = 1 \\ 2x + 5y = 3 \\ 3x + y - 2z = -2. \end{cases}$$

Solution   If we let

$$A = \begin{bmatrix} -1 & 3 & 1 \\ 2 & 5 & 0 \\ 3 & 1 & -2 \end{bmatrix}, \quad X = \begin{bmatrix} x \\ y \\ z \end{bmatrix}, \quad \text{and} \quad B = \begin{bmatrix} 1 \\ 3 \\ -2 \end{bmatrix},$$

then the given system may be written in terms of matrices as $AX = B$. From the preceding discussion this implies that $X = A^{-1}B$. The matrix A^{-1} was found in Example 2. Substituting for X, A^{-1}, and B in the last equation gives us

$$\begin{bmatrix} x \\ y \\ z \end{bmatrix} = \frac{1}{9} \begin{bmatrix} -10 & 7 & -5 \\ 4 & -1 & 2 \\ -13 & 10 & -11 \end{bmatrix} \begin{bmatrix} 1 \\ 3 \\ -2 \end{bmatrix} = \frac{1}{9} \begin{bmatrix} 21 \\ -3 \\ 39 \end{bmatrix} = \begin{bmatrix} \frac{7}{3} \\ -\frac{1}{3} \\ \frac{13}{3} \end{bmatrix}.$$

It follows that $x = \frac{7}{3}, y = -\frac{1}{3}$, and $z = \frac{13}{3}$. Hence the ordered triple $(\frac{7}{3}, -\frac{1}{3}, \frac{13}{3})$ is the solution of the given system. ∎

The method of solution employed in Example 3 is beneficial only if A^{-1} is known, or if many systems with the same coefficient matrix are to be considered. The preferred technique for solving a system of linear equations is still either the method of elimination or the matrix method discussed in Section 8.4.

EXERCISES 8.6

In Exercises 1–12 find the inverse of the given matrix, if it exists.

1 $\begin{bmatrix} 2 & -4 \\ 1 & 3 \end{bmatrix}$

2 $\begin{bmatrix} 3 & 2 \\ 4 & 5 \end{bmatrix}$

3 $\begin{bmatrix} 2 & 4 \\ 4 & 8 \end{bmatrix}$

4 $\begin{bmatrix} 3 & -1 \\ 6 & -2 \end{bmatrix}$

5 $\begin{bmatrix} 3 & -1 & 0 \\ 2 & 2 & 0 \\ 0 & 0 & 4 \end{bmatrix}$

6 $\begin{bmatrix} 3 & 0 & 2 \\ 0 & 1 & 0 \\ -4 & 0 & 2 \end{bmatrix}$

7 $\begin{bmatrix} -2 & 2 & 3 \\ 1 & -1 & 0 \\ 0 & 1 & 4 \end{bmatrix}$

8 $\begin{bmatrix} 1 & 2 & 3 \\ -2 & 1 & 0 \\ 3 & -1 & 1 \end{bmatrix}$

9 $\begin{bmatrix} 2 & 0 & 0 \\ 0 & 4 & 0 \\ 0 & 0 & 6 \end{bmatrix}$

10 $\begin{bmatrix} 1 & 1 & 1 \\ 2 & 2 & 2 \\ 3 & 3 & 3 \end{bmatrix}$

11 $\begin{bmatrix} 1 & -1 & 0 & 1 \\ 0 & 1 & -2 & 0 \\ -1 & 2 & 1 & 2 \\ -2 & 1 & 2 & 0 \end{bmatrix}$

12 $\begin{bmatrix} 1 & 2 & 0 & 1 \\ 0 & -1 & 1 & -2 \\ 0 & 0 & 2 & 0 \\ 0 & 0 & 0 & 1 \end{bmatrix}$

13 State conditions on a and b which guarantee that the matrix $\begin{bmatrix} a & 0 \\ 0 & b \end{bmatrix}$ has an inverse, and find a formula for the inverse when it exists.

14 If $abc \neq 0$, find the inverse of $\begin{bmatrix} a & 0 & 0 \\ 0 & b & 0 \\ 0 & 0 & c \end{bmatrix}$.

15 If $A = \begin{bmatrix} a_{11} & a_{12} & a_{13} \\ a_{21} & a_{22} & a_{23} \\ a_{31} & a_{32} & a_{33} \end{bmatrix}$. prove that $AI_3 = A = I_3 A$.

16 Prove that $AI_4 = A = I_4 A$ for every square matrix A of order 4.

Solve the systems in Exercises 17–20 by the method of Example 3. (Refer to inverses of matrices found in Exercises 1–8.)

17 $\begin{cases} 2x - 4y = 3 \\ x + 3y = 1 \end{cases}$

18 $\begin{cases} 3x + 2y = -1 \\ 4x + 5y = 1 \end{cases}$

19 $\begin{cases} -2x + 2y + 3z = 1 \\ x - y = 3 \\ y + 4z = -2 \end{cases}$

20 $\begin{cases} x + 2y + 3z = -1 \\ -2x + y = 4 \\ 3x - y + z = 2 \end{cases}$

8.7 DETERMINANTS

Throughout this section and the next it is assumed that all matrices under discussion are *square* matrices. Associated with each such matrix A is a number called the **determinant** of A. Determinants can be used to solve systems of linear equations if the number of equations is the same as the number of variables. In this section we shall state the definition and give some basic properties of determinants. It may be difficult to see exactly why the definitions given below are used. One reason will be pointed out in Section 8.9, where it is shown that our definitions arise naturally when solving systems of linear equations.

The determinant of a square matrix A will be denoted by $|A|$. This notation should not be confused with the symbol used for the absolute value of a real number. To avoid any misunderstanding, the expression det A is used in some mathematics texts instead of $|A|$. We shall define $|A|$ by beginning with the case in which A has order 1 and then by increasing the order a step at a time.

If A is a square matrix of order 1, then A has only one element. Thus, $A = [a_{11}]$ and we define $|A| = a_{11}$. If A is a square matrix of order 2, then we may write

$$A = \begin{bmatrix} a_{11} & a_{12} \\ a_{21} & a_{22} \end{bmatrix}$$

and the determinant of A is defined by

$$|A| = a_{11}a_{22} - a_{21}a_{12}.$$

Another notation for $|A|$ is obtained by replacing the brackets in the symbol for A with vertical bars as follows:

$$|A| = \begin{vmatrix} a_{11} & a_{12} \\ a_{21} & a_{22} \end{vmatrix} = a_{11}a_{22} - a_{21}a_{12}.$$

Example 1   Find $|A|$ if $A = \begin{bmatrix} 2 & -1 \\ 4 & -3 \end{bmatrix}$.

Solution   By definition,

$$|A| = \begin{vmatrix} 2 & -1 \\ 4 & -3 \end{vmatrix} = (2)(-3) - (4)(-1) = -6 + 4 = -2. \qquad \blacksquare$$

For matrices of higher order it is convenient to introduce additional terminology as follows.

> If A is a matrix of order 3, then the **minor** M_{ij} of an element a_{ij} is the determinant of the matrix of order 2 obtained by deleting row i and column j of A.

Thus, to determine the minor of an element, we discard the row and column in which the element appears and then find the determinant of the resulting matrix. To illustrate, given the general 3×3 matrix

$$A = \begin{bmatrix} a_{11} & a_{12} & a_{13} \\ a_{21} & a_{22} & a_{23} \\ a_{31} & a_{32} & a_{33} \end{bmatrix}$$

we obtain

$$M_{11} = \begin{vmatrix} a_{22} & a_{23} \\ a_{32} & a_{33} \end{vmatrix} = a_{22}a_{33} - a_{32}a_{23}$$

$$M_{12} = \begin{vmatrix} a_{21} & a_{23} \\ a_{31} & a_{33} \end{vmatrix} = a_{21}a_{33} - a_{31}a_{23}$$

$$M_{13} = \begin{vmatrix} a_{21} & a_{22} \\ a_{31} & a_{32} \end{vmatrix} = a_{21}a_{32} - a_{31}a_{22}$$

$$M_{23} = \begin{vmatrix} a_{11} & a_{12} \\ a_{31} & a_{32} \end{vmatrix} = a_{11}a_{32} - a_{31}a_{12}$$

and likewise for the other minors M_{21}, M_{22}, M_{31}, M_{32}, and M_{33}.

We shall also make use of the following concept.

The **cofactor** A_{ij} of the element a_{ij} is defined by

$$A_{ij} = (-1)^{i+j}M_{ij}.$$

Thus to obtain the cofactor of a_{ij} we find the minor and multiply it by 1 or -1 depending on whether the sum of i and j is even or odd, respectively. An easy way to remember the sign $(-1)^{i+j}$ associated with the cofactor A_{ij} is to consider the following "checkerboard" scheme:

$$\begin{bmatrix} + & - & + \\ - & + & - \\ + & - & + \end{bmatrix}$$

where we may regard the $+$ signs as occurring on red squares and the $-$ signs on black squares.

Example 2 If

$$A = \begin{bmatrix} 1 & -3 & 3 \\ 4 & 2 & 0 \\ -2 & -7 & 5 \end{bmatrix}$$

find M_{11}, M_{21}, M_{22}, A_{11}, A_{21}, and A_{22}.

Solution   By definition

$$M_{11} = \begin{vmatrix} 2 & 0 \\ -7 & 5 \end{vmatrix} = (2)(5) - (-7)(0) = 10$$

$$M_{21} = \begin{vmatrix} -3 & 3 \\ -7 & 5 \end{vmatrix} = (-3)(5) - (-7)(3) = 6$$

$$M_{22} = \begin{vmatrix} 1 & 3 \\ -2 & 5 \end{vmatrix} = (1)(5) - (-2)(3) = 11.$$

All that is necessary to obtain the cofactors is to prefix the corresponding minors with the proper signs. Thus, using the definition of cofactor,

$$A_{11} = (-1)^{1+1} M_{11} = (1)(10) = 10$$
$$A_{21} = (-1)^{2+1} M_{21} = (-1)(6) = -6$$
$$A_{22} = (-1)^{2+2} M_{22} = (1)(11) = 11.$$

The checkerboard scheme could also be used to determine the proper signs. ∎

The determinant $|A|$ of a square matrix of order 3 is defined by

$$|A| = \begin{vmatrix} a_{11} & a_{12} & a_{13} \\ a_{21} & a_{22} & a_{23} \\ a_{31} & a_{32} & a_{33} \end{vmatrix} = a_{11}A_{11} + a_{12}A_{12} + a_{13}A_{13}.$$

Since $A_{11} = (-1)^{1+1} M_{11} = M_{11}$, $A_{12} = (-1)^{1+2} M_{12} = -M_{12}$, and $A_{13} = (-1)^{1+3} M_{13} = M_{13}$, the preceding definition may also be written

$$|A| = a_{11}M_{11} - a_{12}M_{12} + a_{13}M_{13}.$$

If we substitute for M_{11}, M_{12}, and M_{13}, we obtain the following formula for $|A|$ in terms of the elements of A:

$$|A| = a_{11}a_{22}a_{33} - a_{11}a_{32}a_{23} - a_{12}a_{21}a_{33} + a_{12}a_{31}a_{23} + a_{13}a_{21}a_{32} - a_{13}a_{31}a_{22}$$

The definition of $|A|$ for a square matrix A of order 3 displays a pattern of multiplying each element in row 1 by its cofactor and then adding to find $|A|$. This is referred to as *expanding* $|A|$ *by the first row.* By actually carrying out the computations, it is not difficult to show that $|A|$ *can be expanded in similar fashion by using any row or column.* As an illustration, the expansion by the second column is

$$|A| = a_{12}A_{12} + a_{22}A_{22} + a_{32}A_{32}$$

$$= a_{12}\left(- \begin{vmatrix} a_{21} & a_{23} \\ a_{31} & a_{32} \end{vmatrix} \right) + a_{22}\left(+ \begin{vmatrix} a_{11} & a_{13} \\ a_{31} & a_{33} \end{vmatrix} \right) + a_{32}\left(- \begin{vmatrix} a_{11} & a_{13} \\ a_{21} & a_{23} \end{vmatrix} \right).$$

Applying the definition to the determinants in parentheses, multiplying as indicated, and rearranging the terms in the sum, we could arrive at the formula for $|A|$ in terms of the elements of A. Similarly, the expansion by the third row is

$$|A| = a_{31}A_{31} + a_{32}A_{32} + a_{33}A_{33}$$

$$= a_{31}\left(+ \begin{vmatrix} a_{12} & a_{13} \\ a_{22} & a_{23} \end{vmatrix} \right) + a_{32}\left(- \begin{vmatrix} a_{11} & a_{13} \\ a_{21} & a_{23} \end{vmatrix} \right) + a_{33}\left(+ \begin{vmatrix} a_{11} & a_{12} \\ a_{21} & a_{22} \end{vmatrix} \right).$$

Once again it can be shown that this agrees with previous expansions.

Example 3   Find $|A|$ if

$$A = \begin{bmatrix} -1 & 3 & 1 \\ 2 & 5 & 0 \\ 3 & 1 & -2 \end{bmatrix}.$$

Solution   Expanding $|A|$ by the second row gives us

$$|A| = (2)A_{21} + (5)A_{22} + (0)A_{23}.$$

Using the definition of cofactor we have

$$A_{21} = (-1)^3 M_{21} = -\begin{vmatrix} 3 & 1 \\ 1 & -2 \end{vmatrix} = -[(3)(-2) - (1)(1)] = 7$$

$$A_{22} = (-1)^4 M_{22} = \begin{vmatrix} -1 & 1 \\ 3 & -2 \end{vmatrix} = [(-1)(-2) - (3)(1)] = -1.$$

Consequently,

$$|A| = (2)(7) + (5)(-1) + (0)A_{23} = 14 - 5 + 0 = 9. \qquad \blacksquare$$

If A is a matrix of order 4, we define the minor M_{ij} of the element a_{ij} as the determinant of the matrix of order 3 obtained by deleting row i and column j of A. The cofactor A_{ij} is again given by $(-1)^{i+j}M_{ij}$. In a manner analogous to the case for order 3, we define

$$|A| = a_{11}A_{11} + a_{12}A_{12} + a_{13}A_{13} + a_{14}A_{14}.$$

The last formula is called *the expansion of* $|A|$ *by the first row*. In terms of minors, the formula may be written as

$$|A| = a_{11}M_{11} - a_{12}M_{12} + a_{13}M_{13} - a_{14}M_{14}.$$

It can be shown that the same number is obtained if $|A|$ is expanded by any other row or column.

The method of defining determinants of matrices of arbitrary order n should now be apparent. If a_{ij} is an element of a matrix of order $n > 1$, then the **minor** M_{ij} is defined as the determinant of the matrix of order $n - 1$ obtained by deleting row i and column j. The **cofactor** A_{ij} is defined as $(-1)^{i+j}M_{ij}$. The sign $(-1)^{i+j}$ associated with A_{ij} can be remembered by using a checkerboard similar to that used for order 3, extending the rows and columns as far as necessary. We then define the determinant $|A|$ of a matrix A of order n as the expansion by the first row, that is

$$|A| = a_{11}A_{11} + a_{12}A_{12} + \cdots + a_{1n}A_{1n}$$

or, in terms of minors,

$$|A| = a_{11}M_{11} - a_{12}M_{12} + \cdots + a_{1n}(-1)^{1+n}M_{1n}.$$

As was the case with matrices of small order, the number $|A|$ may be found by using *any* row or column. Specifically, we have the following theorem.

The Expansion Theorem for Determinants

> If A is a square matrix of order $n > 1$, then the determinant $|A|$ may be found by multiplying the elements of any row (or column) by their respective cofactors, and adding the resulting products.

The proof of this theorem is difficult and may be found in texts on matrix theory. The theorem is quite useful if many zeros appear in a row or column, as illustrated in the following example.

Example 4   Find $|A|$ if

$$A = \begin{bmatrix} 1 & 0 & 2 & 5 \\ -2 & 1 & 5 & 0 \\ 0 & 0 & -3 & 0 \\ 0 & -1 & 0 & 3 \end{bmatrix}.$$

Solution   Note that all but one of the elements in the third row is zero. Hence if we expand $|A|$ by the third row there will be at most one nonzero term. Specifically,

$$|A| = (0)A_{31} + (0)A_{32} + (-3)A_{33} + (0)A_{34} = -3A_{33}$$

where

$$A_{33} = \begin{vmatrix} 1 & 0 & 5 \\ -2 & 1 & 0 \\ 0 & -1 & 3 \end{vmatrix}.$$

Expanding A_{33} by column 1, we obtain

$$A_{33} = (1)\begin{vmatrix} 1 & 0 \\ -1 & 3 \end{vmatrix} + (-2)\left(-\begin{vmatrix} 0 & 5 \\ -1 & 3 \end{vmatrix} \right) + 0\begin{vmatrix} 0 & 5 \\ 1 & 0 \end{vmatrix} = 3 + 10 + 0 = 13.$$

Therefore,

$$|A| = -3A_{33} = (-3)(13) = -39. \qquad ■$$

In general, if all but one element a in some row (or column) of A is zero and if the determinant $|A|$ is expanded by that row (or column), then all terms drop out except the product of the element a with its cofactor. We will make important use of this fact in the next section.

If *every* element in a row (or column) of a matrix A is zero, then upon expanding $|A|$ by that row (or column) we obtain the number 0. This gives us the following result.

Theorem

> If every element of a row (or column) of a square matrix A is zero, then $|A| = 0$.

EXERCISES 8.7

In each of Exercises 1–4 find all the minors and cofactors of the elements in the given matrix.

1 $\begin{bmatrix} 2 & 4 & -1 \\ 0 & 3 & 2 \\ -5 & 7 & 0 \end{bmatrix}$ **2** $\begin{bmatrix} 5 & -2 & 1 \\ 4 & 7 & 0 \\ -3 & 4 & -1 \end{bmatrix}$

3 $\begin{bmatrix} 7 & -1 \\ 5 & 0 \end{bmatrix}$ **4** $\begin{bmatrix} -6 & 4 \\ 3 & 2 \end{bmatrix}$

5–8 Find the determinants of the matrices given in Exercises 1–4.

Find the determinants of the matrices in Exercises 9–20.

9 $\begin{bmatrix} -5 & 4 \\ -3 & 2 \end{bmatrix}$ **10** $\begin{bmatrix} 6 & 4 \\ -3 & 2 \end{bmatrix}$

11 $\begin{bmatrix} a & -a \\ b & -b \end{bmatrix}$ **12** $\begin{bmatrix} c & d \\ -d & c \end{bmatrix}$

13 $\begin{bmatrix} 3 & 1 & -2 \\ 4 & 2 & 5 \\ -6 & 3 & -1 \end{bmatrix}$ **14** $\begin{bmatrix} 2 & -5 & 1 \\ -3 & 1 & 6 \\ 4 & -2 & 3 \end{bmatrix}$

15 $\begin{bmatrix} -5 & 4 & 1 \\ 3 & -2 & 7 \\ 2 & 0 & 6 \end{bmatrix}$ **16** $\begin{bmatrix} 2 & 7 & -3 \\ 1 & 0 & 4 \\ 4 & -1 & -2 \end{bmatrix}$

17 $\begin{bmatrix} 3 & -1 & 2 & 0 \\ 4 & 0 & -3 & 5 \\ 0 & 6 & 0 & 0 \\ 1 & 3 & -4 & 2 \end{bmatrix}$

18 $\begin{bmatrix} 2 & 5 & 1 & 0 \\ -4 & 0 & -3 & 0 \\ 3 & -2 & 1 & 6 \\ -1 & 4 & 2 & 0 \end{bmatrix}$

19 $\begin{bmatrix} 0 & b & 0 & 0 \\ 0 & 0 & c & 0 \\ a & 0 & 0 & 0 \\ 0 & 0 & 0 & d \end{bmatrix}$ **20** $\begin{bmatrix} a & u & v & w \\ 0 & b & x & y \\ 0 & 0 & c & z \\ 0 & 0 & 0 & d \end{bmatrix}$

Verify the identities in Exercises 21–28 by expanding each determinant.

21 $\begin{vmatrix} a & b \\ c & d \end{vmatrix} = -\begin{vmatrix} c & d \\ a & b \end{vmatrix}$ **22** $\begin{vmatrix} a & b \\ c & d \end{vmatrix} = -\begin{vmatrix} b & a \\ d & c \end{vmatrix}$

23 $\begin{vmatrix} a & kb \\ c & kd \end{vmatrix} = k\begin{vmatrix} a & b \\ c & d \end{vmatrix}$ **24** $\begin{vmatrix} a & b \\ kc & kd \end{vmatrix} = k\begin{vmatrix} a & b \\ c & d \end{vmatrix}$

25 $\begin{vmatrix} a & b \\ c & d \end{vmatrix} = \begin{vmatrix} a & b \\ ka+c & kb+d \end{vmatrix}$

26 $\begin{vmatrix} a & b \\ c & d \end{vmatrix} = \begin{vmatrix} a & ka+b \\ c & kc+d \end{vmatrix}$

27 $\begin{vmatrix} a & b \\ c & d \end{vmatrix} + \begin{vmatrix} a & e \\ c & f \end{vmatrix} = \begin{vmatrix} a & b+e \\ c & d+f \end{vmatrix}$

28 $\begin{vmatrix} a & b \\ c & d \end{vmatrix} + \begin{vmatrix} a & b \\ e & f \end{vmatrix} = \begin{vmatrix} a & b \\ c+e & d+f \end{vmatrix}$

29 Prove that if a square matrix A of order 2 has two identical rows or columns, then $|A| = 0$.

30 Repeat Exercise 29 for a matrix of order 3.

8.8 PROPERTIES OF DETERMINANTS

The method of evaluating a determinant by means of the Expansion Theorem stated in Section 8.7 is not very efficient for matrices of high order. For example, if a determinant of a matrix of order 10 is expanded by any row, a sum of 10 terms is obtained, where each term contains the determinant of a matrix of order 9 (that is, a cofactor of the original matrix). If any of the latter determinants is expanded by a row (or column), a sum of 9 terms is obtained, each containing the determinant of a matrix of order 8. Hence, at this stage there are 90 determinants of matrices of order 8 to evaluate! The process could be continued until only determinants of matrices of order 2 remain. Unless many elements of the original matrix are zero, it is an enormous task to carry out all of the computations.

We shall now consider some rules which make the process of evaluating determinants simpler. These rules are used mainly for introducing zeros into the determinant. They may also be used to change the determinant to echelon form, that is, a form in which the elements below the main diagonal elements a_{ii} are all zero. The transformations on rows stated in the next theorem are the same as the elementary row transformations of a matrix introduced in Section 4. However, for determinants we may also employ similar transformations on columns.

Theorem on Row and Column Transformations of a Determinant

Let A be a matrix of order n.

  (i)   If a matrix B is obtained from A by interchanging two rows (or columns), then $|B| = -|A|$.

 (ii)   If B is obtained from A by multiplying every element of one row (or column) of A by a real number k, then $|B| = k|A|$.

(iii)   If B is obtained from A by adding to any row (or column) of A, k times another row (or column), where k is any real number, then $|B| = |A|$.

When using this theorem to justify manipulations with determinants, we shall refer to the rows (or columns) of the *determinant* in the obvious way. For example, property (iii) may be phrased: "Adding the product of k times another row (or column) to any row (or column) of a determinant does not affect the value of the determinant."

We shall not give a general proof of the preceding theorem. For the case of matrices of orders 2 or 3 the theorem can be proved by evaluating $|B|$. For example, given a matrix $A = (a_{ij})$ of order 3, suppose that B is the matrix obtained by adding to row 2 the product of k times row 1, that is,

$$B = \begin{bmatrix} a_{11} & a_{12} & a_{13} \\ ka_{11} + a_{21} & ka_{12} + a_{22} & ka_{13} + a_{23} \\ a_{31} & a_{32} & a_{33} \end{bmatrix}.$$

To evaluate $|B|$ we expand by row 2, obtaining

$$|B| = (ka_{11} + a_{21})\left(-\begin{vmatrix} a_{12} & a_{13} \\ a_{32} & a_{33} \end{vmatrix}\right) + (ka_{12} + a_{22})\begin{vmatrix} a_{11} & a_{13} \\ a_{31} & a_{33} \end{vmatrix}$$
$$+ (ka_{13} + a_{23})\left(-\begin{vmatrix} a_{11} & a_{12} \\ a_{31} & a_{32} \end{vmatrix}\right).$$

The determinants which appear in this equation are those associated with the cofactors A_{21}, A_{22}, and A_{23} of the original matrix. Hence

$$|B| = (ka_{11} + a_{21})A_{21} + (ka_{12} + a_{22})A_{22} + (ka_{13} + a_{23})A_{23}$$

which may also be written in the form

$$|B| = k(a_{11}A_{21} + a_{12}A_{22} + a_{13}A_{23}) + (a_{21}A_{21} + a_{22}A_{22} + a_{23}A_{23}).$$

The second expression in parentheses equals $|A|$. (Why?) By actually carrying out the computations it is possible to show that the first expression in parentheses is zero. Consequently,

$$|B| = 0 + |A| = |A|.$$

Statement (i) of the last theorem is often phrased: "Interchanging two rows (or columns) changes the sign of a determinant." As illustrations we have

$$\begin{vmatrix} 2 & 0 & 1 \\ 6 & 4 & 3 \\ 0 & 3 & 5 \end{vmatrix} = - \begin{vmatrix} 6 & 4 & 3 \\ 2 & 0 & 1 \\ 0 & 3 & 5 \end{vmatrix} \quad \text{and} \quad \begin{vmatrix} 2 & 0 & 1 \\ 6 & 4 & 3 \\ 0 & 3 & 5 \end{vmatrix} = - \begin{vmatrix} 1 & 0 & 2 \\ 3 & 4 & 6 \\ 5 & 3 & 0 \end{vmatrix}$$

where in the first case rows 1 and 2 were interchanged and in the second case we interchanged columns 1 and 3.

As an illustration of part (iii) of the theorem,

$$\begin{vmatrix} 1 & -3 & 4 \\ 2 & -1 & 0 \\ 3 & 1 & 6 \end{vmatrix} = \begin{vmatrix} 1 & -3 & 4 \\ 0 & 5 & -8 \\ 3 & 1 & 6 \end{vmatrix}$$

where we have added to the second row the product of -2 times the first row. This type of manipulation is very important for evaluating determinants of large order as will be shown below in Example 1. In like manner,

$$\begin{vmatrix} 1 & -3 & 4 \\ 2 & -1 & 0 \\ 3 & 1 & 6 \end{vmatrix} = \begin{vmatrix} -5 & -3 & 4 \\ 0 & -1 & 0 \\ 5 & 1 & 6 \end{vmatrix}$$

where we have added to the first column the product of 2 times the second column. Part (ii) of the theorem will be illustrated in Examples 2 and 3.

Theorem

If two rows (or columns) of a square matrix A are identical, then $|A| = 0$.

Proof

If B is the matrix obtained from A by interchanging the two identical rows (or columns), then B and A are the same and consequently $|B| = |A|$. However, by (i) of the Theorem on Row and Column Transformations of a Determinant, $|B| = -|A|$ and hence $-|A| = |A|$, which implies that $|A| = 0$.

Example 1   Find $|A|$ if

$$A = \begin{bmatrix} 2 & 3 & 0 & 4 \\ 0 & 5 & -1 & 6 \\ 1 & 0 & -2 & 3 \\ -3 & 2 & 0 & -5 \end{bmatrix}.$$

Solution   We plan to use (iii) of the Theorem on Row and Column Transformations to introduce many zeros in some row or column. To do this, it is convenient to work with an element of the matrix which equals 1 or -1, since this enables us to avoid the use of fractions. If no such element appears in the original matrix, it is always possible to introduce the number 1 by using (iii) or (ii) of the theorem. In this example there is no such problem, since 1 appears in row 3 and -1 in row 2. Let us work with the element 1 and introduce zero everywhere else in the first column, as shown below:

$$|A| = \begin{vmatrix} 0 & 3 & 4 & -2 \\ 0 & 5 & -1 & 6 \\ 1 & 0 & -2 & 3 \\ 0 & 2 & -6 & 4 \end{vmatrix}$$

Add to the first row -2 times the third row and then add to the fourth row 3 times the third row.

$$= (1) \begin{vmatrix} 3 & 4 & -2 \\ 5 & -1 & 6 \\ 2 & -6 & 4 \end{vmatrix}$$

Expand by column 1.

$$= \begin{vmatrix} 23 & 4 & 22 \\ 0 & -1 & 0 \\ -28 & -6 & -32 \end{vmatrix}$$

Add to the first column 5 times the second column and then add to the third column 6 times the second column.

$$= (-1) \begin{vmatrix} 23 & 22 \\ -28 & -32 \end{vmatrix}$$

Expand by row 2.

$$= (-1)[(23)(-32) - (-28)(22)]$$ By definition.

$$= 120. \qquad \blacksquare$$

Part (ii) of the Theorem on Row and Column Transformations is useful for finding factors of determinants. To illustrate, for a determinant of a matrix of order 3, we have the following:

$$\begin{vmatrix} a_{11} & a_{12} & a_{13} \\ ka_{21} & ka_{22} & ka_{23} \\ a_{31} & a_{32} & a_{33} \end{vmatrix} = k \begin{vmatrix} a_{11} & a_{12} & a_{13} \\ a_{21} & a_{22} & a_{23} \\ a_{31} & a_{32} & a_{33} \end{vmatrix}.$$

Similar formulas hold if k is a common factor of the elements of some other row or column. When the theorem is used in this way, we often use the phrase "k is a common factor in the row (or column)."

Example 2   Find $|A|$ if

$$A = \begin{bmatrix} 14 & -6 & 4 \\ 4 & -5 & 12 \\ -21 & 9 & -6 \end{bmatrix}.$$

Solution

$$|A| = 2 \begin{vmatrix} 7 & -3 & 2 \\ 4 & -5 & 12 \\ -21 & 9 & -6 \end{vmatrix} \qquad \text{2 is a common factor in row 1.}$$

$$= (2)(-3) \begin{vmatrix} 7 & -3 & 2 \\ 4 & -5 & 12 \\ 7 & -3 & 2 \end{vmatrix} \qquad -3 \text{ is a common factor in row 3.}$$

$$= 0 \qquad \text{Two rows are identical.} \qquad \blacksquare$$

Example 3 Without expanding, show that $a - b$ is a factor of

$$\begin{vmatrix} 1 & 1 & 1 \\ a & b & c \\ a^2 & b^2 & c^2 \end{vmatrix}.$$

Solution

$$\begin{vmatrix} 1 & 1 & 1 \\ a & b & c \\ a^2 & b^2 & c^2 \end{vmatrix} = \begin{vmatrix} 0 & 1 & 1 \\ a-b & b & c \\ a^2-b^2 & b^2 & c^2 \end{vmatrix} \qquad \text{Add to the first column } -1 \text{ times the second column.}$$

$$= (a - b) \begin{vmatrix} 0 & 1 & 1 \\ 1 & b & c \\ a+b & b^2 & c^2 \end{vmatrix} \qquad a - b \text{ is a common factor of column 1.} \qquad \blacksquare$$

EXERCISES 8.8

Without expanding, explain why the statements in Exercises 1–14 are true.

1 $\begin{vmatrix} 1 & 0 & 1 \\ 0 & 1 & 1 \\ 1 & 1 & 0 \end{vmatrix} = - \begin{vmatrix} 1 & 0 & 1 \\ 1 & 1 & 0 \\ 0 & 1 & 1 \end{vmatrix}$

2 $\begin{vmatrix} 1 & 0 & 1 \\ 0 & 1 & 1 \\ 1 & 1 & 0 \end{vmatrix} = - \begin{vmatrix} 1 & 1 & 0 \\ 0 & 1 & 1 \\ 1 & 0 & 1 \end{vmatrix}$

3 $\begin{vmatrix} 1 & 0 & 1 \\ 2 & 1 & 0 \\ 1 & 1 & 2 \end{vmatrix} = \begin{vmatrix} 1 & 0 & 1 \\ 2 & 1 & 0 \\ 0 & 1 & 1 \end{vmatrix}$

4 $\begin{vmatrix} 1 & 1 & 2 \\ 1 & 0 & 1 \\ 2 & 1 & 1 \end{vmatrix} = \begin{vmatrix} 0 & 1 & 1 \\ 1 & 0 & 1 \\ 2 & 1 & 1 \end{vmatrix}$

5 $\begin{vmatrix} 2 & 4 & 2 \\ 1 & 2 & 4 \\ 2 & 6 & 4 \end{vmatrix} = 4 \begin{vmatrix} 1 & 2 & 1 \\ 1 & 2 & 4 \\ 1 & 3 & 2 \end{vmatrix}$

6 $\begin{vmatrix} 2 & 1 & 6 \\ 4 & 3 & 3 \\ 2 & 1 & 3 \end{vmatrix} = 6 \begin{vmatrix} 1 & 1 & 2 \\ 2 & 3 & 1 \\ 1 & 1 & 1 \end{vmatrix}$

7 $\begin{vmatrix} 1 & -1 & 2 \\ 1 & 2 & -1 \\ 1 & -1 & 2 \end{vmatrix} = 0$

8 $\begin{vmatrix} 1 & -1 & 1 \\ 0 & 1 & 0 \\ -1 & 0 & -1 \end{vmatrix} = 0$

9 $\begin{vmatrix} 1 & 5 \\ -3 & 2 \end{vmatrix} = - \begin{vmatrix} 1 & 5 \\ 3 & -2 \end{vmatrix}$

10 $\begin{vmatrix} 2 & -2 \\ 1 & 1 \end{vmatrix} = -\begin{vmatrix} -2 & 2 \\ 1 & 1 \end{vmatrix}$

11 $\begin{vmatrix} 0 & 0 & 1 \\ 1 & 0 & 0 \\ 0 & 0 & 1 \end{vmatrix} = 0$ **12** $\begin{vmatrix} 1 & 0 & 1 \\ 0 & 0 & 0 \\ 1 & 1 & 0 \end{vmatrix} = 0$

13 $\begin{vmatrix} 1 & -1 & -2 \\ -1 & 2 & 1 \\ 0 & 1 & 1 \end{vmatrix} = \begin{vmatrix} 1 & -1 & 0 \\ -1 & 2 & -1 \\ 0 & 1 & 1 \end{vmatrix}$

14 $\begin{vmatrix} a & 0 & 0 \\ 0 & b & 0 \\ 0 & 0 & c \end{vmatrix} = -\begin{vmatrix} 0 & 0 & a \\ 0 & b & 0 \\ c & 0 & 0 \end{vmatrix}$

In each of Exercises 15–24 find the determinant of the matrix after introducing zeros as in Example 1.

15 $\begin{bmatrix} 3 & 1 & 0 \\ -2 & 0 & 1 \\ 1 & 3 & -1 \end{bmatrix}$ **16** $\begin{bmatrix} -3 & 0 & 4 \\ 1 & 2 & 0 \\ 4 & 1 & -1 \end{bmatrix}$

17 $\begin{bmatrix} 5 & 4 & 3 \\ -3 & 2 & 1 \\ 0 & 7 & -2 \end{bmatrix}$ **18** $\begin{bmatrix} 0 & 2 & -6 \\ 5 & 1 & -3 \\ 6 & -2 & 5 \end{bmatrix}$

19 $\begin{bmatrix} 2 & 2 & -3 \\ 3 & 6 & 9 \\ -2 & 5 & 4 \end{bmatrix}$ **20** $\begin{bmatrix} 3 & 8 & 5 \\ 5 & 3 & -6 \\ 2 & 4 & -2 \end{bmatrix}$

21 $\begin{bmatrix} 3 & 1 & -2 & 2 \\ 2 & 0 & 1 & 4 \\ 0 & 1 & 3 & 5 \\ -1 & 2 & 0 & -3 \end{bmatrix}$

22 $\begin{bmatrix} 3 & 2 & 0 & 4 \\ -2 & 0 & 5 & 0 \\ 4 & -3 & 1 & 6 \\ 2 & -1 & 2 & 0 \end{bmatrix}$

23 $\begin{bmatrix} 2 & -2 & 0 & 0 & -3 \\ 3 & 0 & 3 & 2 & -1 \\ 0 & 1 & -2 & 0 & 2 \\ -1 & 2 & 0 & 3 & 0 \\ 0 & 4 & 1 & 0 & 0 \end{bmatrix}$

24 $\begin{bmatrix} 2 & 0 & -1 & 0 & 2 \\ 1 & 3 & 0 & 0 & 1 \\ 0 & 4 & 3 & 0 & -1 \\ -1 & 2 & 0 & -2 & 0 \\ 0 & 1 & 5 & 0 & -4 \end{bmatrix}$

25 Prove that
$$\begin{vmatrix} 1 & 1 & 1 \\ a & b & c \\ a^2 & b^2 & c^2 \end{vmatrix} = (a - b)(b - c)(c - a).$$
(*Hint:* See Example 3.)

26 Prove that
$$\begin{vmatrix} 1 & 1 & 1 \\ a & b & c \\ a^3 & b^3 & c^3 \end{vmatrix} =$$
$$(a - b)(b - c)(c - a)(a + b + c).$$

27 If A is a matrix of order 4 of the form
$$A = \begin{bmatrix} a_{11} & a_{12} & a_{13} & a_{14} \\ 0 & a_{22} & a_{23} & a_{24} \\ 0 & 0 & a_{33} & a_{34} \\ 0 & 0 & 0 & a_{44} \end{bmatrix}$$
show that $|A| = a_{11}a_{22}a_{33}a_{44}$.

28 If
$$A = \begin{bmatrix} a & b & 0 & 0 \\ c & d & 0 & 0 \\ 0 & 0 & e & f \\ 0 & 0 & g & h \end{bmatrix}$$
prove that
$$|A| = \begin{vmatrix} a & b \\ c & d \end{vmatrix} \begin{vmatrix} e & f \\ g & h \end{vmatrix}.$$

29 If $A = (a_{ij})$ and $B = (b_{ij})$ are arbitrary square matrices of order 2, prove that $|AB| = |A|\,|B|$.

30 If $A = (a_{ij})$ is a square matrix of order n and k is any real number, prove that $|kA| = k^n|A|$. (*Hint:* Use (ii) of the Theorem on Row and Column Transformations of a Determinant.)

8.9 CRAMER'S RULE

Determinants arise naturally in the study of solutions of systems of linear equations. To illustrate, let us consider the following case of two general linear equations in two unknowns:

$$\begin{cases} a_{11}x + a_{12}y = k_1 \\ a_{21}x + a_{22}y = k_2 \end{cases}$$

where at least one nonzero coefficient appears in each equation. We may as well assume that $a_{11} \neq 0$, for otherwise $a_{12} \neq 0$ and we could regard y as the "first" variable instead of x. We shall use the matrix method to obtain the matrix of an equivalent system. Changing the matrix of the system by means of elementary row transformations we obtain

$$\begin{bmatrix} a_{11} & a_{12} & k_1 \\ a_{21} & a_{22} & k_2 \end{bmatrix}$$

$$\rightarrow \begin{bmatrix} a_{11} & a_{12} & k_1 \\ 0 & a_{22} - \left(\dfrac{a_{12}a_{21}}{a_{11}}\right) & k_2 - \left(\dfrac{a_{21}k_1}{a_{11}}\right) \end{bmatrix}$$

Add to the second row $-a_{21}/a_{11}$ times the first row.

$$\rightarrow \begin{bmatrix} a_{11} & a_{12} & k_1 \\ 0 & (a_{11}a_{22} - a_{12}a_{21}) & (a_{11}k_2 - a_{21}k_1) \end{bmatrix}$$

Multiply row 2 by a_{11}.

Thus the given system is equivalent to

$$\begin{cases} a_{11}x + a_{12}y = k_1 \\ (a_{11}a_{22} - a_{12}a_{21})y = a_{11}k_2 - a_{21}k_1. \end{cases}$$

Notice that the numbers in the second equation may be written in determinant form as follows:

$$\begin{cases} a_{11}x + a_{12}y = k_1 \\ \begin{vmatrix} a_{11} & a_{12} \\ a_{21} & a_{22} \end{vmatrix} y = \begin{vmatrix} a_{11} & k_1 \\ a_{21} & k_2 \end{vmatrix}. \end{cases}$$

The following results now follow from the discussion in Section 8.2.

(a) If

$$\begin{vmatrix} a_{11} & a_{12} \\ a_{21} & a_{22} \end{vmatrix} = 0 \quad \text{and} \quad \begin{vmatrix} a_{11} & k_1 \\ a_{21} & k_2 \end{vmatrix} = 0$$

then the equations are dependent.

(b) If

$$\begin{vmatrix} a_{11} & a_{12} \\ a_{21} & a_{22} \end{vmatrix} = 0 \quad \text{and} \quad \begin{vmatrix} a_{11} & k_1 \\ a_{21} & k_2 \end{vmatrix} \neq 0$$

then the equations are inconsistent.

(c) If

$$\begin{vmatrix} a_{11} & a_{12} \\ a_{21} & a_{22} \end{vmatrix} \neq 0$$

then the equations are consistent.

If (c) occurs, then we can solve the second equation for y, obtaining

$$y = \frac{\begin{vmatrix} a_{11} & k_1 \\ a_{21} & k_2 \end{vmatrix}}{\begin{vmatrix} a_{11} & a_{12} \\ a_{21} & a_{22} \end{vmatrix}}.$$

The corresponding value for x may be found by substituting for y in the first equation. It can be shown that this leads to

$$x = \frac{\begin{vmatrix} k_1 & a_{12} \\ k_2 & a_{22} \end{vmatrix}}{\begin{vmatrix} a_{11} & a_{12} \\ a_{21} & a_{22} \end{vmatrix}}.$$

This proves that *if the determinant of the coefficient matrix of a system of two linear equations in two variables is not zero, then the system has a unique solution.* The last two formulas for x and y as quotients of certain determinants constitute what is known as **Cramer's Rule**.

There is an easy way to remember Cramer's Rule. Let

$$D = \begin{bmatrix} a_{11} & a_{12} \\ a_{21} & a_{22} \end{bmatrix}$$

be the coefficient matrix of the system and let D_x denote the matrix obtained from D by replacing the coefficients a_{11}, a_{21} of x by the numbers k_1, k_2, respectively. Similarly, let D_y denote the matrix obtained from D by replacing the coefficients a_{12}, a_{22} of y by the numbers k_1, k_2, respectively. Thus

$$D_x = \begin{bmatrix} k_1 & a_{12} \\ k_2 & a_{22} \end{bmatrix}, \quad D_y = \begin{bmatrix} a_{11} & k_1 \\ a_{21} & k_2 \end{bmatrix}.$$

According to the previous discussion, if $|D| \neq 0$, the solution (x, y) is given by

Cramer's Rule

$$x = \frac{|D_x|}{|D|}, \quad y = \frac{|D_y|}{|D|}.$$

Example 1 Use Cramer's Rule to solve the system

$$\begin{cases} 2x - 3y = -4 \\ 5x + 7y = 1. \end{cases}$$

Solution   The determinant of the coefficient matrix is

$$|D| = \begin{vmatrix} 2 & -3 \\ 5 & 7 \end{vmatrix} = 29.$$

Using the notation introduced previously,

$$|D_x| = \begin{vmatrix} -4 & -3 \\ 1 & 7 \end{vmatrix} = -25, \quad |D_y| = \begin{vmatrix} 2 & -4 \\ 5 & 1 \end{vmatrix} = 22.$$

Hence
$$x = \frac{|D_x|}{|D|} = \frac{-25}{29}, \quad y = \frac{|D_y|}{|D|} = \frac{22}{29}.$$

Thus the system has the unique solution $(-25/29, 22/29)$.   ■

Let us briefly consider the case of a general homogeneous system of two linear equations in x and y; that is,

$$\begin{cases} a_{11}x + a_{12}y = 0 \\ a_{21}x + a_{22}y = 0. \end{cases}$$

In this event the determinants $|D_x|$ and $|D_y|$ in Cramer's Rule are both zero. (Why?) Consequently, if

$$|D| = \begin{vmatrix} a_{11} & a_{12} \\ a_{21} & a_{22} \end{vmatrix} \neq 0$$

then $x = 0/|D| = 0$ and $y = 0/|D| = 0$; that is, the only solution is the trivial one $(0,0)$. This proves that *if the given system of homogeneous equations has a nontrivial solution, then*

$$\begin{vmatrix} a_{11} & a_{12} \\ a_{21} & a_{22} \end{vmatrix} = 0.$$

It can be shown, conversely, that if the determinant of the coefficient matrix is zero, then a homogeneous system has a nontrivial solution.

The previous discussion can be extended to systems of n linear equations in n variables. It is possible to show that such a system has a unique solution if and only if the determinant of the coefficient matrix is different from zero. If the system is homogeneous, then nontrivial solutions exist if and only if the determinant of the coefficient matrix is zero.

Cramer's Rule can be extended to systems of n linear equations in n variables x_1, $x_2, \ldots, x_n$, where each equation is written in the form

$$a_1x_1 + a_2x_2 + \cdots + a_nx_n = a.$$

To solve such a system, let D denote the coefficient matrix and let D_{x_i} denote the matrix obtained by replacing the coefficients of x_i in D by the column of numbers $k_1, \ldots, k_n$, which appears to the right of the equals signs in the system. It can be shown that if $|D| \neq 0$, then the system has a unique solution given by the following:

Cramer's Rule (General Form)

$$x_1 = \frac{|D_{x_1}|}{|D|}, \quad x_2 = \frac{|D_{x_2}|}{|D|}, \ldots, \quad x_n = \frac{|D_{x_n}|}{|D|}.$$

It can be shown that if $|D| = 0$, the equations are dependent or inconsistent, depending on whether all the $|D_{x_i}|$ are zero or at least one of them is not zero.

Example 2   Use Cramer's Rule to solve the system

$$\begin{cases} x & -2z = 3 \\ & -y + 3z = 1 \\ 2x & + 5z = 0. \end{cases}$$

Solution   We shall merely list the various determinants which are used, leaving the reader to check the answers:

$$|D| = \begin{vmatrix} 1 & 0 & -2 \\ 0 & -1 & 3 \\ 2 & 0 & 5 \end{vmatrix} = -9, \quad |D_x| = \begin{vmatrix} 3 & 0 & -2 \\ 1 & -1 & 3 \\ 0 & 0 & 5 \end{vmatrix} = -15,$$

$$|D_y| = \begin{vmatrix} 1 & 3 & -2 \\ 0 & 1 & 3 \\ 2 & 0 & 5 \end{vmatrix} = 27, \quad |D_z| = \begin{vmatrix} 1 & 0 & 3 \\ 0 & -1 & 1 \\ 2 & 0 & 0 \end{vmatrix} = 6.$$

By Cramer's Rule, the solution is

$$x = \frac{|D_x|}{|D|} = \frac{-15}{-9} = \frac{5}{3}, \quad y = \frac{|D_y|}{|D|} = \frac{27}{-9} = -3, \quad z = \frac{|D_z|}{|D|} = \frac{6}{-9} = -\frac{2}{3}. \quad \blacksquare$$

Cramer's Rule is an inefficient method to apply if there are a large number of equations, since many determinants of matrices of high order must be evaluated. Note also that Cramer's Rule cannot be used directly if $|D| = 0$ or if the number of equations is not the same as the number of variables. In general, the method of elimination or the matrix method is far superior to Cramer's Rule.

EXERCISES 8.9

1–18 Use Cramer's Rule to solve the systems in Exercises 1–18 of Section 8.2.

19–26 Use Cramer's Rule to solve the systems in Exercises 1–8 of Section 8.3.

27–30 Use Cramer's Rule to solve the systems in Exercises 17–20 of Section 8.3.

8.10 SYSTEMS OF INEQUALITIES

Our previous work with inequalities was restricted to inequalities in one variable. The notion of inequalities in several variables can be developed in a manner similar to our work with equations in several variables. For example, expressions of the form

$$3x + y < 5y^2 + 1$$
$$2x^2 \geq 4 - 3y$$

are called **inequalities in x and y**. As with equations, a **solution** of an inequality in x and y is defined as an ordered pair (a, b) which produces a true statement when a and b are substituted for x and y, respectively. The **graph of an inequality** is the graph of the totality of solutions. Two inequalities are **equivalent** if they have exactly the same solutions. An inequality in x and y can be simplified by adding an expression in x and y to both sides or by multiplying both sides by such an expression, provided we are careful about signs. Similar definitions and remarks hold for inequalities in more than two variables. We shall restrict our discussion, however, to the case of inequalities in two variables.

Example 1 Find the solutions and sketch the graph of the inequality $3x - 3 < 5x - y$.

Solution Adding the expression $y + 3 - 3x$ to both sides, we obtain the equivalent inequality $y < 2x + 3$. Hence the solutions consist of all ordered pairs (x, y) such that $y < 2x + 3$. It is convenient to denote the solutions as follows:

$$\{(x, y): y < 2x + 3\}.$$

There is a close relationship between the graph of the inequality $y < 2x + 3$ and the graph of the equation $y = 2x + 3$. The graph of the equation is the straight line sketched in Figure 8.5. For each real number a, the point on the line with abscissa a has coordinates $(a, 2a + 3)$. A point $P(a, b)$ belongs to the graph of the *inequality* if and only if $b < 2a + 3$; that is, if and only if the point $P(a, b)$ lies directly below the point with coordinates $(a, 2a + 3)$ as shown in Figure 8.5. It

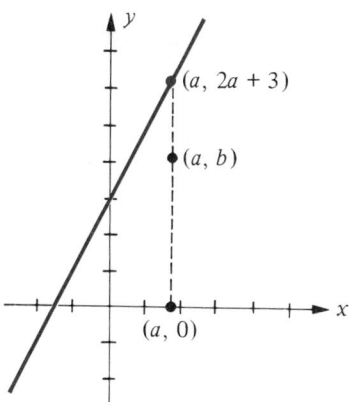

Figure 8.5

follows that the graph of the inequality $y < 2x + 3$ consists of all points in the plane which lie below the line $y = 2x + 3$. In Figure 8.6 we have shaded a portion of the graph. Dashes used for the line indicate that it is not part of the graph of the inequality.

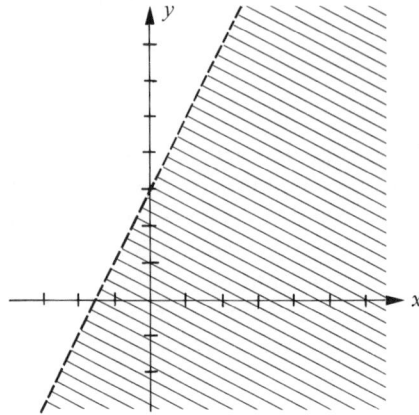

Figure 8.6. $\quad y < 2x + 3$ ∎

A region of the type shown in Figure 8.6 is called a **half-plane**. More precisely, when the line is *not* included, we refer to such a region as an **open half-plane**. If the line *is* included, as would be the case for the graph of the inequality $y \le 2x + 3$, then the region is called a **closed half-plane**.

By an argument similar to that used in Example 1, it can be shown that the graph of the inequality $y > 2x + 3$ is the open half-plane which lies *above* the line with equation $y = 2x + 3$.

If an equality involves only polynomials of the first degree in x and y, as was the case in Example 1, it is called a **linear inequality**.

The procedure used in Example 1 can be generalized to inequalities of the form $y < f(x)$, where f is a function. Specifically, the following can be established.

Theorem

> If f is a function, then the graph of the inequality $y < f(x)$ is the set of points which lie *below* the graph of the equation $y = f(x)$. Similarly, the graph of $y > f(x)$ is the set of points which lie *above* the graph of $y = f(x)$.

Example 2   Find the solutions and sketch the graph of the inequality $x(x + 1) - 2y > 3(x - y)$.

Solution   The given inequality is equivalent to

$$x^2 + x - 2y > 3x - 3y.$$

Adding $3y - x^2 - x$ to both sides we obtain

$$y > 2x - x^2.$$

Hence the solutions are $\{(x,y):y > 2x - x^2\}$. To find the graph, we begin by sketching the graph of $y = 2x - x^2$ (a parabola) with dashes as illustrated in Figure 8.7. Using the preceding theorem, the graph is the region above the parabola, as indicated by the shaded portion of the figure.

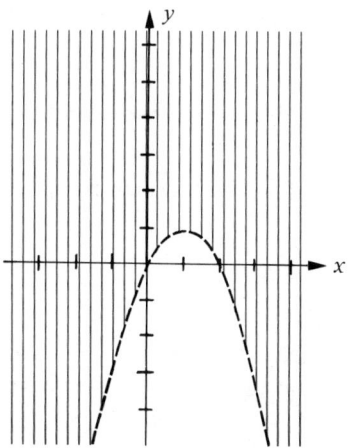

Figure 8.7. $y > 2x - x^2$

The following result can also be proved.

Theorem

> If g is a function, then the graph of the inequality $x < g(y)$ is the set of points to the *left* of the graph of the equation $x = g(y)$. Similarly, the graph of $x > g(y)$ is the set of points to the *right* of the graph of $x = g(y)$.

Example 3   Sketch the graph of $x \geq y^2$.

Solution   The graph of the equation $x = y^2$ is a parabola. By the preceding theorem, the graph of the inequality consists of all points on the parabola together with the points in the region to the right of the parabola (see Figure 8.8).

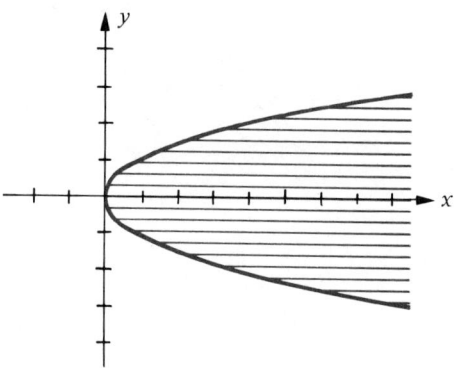

Figure 8.8. $x \geq y^2$

It is sometimes necessary to work simultaneously with several inequalities in two variables. In this case we refer to the given inequalities as a **system of inequalities**. The **solutions of a system** of inequalities are, by definition, the common solutions of all the inequalities in the system. It should be clear how to define **equivalent systems** and the **graph of a system** of inequalities. The following examples illustrate a method for solving systems of inequalities.

Example 4   Find the solutions and sketch the graph of the system $\begin{cases} x + y \leq 4 \\ 2x - y \leq 4. \end{cases}$

Solution   The given system is equivalent to

$$\begin{cases} y \leq 4 - x \\ y \geq 2x - 4. \end{cases}$$

We begin by sketching the graphs of the lines $y = 4 - x$ and $y = 2x - 4$. The lines intersect at the point $(\frac{8}{3}, \frac{4}{3})$ shown in Figure 8.9. The graph of $y \leq 4 - x$ includes the points on the graph of $y = 4 - x$ together with the points which lie below this line. The graph of $y \geq 2x - 4$ includes the points on the graph of $y = 2x - 4$ together with the points which lie above this line. A portion of each of these regions is shown in Figure 8.9. The graph of the system consists of the points that are in *both* regions. This corresponds to the cross-hatched region shown in Figure 8.9.

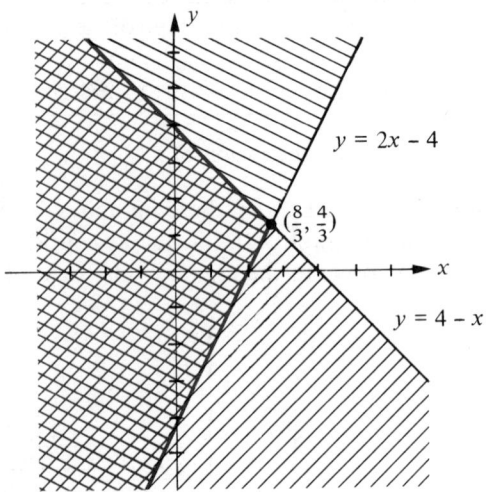

Figure 8.9

Example 5   Sketch the graph of the system

$$\begin{cases} x + y \leq 4 \\ 2x - y \leq 4 \\ x \geq 0 \\ y \geq 0. \end{cases}$$

Solution   The first two inequalities are the same as those considered in Example 4 and hence the points on the graph of the present system must lie within the region shown in Figure 8.10. In addition, the third and fourth inequalities in the system tell us that the points must lie in the first quadrant. This gives us the region shown in Figure 8.10.

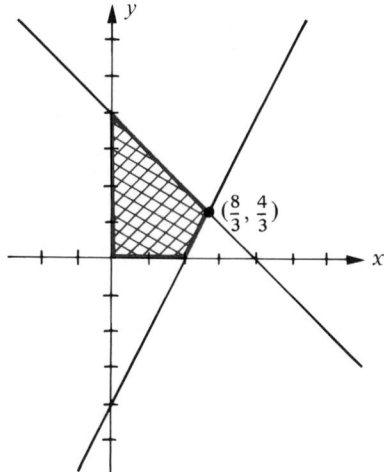

Figure 8.10   ■

Example 6   Sketch the graph of the system

$$\begin{cases} x^2 + y^2 \le 1 \\ (x - 1)^2 + y^2 \le 1. \end{cases}$$

Solution   The graph of the equation $x^2 + y^2 = 1$ is a unit circle with center at the origin, and the graph of $(x - 1)^2 + y^2 = 1$ is a unit circle with center at the point $C(1, 0)$. To find the points of intersection of the two circles, let us solve the equation $x^2 + y^2 = 1$ for y^2, obtaining $y^2 = 1 - x^2$. Substituting for y^2 in $(x - 1)^2 + y^2 = 1$ leads to the following equations:

$$(x - 1)^2 + (1 - x^2) = 1$$
$$x^2 - 2x + 1 + 1 - x^2 = 1$$
$$-2x = -1$$
$$x = 1/2.$$

The corresponding values for y are given by

$$y^2 = 1 - x^2 = 1 - (1/2)^2 = 3/4$$

and hence $y = \pm\sqrt{3}/2$. Thus the points of intersection are $(1/2, \sqrt{3}/2)$ and $(1/2, -\sqrt{3}/2)$ as shown in Figure 8.11. By the distance formula, the graphs of the given inequalities are the regions within and on the two circles. The graph of the system consists of the points common to both regions, as indicated by the shaded portion of the figure.

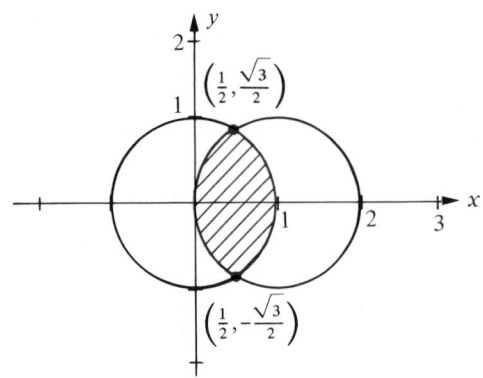

Figure 8.11

EXERCISES 8.10

In each of Exercises 1–10 find the solutions and sketch the graph of the given inequality.

1 $3x - 2y < 6$

2 $4x + 3y < 12$

3 $2x + 3y \geq 2y + 1$

4 $2x - y > 3$

5 $y + 2 < x^2$

6 $y^2 - x \leq 0$

7 $x^2 + 1 \leq y$

8 $y - x^3 < 1$

9 $yx^2 \geq 1$

10 $x^2 + 4 \geq y$

In each of Exercises 11–24 sketch the graph of the given system.

11 $\begin{cases} 3x + y < 3 \\ 4 - y < 2x \end{cases}$

12 $\begin{cases} y + 2 < 2x \\ y - x > 4 \end{cases}$

13 $\begin{cases} y - x < 0 \\ 2x + 5y < 10 \end{cases}$

14 $\begin{cases} 2y - x \leq 4 \\ 3y + 2x < 6 \end{cases}$

15 $\begin{cases} 3x + y \leq 6 \\ y - 2x \geq 1 \\ x \geq -2 \\ y \leq 4 \end{cases}$

16 $\begin{cases} 3x - 4y \geq 12 \\ x - 2y \leq 2 \\ x \geq 9 \\ y \leq 5 \end{cases}$

17 $\begin{cases} x^2 + y^2 \leq 4 \\ x + y \geq 1 \end{cases}$

18 $\begin{cases} x^2 + y^2 > 1 \\ x^2 + y^2 < 4 \end{cases}$

19 $\begin{cases} x^2 \leq 1 - y \\ x \geq 1 + y \end{cases}$

20 $\begin{cases} x - y^2 < 0 \\ x + y^2 > 0 \end{cases}$

21 $\begin{cases} y < 3^x \\ y > 2^x \\ x \geq 0 \end{cases}$

22 $\begin{cases} y \geq \log x \\ y - x \leq 1 \\ x \geq 1 \end{cases}$

23 $\begin{cases} y \leq \log x \\ y + x \geq 1 \\ x \leq 10 \end{cases}$

24 $\begin{cases} y \leq 3^{-x} \\ y \geq 2^{-x} \\ y < 9 \end{cases}$

25 The manager of a baseball team wishes to buy bats and balls, costing $3.50 and $2.50, respectively. If the maximum amount available is $40 and if at least two balls and three bats are required, find a system of inequalities describing all the possibilities and sketch the graph.

26 An office worker wishes to purchase some 40-cent postage stamps and also some 50-cent stamps, totaling not more than $35. Moreover, it is desired to have at least twice as many 40-cent stamps as 50-cent stamps and more than ten 40-cent stamps. Find a system of inequalities describing all the possibilities and sketch the graph.

27 A store sells two brands of television sets. Customer demand indicates that it is necessary to stock at least twice as many sets of brand A as of brand B. It is also necessary to have on hand at least 20 of brand A and 10 of brand B. If there is room for not more than 100 sets in the store, find a system of inequalities describing all possibilities and sketch the graph.

28 An auditorium contains 600 seats. For a certain event it is planned to charge $4.00 for certain seats and $3.00 for others. At least 225 tickets are to be sold for $3.00, and total sales of more than $2,000 is desired. Find a system of inequalities describing all possibilities and sketch the graph.

29 A woman wishes to invest $15,000 in two different savings accounts. She also wants to have at least $2,000 in each account, with the amount in one account being at least three times that in the other. Find a system of inequalities describing all possibilities and sketch the graph.

30 The manager of a college bookstore stocks two types of notebooks, the first wholesaling for 55 cents and the second for 85 cents. If the maximum amount to be spent is $600 and if an inventory of at least 300 of the 85-cent variety and 400 of the 55-cent variety is desired, find a system of inequalities describing all possibilities and sketch the graph.

8.11 LINEAR PROGRAMMING

In applications, problems sometimes arise which require finding solutions of systems of inequalities. A typical problem is that of finding maximum and minimum values of certain expressions involving variables which are subject to various constraints. If all the expressions and inequalities are linear in the variables, then a technique called **linear programming** may be used to help solve such problems. This technique has become very important in businesses where decisions must be made concerning the best use of stock, parts, manufacturing processes, and so on. Usually the objective of management is to maximize profit or to minimize cost. Since there are often many choices, it may be extremely difficult to arrive at a correct decision. A mathematical theory such as that afforded by linear programming can simplify the task considerably. The logical development of the theorems and techniques which are needed would take us beyond the objectives of this text. We shall, therefore, limit ourselves to several examples.

Example 1 A manufacturer of a certain product has two warehouses W_1 and W_2. There are 80 units of his product stored at W_1 and 70 units at W_2. Two customers A and B order 35 units and 60 units, respectively. The shipping cost from each warehouse to A and B is determined according to the following table. How should the order be filled so as to minimize the total shipping cost?

Warehouse	Customer	Shipping cost per unit
W_1	A	$ 8
W_1	B	12
W_2	A	10
W_2	B	13

Solution If we let x denote the number of units to be sent to A from W_1, then $35 - x$ units must be sent from W_2 to A. Similarly, if y denotes the number of units to be sent from W_1 to B, then $60 - y$ units must be sent from W_2 to B. We wish to determine values for x and y which make the total shipping costs minimal. We first note that since x and y are between 35 and 60, respectively, the pair (x, y) must be a solution of the following system of inequalities:

$$0 \leq x \leq 35, \quad 0 \leq y \leq 60.$$

The graph of this system is the rectangular region shown in Figure 8.12.

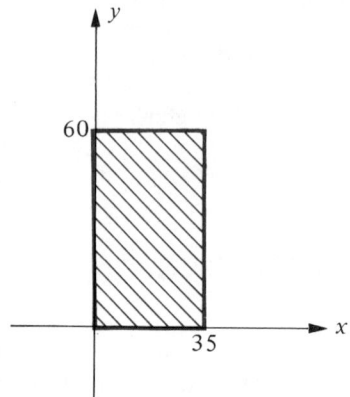

Figure 8.12

There are further constraints on x and y which make it possible to reduce the size of the region above. Since the total number of units shipped from W_1 cannot exceed 80 and the total shipped from W_2 cannot exceed 70, the pair (x, y) must also be a solution of the system

$$\begin{cases} x + y \le 80 \\ (35 - x) + (60 - y) \le 70. \end{cases}$$

This system is equivalent to

$$\begin{cases} x + y \le 80 \\ x + y \ge 25. \end{cases}$$

The graph of the last system is the region between the parallel lines $x + y = 80$ and $x + y = 25$ (see Figure 8.13). Since the pair (x, y) we seek must be a solution of this system and also the system $0 \le x \le 35, 0 \le y \le 60$, the corresponding point must lie in the region shown in Figure 8.14.

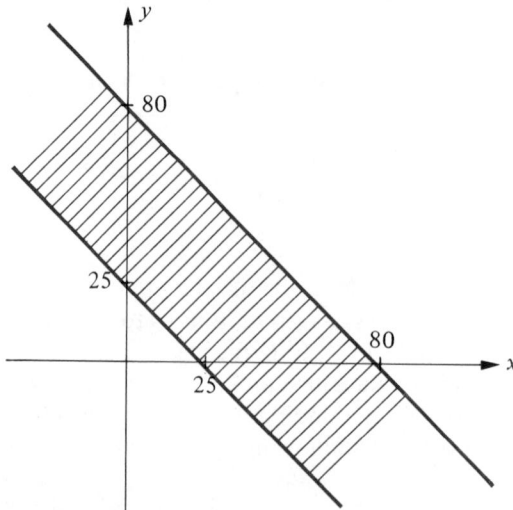

Figure 8.13

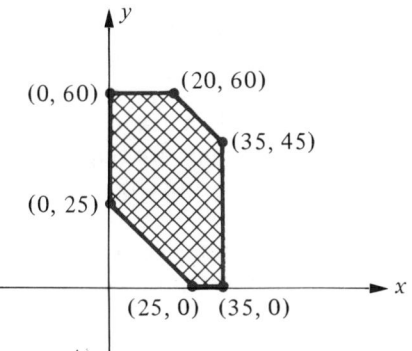

Figure 8.14

Now let C denote the total cost (in dollars) of shipping the merchandise to A and B. We see from the table that the cost of shipping the 35 units to A is $8x + 10(35 - x)$, whereas the cost of shipping the 60 units to B is $12y + 13(60 - y)$. Hence the total cost is

$$C = (8x + 350 - 10x) + (12y + 780 - 13y)$$

or $$C = 1130 - 2x - y.$$

For each point (x, y) of the region shown in Figure 8.14 we obtain a value for C. For example, at $(20, 40)$

$$C = 1130 - 40 - 40 = 1050;$$

and at $(10, 50)$, $$C = 1130 - 20 - 50 = 1060.$$

Since x and y must be integers, there are only a finite number of possible values for C. By checking each possibility, we could find the pair (x, y) which produces the smallest cost. However, since there are a very large number of pairs, the task of checking each one would be very tedious. This is where the theory developed in linear programming is helpful. It can be shown that if we are interested in the value C of a linear expression $ax + by + c$, where each pair (x, y) is a solution of a system of linear inequalities and hence corresponds to a point which is in the intersection of half-planes, then C takes on its maximum and minimum values at a point of intersection of the lines which determine the half-planes. This means that in order to determine the minimum (or maximum) value of C we need only check the points $(0, 25)$, $(0, 60)$, $(20, 60)$, $(35, 45)$, $(35, 0)$, and $(25, 0)$ shown in Figure 8.14. The values are arranged in tabular form below.

Point	$1130 - 2x - y = C$
$(0, 25)$	$1130 - 2(0) \ - 25 = 1105$
$(0, 60)$	$1130 - 2(0) \ - 60 = 1070$
$(20, 60)$	$1130 - 2(20) - 60 = 1030$
$(35, 45)$	$1130 - 2(35) - 45 = 1015$
$(35, 0)$	$1130 - 2(35) - \ \ 0 = 1060$
$(25, 0)$	$1130 - 2(25) - \ \ 0 = 1080$

According to our remarks, the minimal shipping cost $1,015 occurs if $x = 35$ and $y = 45$. This means that the manufacturer should ship all of the units to A from W_1. In addition, the manufacturer should ship 45 units to B from W_1 and 15 units to B from W_2. Note that the *maximum* shipping cost will occur when $x = 0$ and $y = 25$, that is, when all 35 units are shipped to A from W_2 and when B receives 25 units from W_1 and 35 units from W_2. ∎

The preceding example illustrates how linear programming can be used to minimize the cost in a certain situation. The next example has to do with maximization of profit.

Example 2   A firm manufactures two products X and Y. For each product it is necessary to use three different machines A, B, and C. In order to manufacture one unit of product X, machine A must be used for 3 hours, machine B for 1 hour, and machine C for 1 hour. To manufacture one unit of product Y requires 2 hours on A, 2 hours on B, and 1 hour on C. The profit on product X is $500 per unit and the profit on product Y is $350 per unit. Machine A is available for a total of 24 hours per day; however, B can only be used for 16 hours and C for 9 hours. If the machines are available when needed (subject to the noted total hour restrictions), determine the number of units of each product that should be manufactured each day in order to maximize the profit.

Solution   The following table summarizes the data given in the statement of the problem.

Machine	Hours required for 1 unit of X	Hours required for one unit of Y	Hours available
A	3	2	24
B	1	2	16
C	1	1	9

Let x and y denote the number of units of products X and Y, respectively, to be produced per day. Since each unit of product X requires 3 hours on machine A, x units require $3x$ hours. Similarly, since each unit of product Y requires 2 hours on A, y units require $2y$ hours. Hence the total number of hours per day that machine A must be used is $3x + 2y$. Since A can be used for at most 24 hours per day, we have

$$3x + 2y \leq 24.$$

Using the same type of reasoning on rows two and three of the table we see that

$$x + 2y \leq 16$$
$$x + y \leq 9.$$

This system of three linear inequalities, together with the obvious inequalities

$$x \geq 0, \quad y \geq 0$$

states, in mathematical form, the restraints which occur in the manufacturing process. The graph of the preceding system of five linear inequalities is sketched in Figure 8.15. The points shown in the figure are found by solving systems of linear equations. Specifically, $(6, 3)$ is a solution of the system $3x + 2y = 24$, $x + y = 9$, and $(2, 7)$ is a solution of the system $x + 2y = 16$, $x + y = 9$.

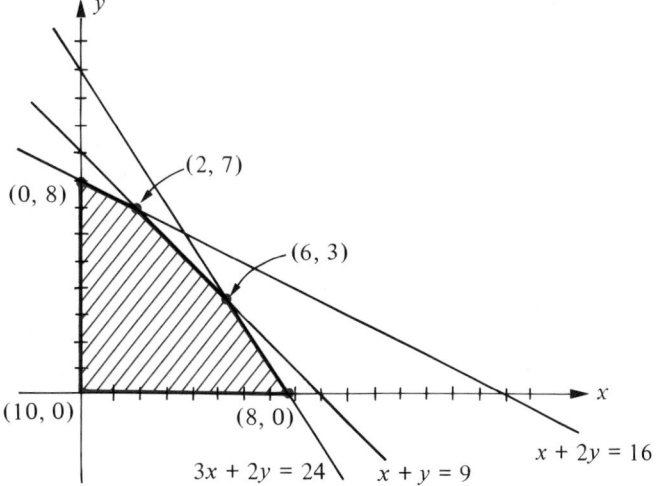

Figure 8.15

Since the production of each unit of product X results in a profit of $500, and each unit of product Y yields a profit of $350, the profit P obtained by producing x units of X together with y units of Y is given by

$$P = 500x + 350y.$$

As pointed out in the solution of Example 1, the maximum and minimum values of P occur at certain of the points shown in Figure 8.15. The values of P at all the points are given in the following table.

(x, y)	$500x + 350y = P$
$(0, 8)$	$500(0) + 350(8) = 2800$
$(2, 7)$	$500(2) + 350(7) = 3450$
$(6, 3)$	$500(6) + 350(3) = 4050$
$(8, 0)$	$500(8) + 350(0) = 4000$
$(0, 0)$	$500(0) + 350(0) = 0$

We see from the table that a maximum profit of $4,050 occurs if there is a daily production of 6 units of product X and 3 units of product Y. ∎

The two illustrations we have given in this section are elementary problems in linear programming which can be solved by rather crude methods. The much more complicated problems that occur in practice are usually solved by employing matrix techniques which are adapted for solutions by computers.

EXERCISES 8.11

1 A manufacturer of tennis rackets makes a profit of $15 on each Set Point racket and $8 on each Double Fault racket. To meet dealer demand, daily production of Double Faults should be between 30 and 80, whereas the number of Set Points should be between 10 and 30. In order to maintain high quality, the total number of rackets produced should not exceed 80 per day. How many of each type should be manufactured daily to maximize the profit?

2 A manufacturer of CB radios makes a profit of $25 on a deluxe model and $30 on a standard model. The company wishes to produce at least 80 deluxe models and at least 100 standard models per day. To maintain high quality, the daily production should not exceed 200 radios. How many of each type should be produced daily in order to maximize the profit?

3 Two substances S and T each contain two types of ingredients I and G. One pound of S contains 2 ounces of I and 4 ounces of G. One pound of T contains 2 ounces of I and 6 ounces of G. It is desired to combine quantities of the two substances to obtain a mixture which contains at least 9 ounces of I and 20 ounces of G. If the cost of S is $3.00 per pound and the cost of T is $4.00 per pound, how much of each substance should be used to keep the cost to a minimum?

4 A stationery company makes two types of notebooks. Type M sells for $1.25 and type N sells for $0.90. It costs the company $1.00 to produce one type M notebook and $0.75 to produce one of type N. The company has the facilities to manufacture between 2,000 and 3,000 of type M and between 3,000 and 6,000 of type N, but not more than 7,000 altogether. How many notebooks of each type should be manufactured to maximize the difference between the selling prices and the costs of production?

5 In Example 1 of this section, if the shipping costs are $12 per unit from W_1 to X, $10 per unit from W_2 to X, $16 per unit from W_1 to Y, and $12 per unit from W_2 to Y, determine how the order should be filled so as to minimize shipping costs.

6 A coffee company purchases mixed lots of coffee beans and then grades them into premium, regular and unusable beans. The company needs at least 280 tons of premium-grade and 200 tons of regular-grade coffee beans. The company can purchase ungraded coffee from two suppliers A and B in any amount desired. Samples from the two suppliers contain the following percentages of premium, regular, and unusable beans:

Supplier	Premium	Regular	Unusable
A	20%	50%	30%
B	40%	20%	40%

If A charges $125 per ton and B charges $200 per ton, how much should the company purchase from each supplier to fulfill its needs at minimum cost?

7 A farmer has 100 acres available for planting two crops A and B. The seed for crop A costs $4 per acre and the seed for crop B costs $6 per acre. The total cost of labor will amount to $20 per acre for crop A and $10 per acre for crop B. The expected income from crop A is $110 per acre, whereas from B it is $150 per acre. If the farmer does not wish to spend more than $480 for seed and $1,400 for labor, how many acres of each crop should be planted in order to obtain the maximum profit?

8 A firm manufactures two products A and B. For each product it is necessary to employ two different machines X and Y. To manufacture product A, machine X must be used for $\frac{1}{2}$ hour and machine Y for 1 hour. To manufacture product B, machine X must be used for 2 hours and machine Y for 2 hours. The profit on product A is $20 per unit and the profit on B is $50 per unit. If machine X can be used for 8 hours per day and machine Y for 12 hours per day, determine how many units of each product should be manufactured each day in order to maximize the profit.

9 Three substances X, Y, and Z each contain four ingredients A, B, C, and D. The percentage of each ingredient and the cost in cents per ounce of the substances are given in the following table.

Substance	A	B	C	D	Cost/ Ounce
X	20%	10%	25%	45%	25¢
Y	20%	40%	15%	25%	35¢
Z	10%	20%	25%	45%	50¢

If the cost is to be minimum, how many ounces of each substance should be combined in order to obtain a mixture of 20 ounces containing at least 14% A, 16% B, and 20% C? What combination would make the cost greatest?

10 A man plans to operate a stand at a one-day fair in which he will sell bags of peanuts and bags of candy. He has $100 available to purchase his stock, which will cost 10¢ per bag of peanuts and 20¢ per bag of candy. He intends to sell the peanuts at 15¢ and the candy at 26¢ per bag. His stand can accommodate up to 500 bags of peanuts and 400 bags of candy. From past experience he knows that he will sell no more than a total of 700 bags. Find the number of bags of each that he should have available in order to maximize his profit. What is the maximum profit?

8.12 REVIEW

Concepts

Define or discuss each of the following.

1 System of equations

2 Solution of a system of equations

3 Equivalent systems of equations

4 System of linear equations

5 Homogeneous system of linear equations

6 An $m \times n$ matrix

7 A square matrix of order n

8 The coefficient matrix of a system of linear equations; the augmented matrix

9 Elementary row transformations

10 The sum and product of two matrices

11 Zero matrix   **12** Identity matrix

13 Inverse of a matrix   **14** Minor

15 Cofactor   **16** Determinant

17 Properties of determinants

18 Cramer's Rule

19 System of inequalities

20 Linear programming

Exercises

Find the solutions of the systems of equations in Exercises 1–16.

1 $\begin{cases} 2x - 3y = 4 \\ 5x + 4y = 1 \end{cases}$

2 $\begin{cases} x - 3y = 4 \\ -2x + 6y = 2 \end{cases}$

3 $\begin{cases} y + 4 = x^2 \\ 2x + y = -1 \end{cases}$

4 $\begin{cases} x^2 + y^2 = 25 \\ x - y = 7 \end{cases}$

5 $\begin{cases} 9x^2 + 16y^2 = 140 \\ x^2 - 4y^2 = 4 \end{cases}$

6 $\begin{cases} 2x = y^2 + 3z \\ x = y^2 + z - 1 \\ x^2 = xz \end{cases}$

7 $\begin{cases} \dfrac{1}{x} + \dfrac{3}{y} = 7 \\ \dfrac{4}{x} - \dfrac{2}{y} = 1 \end{cases}$

8 $\begin{cases} 2^x + 3^{y+1} = 10 \\ 2^{x+1} - 3^y = 5 \end{cases}$

9 $\begin{cases} 3x + y - 2z = -1 \\ 2x - 3y + z = 4 \\ 4x + 5y - z = -2 \end{cases}$

10 $\begin{cases} x + 3y = 0 \\ y - 5z = 3 \\ 2x + z = -1 \end{cases}$ **11** $\begin{cases} 4x - 3y - z = 0 \\ x - y - z = 0 \\ 3x - y + 3z = 0 \end{cases}$

12 $\begin{cases} 2x + y - z = 0 \\ x - 2y + z = 0 \\ 3x + 3y + 2z = 0 \end{cases}$ **13** $\begin{cases} 4x + 2y - z = 1 \\ 3x + 2y + 4z = 2 \end{cases}$

14 $\begin{cases} 2x + y = 6 \\ x - 3y = 17 \\ 3x + 2y = 7 \end{cases}$ **15** $\begin{cases} \dfrac{4}{x} + \dfrac{1}{y} + \dfrac{2}{z} = 4 \\ \dfrac{2}{x} + \dfrac{3}{y} - \dfrac{1}{z} = 1 \\ \dfrac{1}{x} + \dfrac{1}{y} + \dfrac{1}{z} = 4 \end{cases}$

16 $\begin{cases} 2x - y + 3z - w = -3 \\ 3x + 2y - z + w = 13 \\ x - 3y + z - 2w = -4 \\ -x + y + 4z + 3w = 0 \end{cases}$

Find the solutions and sketch the graphs of the systems in Exercises 17–20.

17 $\begin{cases} x^2 + y^2 < 16 \\ y - x^2 > 0 \end{cases}$ **18** $\begin{cases} y - x \le 0 \\ y + x \ge 2 \\ x \le 5 \end{cases}$

19 $\begin{cases} x - 2y \le 2 \\ y - 3x \le 4 \\ 2x + y \le 4 \end{cases}$ **20** $\begin{cases} x^2 - y < 0 \\ y - 2x < 5 \\ xy < 0 \end{cases}$

Find the determinants of the matrices in Exercises 21–30.

21 $[-6]$ **22** $\begin{bmatrix} 3 & 4 \\ -6 & -5 \end{bmatrix}$

23 $\begin{bmatrix} 3 & -4 \\ 6 & 8 \end{bmatrix}$ **24** $\begin{bmatrix} 0 & 4 & -3 \\ 2 & 0 & 4 \\ -5 & 1 & 0 \end{bmatrix}$

25 $\begin{bmatrix} 2 & -3 & 5 \\ -4 & 1 & 3 \\ 3 & 2 & -1 \end{bmatrix}$ **26** $\begin{bmatrix} 3 & 1 & -2 \\ -5 & 2 & -4 \\ 7 & 3 & -6 \end{bmatrix}$

27 $\begin{bmatrix} 5 & 0 & 0 & 0 \\ 6 & -3 & 0 & 0 \\ 1 & 4 & -4 & 0 \\ 7 & 2 & 3 & 2 \end{bmatrix}$

28 $\begin{bmatrix} 1 & 2 & 0 & 3 & 1 \\ -2 & -1 & 4 & 1 & 2 \\ 3 & 0 & -1 & 0 & -1 \\ 2 & -3 & 2 & -4 & 2 \\ -1 & 1 & 0 & 1 & 3 \end{bmatrix}$

29 $\begin{bmatrix} 2 & 0 & 1 & 0 & -1 \\ 0 & 1 & 0 & 1 & 2 \\ 2 & -2 & 1 & -2 & 0 \\ 0 & 0 & -2 & 0 & 1 \\ 1 & -1 & 0 & -1 & 0 \end{bmatrix}$

30 $\begin{bmatrix} 1 & 2 & 0 & 0 & 0 \\ 3 & 4 & 0 & 0 & 0 \\ 0 & 0 & 1 & 2 & 3 \\ 0 & 0 & 2 & -1 & 1 \\ 0 & 0 & 1 & 3 & -1 \end{bmatrix}$

31 Find the determinant of the $n \times n$ matrix (a_{ij}), where $a_{ij} = 0$ if $i \ne j$.

32 Without expanding show that

$$\begin{vmatrix} 1 & a & b+c \\ 1 & b & a+c \\ 1 & c & a+b \end{vmatrix} = 0.$$

Find the inverses of the matrices in Exercises 33–36.

33 $\begin{bmatrix} 5 & -4 \\ -3 & 2 \end{bmatrix}$ **34** $\begin{bmatrix} 2 & -1 & 0 \\ 1 & 4 & 2 \\ 3 & -2 & 1 \end{bmatrix}$

35 $\begin{bmatrix} 3 & -1 & 0 & 0 \\ 1 & 2 & 0 & 0 \\ 0 & 0 & -1 & -2 \\ 0 & 0 & 5 & 3 \end{bmatrix}$

36 $\begin{bmatrix} 2 & 0 & 0 & 0 \\ 0 & 3 & 0 & 0 \\ 0 & 0 & 4 & 0 \\ 0 & 0 & 0 & 5 \end{bmatrix}$

In Exercises 37–46 express as a single matrix.

37 $\begin{bmatrix} 2 & -1 & 0 \\ 3 & 0 & -2 \end{bmatrix} \begin{bmatrix} 2 & -1 & 3 \\ 0 & 3 & 0 \\ 1 & 4 & 2 \end{bmatrix}$

38 $\begin{bmatrix} 4 & 2 \\ 5 & -3 \end{bmatrix} \begin{bmatrix} 3 \\ 7 \end{bmatrix}$

39 $\begin{bmatrix} 2 & 0 \\ 1 & 4 \\ -2 & 3 \end{bmatrix} \begin{bmatrix} 0 & 2 & -3 \\ 4 & 5 & 1 \end{bmatrix}$

40 $\begin{bmatrix} 0 & -2 & 3 \\ 4 & 1 & 2 \end{bmatrix} \begin{bmatrix} 2 & 0 \\ 3 & 8 \\ 2 & -7 \end{bmatrix}$

41 $2\begin{bmatrix} 0 & -1 & -4 \\ 3 & 2 & 1 \end{bmatrix} - 3\begin{bmatrix} 4 & -2 & 1 \\ 0 & 5 & -1 \end{bmatrix}$

42 $\begin{bmatrix} 1 & 3 \\ 2 & 4 \end{bmatrix} \begin{bmatrix} a & 0 \\ 0 & a \end{bmatrix}$

43 $\begin{bmatrix} a & 0 \\ 0 & b \end{bmatrix} \begin{bmatrix} 1 & 3 \\ 2 & 4 \end{bmatrix}$ **44** $\begin{bmatrix} 3 & 2 \\ 0 & 0 \end{bmatrix} \begin{bmatrix} -2 & 0 \\ 3 & 0 \end{bmatrix}$

45 $\begin{bmatrix} 1 & 2 \\ 3 & 4 \end{bmatrix} \left\{ \begin{bmatrix} 2 & -4 \\ 3 & 7 \end{bmatrix} + \begin{bmatrix} 1 & 5 \\ -2 & -3 \end{bmatrix} \right\}$

46 $\begin{bmatrix} 3 & 2 & 5 \\ -3 & 4 & 7 \\ 6 & 5 & 1 \end{bmatrix} \begin{bmatrix} 3 & 2 & 5 \\ -3 & 4 & 7 \\ 6 & 5 & 1 \end{bmatrix}^{-1}$

Verify Exercises 47 and 48 without expanding the determinants.

47 $\begin{vmatrix} 2 & 4 & -6 \\ 1 & 4 & 3 \\ 2 & 2 & 0 \end{vmatrix} = 12 \begin{vmatrix} 1 & 1 & -1 \\ 1 & 2 & 1 \\ 2 & 1 & 0 \end{vmatrix}$

48 $\begin{vmatrix} a & b & c \\ d & e & f \\ g & h & k \end{vmatrix} = \begin{vmatrix} d & e & f \\ g & h & k \\ a & b & c \end{vmatrix}$

49 Suppose that $A = (a_{ij})$ is a square matrix of order n such that $a_{ij} = 0$ if $i < j$. Prove that

$$|A| = a_{11}a_{22}\ldots a_{nn}.$$

50 If $A = (a_{ij})$ is any 2×2 matrix such that $|A| \neq 0$, prove that A has an inverse and find a general formula for A^{-1}

SEQUENCES
AND SERIES

The method of proof called mathematical induction *considered in the first section of this chapter is very important in all branches of mathematics. In particular, in Section 9.2 it is used to prove the famous* Binomial Theorem. *Section 9.3 contains a discussion of* sequences *and* summation notation. *Of special interest to us are* arithmetic *and* geometric *sequences. These are discussed in the final two sections of the chapter.*

9.1 MATHEMATICAL INDUCTION

If n is a positive integer, let P_n denote the statement

$$(xy)^n = x^n y^n$$

where x and y are real numbers. Thus, P_1 represents the statement $(xy)^1 = x^1 y^1$, P_2 denotes $(xy)^2 = x^2 y^2$, P_3 is $(xy)^3 = x^3 y^3$, and so on. It is easy to show that P_1, P_2, and P_3 are *true* statements. However, since the set of positive integers is infinite, it is impossible to check the validity of P_n for every positive integer n. In order to give a proof, the method of mathematical induction is required. This method is based on the following fundamental axiom.

Axiom of Mathematical Induction

Suppose a set S of positive integers has the following two properties:

(i) S contains the integer 1.
(ii) Whenever S contains a positive integer k, S also contains $k + 1$.

Then S contains every positive integer.

If S is a set of positive integers satisfying property (ii), then whenever S contains an arbitrary positive integer k, it must also contain the next positive integer, $k + 1$. If S also satisfies property (i), then S contains 1 and hence by (ii), S contains $1 + 1$, or 2. Applying (ii) again, we see that S contains $2 + 1$, or 3. Once again, S must contain $3 + 1$, or 4. If we continue in this manner, it can be argued that if n is any *specific* positive integer, then n is in S, since we can proceed a step at a time as above, eventually reaching n. Although this argument does not *prove* the axiom, it certainly makes it plausible.

We shall use the preceding axiom to establish the following fundamental principle.

Principle of Mathematical Induction

> If with each positive integer n there is associated a statement P_n, then all the statements P_n are true provided the following two conditions hold:
>
> (i) P_1 is true.
> (ii) Whenever k is a positive integer such that P_k is true, then P_{k+1} is also true.

Proof

Assume that conditions (i) and (ii) of the Principle hold, and let S denote the set of all positive integers n such that P_n is true. By assumption, P_1 is true, and consequently, 1 is in S. Thus S satisfies property (i) of the Axiom of Mathematical Induction. Whenever S contains a positive integer k, then by the definition of S, P_k is true and hence from condition (ii) of the Principle, P_{k+1} is also true. This means that S contains $k + 1$. We have shown that whenever S contains a positive integer k, then S also contains $k + 1$. Consequently, property (ii) of the Axiom of Mathematical Induction is true, and hence S contains every positive integer; that is, P_n is true for every positive integer n.

There are other variations of the Principle of Mathematical Induction. One is stated later in this section. In most of our work the statement P_n will usually be given in the form of an equation involving the arbitrary positive integer n, as in our illustration $(xy)^n = x^n y^n$.

When applying the Principle of Mathematical Induction the following two steps should always be followed:

> Step (i)   Prove that P_1 is true.
> Step (ii) Assume that P_k is true and prove that P_{k+1} is true.

Step (ii) is usually the most confusing for the beginning student. We do not *prove* that P_k is true (except for $k = 1$). Instead, we show that *if* P_k is true, then the statement P_{k+1} is true. That is all that is necessary according to the Principle of Mathematical Induction. The assumption that P_k is true is referred to as the **induction hypothesis**.

Many interesting formulas about positive integers can be established by using mathematical induction, two of which are illustrated in Examples 1 and 2. Others appear in the Exercises.

Example 1   Prove that for every positive integer n, the sum of the first n positive integers is $n(n + 1)/2$.

Solution   If n is any positive integer, let P_n denote the statement

$$1 + 2 + 3 + \cdots + n = \frac{n(n+1)}{2}$$

where, by convention, when $n \leq 4$, the left side is adjusted so that there are precisely n terms in the sum. The following are some special cases of P_n: If $n = 2$, then P_2 is

$$1 + 2 = \frac{2(2+1)}{2}, \quad \text{or} \quad 3 = 3.$$

If $n = 3$, then P_3 is

$$1 + 2 + 3 = \frac{3(3+1)}{2}, \quad \text{or} \quad 6 = 6.$$

If $n = 5$, then P_5 is

$$1 + 2 + 3 + 4 + 5 = \frac{5(5+1)}{2}, \quad \text{or} \quad 15 = 15.$$

Although it is instructive to check the validity of P_n for several values of n as we did above, it is unnecessary to do so. We need only apply the two step process outlined prior to this example. Thus we proceed as follows.

Step (i).   If we substitute $n = 1$ in P_n, then, by convention, the left side collapses to 1 and the right side is $\frac{1(1+1)}{2}$, which also equals 1. This proves that P_1 is true.

Step (ii).   Assume that P_k is true. Thus the induction hypothesis is

$$1 + 2 + 3 + \cdots + k = \frac{k(k+1)}{2}.$$

Our goal is to prove that P_{k+1} is true, that is,

$$1 + 2 + 3 + \cdots + (k+1) = \frac{(k+1)[(k+1)+1]}{2}.$$

By the induction hypothesis we already have a formula for the sum of the first k positive integers. Hence a formula for the sum of the first $k + 1$ positive integers may be found simply by adding $(k + 1)$ to both sides. Doing so and simplifying, we obtain

$$
\begin{aligned}
1 + 2 + 3 + \cdots + k + (k + 1) &= \frac{k(k + 1)}{2} + (k + 1) \\
&= \frac{k(k + 1) + 2(k + 1)}{2} \\
&= \frac{k^2 + 3k + 2}{2} \\
&= \frac{(k + 1)(k + 2)}{2} \\
&= \frac{(k + 1)[(k + 1) + 1]}{2}.
\end{aligned}
$$

We have shown that P_{k+1} is true and, therefore, the proof by mathematical induction is complete.   ∎

Example 2   Prove that for each positive integer n,

$$
1^2 + 3^2 + \cdots + (2n - 1)^2 = \frac{n(2n - 1)(2n + 1)}{3}.
$$

Solution   For each positive integer n, let P_n denote the given statement. Note that this is a formula for the sum of the squares of the first n odd positive integers. We again follow the two-step procedure used in Example 1.

Step (i). Substituting 1 for n in P_n, we obtain

$$
1^2 = \frac{(1)(2 - 1)(2 + 1)}{3} = \frac{3}{3} = 1
$$

which shows that P_1 is true.

Step (ii). Assume that P_k is true. Thus the induction hypothesis is

$$
1^2 + 3^2 + \cdots + (2k - 1)^2 = \frac{k(2k - 1)(2k + 1)}{3}.
$$

We wish to prove that P_{k+1} is true, that is,

$$
\begin{aligned}
1^2 + 3^2 + \cdots &+ [2(k + 1) - 1]^2 \\
&= \frac{(k + 1)[2(k + 1) - 1][2(k + 1) + 1]}{3}.
\end{aligned}
$$

This equation for P_{k+1} simplifies to

$$
1^2 + 3^2 + \cdots + (2k + 1)^2 = \frac{(k + 1)(2k + 1)(2k + 3)}{3}.
$$

Observe that the second from the last term on the left-hand side of this equation is $(2k - 1)^2$. (Why?) In a manner similar to the solution of Example 1, we may obtain the left side of P_{k+1} by adding $(2k + 1)^2$ to both sides of the equation stated in the induction hypothesis. This gives us

$$1^2 + 3^2 + \cdots + (2k - 1)^2 + (2k + 1)^2$$
$$= \frac{k(2k - 1)(2k + 1)}{3} + (2k + 1)^2.$$

We leave it as an exercise for the reader to show that the right side of the preceding equation may be written in the form of the right side of P_{k+1}. This proves that P_{k+1} is true, and hence P_n is true for every n. ■

The Laws of Exponents can be proved by mathematical induction. We shall use the following definition of exponents.

Definition of Exponents

> If x is any real number, then
>
> (i) $x^1 = x$
> (ii) whenever k is a positive integer for which x^k is defined, let $x^{k+1} = x^k \cdot x$.

A definition of this type is called a **recursive definition**. In general, if a concept is defined for every positive integer n in such a way that the case corresponding to $n = 1$ is given, and if it is also stated how any case after the first is obtained from the preceding one, then the definition is a recursive definition. For example, by (i) of the definition we have $x^1 = x$. Next, applying (ii) of the definition we obtain

$$x^2 = x^{1+1} = x^1 \cdot x = x \cdot x.$$

Since x^2 is now defined, we may employ (ii) again (with $k = 2$), obtaining

$$x^3 = x^{2+1} = x^2 \cdot x = (x \cdot x) \cdot x.$$

This defines x^3, and hence (ii) of the definition may be used again to obtain x^4. Thus

$$x^4 = x^{3+1} = x^3 \cdot x = [(x \cdot x) \cdot x] \cdot x.$$

Observe that this agrees with the formulation of x^n as a product of x by itself n times. It can be shown by mathematical induction that x^n is defined for every positive integer n.

Example 3 If x is a real number, prove that $x^m \cdot x^n = x^{m+n}$ for all positive integers m and n.

Solution Let m be an arbitrary positive integer. For each positive integer n, let P_n denote the statement

$$x^m \cdot x^n = x^{m+n}.$$

We shall employ the two step process used in previous examples to prove that P_n is true for every positive integer n.

Step (i).   To show that P_1 is true we may use (i) and (ii) of the Definition of Exponents as follows:

$$x^m \cdot x^1 = x^m \cdot x$$
$$= x^{m+1}$$

which is the statement P_n with $n = 1$. Hence P_1 is true.

Step (ii).   Assume that P_k is true. Thus the induction hypothesis is

$$x^m \cdot x^k = x^{m+k}.$$

We wish to prove that P_{k+1} is true, that is,

$$x^m \cdot x^{k+1} = x^{m+(k+1)}.$$

The proof may be arranged as follows, where reasons are stated to the right of each step.

$$
\begin{aligned}
x^m \cdot x^{k+1} &= x^m \cdot (x^k \cdot x) && \text{((ii) of the definition of exponents)} \\
&= (x^m \cdot x^k) \cdot x && \text{(associative law in } \mathbb{R}) \\
&= x^{m+k} \cdot x && \text{(induction hypothesis)} \\
&= x^{(m+k)+1} && \text{((ii) of the definition of exponents)} \\
&= x^{m+(k+1)} && \text{(associative law for integers).}
\end{aligned}
$$

This completes the proof.   ■

Consider a positive integer j and suppose that with each integer $n \geq j$ there is associated a statement P_n. For example, if $j = 6$, then the statements are numbered $P_6, P_7, P_8, \ldots$. The principle of mathematical induction may be extended to cover this situation. Just as before, two steps are used. Specifically, to prove that the statements S_n are true for $n \geq j$, we use the following two steps.

Extended Principle of Mathematical Induction

(i′)   Prove that S_j is true.

(ii′)   Assume that S_k is true for $k \geq j$ and prove that S_{k+1} is true.

Example 4   Let a be a nonzero real number such that $a > -1$. Prove that $(1 + a)^n > 1 + na$ for every integer $n \geq 2$.

Solution   For each positive integer n, let P_n denote the inequality $(1 + a)^n > 1 + na$. Note that P_1 is *false*, since $(1 + a)^1 = 1 + (1)(a)$. However, we can show that P_n is true for $n \geq 2$ by using the Extended Principle with $j = 2$.

Step (i′). We first note that $(1 + a)^2 = 1 + 2a + a^2$. Since $a \neq 0$, we have $a^2 > 0$ and therefore $1 + 2a + a^2 > 1 + 2a$. This gives us $(1 + a)^2 > 1 + 2a$, and hence P_2 is true.

Step (ii′). Assume that P_k is true. Thus the induction hypothesis is

$$(1 + a)^k > 1 + ka.$$

We wish to show that P_{k+1} is true, that is,

$$(1 + a)^{k+1} > 1 + (k + 1)a.$$

Since $a > -1$, we have $a + 1 > 0$, and hence multiplying both sides of the induction hypothesis by $1 + a$ will not change the inequality sign. Consequently,

$$(1 + a)^k(1 + a) > (1 + ka)(1 + a)$$

which may be rewritten as

$$(1 + a)^{k+1} > 1 + ka + a + ka^2$$

or as

$$(1 + a)^{k+1} > 1 + (k + 1)a + ka^2.$$

Since $ka^2 > 0$, we have

$$1 + (k + 1)a + ka^2 > 1 + (k + 1)a$$

and therefore,

$$(1 + a)^{k+1} > 1 + (k + 1)a.$$

Thus, P_{k+1} is true and the proof is complete. ■

EXERCISES 9.1

In each of Exercises 1–18, prove that the given formula is true for every positive integer n.

1 $2 + 4 + 6 + \cdots + 2n = n(n + 1)$

2 $1 + 4 + 7 + \cdots + (3n - 2) = \dfrac{n(3n - 1)}{2}$

3 $1 + 3 + 5 + \cdots + (2n - 1) = n^2$

4 $3 + 9 + 15 + \cdots + (6n - 3) = 3n^2$

5 $2 + 7 + 12 + \cdots + (5n - 3) = \dfrac{n}{2}(5n - 1)$

6 $2 + 6 + 18 + \cdots + 2 \cdot 3^{n-1} = 3^n - 1$

7 $1 + 2 \cdot 2 + 3 \cdot 2^2 + 4 \cdot 2^3 + \cdots + n \cdot 2^{n-1}$
$= 1 + (n - 1) \cdot 2^n$

8 $(-1)^1 + (-1)^2 + (-1)^3 + \cdots + (-1)^n$
$= \dfrac{(-1)^n - 1}{2}$

9 $1^2 + 2^2 + 3^2 + \cdots + n^2 = \dfrac{n(n + 1)(2n + 1)}{6}$

10 $1^3 + 2^3 + 3^3 + \cdots + n^3 = \left[\dfrac{n(n + 1)}{2}\right]^2$

11 $\dfrac{1}{1 \cdot 2} + \dfrac{1}{2 \cdot 3} + \dfrac{1}{3 \cdot 4} + \cdots + \dfrac{1}{n(n + 1)} = \dfrac{n}{n + 1}$

12 $\dfrac{1}{1\cdot 2\cdot 3}+\dfrac{1}{2\cdot 3\cdot 4}+\dfrac{1}{3\cdot 4\cdot 5}+\cdots$

$\quad+\dfrac{1}{n(n+1)(n+2)}=\dfrac{n(n+3)}{4(n+1)(n+2)}$

13 $3+3^2+3^3+\cdots+3^n=\frac{3}{2}(3^n-1)$

14 $1^3+3^3+5^3+\cdots+(2n-1)^3=n^2(2n^2-1)$

15 $n<2^n$ **16** $1+2n\le 3^n$

17 $1+2+3+\cdots+n<\frac{1}{8}(2n+1)^2$

18 If $0<a<b$, then $\left(\dfrac{a}{b}\right)^{n+1}<\left(\dfrac{a}{b}\right)^{n}$.

Prove that the statements in Exercises 19–22 are true for every positive integer n.

19 3 is a factor of n^3-n+3.

20 2 is a factor of n^2+n.

21 4 is a factor of 5^n-1.

22 9 is a factor of $10^{n+1}+3\cdot 10^n+5$.

23 Use mathematical induction to prove that if a is any real number greater than 1, then $a^n>1$ for every positive integer n.

24 If $a\ne 1$, prove that

$$1+a+a^2+\cdots+a^{n-1}=\dfrac{a^n-1}{a-1}$$

for every positive integer n.

25 If a and b are real numbers, use mathematical induction to prove that $(ab)^n=a^nb^n$ for every positive integer n.

26 If a is a real number, prove that $(a^m)^n=a^{mn}$ for all positive integers m and n.

27 Use mathematical induction to prove that $a-b$ is a factor of a^n-b^n for every positive integer n. (*Hint*: $a^{k+1}-b^{k+1}=a^k(a-b)+(a^k-b^k)b$.)

28 Prove that $a+b$ is a factor of $a^{2n-1}+b^{2n-1}$ for every positive integer n.

29 If z is a complex number and $\bar z$ is its conjugate, prove that $\overline{z^n}=\bar z^n$ for every positive integer n.

30 Prove that for every positive integer n, if $z_1,z_2,\ldots,z_n$ are complex numbers, then $\overline{z_1z_2\cdots z_n}=\bar z_1\bar z_2\cdots\bar z_n$.

31 Prove that

$$\log(a_1a_2\cdots a_n)=\log a_1+\log a_2+\cdots+\log a_n$$

for all $n\ge 2$, where each a_i is a positive real number.

32 Prove the **Generalized Distributive Law**

$$a(b_1+b_2+\cdots+b_n)=ab_1+ab_2+\cdots+ab_n$$

for all $n\ge 2$, where a and each b_i are real numbers.

33 Prove that

$$a+ar+ar^2+\cdots+ar^{n-1}=\dfrac{a(1-r^n)}{1-r}$$

where n is any positive integer and a and r are real numbers with $r\ne 1$.

34 Prove that

$$a+(a+d)+(a+2d)+\cdots+[a+(n-1)d]$$
$$=(n/2)[2a+(n-1)d]$$

where n is any positive integer and a and d are real numbers.

35 If a and b are real numbers and n is any positive integer, prove that

$$(a-b)(a^{n-1}+a^{n-2}b+\cdots+ab^{n-2}+b^{n-1})$$
$$=a^n-b^n.$$

36 Use mathematical induction to prove De Moivre's Theorem:

$$[r(\cos\theta+i\sin\theta)]^n=r^n[\cos n\theta+i\sin n\theta]$$

for every positive integer n.

37 Prove that for every positive integer $n\ge 3$, the sum of the interior angles of a polygon of n sides is $(n-2)\cdot 180°$.

38 Prove, by mathematical induction, that if a finite set S consists of n elements, then the number of subsets of S is 2^n.

Prove that the formulas in Exercises 39 and 40 are true for every positive integer n.

39 $\sin(\theta+n\pi)=(-1)^n\sin\theta$

40 $\cos(\theta+n\pi)=(-1)^n\cos\theta$

9.2 THE BINOMIAL THEOREM

It is often necessary to work with expressions of the form $(a + b)^n$, where a and b are mathematical expressions of some type and n is a large positive integer. There exists a general formula for *expanding* $(a + b)^n$, that is, for expressing it as a sum. The theorem which gives us the formula is called the **Binomial Theorem**. In order to obtain the formula let us first consider cases where n is a small positive integer. By examining the patterns which emerge, we shall make an educated guess about the nature of the general formula. Finally, we shall prove by means of mathematical induction that our guess is correct.

If we actually perform the multiplications, the following expansions of $(a + b)^n$ are obtained for the cases $n = 2$, 3, 4, and 5:

$$(a + b)^2 = a^2 + 2ab + b^2$$
$$(a + b)^3 = a^3 + 3a^2b + 3ab^2 + b^3$$
$$(a + b)^4 = a^4 + 4a^3b + 6a^2b^2 + 4ab^3 + b^4$$
$$(a + b)^5 = a^5 + 5a^4b + 10a^3b^2 + 10a^2b^3 + 5ab^4 + b^5.$$

Let us now make some observations regarding the above expansions of $(a + b)^n$. We see that there are always $n + 1$ terms, the first being a^n and the last b^n. Each intermediate term contains a product of the form a^ib^j, where $i + j = n$. Moreover, as we move from one term to the next, the exponent associated with a decreases by 1 whereas the exponent associated with b increases by 1. The pattern for the coefficients is rather interesting. If we start at one end of the expansion and consider successive terms, the coefficients match those obtained by starting at the other end of the expansion and proceeding in the reverse direction.

It requires some ingenuity to find a formula for the general term in the expansion. The second term of the preceding special cases for $(a + b)^n$ is always $na^{n-1}b$. The third term is seen to be

$$\frac{n(n - 1)}{2} a^{n-2}b^2.$$

If there are more than three terms, the fourth term is

$$\frac{n(n - 1)(n - 2)}{3 \cdot 2} a^{n-3}b^3.$$

It appears that for each term, *if we take the product of the coefficient and the exponent of a and then divide by the number of the term, we obtain the coefficient of the next term* in the expansion. For example, applying this rule to the general fourth term displayed above gives us the following fifth term (if it exists):

$$\frac{n(n - 1)(n - 2)(n - 3)}{4 \cdot 3 \cdot 2} a^{n-4}b^4.$$

The denominators in these terms may be abbreviated by employing the **factorial notation**. If n is any positive integer, then the symbol $n!$ (read "n factorial") is defined by

$$n! = n(n - 1)(n - 2) \cdots 1$$

where there are n factors on the right side. As special cases we have

$$1! = 1, \quad 2! = 2 \cdot 1, \quad 3! = 3 \cdot 2 \cdot 1 = 6, \quad 4! = 4 \cdot 3 \cdot 2 \cdot 1 = 24, \quad 5! = 5 \cdot 4 \cdot 3 \cdot 2 \cdot 1 = 120.$$

To ensure that certain formulas will be true for all *nonnegative* integers, we define $0! = 1$. The factorial notation may also be defined recursively by writing $1! = 1$ and, for any positive integer k, $(k + 1)! = (k + 1)k!$.

If we compare the coefficients with the exponents in the general fourth and fifth terms of $(a + b)^n$, we are led to believe that for a positive integer r, the $(r + 1)$st term in the expansion of $(a + b)^n$ is given by

$$\frac{n(n - 1)(n - 2) \cdots (n - r + 1)}{r!} a^{n-r} b^r.$$

Note that if $r = n$, then the coefficient reduces to $n!/n!$ and we obtain b^n, the last term in the expansion. If $r = n - 1$, then we obtain nab^{n-1}, which is the second from the last term. Hence we *conjecture* that the following formula is true for every positive integer n and all real or complex numbers a and b.

The Binomial Theorem

$$(a + b)^n = a^n + na^{n-1}b + \frac{n(n - 1)}{2!} a^{n-2} b^2 + \cdots$$

$$+ \frac{n(n - 1)(n - 2) \cdots (n - r + 1)}{r!} a^{n-r} b^r$$

$$+ \cdots + nab^{n-1} + b^n$$

Proof

We shall use mathematical induction as follows. For each positive integer n, let P_n denote the statement given in the Binomial Theorem.

Step (i). If $n = 1$, the statement reduces to $(a + b)^1 = a^1 + b^1$. Consequently, P_1 is true.

Step (ii). Assume that P_k is true. Thus the induction hypothesis is

$$(a + b)^k = a^k + ka^{k-1}b + \frac{k(k - 1)}{2!} a^{k-2} b^2 + \cdots$$

$$+ \frac{k(k - 1)(k - 2) \cdots (k - r + 2)}{(r - 1)!} a^{k-r+1} b^{r-1}$$

$$+ \frac{k(k - 1)(k - 2) \cdots (k - r + 1)}{r!} a^{k-r} b^r$$

$$+ \cdots + kab^{k-1} + b^k$$

where we have shown both the rth and the $(r + 1)$st terms in the expansion.

If we multiply both sides of the last equation by $(a + b)$, we obtain

$$(a + b)^{k+1} = \left[a^{k+1} + ka^k b + \frac{k(k-1)}{2!} a^{k-1} b^2 + \cdots \right.$$

$$+ \frac{k(k-1)\cdots(k-r+1)}{r!} a^{k-r+1} b^r + \cdots + ab^k \right]$$

$$+ \left[a^k b + ka^{k-1} b^2 + \cdots + \frac{k(k-1)\cdots(k-r+2)}{(r-1)!} a^{k-r+1} b^r \right.$$

$$\left. + \cdots + kab^k + b^{k+1} \right]$$

where the terms in the first pair of brackets result from multiplying the right side of the induction hypothesis by a and the terms in the second pair of brackets result from multiplying by b. Rearranging and combining terms, we have

$$(a + b)^{k+1} = a^{k+1} + (k+1)a^k b + \left[\frac{k(k-1)}{2!} + k \right] a^{k-1} b^2 + \cdots$$

$$+ \left[\frac{k(k-1)\cdots(k-r+1)}{r!} + \frac{k(k-1)\cdots(k-r+2)}{(r-1)!} \right] a^{k-r+1} b^r$$

$$+ \cdots + (1+k)ab^k + b^{k+1}.$$

It is left to the reader to show that if the coefficients are simplified, then we obtain statement P_n with $k + 1$ substituted for n. Thus, P_{k+1} is true and, therefore, P_n holds for every positive integer n.

Example 1   Find the binomial expansion of $(2x + 3y^2)^4$.

Solution   Using the Binomial Theorem with $a = 2x$, $b = 3y^2$, and $n = 4$, we obtain

$$(2x + 3y^2)^4 = (2x)^4 + 4(2x)^3(3y^2) + \frac{4\cdot 3}{2!}(2x)^2(3y^2)^2$$

$$+ \frac{4\cdot 3\cdot 2}{3!}(2x)(3y^2)^3 + \frac{4\cdot 3\cdot 2\cdot 1}{4!}(3y^2)^4.$$

This simplifies to

$$(2x + 3y^2)^4 = 16x^4 + 96x^3 y^2 + 216x^2 y^4 + 216xy^6 + 81y^8. \qquad \blacksquare$$

The symbol $\binom{n}{r}$ is often used to denote the coefficient of $a^{n-r}b^r$ in the Binomial

Theorem, that is,

$$\binom{n}{r} = \frac{n(n-1)\cdots(n-r+1)}{r!}.$$

An equivalent formula is

$$\binom{n}{r} = \frac{n!}{r!(n-r)!}$$

where r may be assigned any integral value between 0 and n. If we substitute n for r in the first of these formulas we obtain

$$\binom{n}{n} = \frac{n!}{n!} = 1.$$

It is also convenient to *define*

$$\binom{n}{0} = 1.$$

The formulas for $\binom{n}{r}$ may then be used if r is any integer such that $0 \le r \le n$. Employing this notation gives us the following.

The Binomial Theorem (Alternate Form)

$$(a+b)^n = \binom{n}{0}a^n b^0 + \binom{n}{1}a^{n-1}b + \binom{n}{2}a^{n-2}b^2 + \cdots + \binom{n}{r}a^{n-r}b^r$$
$$+ \cdots + \binom{n}{n-1}a^{n-(n-1)}b^{n-1} + \binom{n}{n}a^0 b^n$$

The numbers $\binom{n}{r}$ are called **binomial coefficients**. As a special case, if $n = 4$,

$$(a+b)^4 = \binom{4}{0}a^4 b^0 + \binom{4}{1}a^3 b + \binom{4}{2}a^2 b^2 + \binom{4}{3}ab^3 + \binom{4}{4}a^0 b^4.$$

It is easy to show that this reduces to the expansion of $(a+b)^4$ given at the beginning of this section.

The next example illustrates the fact that if one of a or b is negative, then the terms of the expansion are alternately positive and negative.

Example 2 Expand $\left(\dfrac{1}{x} - 2\sqrt{x}\right)^5$.

Solution Letting $a = 1/x$, $b = -2\sqrt{x}$, and $n = 5$ in the Binomial Theorem, we obtain

$$\left(\frac{1}{x} - 2\sqrt{x}\right)^5 = \left(\frac{1}{x}\right)^5 + 5\left(\frac{1}{x}\right)^4(-2\sqrt{x}) + \frac{5\cdot 4}{2!}\left(\frac{1}{x}\right)^3(-2\sqrt{x})^2$$

$$+ \frac{5\cdot 4\cdot 3}{3!}\left(\frac{1}{x}\right)^2(-2\sqrt{x})^3 + \frac{5\cdot 4\cdot 3\cdot 2}{4!}\left(\frac{1}{x}\right)(-2\sqrt{x})^4$$

$$+ \frac{5\cdot 4\cdot 3\cdot 2\cdot 1}{5!}(-2\sqrt{x})^5.$$

This simplifies to

$$\left(\frac{1}{x} - 2\sqrt{x}\right)^5 = \frac{1}{x^5} - \frac{10}{x^{7/2}} + \frac{40}{x^2} - \frac{80}{x^{1/2}} + 80x - 32x^{5/2}. \qquad \blacksquare$$

For certain problems it is only required to find a specific term in the expansion of $(a + b)^n$. To work such problems we first find the exponent r that is to be assigned to b. Notice that by the Binomial Theorem, *the exponent of b is always one less than the number of the term.* Once r is found, the exponent of a is $n - r$. Referring to the form of the $(r + 1)$st term, we see that the coefficient of the term involving $a^{n-r}b^r$ is of the form $p/r!$, *where p is the product of r factors and where the first factor is n and each factor is one less than the preceding factor.*

Example 3 Find the fifth term in the expansion of $(4x^2 + 3/y)^{13}$.

Solution We let $a = 4x^2$ and $b = 3/y$. The exponent of b in the fifth term is 4 and hence the exponent of a is 9. From the discussion of the preceding paragraph we obtain

$$\frac{13\cdot 12\cdot 11\cdot 10}{4!}(4x^2)^9(3/y)^4. \qquad \blacksquare$$

Finally, there is an interesting triangular array of numbers, called **Pascal's Triangle,** which can be used to obtain the binomial coefficients. The numbers are arranged as follows.

$$
\begin{array}{ccccccccccccc}
 & & & & & & 1 & & & & & & \\
 & & & & & 1 & & 1 & & & & & \\
 & & & & 1 & & 2 & & 1 & & & & \\
 & & & 1 & & 3 & & 3 & & 1 & & & \\
 & & 1 & & 4 & & 6 & & 4 & & 1 & & \\
 & 1 & & 5 & & 10 & & 10 & & 5 & & 1 & \\
1 & & 6 & & 15 & & 20 & & 15 & & 6 & & 1
\end{array}
$$

The numbers in the second row are the coefficients in the expansion of $(a + b)^1$; those in the third row are the coefficients determined by $(a + b)^2$; those in the fourth row are obtained from $(a + b)^3$, and so on. Each number in the array which is different from 1 can be found by adding the two numbers in the previous row which appear above and immediately to the left and right of the number.

EXERCISES 9.2

In each of Exercises 1–12, expand and simplify the given expression.

1 $(a + b)^6$

2 $(a + b)^7$

3 $(a - b)^8$

4 $(a - b)^9$

5 $(3x - 5y)^4$

6 $(2t - s)^5$

7 $(u^2 + 4v)^5$

8 $(\frac{1}{2}c + d^3)^4$

9 $(r^{-2} - 2r)^6$

10 $(x^{1/2} - y^{-1/2})^6$

11 $(1 + x)^{10}$

12 $(1 - x)^{10}$

13 Find the first four terms in the binomial expansion of $(3c^{2/5} + c^{4/5})^{25}$.

14 Find the first three terms and the last three terms in the binomial expansion of $(x^3 + 5x^{-2})^{20}$.

15 Find the last two terms in the expansion of $(4b^{-1} - 3b)^{15}$.

16 Find the last three terms in the expansion of $(s - 2t^3)^{12}$.

Solve Exercises 17–28 without expanding completely.

17 Find the fifth term in the expansion of $(3a^2 + \sqrt{b})^9$.

18 Find the sixth term in the expansion of $\left(\frac{2}{c} + \frac{c^2}{3}\right)^7$.

19 Find the seventh term in the expansion of $(\frac{1}{2}u - 2v)^{10}$.

20 Find the fourth term in the expansion of $(2x^3 - y^2)^6$.

21 Find the middle term in the expansion of $(x^{1/3} + y^{1/3})^{12}$.

22 Find the two middle terms in the expansion of $(rs + t)^7$.

23 Find the term which does not contain x in the expansion of $\left(6x - \dfrac{1}{2x}\right)^{10}$.

24 Find the term involving x^8 in the expansion of $(y + 3x^2)^6$.

25 Find the term containing y^6 in the expansion of $(x - 2y^3)^4$.

26 Find the term containing b^9 in the expansion of $(5a + 2b^3)^4$.

27 Find the term containing c^3 in the expansion of $(\sqrt{c} + \sqrt{d})^{10}$.

28 In the expansion of $(xy - 2y^{-3})^8$ find the term which does not contain y.

29 Use the first four terms in the binomial expansion of $(1 + 0.02)^{10}$ to approximate $(1.02)^{10}$. Also approximate $(1.02)^{10}$ by using logarithms and then compare answers.

30 Use the first four terms in the binomial expansion of $(1 - 0.01)^4$ to approximate $(0.99)^4$. Compare with the answer obtained by approximating $(0.99)^4$ through the use of logarithms.

9.3 INFINITE SEQUENCES AND SUMMATION NOTATION

A function f from a set X to a set Y is a correspondence that associates with each element x of X a unique element $f(x)$ of Y. Up to now the domain X has usually been an interval of real numbers. In this section we shall consider a different class of functions.

Definition

> An **infinite sequence** is a function whose domain is the set of positive integers.

For convenience we sometimes refer to infinite sequences merely as *sequences*. In this book the range of an infinite sequence will be a set of real numbers.

If f is an infinite sequence, then to each positive integer n there corresponds a real number $f(n)$. These numbers in the range of f may be represented by writing

$$f(1), f(2), f(3), \ldots, f(n), \ldots$$

where the dots at the end indicate that the sequence does not terminate. The number $f(1)$ is called the **first term** of the sequence, $f(2)$ the **second term** and, in general, $f(n)$ the **nth term** of the sequence. It is customary to use a subscript notation instead of the functional notation and write these numbers as

$$a_1, a_2, a_3, \ldots, a_n, \ldots$$

where it is understood that for each positive integer n, the symbol a_n denotes the real number $f(n)$. In this way we obtain an infinite collection of real numbers which is *ordered* in the sense that there is a first number, a second number, a forty-fifth number, and so on. Although sequences are functions, an ordered collection of the type displayed above will also be referred to as an infinite sequence. If we wish to convert the collection to a function f we let $f(n) = a_n$ for all positive integers n.

From the definition of equality of functions we see that a sequence

$$a_1, a_2, a_3, \ldots, a_n, \ldots$$

is **equal** to a sequence $\qquad b_1, b_2, b_3, \ldots, b_n, \ldots$

if and only if $a_i = b_i$ for every positive integer i. Infinite sequences are often defined by stating a formula for the nth term, as in the following example.

Example 1   List the first four terms and the tenth term of the sequence whose nth term is as follows.

(a) $a_n = \dfrac{n}{n+1}$

(b) $a_n = 2 + (0.1)^n$

(c) $a_n = (-1)^{n+1} \dfrac{n^2}{3n-1}$

(d) $a_n = 4$

Solutions   To find the first four terms we substitute, successively, $n = 1, 2, 3$, and 4 in the formula for a_n. The tenth term is found by susbstituting 10 for n. Doing this and simplifying gives us the following:

First four terms	*Tenth term*
(a) $\frac{1}{2}, \frac{2}{3}, \frac{3}{4}, \frac{4}{5}$	$\frac{10}{11}$
(b) $2.1, 2.01, 2.001, 2.0001$	2.0000000001
(c) $\frac{1}{2}, -\frac{4}{5}, \frac{9}{8}, -\frac{16}{11}$	$-\frac{100}{29}$
(d) $4, 4, 4, 4$	4 ∎

It is not essential that a formula for a_n be given. Indeed, sometimes that is impossible, as illustrated in the next example.

Example 2   List the first seven terms of the sequence whose nth term a_n is the nth positive prime number.

Solution   The first seven primes are 2, 3, 5, 7, 11, 13, and 17. Hence $a_1 = 2, a_2 = 3$, $a_3 = 5, a_4 = 7, a_5 = 11, a_6 = 13$, and $a_7 = 17$. No one has yet found a formula which yields the nth prime number. Our sequence is perfectly legitimate, however, since a_n is uniquely determined for every positive integer n. ∎

An infinite sequence is sometimes defined by stating a recursive definition for the terms, as in the following example.

Example 3   Find the first four terms and the nth term of the infinite sequence defined by

$$a_1 = 3 \quad \text{and} \quad a_{k+1} = 2a_k, \quad \text{for} \quad k \geq 1.$$

Solution   The sequence is defined recursively since the first term is given and, moreover, whenever a term a_k of the sequence is known, then the next term a_{k+1} can be found. Thus,

$$a_1 = 3$$
$$a_2 = 2a_1 = 2 \cdot 3 = 6$$
$$a_3 = 2a_2 = 2 \cdot 2 \cdot 3 = 2^2 \cdot 3 = 12$$
$$a_4 = 2a_3 = 2 \cdot 2 \cdot 2 \cdot 3 = 2^3 \cdot 3 = 24.$$

We have written the terms as products, so as to gain some insight into the nature of the nth term. Continuing, we obtain $a_5 = 2^4 \cdot 3$ and $a_6 = 2^5 \cdot 3$; and it appears that

$$a_n = 2^{n-1} \cdot 3$$

for every positive integer n. We shall prove that this guess is correct by means of mathematical induction. If we let P_n denote the statement $a_n = 2^{n-1} \cdot 3$, then P_1 is true since $a_1 = 2^0 \cdot 3 = 3$. Next, *assume* that P_k is true, that is, $a_k = 2^{k-1} \cdot 3$. We then have

$$
\begin{aligned}
a_{k+1} &= 2a_k & \text{(definition of } a_{k+1}) \\
&= 2 \cdot 2^{k-1} \cdot 3 & \text{(induction hypothesis)} \\
&= 2^k \cdot 3 & \text{(a law of exponents)} \\
&= 2^{(k+1)-1} \cdot 3 & \text{(Why?)}
\end{aligned}
$$

which shows that P_{k+1} is true. Hence, $a_n = 2^{n-1} \cdot 3$ for every positive integer n. ∎

It is important to observe that if only the first few terms of an infinite sequence are known, then it is impossible to predict additional terms. For example, if we were given $3, 6, 9, \ldots$ and asked to find the fourth term, we could not proceed without further information. The infinite sequence with nth term

$$a_n = 3n + (1-n)^3 (2-n)^2 (3-n)$$

has for its first four terms 3, 6, 9, and 120. It is possible to describe sequences where the first three terms are 3, 6, and 9 and the fourth term is *any* given number. This shows that when we work with infinite sequences it is essential to have specific information about the nth term or to know a general scheme for obtaining each term from the preceding one.

It is often desirable to find the sum of many terms of an infinite sequence. For ease in expressing such sums we use the **summation notation** described below.

Given an infinite sequence

$$a_1, a_2, a_3, \ldots, a_n, \ldots$$

the symbol $\sum_{i=1}^{m} a_i$ represents the sum of the first m terms, that is,

$$\sum_{i=1}^{m} a_i = a_1 + a_2 + a_3 + \cdots + a_m.$$

The Greek capital letter Σ (sigma) indicates a sum and the symbol a_i represents the ith term. The letter i is called the **index of summation** or the **summation variable**, and the numbers 1 and m indicate the extreme values of the summation variable.

Example 4 Find $\sum_{i=1}^{4} i^2 (i-3)$.

Solution In this case, $a_i = i^2 (i-3)$. To find the indicated sum we merely substitute, in succession, the integers 1, 2, 3 and 4 for i and add the resulting terms. Thus,

$$\sum_{i=1}^{4} i^2 (i-3) = 1^2 (1-3) + 2^2 (2-3) + 3^2 (3-3) + 4^2 (4-3)$$

$$= (-2) + (-4) + 0 + 16 = 10.$$ ∎

The letter used for the summation variable is arbitrary. To illustrate, if we use j for the summation variable, then

$$\sum_{j=1}^{m} a_j = a_1 + a_2 + a_3 + \cdots + a_m$$

which is the same as $\sum_{i=1}^{m} a_i$. Other symbols can be used similarly. As a numerical example, the sum in Example 4 can be written as

$$\sum_{k=1}^{4} k^2(k-3).$$

If n is a positive integer, then the sum of the first n terms of an infinite sequence will be denoted by S_n. For example, given $a_1, a_2, a_3, \ldots, a_n, \ldots,$

$$S_1 = a_1$$
$$S_2 = a_1 + a_2$$
$$S_3 = a_1 + a_2 + a_3$$
$$S_4 = a_1 + a_2 + a_3 + a_4$$

and, in general,

$$S_n = \sum_{i=1}^{n} a_i = a_1 + a_2 + \cdots + a_n.$$

The number S_n is called the **nth partial sum** of the sequence $a_1, a_2, a_3, \ldots, a_n, \ldots,$ and the infinite sequence

$$S_1, S_2, S_3, \ldots, S_n, \ldots$$

is called a **sequence of partial sums**. Sequences of partial sums are very important in calculus, where the concept of *infinite series* is introduced. We shall discuss some special types of infinite series in Section 9.5.

Example 5   Find the first four terms and the nth term of the sequence of partial sums associated with the sequence $1, 2, 3, \ldots, n, \ldots$ of positive integers.

Solution   The first four terms of the partial sum sequence are

$$S_1 = 1$$
$$S_2 = 1 + 2 = 3$$
$$S_3 = 1 + 2 + 3 = 6$$
$$S_4 = 1 + 2 + 3 + 4 = 10.$$

From Example 1 of Section 9.1 we see that

$$S_n = 1 + 2 + 3 + \cdots + n = \frac{n(n+1)}{2}.$$ ∎

If a_i is the same for all positive integers i, say $a_i = c$, where c is a real number, then

$$\sum_{i=1}^{n} a_i = a_1 + a_2 + a_3 + \cdots + a_n$$

$$= c + c + c + \cdots + c$$

$$= nc$$

that is,

$$\boxed{\sum_{i=1}^{n} c = nc.}$$

This formula can also be established by mathematical induction (see Exercise 60).

The domain of the summation variable does not have to begin at 1. For example, the following is self-explanatory:

$$\sum_{i=4}^{8} a_i = a_4 + a_5 + a_6 + a_7 + a_8.$$

As another variation, if the first term of an infinite sequence is a_0, as in

$$a_0, a_1, a_2, \ldots, a_n, \ldots$$

then sums of the form

$$\sum_{i=0}^{n} a_i = a_0 + a_1 + a_2 + \cdots + a_n$$

may be considered. Note that this is the sum of the first $n + 1$ terms of the given sequence.

Example 6 Find $\sum_{i=0}^{3} \dfrac{2^i}{(i+1)}$.

Solution

$$\sum_{i=0}^{3} \frac{2^i}{(i+1)} = \frac{2^0}{(0+1)} + \frac{2^1}{(1+1)} + \frac{2^2}{(2+1)} + \frac{2^3}{(3+1)}$$

$$= 1 + 1 + \frac{4}{3} + 2$$

$$= \frac{16}{3}.$$ ∎

Summation notation can be used to denote polynomials compactly. For example, in place of

$$f(x) = a_0 + a_1 x + a_2 x^2 + \cdots + a_n x^n$$

we may write

$$f(x) = \sum_{i=0}^{n} a_i x^i.$$

As another illustration, the rather cumbersome formula for the Binomial Theorem can be written as

$$(a + b)^n = \sum_{r=0}^{n} \binom{n}{r} a^{n-r} b^r.$$

The following theorem concerning sums has many uses in advanced courses in mathematics.

Theorem on Sums

If $a_1, a_2, \ldots, a_n, \ldots$ and $b_1, b_2, \ldots, b_n, \ldots$ are infinite sequences, then for every positive integer n,

(i) $\displaystyle\sum_{i=1}^{n} (a_i + b_i) = \sum_{i=1}^{n} a_i + \sum_{i=1}^{n} b_i$

(ii) $\displaystyle\sum_{i=1}^{n} (a_i - b_i) = \sum_{i=1}^{n} a_i - \sum_{i=1}^{n} b_i$

(iii) $\displaystyle\sum_{i=1}^{n} ca_i = c\left(\sum_{i=1}^{n} a_i\right)$, for every number c.

Proof

Although the theorem can be proved by mathematical induction, we shall use an argument that makes the truth of the formulas transparent. We begin as follows:

$$\sum_{i=1}^{n} (a_i + b_i) = (a_1 + b_1) + (a_2 + b_2) + (a_3 + b_3) + \cdots + (a_n + b_n).$$

Using the Commutative and Associative Properties many times, we may rearrange the terms on the right to produce

$$\sum_{i=1}^{n} (a_i + b_i) = (a_1 + a_2 + a_3 + \cdots + a_n) + (b_1 + b_2 + b_3 + \cdots + b_n).$$

Expressing the right side in summation notation gives us formula (i).

For formula (iii) we have

$$\sum_{i=1}^{n} (ca_i) = ca_1 + ca_2 + ca_3 + \cdots + ca_n$$

$$= c(a_1 + a_2 + a_3 + \cdots + a_n)$$

$$= c\left(\sum_{i=1}^{n} a_i\right).$$

The proof of (ii) is left as an exercise.

EXERCISES 9.3

In Exercises 1–20, find the first five terms and the eighth term of the sequence which has the given nth term.

1 $a_n = 12 - 3n$

2 $a_n = \dfrac{3}{5n - 2}$

3 $a_n = \dfrac{3n - 2}{n^2 + 1}$

4 $a_n = 10 + \dfrac{1}{n}$

5 $a_n = 9$

6 $a_n = (n - 1)(n - 2)(n - 3)$

7 $a_n = 2 + (-0.1)^n$ **8** $a_n = 4 + (0.1)^n$

9 $a_n = (-1)^{n-1} \dfrac{n+7}{2n}$ **10** $a_n = (-1)^n \dfrac{6-2n}{\sqrt{n+1}}$

11 $a_n = 1 + (-1)^{n+1}$

12 $a_n = (-1)^{n+1} + (0.1)^{n-1}$

13 $a_n = \dfrac{2^n}{n^2+2}$ **14** $a_n = \sqrt{2}$

15 a_n is the square of the nth prime.

16 a_n is the nth prime greater than 50.

17 a_n is the largest prime less than $10n$.

18 a_n is the smallest prime greater than $15n$.

19 a_n is the number of decimal places in $(0.1)^n$.

20 a_n is the number of positive integers less than n^3.

Find the first five terms of the infinite sequences defined recursively in Exercises 21–28.

21 $a_1 = 2, a_{k+1} = 3a_k - 5$

22 $a_1 = 5, a_{k+1} = 7 - 2a_k$

23 $a_1 = -3, a_{k+1} = a_k^2$

24 $a_1 = 128, a_{k+1} = a_k/4$

25 $a_1 = 5, a_{k+1} = ka_k$

26 $a_1 = 3, a_{k+1} = 1/a_k$

27 $a_1 = 2, a_{k+1} = (a_k)^k$

28 $a_1 = 2, a_{k+1} = (a_k)^{1/k}$

29 A test question lists the first four terms of a sequence as 2, 4, 6, and 8 and asks for the fifth term. Show that the fifth term can be any real number a by finding the nth term of a sequence which has for its first five terms, 2, 4, 6, 8, and a.

30 The number of bacteria in a certain culture doubles every day. If the initial number of bacteria is 500, how many are present after one day? Two days? Three days? Find a formula for the number of bacteria present after n days.

Find the number given in each of Exercises 31–46.

31 $\displaystyle\sum_{k=1}^{5} (2k - 7)$ **32** $\displaystyle\sum_{k=1}^{6} (10 - 3k)$

33 $\displaystyle\sum_{k=1}^{4} (k^2 - 5)$ **34** $\displaystyle\sum_{k=1}^{10} [1 + (-1)^k]$

35 $\displaystyle\sum_{k=0}^{5} k(k - 2)$ **36** $\displaystyle\sum_{k=0}^{4} (k - 1)(k - 3)$

37 $\displaystyle\sum_{k=3}^{6} \dfrac{k-5}{k-1}$ **38** $\displaystyle\sum_{k=1}^{6} \dfrac{3}{k+1}$

39 $\displaystyle\sum_{k=1}^{5} (-3)^{k-1}$ **40** $\displaystyle\sum_{k=0}^{4} 3(2^k)$

41 $\displaystyle\sum_{k=1}^{100} 100$ **42** $\displaystyle\sum_{k=1}^{1000} 5$

43 $\displaystyle\sum_{k=1}^{n} (k^2 + 3k + 5)$ (*Hint:* Use the Theorem on Sums to write the sum as $\sum_{k=1}^{n} k^2 + 3\sum_{k=1}^{n} k + \sum_{k=1}^{n} 5$. Next employ Exercise 9 and Example 1 of Section 9.1, together with the formula for $\sum_{k=1}^{n} c$.

44 $\displaystyle\sum_{k=1}^{n} (3k^2 - 2k + 1)$

45 $\displaystyle\sum_{k=1}^{n} (2k - 3)^2$

46 $\displaystyle\sum_{k=1}^{n} (k^3 + 2k^2 - k + 4)$ (*Hint:* See Exercise 10 of Section 9.1).

Express the sums in Exercises 47–56 in terms of summation notation.

47 $1 + 5 + 9 + 13 + 17$

48 $2 + 5 + 8 + 11 + 14$

49 $\dfrac{1}{2} + \dfrac{2}{5} + \dfrac{3}{8} + \dfrac{4}{11}$

50 $\dfrac{1}{4} + \dfrac{2}{9} + \dfrac{3}{14} + \dfrac{4}{19}$

51 $1 - \dfrac{x^2}{2} + \dfrac{x^4}{4} - \dfrac{x^6}{6} + \cdots + (-1)^{n+1} \dfrac{x^{2n}}{2n}$

52 $2 - 4 + 8 - 16 + 32 - 64$

53 $1 - \dfrac{1}{2} + \dfrac{1}{3} - \dfrac{1}{4} + \dfrac{1}{5} - \dfrac{1}{6} + \dfrac{1}{7}$

54 $1 + x + \dfrac{x^2}{2} + \dfrac{x^3}{3} + \cdots + \dfrac{x^n}{n}$

55 $\dfrac{1}{1\cdot 2} + \dfrac{1}{2\cdot 3} + \dfrac{1}{3\cdot 4} + \cdots + \dfrac{1}{99\cdot 100}$

56 $\dfrac{1}{1 \cdot 2 \cdot 3} + \dfrac{1}{2 \cdot 3 \cdot 4} + \dfrac{1}{3 \cdot 4 \cdot 5} + \cdots + \dfrac{1}{98 \cdot 99 \cdot 100}$

57   Prove (ii) of the Theorem on Sums.

58   Extend (i) of the Theorem on Sums to $\sum_{i=1}^{n}(a_i + b_i + c_i)$.

59   Prove the Theorem on Sums by mathematical induction.

60   Prove $\sum_{i=1}^{n} c = nc$ by mathematical induction.

9.4 ARITHMETIC SEQUENCES

In this section and the next we shall concentrate on two special types of sequences. The first type may be defined as follows.

Definition

> An **arithmetic sequence** is a sequence such that successive terms differ by the same real number.

Arithmetic sequences are also called **arithmetic progressions**. By definition, a sequence

$$a_1, a_2, a_3, \ldots, a_n, \ldots$$

is arithmetic if and only if there is a real number d such that

$$a_{k+1} - a_k = d$$

for every positive integer k. The number d is called the **common difference** associated with the arithmetic sequence.

Example 1   Show that the sequence

$$1, 4, 7, 10, \ldots, 3n - 2, \ldots$$

is arithmetic and find the common difference.

Solution   Since $a_n = 3n - 2$ it follows that for every positive integer k,

$$a_{k+1} - a_k = [3(k+1) - 2] - (3k - 2)$$
$$= 3k + 3 - 2 - 3k + 2 = 3.$$

Hence, by definition, the given sequence is arithmetic with common difference 3. ∎

Given an arithmetic sequence we know that

$$a_{k+1} = a_k + d$$

for every positive integer k. This provides a recursive formula for obtaining successive terms. Beginning with any real number a_1, we can generate an arithmetic sequence with common difference d simply by adding d to a_1, then to $a_1 + d$, and so on, obtaining

$$a_1, \quad a_1 + d, \quad a_1 + 2d, \quad a_1 + 3d, \quad a_1 + 4d, \quad \dots .$$

It is evident that the nth term a_n of this sequence is given by

$$\boxed{a_n = a_1 + (n-1)d.}$$

This formula can be proved by mathematical induction.

Example 2 Find the fifteenth term of the arithmetic sequence whose first three terms are 20, 16.5, and 13.

Solution The common difference is -3.5. (Why?) Substituting $a_1 = 20$, $d = -3.5$, and $n = 15$ in the formula $a_n = a_1 + (n-1)d$,

$$a_{15} = 20 + (15-1)(-3.5) = 20 - 49 = -29. \qquad \blacksquare$$

Example 3 If the fourth term of an arithmetic sequence is 5 and the ninth term is 20, find the sixth term.

Solution Substituting $n = 4$ and $n = 9$ in the formula $a_n = a_1 + (n-1)d$, and using the fact that $a_4 = 5$ and $a_9 = 20$, we obtain the following system of linear equations in the variables a_1 and d:

$$\begin{cases} 5 = a_1 + (4-1)d \\ 20 = a_1 + (9-1)d. \end{cases}$$

This system has the unique solution $d = 3$ and $a_1 = -4$. (Verify!) Substitution in the formula $a_n = a_1 + (n-1)d$ gives us

$$a_6 = (-4) + (6-1)(3) = 11. \qquad \blacksquare$$

Theorem

If $a_1, a_2, \dots, a_n, \dots$ is an arithmetic sequence with common difference d, then the nth partial sum S_n is given by both

$$S_n = \frac{n}{2}[2a_1 + (n-1)d] \quad \text{and} \quad S_n = \frac{n}{2}(a_1 + a_n).$$

Proof We may write

$$S_n = a_1 + a_2 + a_3 + \cdots + a_n$$
$$= a_1 + (a_1 + d) + (a_1 + 2d) + \cdots + [a_1 + (n-1)d].$$

Employing the Commutative and Associative Properties many times we obtain

$$S_n = (a_1 + a_1 + a_1 + \cdots + a_1) + [d + 2d + \cdots + (n-1)d]$$

where a_1 appears n times within the first parentheses. It follows that

$$S_n = na_1 + d[1 + 2 + \cdots + (n-1)].$$

The expression within brackets is the sum of the first $n-1$ positive integers. From Example 1 of Section 9.1 (with $n-1$ in place of n), that sum equals

$$\frac{(n-1)n}{2}.$$

Substituting in the last equation for S_n and factoring gives us

$$S_n = na_1 + d\frac{n(n-1)}{2}$$

$$= \frac{n}{2}[2a_1 + (n-1)d].$$

Since $a_n = a_1 + (n-1)d$, we get

$$S_n = \frac{n}{2}(a_1 + a_n).$$

Example 4   Find the sum of all the even integers from 2 through 100.

Solution   The given problem is equivalent to finding the sum of the first fifty terms of the arithmetic sequence $2, 4, 6, \ldots, 2n, \ldots$. Substituting $n = 50$, $a_1 = 2$, and $a_{50} = 100$ in the second formula of the preceding theorem,

$$S_{50} = \frac{50}{2}(2 + 100) = 2550.$$

In order to check our work we use the first formula given in the theorem. Thus

$$S_{50} = \frac{50}{2}[2 \cdot 2 + (50-1)2] = 25[4 + 98] = 2550. \qquad \blacksquare$$

The **arithmetic mean** of two numbers a and b is defined as $(a+b)/2$. This is also called the **average** of a and b. Note that the numbers

$$a, \quad \frac{a+b}{2}, \quad b$$

are terms of an arithmetic sequence. This concept may be generalized as follows.

If $c_1, c_2, \ldots, c_k$ are real numbers such that

$$a, c_1, c_2, \ldots, c_k, b$$

is an arithmetic sequence, then $c_1, c_2, \ldots, c_k$ are called the **k arithmetic means** of the numbers a and b. The process of determining such numbers is referred to as *inserting k arithmetic means between a and b.*

Example 5   Insert three arithmetic means between 2 and 9.

Solution   We wish to find three numbers c_1, c_2, and c_3 such that $2, c_1, c_2, c_3, 9$ is an arithmetic sequence. The common difference d may be found by using the formula $a_n = a_1 + (n - 1)d$ with $n = 5$, $a_5 = 9$, and $a_1 = 2$. This gives us

$$9 = 2 + (5 - 1)d, \quad \text{or} \quad d = \tfrac{7}{4}.$$

The three arithmetic means are

$$c_1 = a_1 + d = 2 + \tfrac{7}{4} = \tfrac{15}{4}$$
$$c_2 = c_1 + d = \tfrac{15}{4} + \tfrac{7}{4} = \tfrac{22}{4} = \tfrac{11}{2}$$
$$c_3 = c_2 + d = \tfrac{11}{2} + \tfrac{7}{4} = \tfrac{29}{4}.$$ ∎

EXERCISES 9.4

In each of Exercises 1–8 find the fifth term, the tenth term, and the nth term of the given arithmetic sequence.

1 $2, 6, 10, 14, \ldots$

2 $16, 13, 10, 7, \ldots$

3 $3, 2.7, 2.4, 2.1, \ldots$

4 $-6, -4.5, -3, -1.5, \ldots$

5 $-7, -3.9, -0.8, 2.3, \ldots$

6 $x - 8, x - 3, x + 2, x + 7, \ldots$

7 $\log 3, \log 9, \log 27, \log 81, \ldots$

8 $\log 1000, \log 100, \log 10, 0, \ldots$

9 Find the twelfth term of the arithmetic sequence whose first two terms are 9.1 and 7.5.

10 Find the eleventh term of the arithmetic sequence whose first two terms are $2 + \sqrt{2}$ and 3.

11 The sixth and seventh terms of an arithmetic sequence are 2.7 and 5.2. Find the first term.

12 Given an arithmetic sequence with $a_3 = 7$ and $a_{20} = 43$, find a_{15}.

In each of Exercises 13–16 find the sum of the arithmetic sequence which satisfies the given conditions.

13 $a_1 = 40, d = -3, n = 30$

14 $a_1 = 5, d = 0.1, n = 40$

15 $a_1 = -9, a_{10} = 15, n = 10$

16 $a_7 = 7/3, d = -2/3, n = 15$

Find the sums in Exercises 17–20.

17 $\displaystyle\sum_{k=1}^{20} (3k - 5)$

18 $\displaystyle\sum_{k=1}^{12} (7 - 4k)$

19 $\displaystyle\sum_{k=1}^{18} \left(\frac{1}{2}k + 7\right)$

20 $\displaystyle\sum_{k=1}^{10} \left(\frac{1}{4}k + 3\right)$

21 How many integers between 32 and 395 are divisible by 6? Find their sum.

22 How many negative integers greater than -500 are divisible by 33? Find their sum.

23 How many terms are in an arithmetic sequence with first term -2, common difference $\frac{1}{4}$, and sum 21?

24 How many terms are in an arithmetic sequence with sixth term -3, common difference 0.2, and sum -33?

25 Insert five arithmetic means between 2 and 10.

26 Insert three arithmetic means between 3 and -5.

27 A pile of logs has 24 logs in the first layer, 23 in the second, 22 in the third, and so on. The last layer contains 10 logs. Find the total number of logs in the pile.

28 A seating section in a certain athletic stadium has 30 seats in the first row, 32 seats in the second, 34 in the third, and so on, until the tenth row is reached, after which there are 10 more rows, each containing 50 seats. Find the total number of seats in the section.

29 A man wishes to construct a ladder with nine rungs which diminish uniformly from 24 inches at the base to 18 inches at the top. Determine the lengths of the seven intermediate rungs.

30 A boy on a bicycle coasts down a hill, covering 4 feet the first second and in each succeeding second 5 feet more than in the preceding second. If he reaches the bottom of the hill in 11 seconds, find the total distance traveled.

9.5 GEOMETRIC SEQUENCES

Another important type of infinite sequence is defined as follows.

Definition

> A **geometric sequence** is a sequence such that the quotient of any term after the first by the preceding term is the same nonzero real number.

Geometric sequences are also called **geometric progressions**. By definition, a sequence

$$a_1, a_2, \ldots, a_n, \ldots$$

is geometric if and only if there is a real number $r \neq 0$ such that

$$\frac{a_{k+1}}{a_k} = r$$

for every positive integer k. The number r is called the **common ratio** associated with the geometric sequence. We see from the preceding formula that the indicated sequence is geometric (with common ratio r) if and only if

$$a_{k+1} = a_k r$$

for every positive integer k. As with arithmetic sequences, this provides a recursive method for obtaining terms. Beginning with any nonzero real number a_1, we *multiply* by the number r successively, obtaining

$$a_1, a_1 r, a_1 r^2, a_1 r^3, \ldots.$$

It appears that the nth term a_n of this sequence is given by

$$a_n = a_1 r^{n-1}.$$

This can be proved by mathematical induction.

Example 1 Find the first five terms and the tenth term of the geometric sequence having first term 3 and common ratio $-1/2$.

Solution If we let $a_1 = 3$ and $r = -1/2$, then the first five terms are,

$$3, -\frac{3}{2}, \frac{3}{4}, -\frac{3}{8}, \frac{3}{16}.$$

Using the formula $a_n = a_1 r^{n-1}$ with $n = 10$, we obtain

$$a_{10} = 3\left(-\frac{1}{2}\right)^9 = -\frac{3}{512}. \qquad\blacksquare$$

Example 2 If the third term of a geometric sequence is 5 and the sixth term is -40, find the eighth term.

Solution We are given $a_3 = 5$ and $a_6 = -40$. Substituting $n = 3$ and $n = 6$ in the formula $a_n = a_1 r^{n-1}$ leads to the following system of equations:

$$\begin{cases} 5 = a_1 r^2 \\ -40 = a_1 r^5. \end{cases}$$

Since $r \neq 0$, the first equation is equivalent to $a_1 = 5/r^2$. Substituting for a_1 in the second equation we obtain

$$-40 = \left(\frac{5}{r^2}\right) \cdot r^5 = 5r^3.$$

Hence $r^3 = -8$ and $r = -2$. If we now substitute -2 for r in the equation $5 = a_1 r^2$, we obtain $a_1 = 5/4$. Finally, using $a_n = a_1 r^{n-1}$ with $n = 8$ gives us

$$a_8 = \left(\frac{5}{4}\right)(-2)^7 = -160. \qquad\blacksquare$$

Let us find a formula for S_n, the nth partial sum of a geometric sequence. If the first term is a_1 and the common ratio is r, then

$$S_n = a_1 + a_1 r + a_1 r^2 + \cdots + a_1 r^{n-2} + a_1 r^{n-1}.$$

If $r = 1$, this gives us $S_n = na_1$. Next suppose that $r \neq 1$. Multiplying both sides of the general formula for S_n by r, we obtain

$$rS_n = a_1 r + a_1 r^2 + a_1 r^3 + \cdots + a_1 r^{n-1} + a_1 r^n.$$

If we subtract the preceding equation from that for S_n, then many terms on the right side drop out, leaving

$$S_n - rS_n = a_1 - a_1 r^n \quad \text{or} \quad (1 - r)S_n = a_1(1 - r^n).$$

Since $r \neq 1$, we have $1 - r \neq 0$, and dividing both sides by $1 - r$ gives us

$$S_n = \frac{a_1(1 - r^n)}{1 - r}.$$

We have proved the following theorem.

Theorem

> The nth partial sum of a geometric sequence with first term a_1 and common ratio $r \neq 1$ is
>
> $$S_n = a_1 \frac{(1 - r^n)}{1 - r}.$$

Example 3 Find the sum of the first five terms of the geometric sequence which begins as follows: 1, 0.3, 0.09, 0.027, ...

Solution We let $a_1 = 1$, $r = 0.3$, and $n = 5$ in the formula for S_n, obtaining

$$S_5 = 1\left(\frac{1 - (0.3)^5}{1 - 0.3}\right)$$

which reduces to $S_5 = 1.4251$. ∎

Example 4 A man wishes to save money by setting aside 1 cent the first day, 2 cents the second day, 4 cents the third day and so on, doubling the amount each day. If this is continued, how much must be set aside on the fifteenth day? Assuming he does not run out of money, what is the total amount saved at the end of 30 days?

Solution The amount (in cents) set aside on successive days forms a geometric sequence

$$1, 2, 4, 8, \ldots$$

with first term 1 and common ratio 2. The amount needed for the fifteenth day is found by using $a_n = a_1 r^{n-1}$ with $a_1 = 1$ and $n = 15$. This gives us $1 \cdot 2^{14}$, or

$163.84. To find the total amount set aside after 30 days, we use the formula for S_n with $n = 30$. Thus

$$S_{30} = 1\frac{(1 - 2^{30})}{1 - 2},$$

which simplifies to $10,737,418.23. ∎

Given the geometric series with first term a_1 and common ratio $r \neq 1$, we may write the formula for S_n of the last theorem in the form

$$S_n = \frac{a_1}{1 - r} - \frac{a_1}{1 - r}r^n.$$

Although we shall not do it here, it can be shown that if $|r| < 1$, then r^n *approaches the number 0 as n increases* without bound; that is, we can make r^n as close as we wish to 0 by taking n sufficiently large. It follows that S_n approaches $a_1/(1 - r)$ as n increases without bound. Using the notation introduced in our work with rational functions in Chapter Three, this may be expressed symbolically as

$$S_n \to \frac{a_1}{1 - r} \quad \text{as } n \to \infty.$$

The number $a_1/(1 - r)$ is called *the sum of the infinite geometric series*

$$a_1 + a_1 r + a_1 r^2 + \cdots + a_1 r^{n-1} + \cdots.$$

This gives us the following result.

Theorem

> If $|r| < 1$, then the infinite geometric series
>
> $$a_1 + a_1 r + a_1 r^2 + \cdots + a_1 r^{n-1} + \cdots$$
>
> has the sum
>
> $$\frac{a_1}{1 - r}.$$

Intuitively, the preceding theorem implies that if we add more and more terms of the indicated infinite geometric series, the sums get closer and closer to $a_1/(1 - r)$. The next example illustrates how the theorem can be used to show that every real number represented by a repeating decimal is rational.

Example 5 Find the rational number which corresponds to the infinite repeating decimal $5.4\overline{27}$, where the bar means that the block of digits underneath is to be repeated indefinitely.

Solution   From the decimal expression 5.4272727... we obtain the infinite series

$$5.4 + 0.027 + 0.00027 + 0.0000027 + \cdots.$$

The part of the expression after the first term is

$$0.027 + 0.00027 + 0.0000027 + \cdots$$

which has the form given in the last theorem with $a_1 = 0.027$ and $r = 0.01$. Hence the sum of this infinite geometric series is

$$\frac{0.027}{1 - 0.01} = \frac{0.027}{0.99} = \frac{27}{990} = \frac{3}{110}.$$

Thus it appears that the desired number is $5.4 + 3/110$, or $597/110$. A check by long division shows that $597/110$ does equal the given repeating decimal.   ■

In general, given any infinite sequence $a_1, a_2, \ldots, a_n, \ldots$, an expression of the form

$$a_1 + a_2 + \cdots + a_n + \cdots$$

is called an **infinite series**, or simply a **series**. In summation notation, this series is denoted by either

$$\sum_{n=1}^{\infty} a_n \quad \text{or} \quad \sum a_n$$

where it is understood that the summation variable in the last sum is n. Each number a_i is called a **term** of the series and a_n is called the **nth term**. Since only finite sums may be added algebraically, it is necessary to *define* what is meant by an infinite sum. A natural way to do so is to use the sequence of partial sums

$$S_1, S_2, \ldots, S_n, \ldots.$$

If there is a number S such that $S_n \to S$ as $n \to \infty$, then, as in our discussion of infinite geometric series, we call S the **sum** of the series and write

$$S = a_1 + a_2 + \cdots + a_n + \cdots.$$

Thus, in Example 5 we may write

$$\frac{597}{110} = 5.4 + 0.027 + 0.00027 + 0.0000027 + \cdots.$$

The next example provides an illustration of a non-geometric infinite series. Since a complete discussion is beyond the scope of this text, the solution is presented in an intuitive manner.

Example 6 Show that the infinite series

$$\frac{1}{1 \cdot 2} + \frac{1}{2 \cdot 3} + \frac{1}{3 \cdot 4} + \cdots + \frac{1}{n(n + 1)} + \cdots$$

has a sum.

Solution The nth term $a_n = 1/n(n + 1)$ has the partial fraction decomposition

$$a_n = \frac{1}{n(n + 1)}$$

$$= \frac{1}{n} - \frac{1}{n + 1}.$$

Consequently, the nth partial sum of the series may be written

$$S_n = a_1 + a_2 + a_3 + \cdots + a_n$$

$$= \left(1 - \frac{1}{2}\right) + \left(\frac{1}{2} - \frac{1}{3}\right) + \left(\frac{1}{3} - \frac{1}{4}\right) + \cdots + \left(\frac{1}{n} - \frac{1}{n + 1}\right)$$

$$= 1 - \frac{1}{n + 1}.$$

Since $1/(n + 1) \to 0$ as $n \to \infty$ it follows that $S_n \to 1$ and we may write

$$1 = \frac{1}{1 \cdot 2} + \frac{1}{2 \cdot 3} + \frac{1}{3 \cdot 4} + \cdots + \frac{1}{n(n + 1)} + \cdots.$$

This means that, as we add more and more terms of the series, the sums get closer and closer to 1. ∎

If the terms of an infinite sequence are alternately positive and negative and if we consider the expression

$$a_1 + (-a_2) + a_3 + (-a_4) + \cdots + [(-1)^{n+1}a_n] + \cdots$$

where all the a_i are positive real numbers, then this expression is referred to as an *alternating infinite series* and we write it in the form

$$a_1 - a_2 + a_3 - a_4 + \cdots + (-1)^{n+1}a_n + \cdots.$$

Illustrations of alternating infinite series can be obtained by using infinite geometric series with negative common ratio.

Infinite series have many applications in mathematics and the sciences. The reader is referred to texts on calculus for a rigorous treatment of this important concept.

EXERCISES 9.5

In each of Exercises 1–12 find the fifth term, the eighth term, and the nth term of the given geometric sequence.

1   8, 4, 2, 1, ...

2   4, 1.2, 0.36, 0.108, ...

3   300, -30, 3, -0.3, ...

4   1, $-\sqrt{3}$, 3, $-\sqrt{27}$, ...

5   5, 25, 125, 625, ...

6   2, 6, 18, 54, ...

7   4, -6, 9, -13.5, ...

8   162, -54, 18, -6, ...

9   1, $-x^2$, x^4, $-x^6$, ...

10   1, $-\dfrac{x}{3}$, $\dfrac{x^2}{9}$, $-\dfrac{x^3}{27}$, ...

11   2, 2^{x+1}, 2^{2x+1}, 2^{3x+1}, ...

12   10, 10^{2x-1}, 10^{4x-3}, 10^{6x-5}, ...

13   Find the sixth term of the geometric sequence whose first two terms are 4 and 6.

14   Find the seventh term of the geometric sequence that has 2 and $-\sqrt{2}$ for its second and third terms, respectively.

15   In a certain geometric sequence $a_5 = 1/16$ and $r = 3/2$. Find a_1 and S_5.

16   Given a geometric sequence in which $a_4 = 4$ and $a_7 = 12$, find r and a_{10}.

Find the sum in each of Exercises 17–20.

17    $\displaystyle\sum_{k=1}^{10} 3^k$

18    $\displaystyle\sum_{k=1}^{9} (-\sqrt{5})^k$

19    $\displaystyle\sum_{k=0}^{9} (-\tfrac{1}{2})^{k+1}$

20    $\displaystyle\sum_{k=1}^{7} (3^{-k})$

21   A vacuum pump removes one-half of the air in a container at each stroke. After 10 strokes, what percentage of the original amount of air remains in the container?

22   The yearly depreciation of a certain machine is 25% of its value at the beginning of the year. If the original cost of the machine is $20,000, find its value after 6 years.

23   A culture of bacteria increases 20% every hour. If the original culture contains 10,000 bacteria, find a formula for the number of bacteria present after t hours. How many bacteria are in the culture at the end of 10 hours?

24   If an amount P of money is deposited in a savings account which pays interest at a rate of r percent per year compounded quarterly and if the principal and accumulated interest are left in the account, find a formula for the total amount in the account after n years.

Find the sums of the infinite geometric series in Exercises 25–30, whenever they exist.

25    $1 - \dfrac{1}{2} + \dfrac{1}{4} - \dfrac{1}{8} + \cdots$

26    $2 + \dfrac{2}{3} + \dfrac{2}{9} + \dfrac{2}{27} + \cdots$

27    $1.5 + 0.015 + 0.00015 + \cdots$

28    $1 - 0.1 + 0.01 - 0.001 + \cdots$

29    $\sqrt{2} - 2 + \sqrt{8} - 4 + \cdots$

30    $250 - 100 + 40 - 16 + \cdots$

In each of Exercises 31–38 find the rational number represented by the given repeating decimal.

31    $0.\overline{23}$ 32 $0.\overline{071}$

33    $2.4\overline{17}$ 34 $10.\overline{55}$

35    $5.1\overline{46}$ 36 $3.2\overline{394}$

37    $1.6\overline{124}$ 38 $123.6\overline{183}$

39   A rubber ball is dropped from a height of 10 meters. If it rebounds approximately one-half the distance after each fall, use an infinite geometric series to approximate the total distance the ball travels before coming to rest.

40   The bob of a pendulum swings through an arc 24 cm long on its first swing. If each successive swing is approximately five-sixths the length of the preceding swing, use an infinite geometric series to approximate the total distance it travels before coming to rest.

9.6 REVIEW

Concepts

Define or discuss each of the following.

1 The axiom of mathematical induction

2 The principle of mathematical induction

3 The Binomial Theorem

4 Binomial coefficients

5 Infinite sequence

6 Summation notation

7 The nth partial sum of an infinite sequence

8 Arithmetic sequence

9 Arithmetic mean of two numbers

10 Geometric sequence

11 Infinite geometric series

12 Infinite series

Exercises

Prove that the statements in Exercises 1–5 are true for every positive integer n.

1   $2 + 5 + 8 + \cdots + (3n - 1) = \dfrac{n(3n + 1)}{2}$

2   $2^2 + 4^2 + 6^2 + \cdots + (2n)^2$
$$= \frac{2n(2n + 1)(n + 1)}{3}$$

3   $\dfrac{1}{1 \cdot 3} + \dfrac{1}{3 \cdot 5} + \dfrac{1}{5 \cdot 7} + \cdots + \dfrac{1}{(2n - 1)(2n + 1)}$
$$= \frac{n}{2n + 1}$$

4   $1 \cdot 2 + 2 \cdot 3 + 3 \cdot 4 + \cdots + n(n + 1)$
$$= \frac{n(n + 1)(n + 2)}{3}$$

5   3 is a factor of $n^3 + 2n$.

6   Prove that $2^n > n^2$ for every positive integer $n \geq 5$.

7   Expand and simplify $(x^2 - 3y)^6$.

8   Find the first four terms in the binomial expansion of $(a^{2/5} + 2a^{-3/5})^{20}$.

9   Find the sixth term in the expansion of $(b^3 - \frac{1}{2}c^2)^9$.

10   In the expansion of $(2c^3 + 5c^{-2})^{10}$ find the term which does not contain c.

In each of Exercises 11–14 find the first four terms and the seventh term of the sequence that has the given nth term.

11   $a_n = \dfrac{5n}{3 - 2n^2}$

12   $a_n = (-1)^{n+1} - (0.1)^n$

13   $a_n = 1 + (-\frac{1}{2})^{n-1}$

14   $a_n = \dfrac{2^n}{(n + 1)(n + 2)(n + 3)}$

Find the first five terms of the infinite sequence defined recursively in each of Exercises 15–18.

15   $a_1 = 10$, $a_{k+1} = 1 + (1/a_k)$

16   $a_1 = 2$, $a_{k+1} = a_k!$

17   $a_1 = 9$, $a_{k+1} = \sqrt{a_k}$

18   $a_1 = 1$, $a_{k+1} = (1 + a_k)^{-1}$

Find the number represented by the sum in each of Exercises 19–22.

19   $\displaystyle\sum_{k=1}^{5} (k^2 + 4)$ 20 $\displaystyle\sum_{k=2}^{6} \frac{2k - 8}{k - 1}$

21 $\displaystyle\sum_{k=1}^{100} 10$

22 $\displaystyle\sum_{i=1}^{4} (2^i - 10)$

In Exercises 23–26 use summation notation to represent the sums.

23 $3 + 6 + 9 + 12 + 15$

24 $2 + 4 + 8 + 16 + 32 + 64 + 128$

25 $100 - 95 + 90 - 85 + 80$

26 $a_0 + a_4 x^4 + a_8 x^8 + \cdots + a_{100} x^{100}$

27 Find the tenth term and the sum of the first ten terms of the arithmetic sequence whose first two terms are $4 + \sqrt{3}$ and 3.

28 Find the sum of the first eight terms of an arithmetic sequence in which the fourth term is 9 and the common difference is -5.

29 The fifth and thirteenth terms of an arithmetic sequence are 5 and 77, respectively. Find the first term and the tenth term.

30 Insert four arithmetic means between 20 and -10.

31 Find the tenth term of the geometric sequence whose first two terms are 1/8 and 1/4.

32 If a geometric sequence has 3 and -0.3 as its third and fourth terms, find the eighth term.

33 Find a positive number c such that 4, c, 8 are successive terms of a geometric sequence.

34 In a certain geometric sequence the eighth term is 100 and the common ratio is $-3/2$. Find the first term.

Find the sums in Exercises 35–38.

35 $\displaystyle\sum_{k=1}^{20} (3k - 5)$

36 $\displaystyle\sum_{k=1}^{10} (6 - \tfrac{1}{2}k)$

37 $\displaystyle\sum_{k=1}^{10} (2^k - \tfrac{1}{2})$

38 $\displaystyle\sum_{k=1}^{8} (\tfrac{1}{2} - 2^k)$

39 Find the sum of the infinite geometric series

$$1 - \frac{2}{5} + \frac{4}{25} - \frac{8}{125} + \cdots.$$

40 Find the rational number whose decimal representation is $6.\overline{274}$.

10 TOPICS IN ANALYTIC GEOMETRY

Plane geometry includes the study of figures such as lines, circles, and triangles which lie in a plane. Theorems are proved by reasoning deductively from postulates. In analytic geometry, plane geometric figures are investigated by introducing a coordinate system and then using equations and formulas of various types. If the study of analytic geometry were to be summarized by means of one statement, perhaps the following would be appropriate: "Given an equation, find its graph and, conversely, given a graph, find its equation." In this chapter we shall apply coordinate methods to several basic plane figures.

10.1 CIRCLES

In Chapter One we proved that the circle in a coordinate plane with center $C(h, k)$ and radius r is the graph of the equation

$$(x - h)^2 + (y - k)^2 = r^2.$$

Squaring and simplifying, we obtain an equation of the form

$$x^2 + y^2 + ax + by + c = 0$$

where a, b, and c are real numbers. Conversely, if we begin with such an equation, it is always possible, by completing the squares in x and y, to obtain an equation of the form

$$(x - h)^2 + (y - k)^2 = d.$$

The method will be illustrated in Example 1. If $d > 0$, then the graph is a circle with center (h, k) and radius $r = \sqrt{d}$. If $d = 0$, then since $(x - h)^2 \geq 0$ and $(y - k)^2 \geq 0$, the only solution of the equation is (h, k) and hence the graph consists of only one point. Finally, if $d < 0$, the equation has no real solutions and there is no graph.

427

Example 1   Find the center and radius of the circle with equation

$$x^2 + y^2 - 4x + 6y - 3 = 0.$$

Solution   We begin by arranging the equation as follows:

$$(x^2 - 4x) + (y^2 + 6y) = 3.$$

Next we complete the squares by adding appropriate numbers within the parentheses. Of course, to obtain equivalent equations we must add the numbers to *both* sides of the equation. In order to complete the square for an expression of the form $x^2 + ax$, we add the square of half the coefficient of x, that is, $(a/2)^2$, to both sides of the equation. Similarly, for $y^2 + by$ we add $(b/2)^2$ to both sides. This leads to

$$(x^2 - 4x + 4) + (y^2 + 6y + 9) = 3 + 4 + 9$$

or
$$(x - 2)^2 + (y + 3)^2 = 16.$$

Hence, by the preceding discussion, the center is $(2, -3)$ and the radius is 4. ■

In the next example we use the following fundamental principle of analytic geometry:

> A point $P(x_1, y_1)$ is on the graph of an equation in x and y if and only if the pair (x_1, y_1) is a solution of the equation.

Example 2   Find an equation of the circle which contains the points $A(-1, 7)$, $B(3, 9)$, and the origin $O(0, 0)$. What is the center and radius of the circle?

Solution   The circle is sketched in Figure 10.1. If we choose a, b, and c appropriately, then the equation has the form $x^2 + y^2 + ax + by + c = 0$. Moreover, since the circle passes through the origin, $(0, 0)$ is a solution, that is,

$$0^2 + 0^2 + a \cdot 0 + b \cdot 0 + c = 0,$$

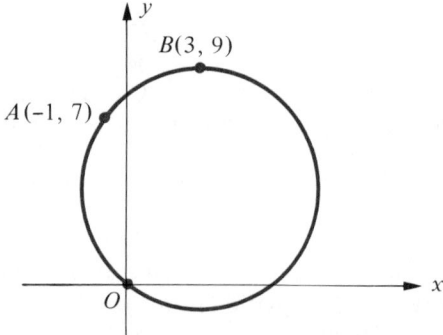

Figure 10.1

and consequently, $c = 0$. Therefore, an equation for the circle is of the form

$$x^2 + y^2 + ax + by = 0.$$

Since A and B are on the circle, $(-1, 7)$ and $(3, 9)$ are solutions of this equation; that is,

$$1 + 49 - a + 7b = 0 \quad \text{and} \quad 9 + 81 + 3a + 9b = 0.$$

Simplifying, we see that the pair (a, b) must be a solution of *both* of the equations

$$a - 7b = 50 \quad \text{and} \quad a + 3b = -30.$$

Solving the first equation for a and substituting in the second equation, we obtain $(7b + 50) + 3b = -30$, or $10b = -80$. Consequently, $b = -8$. Since $a = 7b + 50$, we have $a = 7(-8) + 50 = -6$. Substitution of these numbers for a and b in $x^2 + y^2 + ax + by = 0$ gives us

$$x^2 + y^2 - 6x - 8y = 0.$$

Completing the squares we obtain

$$(x^2 - 6x + 9) + (y^2 - 8y + 16) = 9 + 16$$

or
$$(x - 3)^2 + (y - 4)^2 = 25.$$

Hence the center is $C(3, 4)$ and the radius is 5. ∎

EXERCISES 10.1

In each of Exercises 1–6 find an equation of the circle which satisfies the stated conditions.

1 Center $C(3, -4)$, radius 2

2 Center $C(5, 0)$, radius 5

3 Center $C(6, -2)$, tangent to the line $x = 3$

4 Endpoints of diameter $A(-1, 1)$, $B(4, -6)$

5 Circumscribed about the right triangle with vertices $A(-1, 2)$, $B(2, -3)$, and $C(7, 0)$ (*Hint:* The center of the circle is the midpoint of the hypotenuse.)

6 Passing through the three points $A(-2, 1)$ $B(5, 1)$, and $C(5, -3)$

In each of Exercises 7–16 find the center and radius of the circle with the given equation.

7 $x^2 + y^2 + 2x - 10y + 10 = 0$

8 $x^2 + y^2 - 8x + 4y + 15 = 0$

9 $x^2 + y^2 - 6y + 5 = 0$

10 $x^2 + y^2 + 14x + 46 = 0$

11 $x^2 + y^2 - 10x + 8y = 0$

12 $x^2 + y^2 + 2y = 0$

13 $4x^2 + 4y^2 + 8x - 8y + 7 = 0$

14 $2x^2 + 2y^2 + 8x + 7 = 0$

15 $3x^2 + 3y^2 - 3x + 2y + 1 = 0$

16 $9x^2 + 9y^2 + 12x - 6y + 4 = 0$

In Exercises 17 and 18 use the method of Example 2 to find an equation for the circle which circumscribes the triangle with the given vertices.

17 $A(-2,1)$, $B(-4,-1)$, $C(1,2)$

18 $A(-3,1)$, $B(5,5)$, $C(-3,9)$

19   Show that the center of the circle given by $x^2 + y^2 + ax + by + c = 0$ is on the line $y = x$ if and only if $a = b$.

20   Show that the graph of $x^2 + y^2 + ax + by + c = 0$ contains the origin if and only if $c = 0$.

10.2   LINES

If l is a line which is not parallel to the y-axis and if $P_1(x_1, y_1)$ and $P_2(x_2, y_2)$ are distinct points on l, then the *slope m* of l is

$$m = \frac{y_2 - y_1}{x_2 - x_1}.$$

If l is parallel to the y-axis, then it has no slope.

It was shown in Chapter Two that an equation of a line of slope m through a point $P_1(x_1, y_1)$ is given by the *point-slope form*

$$y - y_1 = m(x - x_1).$$

The *slope-intercept form*

$$y = mx + b$$

is an equation of a line having slope m and y-intercept b. A consequence of these facts is that the graph of every linear equation

$$Ax + By + C = 0,$$

where A, B, and C are real numbers, is a line and, conversely, every line is the graph of a linear equation.

The concept of *inclination*, introduced in the next definition, is fundamental for the further study of lines.

Definition

> If l is a line that is not parallel to the x-axis and if P_1 is the point of intersection of l and the x-axis, then the **inclination** of l is the smallest angle α through which the x-axis must be rotated in a counterclockwise direction about P_1 in order to coincide with l. If l is parallel to the x-axis, then $\alpha = 0°$.

If l is not horizontal, then $0° < \alpha < 180°$. The line shown in (i) of Figure 10.2 illustrates the case $0° < \alpha < 90°$ and that in (ii) illustrates $90° < \alpha < 180°$. It follows from plane geometry that two lines are parallel if and only if they have the same inclination. The next theorem provides the connection between slope and inclination.

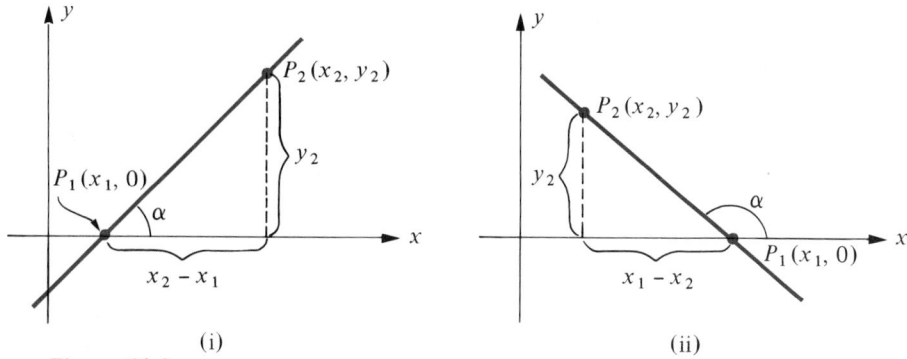

Figure 10.2

Theorem

If a line has slope m and inclination α, then $m = \tan \alpha$.

Proof

If the line is horizontal, then $m = 0$, $\alpha = 0°$ and, since $0 = \tan 0°$, the theorem is true.

If the line is not horizontal, let x_1 be the x-intercept and consider any point $P_2(x_2, y_2)$ on the line, where $y_2 > 0$. If $x_1 < x_2$, then referring to the triangle in (i) of Figure 10.2,

$$\tan \alpha = \frac{y_2}{x_2 - x_1} = m.$$

If $x_2 < x_1$, then from (ii) of Figure 10.2,

$$\tan (180° - \alpha) = \frac{y_2}{x_1 - x_2}.$$

However, by Exercise 35 in Section 6.6, $\tan (180° - \alpha) = -\tan \alpha$, and hence we again obtain $\tan \alpha = m$. This completes the proof.

Corollary

Two lines with slopes m_1 and m_2 are parallel if and only if $m_1 = m_2$.

Proof

If the lines have inclinations α_1 and α_2, then they are parallel if and only if $\alpha_1 = \alpha_2$ or, equivalently, $\tan \alpha_1 = \tan \alpha_2$. The corollary now follows from the theorem.

Example 1
(a) Find the slope m of a line whose inclination is $120°$.
(b) Find the inclination α of a line whose slope is $7/10$.

Solutions
(a) By the preceding theorem, $m = \tan 120° = -\sqrt{3}$.
(b) Since $\tan \alpha = 0.7$, $\alpha = \tan^{-1} (0.7)$. If an approximation is desired, then using Table 3, $\alpha \approx 35°$. ∎

The next theorem provides conditions for testing perpendicularity of lines.

Theorem

> Two nonvertical lines with slopes m_1 and m_2 are perpendicular if and only if $m_1 m_2 = -1$.

Proof   If α_1 and α_2 denote the inclinations of the lines, then

$$m_1 = \tan \alpha_1 \quad \text{and} \quad m_2 = \tan \alpha_2.$$

We may assume, without loss of generality, that $\alpha_2 > \alpha_1$ as illustrated in Figure 10.3. If θ is the angle indicated in the figure, then

$$\alpha_2 = \alpha_1 + \theta, \quad \text{or} \quad \theta = \alpha_2 - \alpha_1.$$

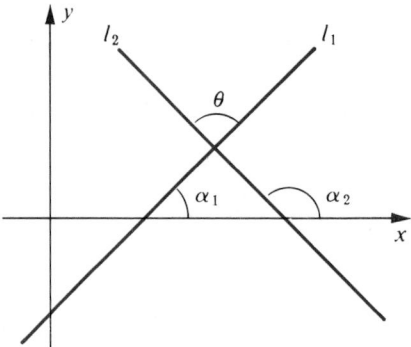

Figure 10.3

Applying the subtraction formula for the tangent function,

$$\tan \theta = \tan (\alpha_2 - \alpha_1) = \frac{\tan \alpha_2 - \tan \alpha_1}{1 + \tan \alpha_1 \tan \alpha_2}$$

or

$$\tan \theta = \frac{m_2 - m_1}{1 + m_1 m_2}.$$

By definition, the lines are perpendicular if and only if $\theta = 90°$, or equivalently, $\tan \theta$ is undefined. However, by the preceding formula, $\tan \theta$ is undefined if and only if $1 + m_1 m_2 = 0$, or $m_1 m_2 = -1$, which is what we wished to prove. Although we have referred to Figure 10.3 a similar argument may be used regardless of the magnitudes of α_1 and α_2, or where the lines intersect.

Another way of stating this theorem is to say that l_1 and l_2 are perpendicular if and only if $m_1 = -1/m_2$; that is, m_1 and m_2 are negative reciprocals of one another.

Example 2   Prove that the triangle with vertices $A(-1, -3)$, $B(6, 1)$, and $C(2, -5)$ is a right triangle.

Solution   If m_1 denotes the slope of the line through B and C and if m_2 denotes the slope of the line through A and C, then

$$m_1 = \frac{1-(-5)}{6-2} = \frac{3}{2}, \qquad m_2 = \frac{-3-(-5)}{-1-2} = -\frac{2}{3}.$$

Since $m_1 m_2 = -1$, the angle at C is a right angle.   ■

Example 3   Find an equation for the perpendicular bisector l.of the line segment from $A(1, 7)$ to $B(-3, 2)$.

Solution   The given points and the perpendicular bisector are shown in Figure 10.4. The slope of the line through A and B is

$$m = \frac{7-2}{1-(-3)} = \frac{5}{4}$$

and hence, by the last theorem, the slope of l is $-4/5$. Applying the midpoint formula we see that the midpoint of AB is $(-1, 9/2)$. Using the point-slope form, an equation for l is

$$y - \frac{9}{2} = -\frac{4}{5}(x + 1).$$

Multiplying both sides by 10 and simplifying gives us

$$8x + 10y - 37 = 0.$$

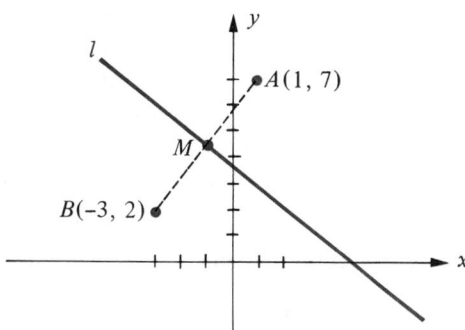

Figure 10.4   ■

We shall conclude this section by deriving a formula which may be used to find the angle between two lines in a coordinate plane. If the lines are parallel, then by definition the angle between them is $0°$. If the lines intersect, label the line of larger inclination l_2 and the line of smaller inclination l_1. Denoting the inclinations of l_1 and l_2 by α_1 and α_2, respectively, it follows that $\alpha_2 > \alpha_1$. The angle θ between l_1 and l_2 is defined by $\theta = \alpha_2 - \alpha_1$, as illustrated in Figure 10.5.

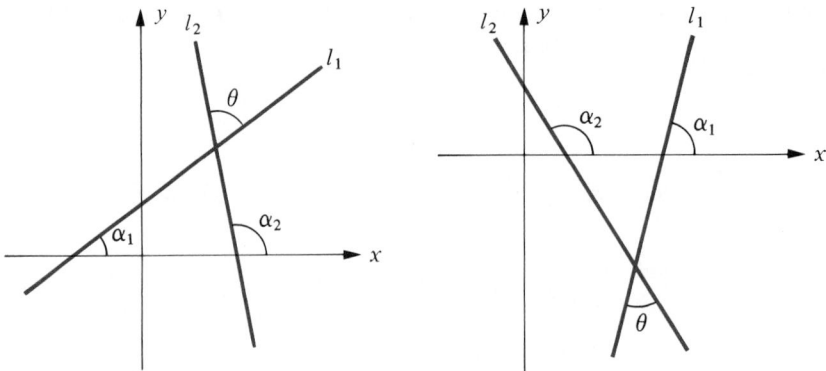

Figure 10.5

Assuming that none of the angles are right angles,

$$\tan \theta = \tan (\alpha_2 - \alpha_1) = \frac{\tan \alpha_2 - \tan \alpha_1}{1 + \tan \alpha_2 \tan \alpha_1}$$

or

$$\tan \theta = \frac{m_2 - m_1}{1 + m_2 m_1}$$

where m_1 and m_2 are the slopes of l_1 and l_2, respectively. If $\theta = 90°$ the lines are perpendicular which, for nonvertical lines, is equivalent to $m_1 m_2 = -1$. If either line is vertical, then the formula for $\tan \theta$ cannot be employed. In the latter case the fact that $\theta = \alpha_2 - \alpha_1$ should be used to find θ.

Example 4   Find the tangent of the angle between the lines $3x - 4y = 8$ and $5x + y - 7 = 0$.

Solution   The slopes of the lines are $3/4$ and -5. (Why?) Setting $m_2 = -5$, $m_1 = 3/4$, and applying the last formula gives us

$$\tan \theta = \frac{-5 - (3/4)}{1 + (-5)(3/4)} = \frac{23}{11}.$$

An approximation to θ may be found by means of Table 3. To the nearest degree $\theta \approx 64°$.   ■

EXERCISES 10.2

In Exercises 1 and 2 find the inclination of the line which has the given slope.

1 (a) -1 (b) $\sqrt{3}/3$

    (c) 2 (d) $\tan 20°$

2 (a) 1 (b) $\sqrt{3}$

    (c) -4 (d) $\cot 55°$

In Exercises 3 and 4 find the slope of the line with the given inclination.

3 (a) $0°$ (b) $120°$

   (c) $\arctan 1$ (d) $\arctan 4/3$

4 (a) $135°$ (b) $5\pi/6$

   (c) $\arcsin 1/2$ (d) $\arctan 1/2$

5 Prove that the following points are vertices of a rectangle: $A(6, 15)$, $B(11, 12)$, $C(-1, -8)$, $D(-6, -5)$.

6 Prove that the points $A(1, 4)$, $B(6, -4)$, and $C(-15, -6)$ are vertices of a right triangle.

In each of Exercises 7–10, find an equation for the lines satisfying the given conditions.

7 Through $A(7, -3)$, perpendicular to the line with equation $2x - 5y = 8$.

8 Through $A(-4, 8)$, perpendicular to the line through $B(5, -1)$ and $C(-2, -3)$.

9 Through $A(-7, 2)$, parallel to the line through $B(0, 4)$ and $C(-6, -6)$.

10 Through $P(-3/4, -1/2)$, parallel to the line with equation $x + 3y = 1$.

11 Find an equation for the perpendicular bisector of the line segment from $A(3, -1)$ to $B(-2, 6)$.

12 Find an equation for the perpendicular bisector of the line segment from the origin to $P(-5, 6)$.

13 Find equations for the altitudes of the triangle with vertices $A(-3, 2)$, $B(5, 4)$, $C(3, -8)$, and find the point at which they intersect.

14 Find equations for the medians of the triangle in Exercise 13, and find their point of intersection.

15 Find the tangent of the angle between lines whose slopes are

(a) 4 and 2/3   (b) -3 and -5.

16 Approximate, to the nearest degree, the angle between the lines having equations $6x - 2y - 3 = 0$ and $3x + 8y + 1 = 0$.

17 Approximate, to the nearest degree, the interior angles of the triangle with vertices $A(1, 6)$, $B(-3, -2)$, and $C(4, 4)$.

18 Given the triangle of Exercise 17, determine the slope of the bisector of the angle at vertex A.

19 Find the slope of a line through the point $P(4, -5)$ which makes an angle of $30°$ with the line having equation $3x + y - 18 = 0$.

20 If the slope of a line l_1 is 1/2, find the slope of a line l_2 such that the angle between l_1 and l_2 is $135°$.

10.3 CONIC SECTIONS

Each of the geometric figures to be discussed in the next five sections can be obtained by intersecting a double-napped right circular cone with a plane. For this reason they are called **conic sections**, or simply **conics**. If, as in (i) of Figure 10.6, the plane cuts entirely across one nappe of the cone and is not perpendicular to the axis, then the curve of intersection is called an **ellipse**. If the plane is perpendicular to the axis of the cone, a *circle* results. If the plane does not cut across one entire nappe and does not intersect both nappes, as illustrated in (ii) of Figure 10.6, then the curve of intersection is a **parabola**. If the plane cuts through both nappes of the cone, as in (iii) of Figure 10.6, then the resulting figure is called a **hyperbola**.

By changing the position of the plane and the shape of the cone, conics can be made to vary considerably. For certain positions of the plane there result what are called **degenerate conics**. For example, if the plane intersects the cone only at the vertex, then the conic consists of one point. If the axis of the cone lies on the plane, then a pair of intersecting lines is obtained. Finally, if we begin with the parabolic case, as in (ii) of

Figure 10.6, and move the plane parallel to its initial position until it coincides with one of the generators of the cone, a line results.

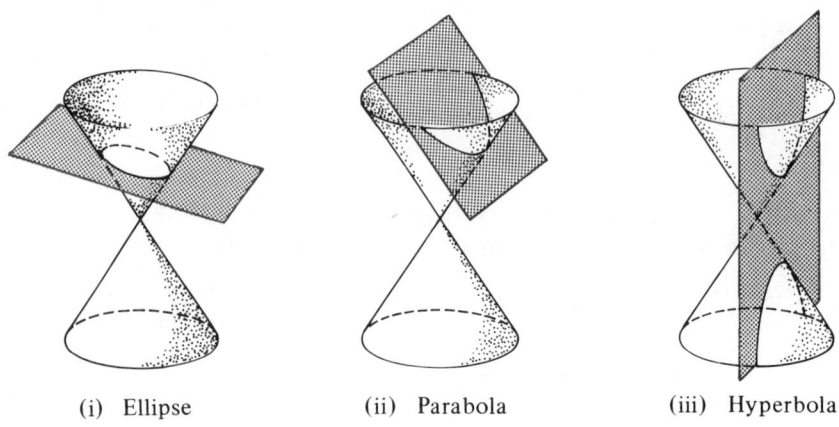

(i) Ellipse   (ii) Parabola   (iii) Hyperbola

Figure 10.6

The conic sections were studied extensively by the early Greek mathematicians, who used the methods of Euclidean geometry. They discovered the properties which enable us to define conics in terms of points (foci) and lines (directrices) in the plane of the conic. Reconciliation of the latter definitions with the previous discussion requires proofs which we shall not go into here.

A remarkable fact about conic sections is that although they were studied thousands of years ago, they are far from obsolete. Indeed, they are important tools for present-day investigations in outer space and for the study of the behavior of atomic particles. It is shown in physics that if a mass moves under the influence of what is called an *inverse square force field*, then its path may be described by means of a conic section. Examples of inverse square fields are gravitational and electromagnetic fields. Planetary orbits are elliptical. If the ellipse is very "flat," the curve resembles the path of a comet. The hyperbola is useful for describing the path of an alpha particle in the electric field of the nucleus of an atom. The interested person can find many other applications of conic sections.

10.4 PARABOLAS

Parabolas are very useful in applications of mathematics to the physical world. For example, it can be shown that if a projectile is fired and it is assumed that it is acted upon only by the force of gravity (that is, air resistance and other outside factors are ignored), then the path of the projectile is parabolic. Properties of parabolas are used in the design of mirrors for telescopes and searchlights. They are also employed in the design of field microphones used in television broadcasts of football games. These are only a few of many physical applications.

Definition

> A **parabola** is the set of all points in a plane equidistant from a fixed point F (the **focus**) and a fixed line l (the **directrix**) in the plane.

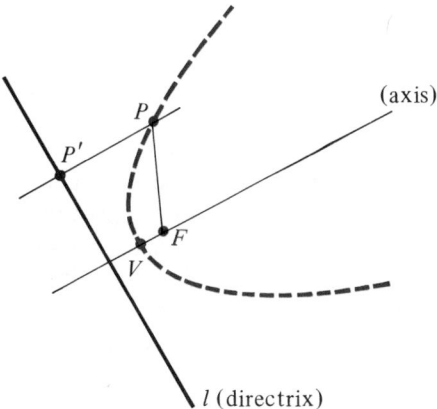

(axis)

l (directrix)

Figure 10.7

We shall assume that F is not on l, for otherwise the parabola degenerates into a line. If P is any point in the plane and P' is the point on l determined by a line through P which is perpendicular to l, then by definition, P is on the parabola if and only if $d(P, F) = d(P, P')$. A typical situation is illustrated in Figure 10.7, where the dashes indicate possible positions of P. The line through F, perpendicular to the directrix, is called the **axis** of the parabola. The point V on the axis, half-way from F to l, is called the **vertex** of the parabola.

In order to obtain a simple equation for a parabola, let us choose the y-axis along the axis of the parabola, with the origin at the vertex V, as illustrated in Figure 10.8. In this case, the focus F has coordinates $(0, p)$ for some real number $p \neq 0$, and the equation of the directrix is $y = -p$. By the distance formula, a point $P(x, y)$ is on the parabola if and only if

$$\sqrt{(x - 0)^2 + (y - p)^2} = \sqrt{(x - x)^2 + (y + p)^2}.$$

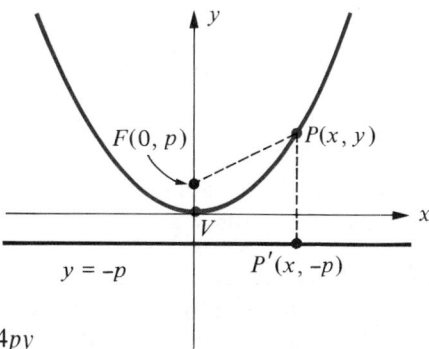

Figure 10.8. $x^2 = 4py$

Squaring both sides gives us

$$(x - 0)^2 + (y - p)^2 = (y + p)^2$$

or

$$x^2 + y^2 - 2py + p^2 = y^2 + 2py + p^2$$

which simplifies to

$$x^2 = 4py.$$

The last equation is called the **standard form** for the equation of a parabola with focus at $F(0, p)$ and directrix $y = -p$. If $p > 0$, the parabola **opens upward**, as in Figure 10.8, whereas if $p < 0$, the parabola **opens downward**.

An analogous situation exists if the axis of the parabola is taken along the x-axis. If the vertex is $V(0, 0)$, the focus $F(p, 0)$, and the directrix has equation $x = -p$ (see Figure 10.9), then using the same type of argument we obtain the **standard form**

$$y^2 = 4px.$$

If $p > 0$, the parabola opens to the right, as in Figure 10.9, whereas if $p < 0$, it opens to the left.

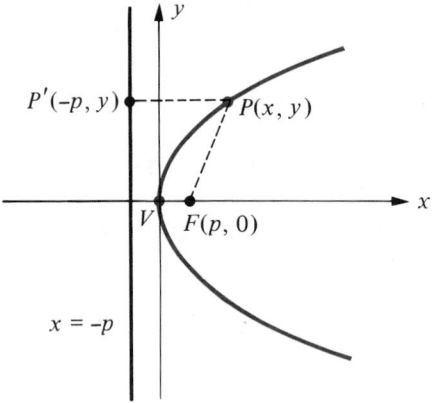

Figure 10.9. $y^2 = 4px$

Example 1   Find the focus and directrix of the parabola having equation $y^2 = -6x$, and sketch the graph.

Solution   The equation is in the standard form $y^2 = 4px$ with $4p = -6$, and hence $p = -3/2$. Consequently, the focus is $F(-3/2, 0)$ and the equation of the directrix is $x = 3/2$. The graph is sketched in Figure 10.10.

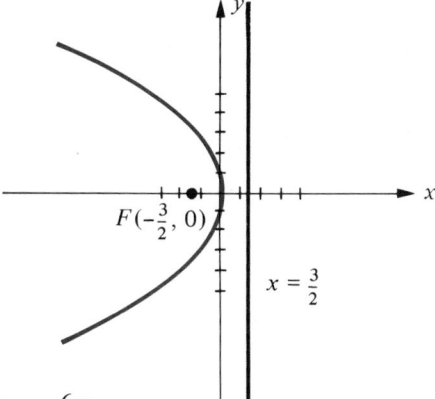

Figure 10.10. $y^2 = -6x$ ■

Example 2 Find an equation of the parabola that has vertex at the origin, opens upward, and passes through the point $P(-3, 7)$.

Solution According to our previous discussion, the general form of the equation is given by $x^2 = 4py$. If P is on the parabola, then $(-3, 7)$ is a solution of the equation. Hence we must have $(-3)^2 = 4p(7)$, or $p = 9/28$. Substitution for p leads to the desired equation $x^2 = (9/7)y$, or $7x^2 = 9y$. ■

It is worth noting that each of the graphs we have discussed is symmetric to one of the coordinate axes. For example, the graph of $x^2 = 4py$ in Figure 10.8 is symmetric with respect to the y-axis, since the equation is unchanged if x is replaced by $-x$. Similarly, the graph of $y^2 = 4px$ in Figure 10.9 is symmetric with respect to the x-axis, since the equation is unchanged if y is replaced by $-y$.

It is not difficult to extend our work to the case in which the axis of the parabola is parallel to one of the coordinate axes. In Figure 10.11 we have taken the vertex at the point $V(h, k)$, the focus at $F(h, k + p)$, and the directrix $y = k - p$. As before, the point $P(x, y)$ is on the parabola if and only if $d(P, F) = d(P, P')$, that is, if and only if

$$\sqrt{(x - h)^2 + (y - k - p)^2} = \sqrt{(x - x)^2 + (y - k + p)^2}.$$

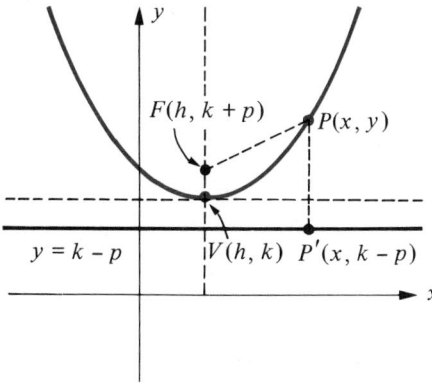

Figure 10.11. $(x - h)^2 = 4p(y - k)$

We leave it to the reader to show that the equation simplifies to

$$(x - h)^2 = 4p(y - k)$$

which is called the **standard form** for the equation of a parabola with vertex (h, k) and axis parallel to the y-axis. Squaring the left side and simplifying leads to an equation of the form

$$y = ax^2 + bx + c$$

where a, b, and c are real numbers. Conversely, given such an equation, we may complete the square in x to arrive at the standard form. Consequently, if $a \neq 0$, then the graph of $y = ax^2 + bx + c$ is a parabola with a vertical axis.

In similar fashion,

$$(y - k)^2 = 4p(x - h)$$

is called the **standard form** for the equation of a parabola with vertex (h, k) and axis parallel to the x-axis. This parabola opens to the right or left as p is positive or negative, respectively.

Example 3   Find an equation of the parabola with vertex $(4, -1)$, with axis parallel to the y-axis, and which passes through the origin.

Solution   The standard form of the equation is

$$(x - 4)^2 = 4p(y + 1).$$

If the origin is on the parabola, then $(0, 0)$ is a solution of this equation, and hence $(0 - 4)^2 = 4p(0 + 1)$. Consequently, $16 = 4p$ and $p = 4$. The desired equation is, therefore,

$$(x - 4)^2 = 16(y + 1). \qquad \blacksquare$$

Example 4   Discuss and sketch the graph of the equation

$$2x = y^2 + 8y + 22.$$

Solution   By our previous remarks, the graph of the equation is a parabola with a horizontal axis. Writing

$$y^2 + 8y = 2x - 22$$

we complete the square on the left by adding 16 to both sides. This gives us

$$y^2 + 8y + 16 = 2y - 6.$$

The latter equation may be written

$$(y + 4)^2 = 2(x - 3)$$

which is in standard form with $h = 3$, $k = -4$, and $p = 1/2$. Hence the vertex is $V(3, -4)$. Since $p = 1/2 > 0$, the parabola opens to the right with focus at $F(7/2, -4)$. The equation of the directrix is $x = h - p$, or $x = 5/2$. The parabola is sketched in Figure 10.12.

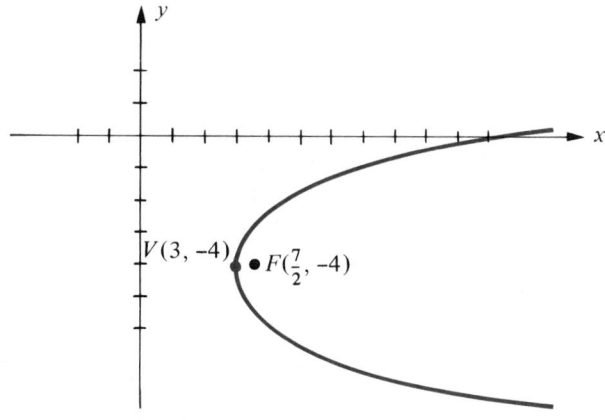

Figure 10.12. $2x = y^2 + 8y + 22$ ∎

EXERCISES 10.4

In each of Exercises 1–16 find the focus and directrix of the parabola with the given equation, and sketch the graph.

1 $y^2 = 12x$

2 $y^2 = -20x$

3 $x^2 = -8y$

4 $x^2 = 5y$

5 $4y^2 = -5x$

6 $4y^2 = x$

7 $9x^2 = y$

8 $7x^2 = 3y$

9 $(y + 3)^2 = 4(x - 1)$

10 $6(x - 5)^2 = 8 - y$

11 $y^2 - 4y - 2x - 4 = 0$

12 $y^2 + 14y + 4x + 45 = 0$

13 $4x^2 + 40x + y + 106 = 0$

14 $y^2 - 20y + 100 = 6x$

15 $y^2 + 20x = 10$

16 $4x^2 + 4x + 4y + 1 = 0$

In each of Exercises 17–22 find an equation for the parabola that satisfies the given conditions.

17 Focus $(-6, 0)$, directrix $x = 6$

18 Focus $(0, 3)$, directrix $y = -3$

19 Focus $(2, 5)$, directrix $y = -1$

20 Focus $(-4, 1)$, directrix $y = 5$

21 Vertex at the origin, symmetric to the y-axis, and passing through the point $A(3, -2)$

22 Vertex $V(1, -4)$, axis parallel to the x-axis, and passing through $A(-6, 7)$

23 A searchlight reflector is designed so that a cross section through its axis is a parabola and the light source is at the focus. Find the focus if the reflector is 3 feet across at the opening and 1 foot deep.

24 Prove that the point on a parabola that is closest to the focus is the vertex.

10.5 ELLIPSES

An ellipse may be defined as follows:

Definition

> An **ellipse** is the set of all points in a plane, the sum of whose distances from two fixed points in the plane (the **foci**) is constant.

It is known that the orbits of planets in the solar system are elliptical, with the sun at one of the foci. This is only one of many important applications of ellipses.

There is an easy way to construct an ellipse on paper. We may begin by inserting two thumbtacks in the paper at points labeled F and F' and fastening the ends of a piece of string to the thumbtacks. If the string is now looped around a pencil and drawn taut at point P, as in Figure 10.13, then moving the pencil and at the same time keeping the string taut, the sum of the distances $d(F, P)$ and $d(F', P)$ is the length of the string, and hence is constant. The pencil will, therefore, trace out a figure which resembles an ellipse with foci at F and F'. By varying the positions of F and F' but keeping the length of string fixed, the shape of the ellipse can be made to change considerably. If F and F' are far apart, in the sense that $d(F, F')$ is almost the same as the length of the string, then the ellipse is quite flat. On the other hand, if $d(F, F')$ is close to zero, the ellipse is almost circular. Indeed, if $F = F'$, a circle is obtained.

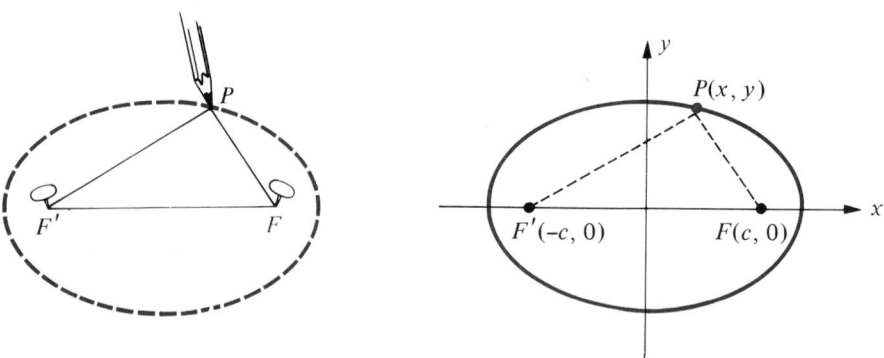

Figure 10.13   **Figure 10.14**

By introducing suitable coordinate systems we may derive simple equations for ellipses. Let us choose the x-axis as the line through the two foci F and F', with the origin at the midpoint of the segment $F'F$. This point is called the **center** of the ellipse. If F has coordinates $(c, 0)$, where $c > 0$, then, as shown in Figure 10.14, F' has coordinates $(-c, 0)$ and hence the distance between F and F' is $2c$. Let the constant sum of the distances of P from F and F' be denoted by $2a$, where in order to get points that are not on the x-axis we must have $2a > 2c$, that is, $a > c$. (Why?) By definition, $P(x, y)$ is on the ellipse if and only if

$$d(P, F) + d(P, F') = 2a$$

or, by the distance formula,

$$\sqrt{(x - c)^2 + (y - 0)^2} + \sqrt{(x + c)^2 + (y - 0)^2} = 2a.$$

Writing the above equation as

$$\sqrt{(x - c)^2 + y^2} = 2a - \sqrt{(x + c)^2 + y^2}$$

and squaring both sides, we obtain

$$x^2 - 2cx + c^2 + y^2 = 4a^2 - 4a\sqrt{(x + c)^2 + y^2} + x^2 + 2cx + c^2 + y^2$$

which simplifies to

$$a\sqrt{(x + c)^2 + y^2} = a^2 + cx.$$

Squaring both sides gives us

$$a^2(x^2 + 2cx + c^2 + y^2) = a^4 + 2a^2cx + c^2x^2$$

which may be written in the form

$$x^2(a^2 - c^2) + a^2y^2 = a^2(a^2 - c^2).$$

Dividing both sides by $a^2(a^2 - c^2)$ leads to

$$\frac{x^2}{a^2} + \frac{y^2}{a^2 - c^2} = 1.$$

For convenience, we let

$$\boxed{b^2 = a^2 - c^2 \qquad \text{where } b > 0}$$

obtaining

$$\boxed{\frac{x^2}{a^2} + \frac{y^2}{b^2} = 1.}$$

Since $c > 0$ and $b^2 = a^2 - c^2$, it follows that $a^2 > b^2$ and hence $a > b$. The equation we have derived is called the **standard form** for the equation of an ellipse with foci on the x-axis and center at the origin. The x-intercepts of the graph may be found by setting $y = 0$. Doing so gives us $x^2/a^2 = 1$, or $x^2 = a^2$, and consequently, the x-intercepts are a and $-a$. The corresponding points $V(a, 0)$ and $V'(-a, 0)$ on the graph are called the **vertices** of the ellipse, and the line segment $V'V$ is referred to as the **major axis**. The y-intercepts are b and $-b$. The segment from $M'(0, -b)$ to $M(0, b)$ is called the **minor axis** of the ellipse. Note that the major axis is longer than the minor axis, since $a > b$.

Applying tests for symmetry we see that the ellipse is symmetric to both the x-axis and the y-axis. It is also symmetric with respect to the origin since substitution of $-x$ for x and $-y$ for y does not change the equation.

Example 1 Discuss and sketch the graph of the equation

$$4x^2 + 18y^2 = 36.$$

Solution To obtain the standard form, we divide both sides of the given equation by 36 and simplify. This leads to

$$\frac{x^2}{9} + \frac{y^2}{2} = 1$$

which is in standard form with $a^2 = 9$ and $b^2 = 2$. Thus $a = 3, b = \sqrt{2}$, and hence the endpoints of the major axis are $(\pm 3, 0)$ and the endpoints of the minor axis are $(0, \pm \sqrt{2})$. Since $b^2 = a^2 - c^2$, we have

$$c^2 = a^2 - b^2 = 9 - 2 = 7, \quad \text{or} \quad c = \sqrt{7}.$$

Consequently, the foci are $(\pm \sqrt{7}, 0)$. The graph is sketched in Figure 10.15.

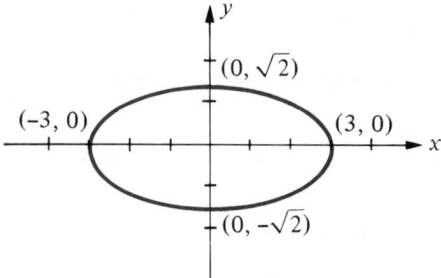

Figure 10.15. $4x^2 + 18y^2 = 36$

Example 2 Find an equation of the ellipse with vertices $(\pm 4, 0)$ and foci $(\pm 2, 0)$.

Solution Since $a = 4$ and $c = 2$,

$$b^2 = a^2 - c^2 = 16 - 4 = 12.$$

Substitution in the standard form gives us

$$\frac{x^2}{16} + \frac{y^2}{12} = 1.$$

Multiplying both sides by 48 leads to $3x^2 + 4y^2 = 48$. ∎

It is sometimes convenient to choose the major axis of an ellipse along the y-axis. If the foci are $(0, \pm c)$, then by the same type of argument used previously we obtain the following **standard form** for the equation of an ellipse with foci on the y-axis and center at the origin:

$$\frac{x^2}{b^2} + \frac{y^2}{a^2} = 1$$

where $a > b$. As before, the connection between a, b, and c is given by $b^2 = a^2 - c^2$ or, equivalently, by $c^2 = a^2 - b^2$. In this case the vertices are $V(0, a)$ and $V'(0, -a)$. The endpoints of the minor axis are $M(b, 0)$ and $M'(-b, 0)$. A typical graph is sketched in Figure 10.16.

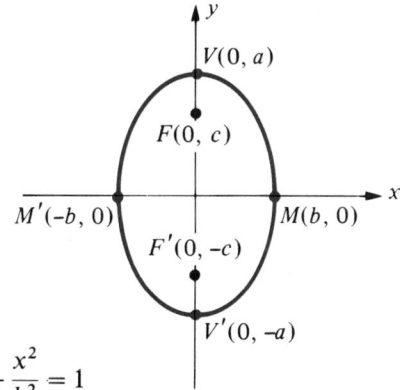

Figure 10.16. $\dfrac{y^2}{a^2} + \dfrac{x^2}{b^2} = 1$

The preceding discussion shows that an equation of an ellipse with center at the origin and foci on a coordinate axis may always be written in the form

$$\frac{x^2}{p} + \frac{y^2}{q} = 1 \quad \text{or} \quad qx^2 + py^2 = pq$$

where p and q are positive. If $p > q$, then the major axis lies on the x-axis, whereas if $q > p$, then the major axis is on the y-axis. It is unnecessary to memorize these facts, since in any given problem the major axis can be determined by examining the x- and y-intercepts.

Example 3 Sketch the graph of the equation $9x^2 + 4y^2 = 25$.

Solution The graph is an ellipse with center at the origin and foci on one of the coordinate axes. To find the x-intercepts, we let $y = 0$, obtaining $9x^2 = 25$, or $x = \pm 5/3$. Similarly, to find the y-intercepts, we let $x = 0$, obtaining $4y^2 = 25$, or $y = \pm 5/2$.

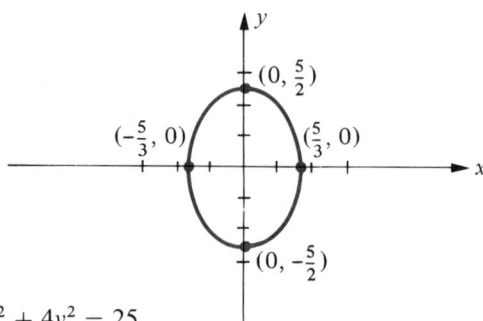

Figure 10.17. $9x^2 + 4y^2 = 25$

This enables us to sketch the ellipse (see Figure 10.17). Since $5/3 < 5/2$, the major axis is on the y-axis. ∎

EXERCISES 10.5

In each of Exercises 1–8 sketch the graph of the equation and give coordinates of the vertices and foci.

1 $\dfrac{x^2}{49} + \dfrac{y^2}{25} = 1$ **2** $\dfrac{x^2}{16} + \dfrac{y^2}{9} = 1$

3 $9x^2 + y^2 = 36$ **4** $y^2 + 25x^2 = 25$

5 $3x^2 + 4y^2 = 12$ **6** $x^2 + 4y^2 = 16$

7 $4x^2 + 9y^2 = 1$ **8** $3y^2 + x^2 = 9$

In each of Exercises 9–14 find an equation for the ellipse satisfying the given conditions.

9 Vertices $V(0, \pm 8)$, foci $F(0, \pm 5)$

10 Vertices $V(\pm 10, 0)$, foci $F(\pm 6, 0)$

11 Vertices $V(\pm 5, 0)$, length of minor axis 3

12 Foci $F(0, \pm 3)$, length of minor axis 2

13 Vertices $V(0, \pm 10)$, passing through $(4, 3)$

14 Center at the origin, symmetric with respect to both axes, passing through $A(-6, 1)$ and $B(2, -3)$.

In each of Exercises 15 and 16 find the points of intersection of the graphs of the given equations. Sketch both graphs on the same coordinate axes, showing points of intersection.

15 $\begin{cases} x^2 + 4y^2 = 20 \\ x + 2y = 6 \end{cases}$

16 $\begin{cases} x^2 + 4y^2 = 36 \\ x^2 + y^2 = 12 \end{cases}$

17 An arch of a bridge is semielliptical with its major axis horizontal. The base of the arch is 30 feet across and the highest part of the arch is 10 feet above the horizontal roadway. Find the height of the arch 6 feet from the center of the base.

18 If a square with sides parallel to the coordinate axes is inscribed in the ellipse with equation $x^2/a^2 + y^2/b^2 = 1$, express the area A of the square in terms of a and b.

19 The **eccentricity** of an ellipse is defined as the ratio $(\sqrt{a^2 - b^2})/a$. If a is fixed and b varies, describe the general shape of the ellipse when the eccentricity is close to 1 and when it is close to zero.

20 Derive the standard form $x^2/b^2 + y^2/a^2 = 1$.

10.6 HYPERBOLAS

The definition of hyperbola is similar to that of an ellipse. The only change is that instead of using the *sum* of distances from two fixed points we use the *difference*.

Definition

> A **hyperbola** is the set of all points in a plane, the difference of whose distances from two fixed points in the plane (the **foci**) is a positive constant.

To find a simple equation for a hyperbola, we choose a coordinate system with foci at $F(c, 0)$ and $F'(-c, 0)$, and denote the (constant) distance by $2a$. Referring to Figure 10.18, we see that a point $P(x, y)$ is on the hyperbola if and only if either one of the following is true:

(∗)
$$d(P, F) - d(P, F') = 2a$$
$$d(P, F') - d(P, F) = 2a.$$

For hyperbolas (unlike ellipses) we need $a < c$ in order to obtain points on the hyperbola which are not on the x-axis, for if P is such a point, then from Figure 10.18 we see that

$$d(P, F) < d(F', F) + d(P, F')$$

since the length of one side of a triangle is always less than the sum of the lengths of the other two sides. Similarly,

$$d(P, F') < d(F', F) + d(P, F).$$

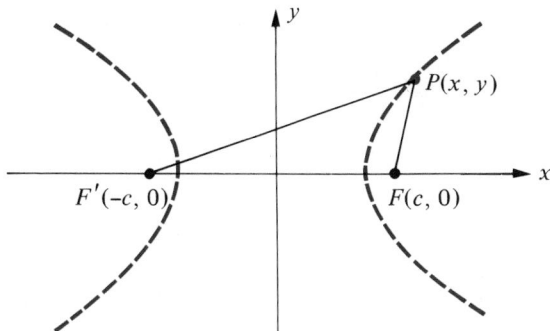

Figure 10.18

Equivalent forms for the previous two inequalities are

$$d(P, F) - d(P, F') < d(F', F)$$
$$d(P, F') - d(P, F) < d(F', F).$$

From the equations labelled (∗) and the fact that $d(F', F) = 2c$, the latter inequalities imply that $2a < 2c$, or $a < c$.

Equations (∗) may be replaced by the single equation

$$|d(P, F) - d(P, F')| = 2a.$$

It then follows from the distance formula that an equation of the hyperbola is given by

$$\left|\sqrt{(x - c)^2 + (y - 0)^2} - \sqrt{(x + c)^2 + (y - 0)^2}\right| = 2a.$$

Employing the type of simplification procedure used to derive an equation for an ellipse, we arrive at the equivalent equation

$$\frac{x^2}{a^2} - \frac{y^2}{c^2 - a^2} = 1.$$

For convenience, we let

$$\boxed{b^2 = c^2 - a^2 \qquad \text{where } b > 0}$$

in the preceding equation, obtaining

$$\boxed{\frac{x^2}{a^2} - \frac{y^2}{b^2} = 1.}$$

The latter equation is called the **standard form** for the equation of a hyperbola with foci on the x-axis and center at the origin. By the tests for symmetry we see that this hyperbola is symmetric with respect to both axes and the origin. The x-intercepts are $\pm a$. The corresponding points $V(a, 0)$ and $V'(-a, 0)$ are called the **vertices**, and the line segment $V'V$ is known as the **transverse axis** of the hyperbola. There are no y-intercepts, since the equation $-y^2/b^2 = 1$ has no solution.

If the equation $x^2/a^2 - y^2/b^2 = 1$ is solved for y, we obtain

$$y = \pm \frac{b}{a}\sqrt{x^2 - a^2}.$$

Hence there are no points (x, y) on the graph if $x^2 - a^2 < 0$, that is, if $-a < x < a$. However, there *are* points $P(x, y)$ on the graph if $x \geq a$ or $x \leq -a$. In order to arrive at a precise description of the graph, it is necessary to investigate the position of the point $P(x, y)$ on the hyperbola when x is numerically very large. If $x \geq a$, we may write the last equation in the form

$$y = \pm \frac{b}{a}x\sqrt{1 - \frac{a^2}{x^2}}.$$

It follows that if x is large (in comparison to a), the radicand is close to 1, and hence the ordinate y of the point $P(x, y)$ on the hyperbola is close to either $(b/a)x$ or $-(b/a)x$. This means that the point $P(x, y)$ is close to the line with equation $y = (b/a)x$ when y is positive, or the line with equation $y = -(b/a)x$ when y is negative. As x increases (or decreases), we say that the point $P(x, y)$ *approaches* one of these lines. A corresponding situation exists when $x \leq -a$. The lines with equations

$$y = \pm \frac{b}{a} x$$

are called the **asymptotes** of the hyperbola under discussion. The asymptotes serve as an excellent guide for sketching the graph. This is illustrated in Figure 10.19, where we have represented the asymptotes by dashed lines and indicated the manner in which the points on the hyperbola approach the asymptotes as x increases or decreases. The two curves that make up the hyperbola are called the **branches** of the hyperbola.

A convenient way to sketch the asymptotes is first to plot the vertices $V(a, 0)$ and $V'(-a, 0)$ and the points $W(0, b)$ and $W'(0, -b)$ (see Figure 10.20). The line segment $W'W$ of length $2b$ is called the **conjugate axis** of the hyperbola. If horizontal and vertical lines are drawn through the endpoints of the conjugate and transverse axes, respectively, then the diagonals of the resulting rectangle have slopes b/a and $-b/a$. Consequently, by extending these diagonals, we obtain the asymptotes. The hyperbola is then sketched as in Figure 10.20 using the asymptotes as a guide.

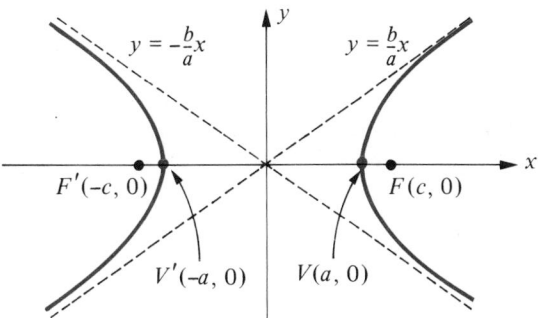

Figure 10.19

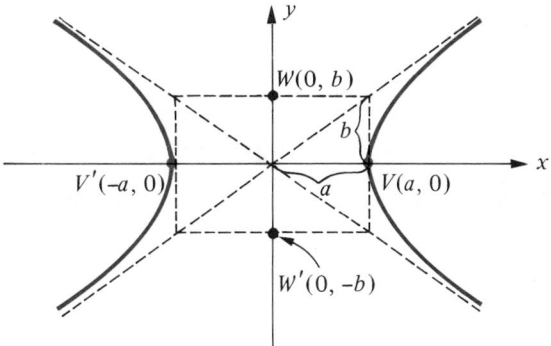

Figure 10.20. $\dfrac{x^2}{a^2} - \dfrac{y^2}{b^2} = 1$

Example 1   Discuss and sketch the graph of the equation

$$9x^2 - 4y^2 = 36.$$

Solution   Dividing both sides by 36, we have

$$\frac{x^2}{4} - \frac{y^2}{9} = 1$$

which is in the preceding standard form with $a^2 = 4$ and $b^2 = 9$. Hence $a = 2$ and $b = 3$. The vertices $(\pm 2, 0)$ and the endpoints $(0, \pm 3)$ of the conjugate axis determine a rectangle whose diagonals (extended) give us the asymptotes. The graph of the equation is sketched in Figure 10.21. The equations of the asymptotes, $y = \pm\frac{3}{2}x$, may be found by referring to the graph, or to the equations $y = \pm (b/a)x$. Since

$$c^2 = a^2 + b^2 = 4 + 9 = 13$$

the foci are $(\pm\sqrt{13}, 0)$.

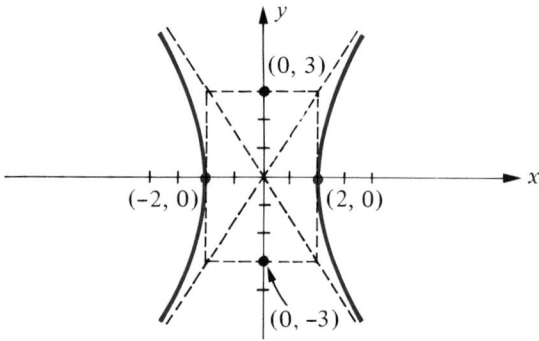

Figure 10.21. $\dfrac{x^2}{4} - \dfrac{y^2}{9} = 1$ ∎

The preceding example indicates that for hyperbolas it is not always true that $a > b$, as was the case for ellipses. Indeed, we may have $a < b$, $a > b$, or $a = b$.

Example 2   Find an equation, the foci, and the asymptotes of the hyperbola which has vertices $(\pm 3, 0)$ and passes through the point $P(5, 2)$.

Solution   Substituting $a = 3$ in the standard form we obtain the equation

$$\frac{x^2}{9} - \frac{y^2}{b^2} = 1.$$

If $(5, 2)$ is a solution of this equation, then

$$\frac{25}{9} - \frac{4}{b^2} = 1.$$

Solving for b^2 gives us $b^2 = 9/4$, and hence the desired equation is

$$\frac{x^2}{9} - \frac{4y^2}{9} = 1$$

or equivalently, $x^2 - 4y^2 = 9$.

Since $c^2 = a^2 + b^2 = 9 + \frac{9}{4} = \frac{45}{4}$ it follows that $c = \sqrt{45/4} = 3\sqrt{5}/2$. Hence the foci are $(\pm\frac{3}{2}\sqrt{5}, 0)$. Substituting for b and a in $y = \pm(b/a)x$ and simplifying, we obtain equations $y = \pm\frac{1}{2}x$ of the asymptotes. ■

If the foci of a hyperbola are the points $(0, \pm c)$ on the y-axis, then it can be shown that an equation for the hyperbola is

$$\boxed{\frac{y^2}{a^2} - \frac{x^2}{b^2} = 1.}$$

This equation is called the **standard form** for the equation of a hyperbola with foci on the y-axis and center at the origin. The numbers a, b and c are again related by means of the equation $b^2 = c^2 - a^2$. The points $V(0, a)$ and $V'(0, -a)$ are the vertices of the hyperbola, and the endpoints of the conjugate axis are now $W(b, 0)$ and $W'(-b, 0)$. The asymptotes are found, as before, by using the diagonals of the rectangle determined by these points and lines parallel to the coordinate axes. The graph is sketched in Figure 10.22. The equations of the asymptotes are

$$\boxed{y = \pm\frac{a}{b}x.}$$

Note the difference between these equations and the equations $y = \pm(b/a)x$ used previously.

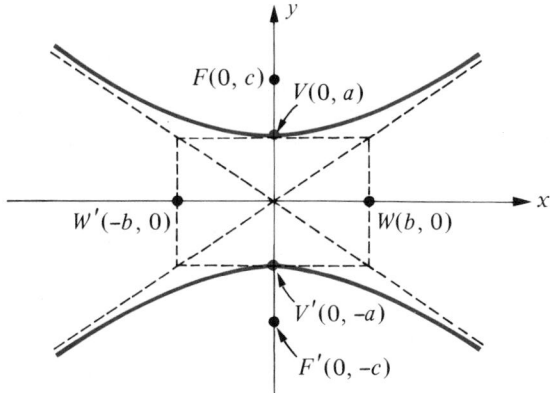

Figure 10.22. $\dfrac{y^2}{a^2} - \dfrac{x^2}{b^2} = 1$

Example 3   Discuss and sketch the graph of the equation

$$4y^2 - 2x^2 = 1.$$

Solution   The standard form is

$$\frac{y^2}{1/4} - \frac{x^2}{1/2} = 1.$$

Thus $a^2 = 1/4$, $b^2 = 1/2$, and $c^2 = 1/4 + 1/2 = 3/4$. Consequently, $a = 1/2$, $b = \sqrt{2}/2$, and $c = \sqrt{3}/2$. The vertices are $(0, \pm 1/2)$ and the foci are $(0, \pm \sqrt{3}/2)$. The graph is sketched in Figure 10.23. Equations of the asymptotes are $y = \pm(\sqrt{2}/2)x$.

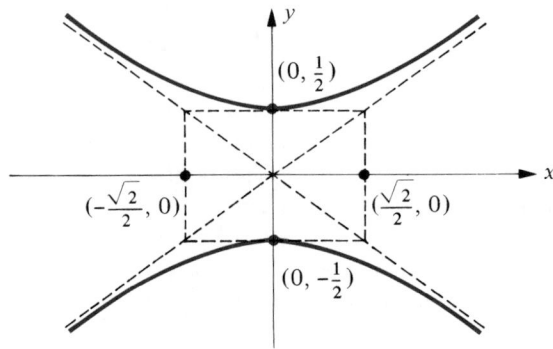

Figure 10.23. $4y^2 - 2x^2 = 1$ ■

EXERCISES 10.6

In Exercises 1–10 sketch the graph of the equation, find the coordinates of the vertices and foci, and write equations for the asymptotes.

1 $\dfrac{x^2}{49} - \dfrac{y^2}{25} = 1$ **2** $\dfrac{y^2}{16} - \dfrac{x^2}{9} = 1$

3 $\dfrac{y^2}{49} - \dfrac{x^2}{25} = 1$ **4** $\dfrac{x^2}{16} - \dfrac{y^2}{9} = 1$

5 $y^2 - 9x^2 = 36$ **6** $x^2 - 5y^2 = 25$

7 $x^2 - y^2 = 1$ **8** $4y^2 - 4x^2 = 1$

9 $16x^2 - 36y^2 = 1$ **10** $3x^2 - y^2 = -3$

In Exercises 11–16 find an equation for the hyperbola satisfying the given conditions.

11 Foci $F(\pm 8, 0)$, vertices $V(\pm 5, 0)$

12 Foci $F(0, \pm 4)$, vertices $V(0, \pm 1)$

13 Foci $F(0, \pm 3)$, length of conjugate axis 2

14 Vertices $V(\pm 6, 0)$ passing through $P(10, 4)$

15 Vertices $V(\pm 10, 0)$, equations of asymptotes $y = \pm \frac{1}{2}x$

16 Foci $F(0, \pm 9)$, equations of asymptotes $y = \pm \frac{1}{5}x$

In each of Exercises 17 and 18 find the points of intersection of the graphs of the given equations and sketch both graphs on the same coordinate axes, showing points of intersection.

17 $\begin{cases} y^2 - 4x^2 = 16 \\ y - x = 4 \end{cases}$ **18** $\begin{cases} x^2 - y^2 = 4 \\ y^2 - 3x = 0 \end{cases}$

19 The graphs of the equations

$$\frac{x^2}{a^2} - \frac{y^2}{b^2} = 1 \quad \text{and} \quad \frac{x^2}{a^2} - \frac{y^2}{b^2} = -1$$

are called **conjugate hyperbolas**. Sketch the graphs of both equations on the same

coordinate system with $a = 2$ and $b = 5$. Describe the relationship between the two graphs.

20 Show how to obtain the equation

$$\frac{x^2}{a^2} - \frac{y^2}{c^2 - a^2} = 1$$

from the equation

$$\left| \sqrt{(x - c)^2 + (y - 0)^2} - \sqrt{(x + c)^2 + (y - 0)^2} \right| = 2a.$$

10.7 TRANSLATION OF AXES

If a and b are coordinates of two points A and B, respectively, on a coordinate line l, the distance between A and B was defined in Chapter One by $d(A, B) = |b - a|$. If we wish to take into account the direction of l, then we use the **directed distance** $\overline{AB}$ from A to B where, by definition,

$$\boxed{\overline{AB} = b - a.}$$

Since $\overline{BA} = a - b$, we have $\overline{AB} = -\overline{BA}$. If the positive direction on l is to the right, then B is to the right of A if and only if $\overline{AB} > 0$, and is to the left of A if and only if $\overline{AB} < 0$. If C is any other point on l with coordinate c, it follows that

$$\boxed{\overline{AC} = \overline{AB} + \overline{BC}}$$

since $(c - a) = (b - a) + (c - b)$. We shall use this formula for directed distances to develop formulas for **translation of axes** in two dimensions.

Suppose that $C(h, k)$ is an arbitrary point in an xy-coordinate plane. Let us introduce a new $x'y'$-coordinate system with origin O' at C such that the x'- and y'- axes are parallel to, and have the same unit lengths and positive directions as, the x- and y-axes, respectively. A typical situation of this type is illustrated in Figure 10.24 where, for simplicity, we have placed C in the first quadrant. We shall use primes on letters to denote coordinates of points in the $x'y'$-coordinate system in order to distinguish them from coordinates with respect to the xy-coordinate system. Thus the point $P(x, y)$ in the xy-system will be denoted by $P(x', y')$ in the $x'y'$-system. If we label projections of P on the various axes as indicated in Figure 10.24, and let A and B denote projections of C on the x- and y-axes, respectively, then, using formulas for directed distances,

$$x = \overline{OQ} = \overline{OA} + \overline{AQ} = \overline{OA} + \overline{O'Q'} = h + x'$$
$$y = \overline{OR} = \overline{OB} + \overline{BR} = \overline{OB} + \overline{O'R'} = k + y'.$$

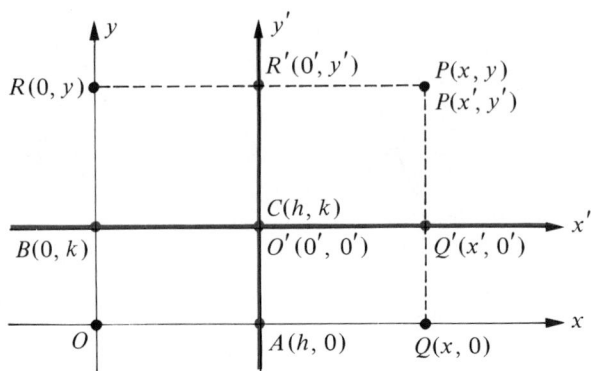

Figure 10.24

To summarize, if (x, y) are the coordinates of a point P relative to the xy-coordinate system, and if (x', y') are the coordinates of P relative to an $x'y'$-coordinate system with origin at the point $C(h, k)$ of the xy-system, then we have the following:

Translation of Axes Formulas

(a) $x = x' + h, \quad y = y' + k$

(b) $x' = x - h, \quad y' = y - k$

These formulas enable us to go from either coordinate system to the other. Their major use is to change the form of equations of graphs. To be specific, if, in the xy-plane, a certain collection of points is the graph of an equation in x and y, then, to find an equation in x' and y' which has the same graph in the $x'y'$-plane, we may substitute $x' + h$ for x and $y' + k$ for y in the given equation. Conversely, if a set of points in the $x'y'$-plane is the graph of an equation in x' and y', then, to find the corresponding equation in x and y, we substitute $x - h$ for x' and $y - k$ for y'.

As a simple illustration of the preceding remarks, the equation

$$(x')^2 + (y')^2 = r^2$$

has, for its graph in the $x'y'$-plane, a circle of radius r with center at the origin O'. Using the translation of axes formulas $x' = x - h, y' = y - k$ we see that an equation for this circle in the xy-plane is

$$(x - h)^2 + (y - k)^2 = r^2,$$

which is in agreement with the formula for a circle of radius r with center at $C(h, k)$ in the xy-plane.

As another illustration, we know that

$$(x')^2 = 4py'$$

is an equation of a parabola with vertex at the origin O' of the $x'y'$-plane. Applying translation of axes formulas, it follows that

$$(x - h)^2 = 4p(y - k)$$

is an equation of the same parabola in the xy-plane. This checks with the formula in Section 10.4 which was derived for a parabola with vertex at the point $V(h, k)$ in the xy-plane.

It should now be evident how this technique can be applied to all the conics. For example, the graph of

$$\frac{(x')^2}{a^2} + \frac{(y')^2}{b^2} = 1$$

is an ellipse with center at O' in the $x'y'$-plane, as illustrated in Figure 10.25. According to translation of axes formulas, its equation relative to the xy-coordinate system is

$$\frac{(x - h)^2}{a^2} + \frac{(y - k)^2}{b^2} = 1.$$

This is called a **standard form** for the equation of an ellipse with center (h, k). A similar situation exists for hyperbolas.

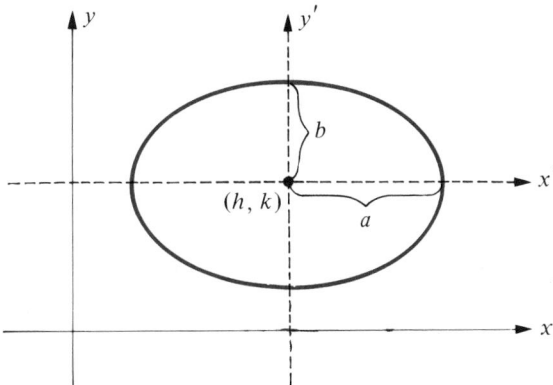

Figure 10.25. $\dfrac{(x')^2}{a^2} + \dfrac{(y')^2}{b^2} = 1$ or $\dfrac{(x - h)^2}{a^2} + \dfrac{(y - k)^2}{b^2} = 1$

In certain cases, given an equation in x and y we may, by a proper translation of axes, obtain a simpler equation in x' and y' which has the same graph. In particular, this is true for an equation in x and y of the form

$$Ax^2 + Cy^2 + Dx + Ey + F = 0$$

where the coefficients are real numbers. The graph of this equation is a conic, except for the degenerate cases in which points, lines, or no graphs are obtained. We shall not give a general proof of this fact but will, instead, illustrate the procedure by means of examples.

Example 1   Discuss and sketch the graph of the equation

$$16x^2 + 9y^2 + 64x - 18y - 71 = 0.$$

Solution   In order to determine the origin of a new $x'y'$-coordinate system which will enable us to simplify the given equation, we begin by writing the given equation in the form

$$16(x^2 + 4x) + 9(y^2 - 2y) = 71.$$

Next, we complete the squares for the expressions within parentheses, obtaining

$$16(x^2 + 4x + 4) + 9(y^2 - 2y + 1) = 71 + 64 + 9.$$

Note that by adding 4 to the expression within the first parentheses we have added 64 to the left side of the equation and hence must compensate by adding 64 to the right side. Similarly, by adding 1 to the expression within the second parentheses, 9 is added to the left side and consequently 9 must also be added to the right side. The last equation may be written

$$16(x + 2)^2 + 9(y - 1)^2 = 144.$$

Dividing by 144 we obtain

$$\frac{(x + 2)^2}{9} + \frac{(y - 1)^2}{16} = 1$$

which is of the form

$$\frac{(x')^2}{9} + \frac{(y')^2}{16} = 1$$

where $x' = x + 2$ and $y' = y - 1$. This shows that if we let $h = -2$ and $k = 1$ in the translation of axes formulas we obtain the given equation. Since the graph of $(x')^2/9 + (y')^2/16 = 1$ is an ellipse with center at the origin O' in the $x'y'$-plane it follows that the given equation is an ellipse with center $C(-2, 1)$ in the xy-plane and with axes parallel to the coordinate axes. The graph is sketched in Figure 10.26.   ■

Example 2   Discuss and sketch the graph of the equation

$$9x^2 - 4y^2 - 54x - 16y + 29 = 0.$$

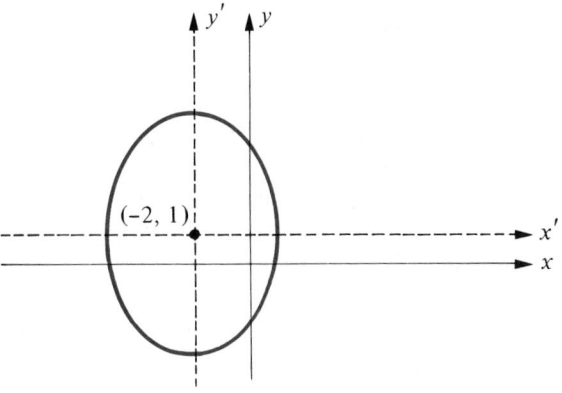

Figure 10.26. $\dfrac{(x + 2)^2}{9} + \dfrac{(y - 1)^2}{16} = 1$

Solution   As in Example 1, we arrange our work as follows:

$$9(x^2 - 6x) - 4(y^2 + 4y) = -29$$
$$9(x^2 - 6x + 9) - 4(y^2 + 4y + 4) = -29 + 81 - 16$$
$$9(x - 3)^2 - 4(y + 2)^2 = 36$$
$$\frac{(x - 3)^2}{4} - \frac{(y + 2)^2}{9} = 1.$$

If we substitute $h = 3$ and $k = -2$ in the translation of axes formulas, then the given equation reduces to a standard form for the equation of a hyperbola, namely,

$$\frac{(x')^2}{4} - \frac{(y')^2}{9} = 1.$$

By translating the x- and y-axes to the new origin $C(3, -2)$ we obtain the sketch shown in Figure 10.27.

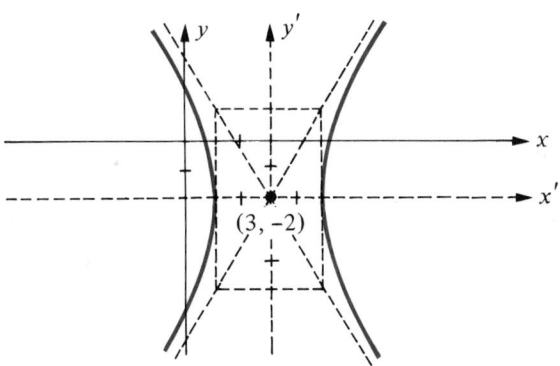

Figure 10.27. $\dfrac{(x - 3)^2}{4} - \dfrac{(y + 2)^2}{9} = 1$ ■

Example 3   Discuss and sketch the graph of the equation

$$2x = y^2 + 8y + 22.$$

Solution   This example is the same as Example 4 of Section 10.4, where we completed the square in y and obtained the equation $(y + 4)^2 = 2(x - 3)$. By the methods of the present section, if we let $h = 3$ and $k = -4$ in the translation of axes formulas, then the given equation reduces to the standard form $(y')^2 = 2x'$ of a parabola with vertex at O'. Consequently, a translation of axes to the new origin $C(3, -4)$ leads to the sketch shown in Figure 10.12.   ∎

Although we have only considered special examples, our methods are perfectly general. If, in the equation

$$Ax^2 + Cy^2 + Dx + Ey + F = 0,$$

A and C are equal and not zero, then the graph, when it exists, is a circle or, in exceptional cases, a point. If A and C are unequal but have the same sign, then by completing squares and properly translating axes we obtain an equation whose graph, when it exists, is an ellipse (or a point). If A and C have opposite signs, the equation of a hyperbola is obtained, or possibly, in the degenerate case, two intersecting straight lines. Finally, if either A or C (but not both) is zero, the graph is a parabola, or, in certain cases, a pair of parallel straight lines.

EXERCISES 10.7

Discuss and sketch the graph of each of the equations in Exercises 1–26 after making a suitable translation of axes.

1 $(y + 3)^2 = 12(x - 1)$

2 $(x - 8)^2 = -4(y + 2)$

3 $\dfrac{(x + 5)^2}{16} + \dfrac{(y - 4)^2}{9} = 1$

4 $8(x + 3)^2 + (y - 6)^2 = 32$

5 $\dfrac{(x - 2)^2}{36} - \dfrac{(y + 7)^2}{49} = 1$

6 $\dfrac{(y - 2)^2}{9} - (x + 4)^2 = 1$

7 $4(x + 2)^2 + (y - 2)^2 = 1$

8 $(x + 1)^2 + 9y^2 = 36$

9 $16y^2 - 100(x - 3)^2 = 1600$

10 $(x - 7)^2 - (y + 5)^2 = 1$

11 $4x^2 + y^2 + 24x - 10y + 45 = 0$

12 $9x^2 + 16y^2 + 36x + 96y + 36 = 0$

13 $9x^2 + y^2 - 108x - 4y + 319 = 0$

14 $x^2 + 4y^2 - 2x = 0$

15 $2x^2 - 5y + 8x + 58 = 0$

16 $y^2 + 2x - 16y + 66 = 0$

17 $y^2 - 4x^2 + 6y - 40x - 107 = 0$

18 $25y^2 - 9x^2 - 100y - 54x + 10 = 0$

19 $9x^2 - y^2 - 36x + 12y - 9 = 0$

20 $4y^2 - x^2 + 32y - 8x + 49 = 0$

21 $y = |x - 5| - 4$

22 $x + 2 = \sqrt{(y - 1)^2}$

23 $x = 3 + (y - 6)^3$

24 $(x - 7)^3 - y - 2 = 0$

25 $2y + 10 - (x + 2)^4 = 0$

26 $(y + 7)(x - 5) = 1$

27 Find an equation of the hyperbola with foci $(h \pm c, k)$ and vertices $(h \pm a, k)$, where $0 < a < c$ and $c^2 = a^2 + b^2$.

28 Find an equation of the hyperbola with vertices $(h, k \pm a)$ and asymptotes $y - k = \pm (a/b)(x - h)$, where $b > 0$.

10.8 ROTATION OF AXES

The $x'y'$-coordinate system discussed in Section 10.7 may be thought of as having been obtained by moving the origin O of the xy-system to a new position $C(h, k)$ while, at the same time, not changing the positive directions of the axes or the units of length. We shall now introduce a new coordinate system by keeping the origin O fixed and rotating the x- and y-axes about O to another position denoted by x' and y'. A transformation of this type will be referred to as a **rotation of axes**.

Let us consider a rotation of axes and, as shown in Figure 10.28, let ϕ denote the angle through which the positive x-axis must be rotated in order to coincide with the positive x'-axis. If (x, y) are the coordinates of a point P relative to the xy-plane, then, as before, (x', y') will denote its coordinates relative to the new $x'y'$-coordinate system. Let the projections of P on the various axes be denoted as in Figure 10.28 and let θ denote angle POQ'. If $p = d(O, P)$, then

$$x' = p \cos \theta, \qquad y' = p \sin \theta,$$
$$x = p \cos (\theta + \phi), \qquad y = p \sin (\theta + \phi).$$

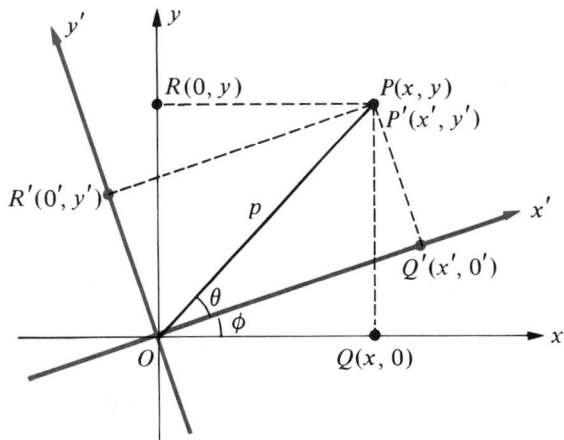

Figure 10.28

Applying addition formulas gives us

$$x = p \cos \theta \cos \phi - p \sin \theta \sin \phi$$
$$y = p \sin \theta \cos \phi + p \cos \theta \sin \phi.$$

Since $p \cos \theta = x'$ and $p \sin \theta = y'$, these equations reduce to formulas (a) in the following.

**Rotation
of Axes
Formulas**

> (a) $x = x' \cos \phi - y' \sin \phi, \quad y = x' \sin \phi + y' \cos \phi$
>
> (b) $x' = x \cos \phi + y \sin \phi, \quad y' = -x \sin \phi + y \cos \phi$

The equations in (b) may be obtained by solving those in (a) for x' and y'.

Example 1   The graph of the equation $xy = 1$, or equivalently $y = 1/x$, is sketched in Figure 3.16. If the coordinate axes are rotated through an angle of $45°$, find the equation of the graph relative to the new $x'y'$-coordinate system.

Solution   Letting $\phi = 45°$ in the rotation of axes formulas,

$$x = x'(\sqrt{2}/2) - y'(\sqrt{2}/2) = (\sqrt{2}/2)(x' - y')$$

$$y = x'(\sqrt{2}/2) + y'(\sqrt{2}/2) = (\sqrt{2}/2)(x' + y').$$

Substituting for x and y in the equation $xy = 1$ gives us

$$(\sqrt{2}/2)(x' - y') \cdot (\sqrt{2}/2)(x' + y') = 1.$$

This reduces to
$$\frac{(x')^2}{2} - \frac{(y')^2}{2} = 1$$

which is the standard equation of a hyperbola with vertices $(\pm\sqrt{2}, 0)$ on the x'-axis. Figure 10.29 shows the graph, together with the new coordinate axes. Note that the asymptotes for the hyperbola have equations $y' = \pm x'$ in the new system. These correspond to the original x- and y-axes. ∎

Example 1 illustrates a method for eliminating a term of an equation which contains the product xy. This method can be used to transform the following **General Quadratic Equation in x and y**:

> $$Ax^2 + Bxy + Cy^2 + Dx + Ey + F = 0$$

where $B \neq 0$, into an equation in x' and y' which contains no $x'y'$ term. Let us prove that this may always be done. If we rotate the axes through an angle ϕ, then substituting

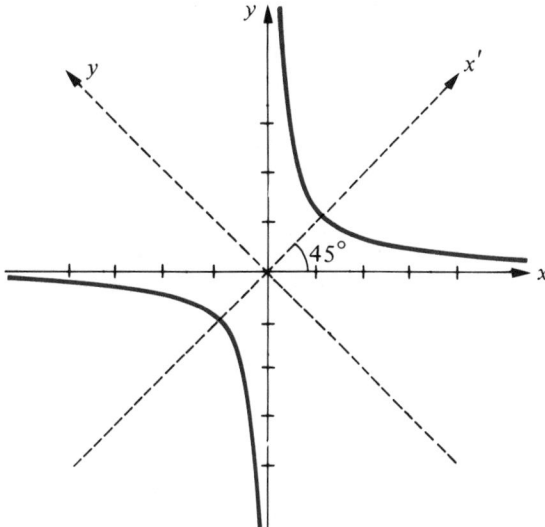

Figure 10.29

the expressions in (a) of the rotation of axes formulas for x and y gives us

$$A(x' \cos \phi - y' \sin \phi)^2$$
$$+ B(x' \cos \phi - y' \sin \phi)(x' \sin \phi + y' \cos \phi)$$
$$+ C(x' \sin \phi + y' \cos \phi)^2 + D(x' \cos \phi - y' \sin \phi)$$
$$+ E(x' \sin \phi + y' \cos \phi) + F = 0.$$

The last equation may be written in the form

$$A'(x')^2 + B'x'y' + C'(y')^2 + D'x' + E'y' + F = 0$$

where the coefficient B' of $x'y'$ is given by

$$B' = 2(C - A) \sin \phi \cos \phi - B(\cos^2 \phi - \sin^2 \phi).$$

In order to eliminate the $x'y'$ term we must select ϕ such that

$$2(C - A) \sin \phi \cos \phi + B(\cos^2 \phi - \sin^2 \phi) = 0.$$

Using the double-angle formulas of Chapter Six, the last equation may be written

$$(C - A) \sin 2\phi + B \cos 2\phi = 0$$

which is equivalent to

$$\boxed{\cot 2\phi = \frac{A - C}{B}, \quad B \neq 0.}$$

Thus, to eliminate the xy term in the given equation we may choose ϕ such that $\cot 2\phi = (A - C)/B$ and then employ the rotation of axes formulas. The resulting equation will contain no $x'y'$ term and, therefore, can be analyzed by previous methods. This proves that if the graph of the general quadratic equation in x and y exists, it is a conic (except for degenerate cases).

Example 2 Discuss and sketch the graph of the equation

$$41x^2 - 24xy + 34y^2 - 25 = 0.$$

Solution Using the notation preceding this example, we have $A = 41$, $B = -24$, and $C = 34$. We wish to determine ϕ such that

$$\cot 2\phi = \frac{A - C}{B} = \frac{41 - 34}{-24} = -\frac{7}{24}.$$

Since $\cot 2\phi$ is negative we may choose 2ϕ such that $90° < 2\phi < 180°$, and consequently, $\cos 2\phi = -7/25$. (Why?) We now use the half-angle formulas of Chapter Six with $v = 2\phi$. Since $45° < \phi < 90°$, this gives us

$$\sin \phi = \sqrt{\frac{1 - \cos 2\phi}{2}} = \sqrt{\frac{1 - (-7/25)}{2}} = \frac{4}{5}$$

$$\cos \phi = \sqrt{\frac{1 + \cos 2\phi}{2}} = \sqrt{\frac{1 + (-7/25)}{2}} = \frac{3}{5}.$$

It follows that the desired rotation of axes formulas are

$$x = \tfrac{3}{5}x' - \tfrac{4}{5}y', \quad y = \tfrac{4}{5}x' + \tfrac{3}{5}y'.$$

We leave it to the reader to show that, after substituting for x and y in the given equation and simplifying, we obtain the equation

$$(x')^2 + 2(y')^2 = 1.$$

Thus the graph is an ellipse with vertices at $(\pm 1, 0)$ on the x'-axis. Since $\tan \phi = \sin \phi / \cos \phi = (4/5)/(3/5) = 4/3$, we obtain $\phi = \tan^{-1}(4/3)$. If an approximation is desired, then, by Table 3, $\phi \approx 53° 8'$. The graph is sketched in Figure 10.30.

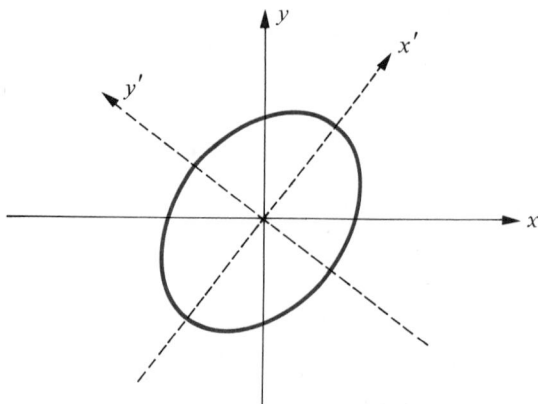

Figure 10.30

EXERCISES 10.8

After a suitable rotation of axes, describe and sketch the graph of each of the equations in Exercises 1–8.

1 $32x^2 - 72xy + 53y^2 = 80$

2 $7x^2 - 48xy - 7y^2 = 225$

3 $11x^2 + 10\sqrt{3}xy + y^2 = 4$

4 $x^2 - xy + y^2 = 3$

5 $5x^2 - 8xy + 5y^2 = 9$

6 $11x^2 - 10\sqrt{3}xy + y^2 = 20$

7 $16x^2 - 24xy + 9y^2 - 60x - 80y + 100 = 0$

8 $64x^2 - 240xy + 225y^2 + 1020x - 544y = 0$

9 Prove that, except for degenerate cases, the graph of the general quadratic equation in x and y is
  (a) a parabola if $B^2 - 4AC = 0$.
  (b) an ellipse if $B^2 - 4AC < 0$.
  (c) a hyperbola if $B^2 - 4AC > 0$.

10 Use the results of Exercise 9 to determine the type of conic in Exercises 1–8.

10.9 POLAR COORDINATES

We have previously specified points in a plane in terms of rectangular coordinates, using the ordered pair (a, b) to denote the point whose directed distances from the x- and y- axes are b and a, respectively. Another important method for representing points is by means of **polar coordinates**. In this case we again use ordered pairs; however, one of the numbers represents the measure of an angle. In order to introduce a system of polar coordinates in a plane we begin with a fixed point O (called the **origin**, or **pole**) and a directed half-line (called the **polar axis**) with endpoint O. Next we consider any point P in the plane different from O. If, as illustrated in Figure 10.31, $r = d(O, P)$, and θ denotes the measure of any angle determined by the polar axis and OP, then r and θ are called **polar coordinates** of P and the symbols (r, θ) or $P(r, \theta)$ are used to denote P. As usual, θ is considered positive if the angle is generated by a counterclockwise rotation of the polar axis and negative if the rotation is clockwise. Either radians or degrees may be used for the measure of θ. Since there are many angles with the same terminal side, the polar coordinates of a point are not unique. For example, $(3, \pi/4)$, $(3, 9\pi/4)$, and $(3, -7\pi/4)$ all represent the same point (see Figure 10.32). We shall also allow r to be negative. In this event, instead of measuring $|r|$ units along the terminal side of the angle θ, we measure along the half-line with endpoint O which has direction opposite to that of the terminal

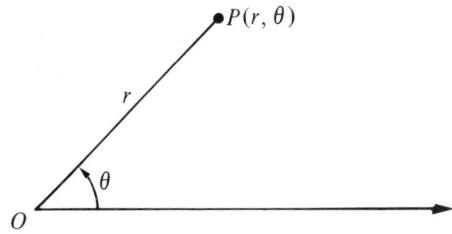

Figure 10.31

side. Figure 10.33 contains illustrations for the pairs $(-3, 5\pi/4)$ and $(-3, -3\pi/4)$. Finally, we agree that the pole O has polar coordinates $(0, \theta)$ for *any* θ. An assignment of ordered pairs of the form (r, θ) to points in a plane will be referred to as a **polar coordinate system** and the plane will be called an $r\theta$-plane.

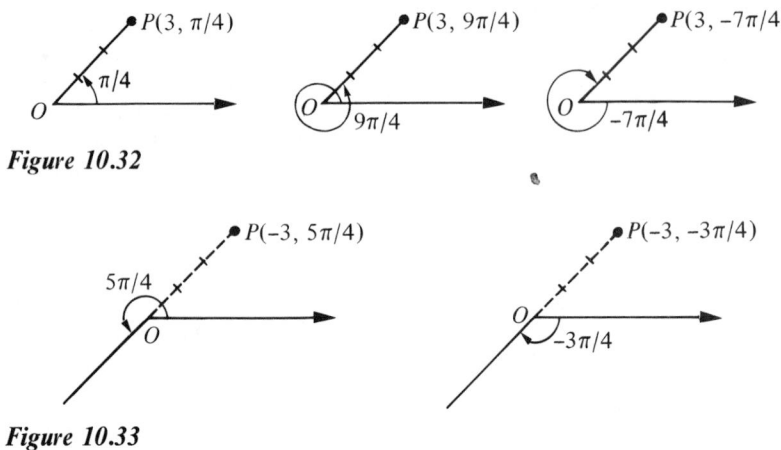

Figure 10.32

Figure 10.33   ∎

In a manner analogous to our work with equations in x and y, we now consider **polar equations**, that is, equations in r and θ. A solution of such an equation is an ordered pair (a, b) which leads to equality when a is substituted for r and b for θ. The graph is the set of all points (in an $r\theta$-plane) which correspond to the solutions. Although many polar equations contain trigonometric expressions, their graphs will differ from those discussed in Chapter Five, since points are plotted in a *polar* coordinate system instead of a rectangular coordinate system.

Example 1   Sketch the graph of the equation $r = 4\sin\theta$.

Solution   The following table contains some solutions of the equation.

θ	0	$\pi/6$	$\pi/4$	$\pi/3$	$\pi/2$	$2\pi/3$	$3\pi/4$	$5\pi/6$	π
r	0	2	$2\sqrt{2}$	$2\sqrt{3}$	4	$2\sqrt{3}$	$2\sqrt{2}$	2	0

We know that in rectangular coordinates, the graph of the given equation consists of sine waves of amplitude 4 and period 2π. However, if polar coordinates are used, then the points which correspond to the pairs in the table appear to lie on a circle of radius 2, and we draw the graph accordingly (see Figure 10.34). The proof that the graph is a circle will be given in Example 4. Additional points obtained by letting θ vary from π to 2π lie on the same circle. For example, the solution $(-2, 7\pi/6)$ gives us the same point as $(2, \pi/6)$; the point corresponding to $(-2\sqrt{2}, 5\pi/4)$ is the same as that obtained from $(2\sqrt{2}, \pi/4)$; and so on. Due to the periodicity of the sine function, if we let θ increase through all real numbers we obtain the same points over and over.

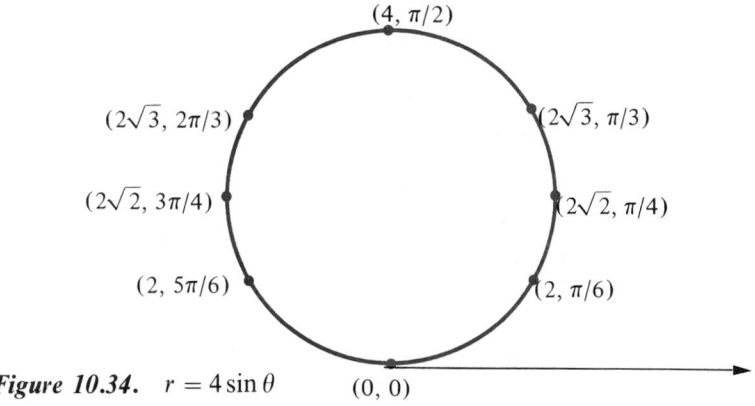

$(4, \pi/2)$

$(2\sqrt{3}, 2\pi/3)$　　　　$(2\sqrt{3}, \pi/3)$

$(2\sqrt{2}, 3\pi/4)$　　　　$(2\sqrt{2}, \pi/4)$

$(2, 5\pi/6)$　　　　$(2, \pi/6)$

Figure 10.34.　$r = 4\sin\theta$　　$(0, 0)$

Example 2　Sketch the graph of the equation $r = 2 + 2\cos\theta$.

Solution　Since the cosine function decreases from 1 to -1 as θ varies from 0 to π, it follows that r decreases from 4 to 0 in this θ-interval. The following table exhibits some solutions of the given equation.

θ	0	$\pi/6$	$\pi/4$	$\pi/3$	$\pi/2$	$2\pi/3$	$3\pi/4$	$5\pi/6$	π
r	4	$2+\sqrt{3}$	$2+\sqrt{2}$	3	2	1	$2-\sqrt{2}$	$2-\sqrt{3}$	0

If θ increases from π to 2π, then $\cos\theta$ increases from -1 to 1, and consequently, r increases from 0 to 4. Plotting points and connecting them with a smooth curve leads to the sketch shown in Figure 10.35, where we have used polar coordinate graph paper which displays lines through O at various angles, and circles with centers at the pole. The graph is called a **cardioid**. The same graph may be obtained by taking other intervals for θ.

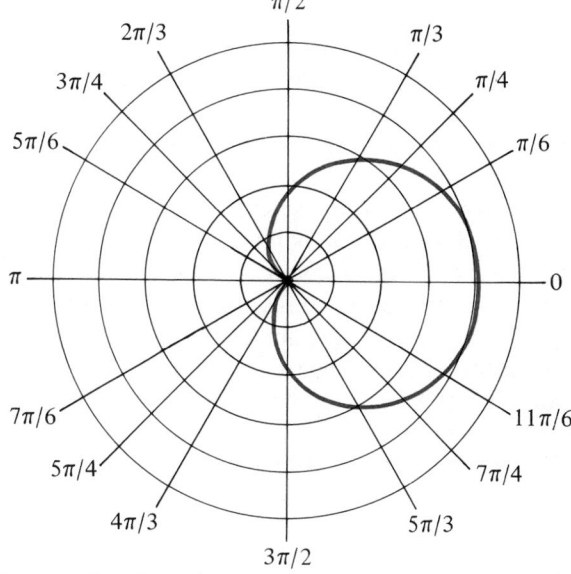

Figure 10.35.　$r = 2 + 2\cos\theta$

Example 3   Sketch the graph of the equation $r = a \sin 2\theta$, where $a > 0$.

Solution   Instead of tabulating solutions, let us reason as follows. If θ increases from 0 to $\pi/4$, then 2θ varies from 0 to $\pi/2$ and hence $\sin 2\theta$ increases from 0 to 1. It follows that r increases from 0 to a in the θ-interval $[0, \pi/4]$. If we next let θ increase from $\pi/4$ to $\pi/2$, then 2θ changes from $\pi/2$ to π. Consequently, r decreases from a to 0 in the θ-interval $[\pi/4, \pi/2]$. (Why?) The corresponding points on the graph constitute a "loop," as illustrated in Figure 10.36. We shall leave it to the reader to show that as θ increases from $\pi/2$ to π, a similar loop is obtained directly *below* the first loop. (Note that for this range of θ we have $\pi < 2\theta < 2\pi$ and hence $\sin 2\theta$ is negative.) Similar loops are obtained for the θ-intervals $[\pi, 3\pi/2]$ and $[3\pi/2, 2\pi]$. We have plotted only those points on the graph which correspond to the largest numerical values of r. The graph is called a **four-leaved rose**.

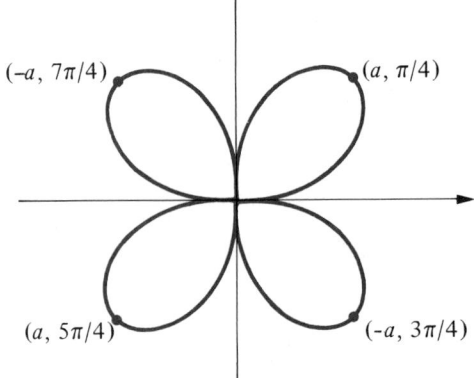

Figure 10.36. $r = a \sin 2\theta$ ■

Many other interesting graphs result from polar equations. Some are included in the exercises at the end of this section. Polar coordinates are very useful in applications involving circles with centers at the origin or lines that pass through the origin, since the equations which have these graphs may be written in the simple forms $r = k$ or $\theta = k$ for some fixed number k. (Verify!)

Let us now superimpose an xy-plane on an $r\theta$-plane in such a way that the positive x-axis coincides with the polar axis. Any point P in the plane may then be assigned rectangular coordinates (x, y) or polar coordinates (r, θ). It is not difficult to obtain formulas which specify the relationship between the two coordinate systems. Thus, if $r > 0$ we have a situation similar to that illustrated in (i) of Figure 10.37, whereas if $r < 0$ we have that shown in (ii) where, for later purposes, we have also plotted the point P' which has polar coordinates $(|r|, \theta)$ and rectangular coordinates $(-x, -y)$. Although we have pictured θ as an acute angle, the discussion which follows is valid for all angles. On the one hand, if $r > 0$ as in (i) of Figure 10.37, then we obtain

$$x = r \cos \theta, \quad y = r \sin \theta.$$

On the other hand, if $r < 0$, then referring to (ii) of Figure 10.37 and using the fact that $|r| = -r$, we have

$$\cos\theta = \frac{-x}{|r|} = \frac{-x}{-r} = \frac{x}{r}$$

$$\sin\theta = \frac{-y}{|r|} = \frac{-y}{-r} = \frac{y}{r}.$$

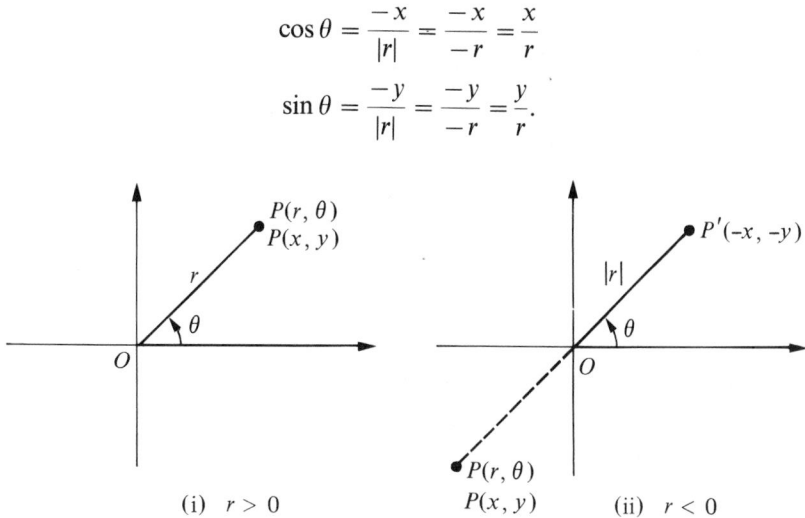

(i) $r > 0$ (ii) $r < 0$

Figure 10.37

Multiplication by r produces $x = r\cos\theta$, $y = r\sin\theta$, and hence these formulas hold whether r is positive or r is negative. If $r = 0$, then the point is the pole and we again see that the formulas are true. The following are further consequences of our discussion.

$$\tan\theta = \frac{y}{x}, \quad r^2 = x^2 + y^2$$

It is possible to use the formulas we have obtained to change the rectangular coordinates of a point to polar coordinates and vice versa. A more important use is for transforming a polar equation to an equation in x and y and vice versa, as illustrated in the next two examples.

Example 4   Find an equation in x and y which has the same graph as $r = 4\sin\theta$.

Solution   The given equation was considered in Example 1. It is convenient to multiply both sides by r, obtaining $r^2 = 4r\sin\theta$. Since $r^2 = x^2 + y^2$, and $r\sin\theta = y$, we obtain $x^2 + y^2 = 4y$. The last equation is equivalent to $x^2 + (y-2)^2 = 4$, whose graph is a circle of radius 2 with center at $(0,2)$ in the xy-plane.   ∎

Example 5   Find the general polar equation of a straight line.

Solution   We know that every straight line in an xy-coordinate system is the graph of a linear equation $Ax + By + C = 0$. Substituting $r\cos\theta$ for x and $r\sin\theta$ for y

leads to the polar equation

$$r(A \cos \theta + B \sin \theta) + C = 0.$$ ■

The graph of a polar equation may be symmetric with respect to the x-axis, the y-axis, or the origin. It is left to the reader to show that if a substitution listed in the following table does not change the solutions of a polar equation, then the graph has the indicated symmetry.

Substitution	Symmetry
$-\theta$ for θ	x-axis
$-r$ for r	origin
$\pi - \theta$ for θ	y-axis

To illustrate, since $\cos(-\theta) = \cos\theta$, the graph of the equation in Example 2 (see Figure 10.35) is symmetric with respect to the x-axis. Since $\sin(\pi - \theta) = \sin\theta$, the graph in Example 1 is symmetric with respect to the y-axis. The graph in Example 3 is symmetric to both axes and the origin. Other tests for symmetry may be stated; however, those listed above are among the easiest to apply.

EXERCISES 10.9

Sketch the graph of each of the equations in Exercises 1–18.

1 $r = 5$

2 $\theta = \pi/4$

3 $\theta = -\pi/6$

4 $r = -2$

5 $r = 4\cos\theta$

6 $r = -2\sin\theta$

7 $r = 4(1 - \sin\theta)$ (cardioid)

8 $r = 1 + 2\cos\theta$ (limaçon)

9 $r = a\cos 3\theta$ (three-leaved rose)

10 $r = a\sin 4\theta$ (eight-leaved rose)

11 $r^2 = a^2 \cos 2\theta$ (lemniscate)

12 $r = a\sin^2(\frac{1}{2}\theta)$ (cardioid)

13 $r = 4\csc\theta$

14 $r = -3\sec\theta$

15 $r = 2 - \cos\theta$ (limaçon)

16 $r = 2 + 2\sec\theta$ (conchoid)

17 $r = 2^\theta, \theta \geq 0$ (spiral)

18 $r\theta = 1, \theta > 0$ (spiral)

In Exercises 19–26 find a polar equation which has the same graph as the given equation.

19 $x = -3$

20 $y = 2$

21 $x^2 + y^2 = 16$

22 $x^2 = 8y$

23 $y = 6$

24 $y = 6x$

25 $x^2 - y^2 = 16$

26 $9x^2 + 4y^2 = 36$

In Exercises 27–34 find an equation in x and y which has the same graph as the given polar equation.

27 $r\cos\theta = 5$

28 $r\sin\theta = -2$

29 $r - 6\sin\theta = 0$

30 $r = 2(1 + \cos\theta)$

31 $r = a$

32 $\theta = \pi/4$

33 $r = \tan\theta$

34 $r = 4\sec\theta$

35 If $P_1(r_1, \theta_1)$ and $P_2(r_2, \theta_2)$ are points in an $r\theta$-plane, use the Law of Cosines to prove that $d(P_1, P_2)^2 = r_1^2 + r_2^2 - 2r_1 r_2 \cos(\theta_2 - \theta_1)$.

36 Prove that the graph of the polar equation $r = a \sin \theta + b \cos \theta$ is a circle and find its center and radius.

10.10 POLAR EQUATIONS OF CONICS

The following theorem provides another means for describing the conic sections.

Theorem

Let F be a fixed point and l a fixed line in a plane. The set of all points P in the plane such that the ratio $d(P, F)/d(P, Q)$ is a positive constant e, where $d(P, Q)$ is the distance from P to l, is a conic section. Moreover, the conic is a parabola if $e = 1$, an ellipse if $0 < e < 1$, or a hyperbola if $e > 1$.

The constant e is called the **eccentricity** of the conic. It will be seen that the point F is a focus of the conic. The line l is called a **directrix**. We shall prove the theorem for $e \leq 1$ and leave the case $e > 1$ as an exercise.

If $e = 1$, then $d(P, F) = d(P, Q)$ and, by definition, a parabola with focus F and directrix l is obtained.

Suppose next that $0 < e < 1$. It is convenient to introduce a polar coordinate system in the plane with F as the pole and with l perpendicular to the polar axis at the point $D(d, 0)$, where $d > 0$. If $P(r, \theta)$ is a point in the plane such that $d(P, F)/d(P, Q) = e < 1$, then referring to Figure 10.38 we see that P lies to the left of l.

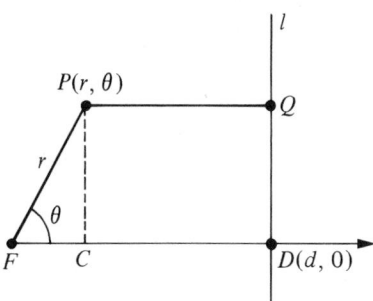

Figure 10.38

Let C be the projection of P on the polar axis. Since $d(P, F) = r$ and $d(P, Q) = \overline{FD} - \overline{FC} = d - r \cos \theta$, it follows that P satisfies the condition in the theorem if and only if

$$\frac{r}{d - r \cos \theta} = e$$

or equivalently,

$$r = de - er \cos \theta.$$

Solving for r gives us

$$r = \frac{de}{1 + e \cos \theta}.$$

This is a polar equation of the graph. Actually, the same equation is obtained if $e = 1$; however, in this event there is no point (r, θ) on the graph if $1 + \cos \theta = 0$. The rectangular equation corresponding to $r = de - er \cos \theta$ is

$$\pm\sqrt{x^2 + y^2} = de - ex.$$

Squaring both sides and rearranging terms leads to

$$(1 - e^2)x^2 + 2de^2x + y^2 = d^2e^2.$$

Completing the square in the previous equation and simplifying, we obtain

$$\left(x + \frac{de^2}{1 - e^2}\right)^2 + \frac{y^2}{1 - e^2} = \frac{d^2e^2}{(1 - e^2)^2}.$$

Finally, dividing both sides by $d^2e^2/(1 - e^2)^2$ gives us a standard form for the equation of an ellipse with center at the point $(-de^2/(1 - e^2), 0)$ and where

$$a^2 = \frac{d^2e^2}{(1 - e^2)^2}, \quad b^2 = \frac{d^2e^2}{1 - e^2}.$$

Since

$$c^2 = a^2 - b^2 = \frac{d^2e^4}{(1 - e^2)^2} = \left(\frac{de^2}{1 - e^2}\right)^2$$

we see that $c = de^2/(1 - e^2)$. This proves that F is a focus of the ellipse. It also follows that $e = c/a$. A similar proof may be given for the case $e > 1$.

It can be shown, conversely, that every conic which is not a circle may be described by means of the statement given in the theorem. This gives us a formulation of conic sections that is equivalent to the approach used previously. Since the theorem includes all three types of conics, it is sometimes regarded as a definition for the conic sections.

If, instead of having l intersect the polar axis at $D(d, 0)$, we had used the point $(-d, 0)$ to the *left* of F, then the resulting polar equation would be

$$r = \frac{de}{1 - e \cos \theta}.$$

If l is taken *parallel* to the polar axis through one of the points $(d, \pi/2)$ or $(d, 3\pi/2)$, then the corresponding equations would contain $\sin \theta$ instead of $\cos \theta$. The proofs of these facts are left to the reader. To summarize, *a polar equation having one of the forms*

$$r = \frac{de}{1 \pm e \cos \theta}, \quad r = \frac{de}{1 \pm e \sin \theta}$$

is a conic section. Moreover, the conic is a parabola if $e = 1$, an ellipse if $0 < e < 1$, or a hyperbola if $e > 1$.

Example 1   Describe and sketch the graph of the equation

$$r = \frac{10}{3 + 2\cos\theta}.$$

Solution   Dividing numerator and denominator of the given fraction by 3 gives us

$$r = \frac{\dfrac{10}{3}}{1 + \dfrac{2}{3}\cos\theta}$$

which has one of the forms discussed previously with $e = 2/3$. Thus the graph is an ellipse with focus F at the pole and major axis along the polar axis. The endpoints of the major axis may be found by setting θ equal to 0 and π. This gives us $V(2,0)$ and $V'(10,\pi)$. Hence $2a = d(V',V) = 12$, or $a = 6$. The center of the ellipse is the midpoint of the segment $V'V$, namely $(4,\pi)$. Using the fact that $e = c/a$ we obtain $c = ae = 6(2/3) = 4$. Hence $b^2 = a^2 - c^2 = 36 - 16 = 20$; that is, the semi-minor axis has length $\sqrt{20}$. The graph is sketched in Figure 10.39 where, for reference, we have superimposed a rectangular coordinate system on the polar system.

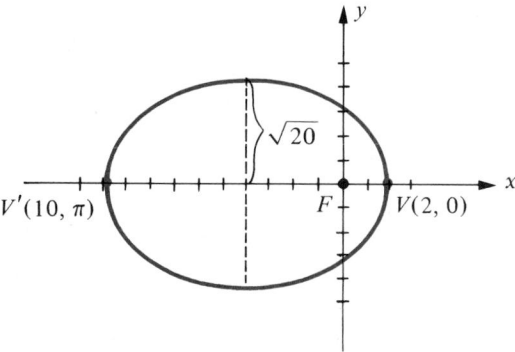

Figure 10.39. $r = \dfrac{10}{3 + 2\cos\theta}$ ∎

Example 2   Describe and sketch the graph of the equation

$$r = \frac{10}{2 + 3\sin\theta}.$$

Solution   To express the equation in one of the proper forms we divide numerator and denominator of the given fraction by 2, obtaining

$$r = \frac{5}{1 + \dfrac{3}{2}\sin\theta}.$$

Thus $e = 3/2$ and the graph is a hyperbola with a focus at the pole. The expression $\sin\theta$ tells us that the transverse axis is perpendicular to the polar axis. To find the vertices we let θ equal $\pi/2$ and $3\pi/2$ in the given equation. This gives us the points $V(2, \pi/2)$ and $V'(-10, 3\pi/2)$, and hence $2a = d(V, V') = 8$, or $a = 4$. The points $(5, 0)$ and $(5, \pi)$ on the graph can be used to get a rough estimate of the lower branch of the hyperbola. The upper branch is obtained by symmetry, as illustrated in Figure 10.40. If more accuracy or additional information is desired, we may calculate

$$c = ae = 4\left(\frac{3}{2}\right) = 6$$

and

$$b^2 = c^2 - a^2 = 36 - 16 = 20.$$

Asymptotes may then be constructed in the usual way.

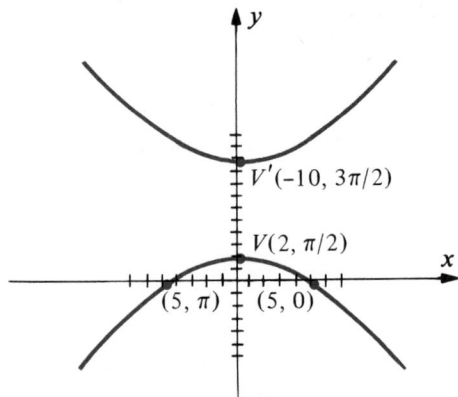

Figure 10.40. $r = \dfrac{10}{2 + 3\sin\theta}$

■

EXERCISES 10.10

In each of Exercises 1–10, identify and sketch the graph of the given equation.

1 $r = \dfrac{12}{6 + 2\sin\theta}$

2 $r = \dfrac{12}{6 - 2\sin\theta}$

3 $r = \dfrac{12}{2 - 6\cos\theta}$

4 $r = \dfrac{12}{2 + 6\cos\theta}$

5 $r = \dfrac{3}{2 + 2\cos\theta}$

6 $r = \dfrac{3}{2 - 2\sin\theta}$

7 $r = \dfrac{4}{\cos\theta - 2}$

8 $r = \dfrac{4\sec\theta}{2\sec\theta - 1}$

9 $r = \dfrac{6\csc\theta}{2\csc\theta + 3}$

10 $r = \csc\theta(\csc\theta - \cot\theta)$

11–20 Find rectangular equations for the graphs in Exercises 1–10.

In each of Exercises 21–26 express the given equation in polar form and then find the eccentricity and an equation for the directrix.

21 $y^2 = 4 - 4x$

22 $x^2 = 1 - 2y$

23 $3y^2 - 16y - x^2 + 16 = 0$

24 $5x^2 + 9y^2 = 32x + 64$

25 $8x^2 + 9y^2 + 4x = 4$

26 $4x^2 - 5y^2 + 36y - 36 = 0$

In each of Exercises 27–32 find a polar equation of the conic with focus at the pole and the given eccentricity and equation of directrix.

27 $e = 1/3, r = 2 \sec \theta$ **28** $e = 2/5, r = 4 \csc \theta$

29 $e = 4, r = -3 \csc \theta$ **30** $e = 3, r = -4 \sec \theta$

31 $e = 1, r \cos \theta = 5$ **32** $e = 1, r \sin \theta = -2$

33 Find a polar equation of the parabola with focus at the pole and vertex $(4, \pi/2)$.

34 Find a polar equation of the ellipse with eccentricity $2/3$, a vertex at $(1, 3\pi/2)$, and a focus at the pole.

35 Prove the theorem of this section for the case $e > 1$.

36 Derive the formulas $r = de/(1 \pm e \sin \theta)$ discussed in this section.

10.11   PLANE CURVES AND PARAMETRIC EQUATIONS

The graph of an equation $y = f(x)$, where the domain of the function f is an interval I, is often called a *plane curve*. However, to use this as a definition is unnecessarily restrictive, since it rules out most of the conic sections and many other useful graphs. The following statement is satisfactory for most applications.

Definition

> A **plane curve** is a set C of ordered pairs of the form
>
> $$(f(t), g(t))$$
>
> where f and g are functions defined on an interval I.

For simplicity, we often refer to a plane curve as a **curve**. The **graph** of C is the set of all points $P(t) = (f(t), g(t))$ in a rectangular coordinate system obtained by letting t vary through I. Each $P(t)$ is referred to as a point on the curve. We shall use the term *curve* interchangeably with *graph of a curve*. Sometimes it is convenient to think that the point $P(t)$ traces the curve C as t varies through the interval I. This is especially true in applications for which t represents time and $P(t)$ is the position of a moving particle at time t.

The graphs of several curves are sketched in Figure 10.41 for the case where I is a closed interval $[a, b]$. If, as in (i) of the figure, $P(a) \neq P(b)$, then $P(a)$ and $P(b)$ are called the **endpoints** of C. Note that the curve illustrated in (i) intersects itself in the sense that two different values of t give rise to the same point. If $P(a) = P(b)$, as illustrated in (ii) of Figure 10.41, then C is called a **closed curve**. If $P(a) = P(b)$ and C does not intersect itself at any other point, as illustrated in (iii) of the figure, then C is called a **simple closed curve**.

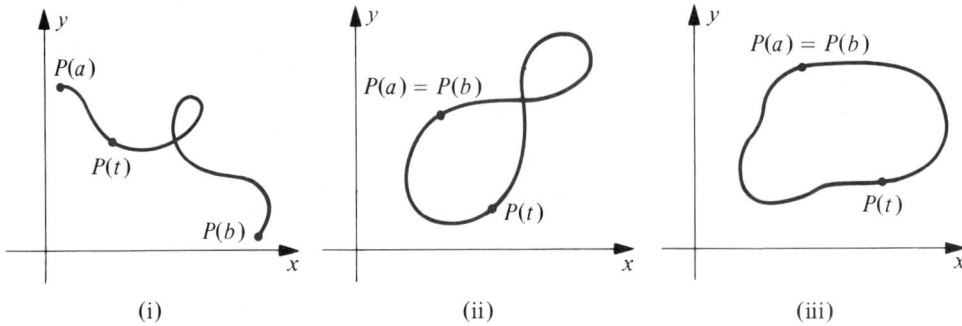

(i)   (ii)   (iii)

Figure 10.41

If C is the curve of the previous definition, then the equations

$$x = f(t), \quad y = g(t)$$

where t is in I are called **parametric equations** for C, and t is called a **parameter**. As t varies through I, the point $P(x, y)$ traces the curve. It is sometimes possible to eliminate the parameter and obtain a rectangular equation for C.

Example 1   Describe and sketch the graph of the curve $C = \{(2t, t^2 - 1):$ $-1 \le t \le 2\}$.

Solution   In this example $f(t) = 2t$, $g(t) = t^2 - 1$, and parametric equations for C are

$$x = 2t, \quad y = t^2 - 1, \quad \text{where } -1 \le t \le 2.$$

These equations can be used to tabulate coordinates for points $P(x, y)$ on C as in the following table.

t	-1	$-\frac{1}{2}$	0	$\frac{1}{2}$	1	$\frac{3}{2}$	2
x	-2	-1	0	1	2	3	4
y	0	$-\frac{3}{4}$	-1	$-\frac{3}{4}$	0	$\frac{5}{4}$	3

Plotting points leads to the sketch in Figure 10.42. A precise description of the graph may be obtained by eliminating the parameter. Solving the first parametric equation for t we obtain $t = x/2$, and substitution in the second equation yields

$$y = \left(\frac{x}{2}\right)^2 - 1, \quad \text{or} \quad y + 1 = \frac{1}{4}x^2.$$

The graph of the last equation is a parabola with vertical axis and vertex at the point $(0, -1)$. The curve C is that part of the parabola shown in Figure 10.42.

Parametric equations of curves are not unique. The curve C of this example is also given by

$$C = \{(t^3, \tfrac{1}{4}t^6 - 1): \quad \sqrt[3]{-2} \le t \le \sqrt[3]{4}\}$$

or by many other expressions.

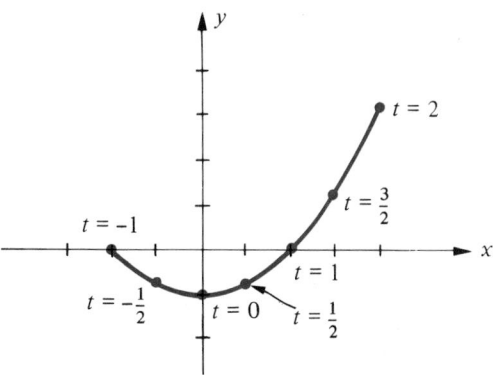

Figure 10.42. $x = 2t, y = t^2 - 1; 1 \le t \le 2$ ■

Example 2 Describe the graph of the curve C having parametric equations $x = \cos t$, $y = \sin t$, where $0 \le t \le 2\pi$.

Solution Eliminating the parameter gives us $x^2 + y^2 = 1$ and hence points on C are on the unit circle with center at the origin. As t increases from 0 to 2π, $P(t)$ starts at the point $A(1, 0)$ and traverses the circle once in the counterclockwise direction. In this example the parameter may be interpreted geometrically as the length of arc from A to P. ■

Example 3 Find parametric equations for the line of slope m through the point (x_1, y_1).

Solution By the point-slope form, an equation for the line is

$$y - y_1 = m(x - x_1).$$

If we let $x - x_1 = t$, then $P(x, y)$ is on the line if and only if $y - y_1 = mt$. It follows that parametric equations for the line are

$$x = x_1 + t, \quad y = y_1 + mt$$

where t varies through $\mathbb{R}$. As in Example 2, other equations could be used. For example, if we use $5t^3$ in place of t above, then the parametric equations $x = x_1 + 5t^3, y = y_1 + 5mt^3$ would describe the line. ■

Example 4 The curve traced by a fixed point P on the circumference of a circle as the circle rolls along a straight line in a plane is called a **cycloid**. Find parametric equations for a cycloid.

Solution   Suppose that the circle has radius a and that it rolls along (and above) the x-axis in the positive direction. If one position of P is the origin, then Figure 10.43 displays part of the curve and a possible position of the circle. Let C denote the center of the circle and T the point of tangency with the x-axis. We introduce a parameter t as the radian measure of angle TCP. Since $\overline{OT}$ is the distance the circle has rolled, $\overline{OT} = at$. Consequently, the coordinates of C are (at, a). If we consider an $x'y'$-coordinate system with origin at $C(at, a)$ and if $P(x', y')$ denotes the point P relative to this system, then by the translation of axes formulas of Section 10.7 with $h = at$ and $k = a$,

$$x = at + x', \quad y = a + y'.$$

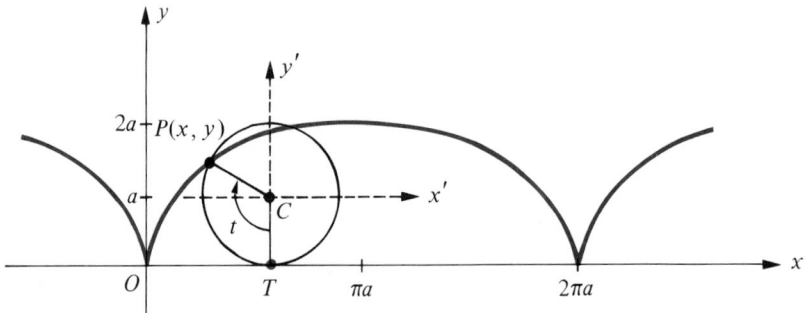

Figure 11.43

If, as in Figure 10.44, θ denotes an angle in standard position on the $x'y'$-system, then $\theta = (3\pi/2) - t$. Hence

$$x' = a\cos\theta = a\cos(3\pi/2 - t) = -a\sin t$$
$$y' = a\sin\theta = a\sin(3\pi/2 - t) = -a\cos t$$

and substitution in $x = at + x'$, $y = a + y'$ gives us parametric equations for the cycloid, namely,

$$x = a(t - \sin t), \quad y = a(1 - \cos t)$$

where t varies through $\mathbb{R}$.

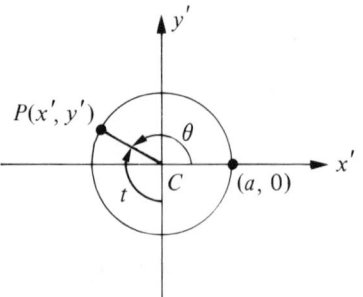

Figure 10.44   ∎

If $a < 0$, then the graph is the inverted cycloid that results if the circle rolls *below* the x-axis. This curve has a number of important physical properties. In particular, suppose that a thin wire passes through two fixed points A and B, as illustrated in Figure 10.45, and that the shape of the wire can be changed by bending it in any manner. Suppose further, that a bead is allowed to slide along the wire and the only force acting on the bead is gravity. We now ask which of all the possible paths will allow the bead to slide from A to B in the least amount of time. It is natural to conjecture that the desired path is the straight line segment from A to B; however, this is not the correct answer. It can be proved, using methods of advanced calculus, that the path which requires the least time coincides with the graph of an inverted cycloid. To cite another interesting property of this curve, suppose that A is the origin and B is the point with abscissa $\pi|a|$, that is, the lowest point on the cycloid occurring in the first arc to the right of A. It can be shown that if the bead is released at *any* point between A and B, the time required for it to reach B is always the same!

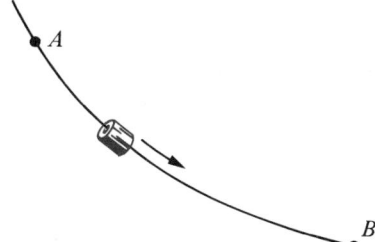

Figure 10.45

Variations of the cycloid occur in practical problems. For example, if a motorcycle wheel rolls along a straight road, then the curve traced by a fixed point on one of the spokes is a cycloid-like curve. In this case the curve does not have sharp corners, nor does it intersect the road (the x-axis) as does the cycloid. In like manner, if the wheel of a train rolls along a railroad track, then the curve traced by a fixed point on the circumference of the wheel (which extends below the track) contains loops at regular intervals. Observe in this event that as the train moves forward there are always points on the wheel that move backward! Several other cycloids are defined in Exercises 21 and 22.

EXERCISES 10.11

In each of Exercises 1–16, (a) sketch the graph of the curve C having the indicated parametric equations, and (b) find a rectangular equation of a graph which contains the points on C.

1   $x = t - 2, \ y = 2t + 3; \ 0 \le t \le 5$

2   $x = 1 - 2t, \ y = 1 + t; \ -1 \le t \le 4$

3   $x = t^2 + 1, \ y = t^2 - 1; \ -2 \le t \le 2$

4   $x = t^3 + 1, \ y = t^3 - 1; \ -2 \le t \le 2$

5   $x = 4t^2 - 5, \ y = 2t + 3; \ t$ in $\mathbb{R}$

6   $x = t^3, \ y = t^2; t$ in $\mathbb{R}$

7   $x = 2^t, \ y = 2^{-2t}; \ t$ in $\mathbb{R}$

8 $x = \sqrt{t}, y = 3t + 4; t \geq 0$

9 $x = 2 \sin t, y = 3 \cos t; 0 \leq t \leq 2\pi$

10 $x = \cos t - 2, y = \sin t + 3; 0 \leq t \leq 2\pi$

11 $x = \sec t, y = \tan t; -\pi/2 < t < \pi/2$

12 $x = \cos 2t, y = \sin t; -\pi \leq t \leq \pi$

13 $x = t^2, y = 2 \log t; t > 0$

14 $x = \cos^3 t, y = \sin^3 t; 0 \leq t \leq 2\pi$

15 $x = \sin t, y = \csc t; 0 < t \leq \pi/2$

16 $x = 2^t, y = 2^{-t}; t$ in $\mathbb{R}$

17 If $P_1(x_1, y_1)$ and $P_2(x_2, y_2)$ are distinct points, show that

$$x = (x_2 - x_1)t + x_1, \quad y = (y_2 - y_1)t + y_1,$$

where t varies through $\mathbb{R}$, are parametric equations of the line l through P_1 and P_2. Find three other pairs of parametric equations for the line l. Show that there are an infinite number of different pairs of parametric equations for l.

18 What is the difference between the graph of the parabola $y = x^2$ and the graph of $x = t^2, y = t^4$, where t varies through $\mathbb{R}$?

19 Show that

$$x = a \cos t + h, \quad y = b \sin t + k,$$

where t is in $\mathbb{R}$ are parametric equations of an ellipse with center at the point (h, k) and semi-axes of lengths a and b.

20 Find parametric equations for the parabola with (a) vertex $V(0,0)$ and focus $F(0, p)$; (b) vertex $V(h, k)$ and focus $F(h, k + p)$.

21 A circle C of radius b rolls on the inside of a second circle having equation $x^2 + y^2 = a^2$, where $b < a$. Let P be a fixed point on C and let the initial position of P be $A(a, 0)$. If the parameter t is the angle from the positive x-axis to the line segment from O to the center of C, show that parametric equations for the curve traced by P (called a **hypocycloid**) are

$$x = (a - b) \cos t + b \cos \frac{a - b}{b} t$$

$$y = (a - b) \sin t - b \sin \frac{a - b}{b} t$$

where t is in $\mathbb{R}$. If $b = a/4$, show that

$$x = a \cos^3 t, \quad y = a \sin^3 t$$

and sketch the graph of the curve.

22 If the circle C of Exercise 21 rolls on the outside of the second circle, find parametric equations for the curve traced by P. (This curve is called an **epicycloid**.)

10.12 REVIEW

Concepts

Define or discuss each of the following:

1 Equations of circles

2 Equations of lines

3 Inclination of a line

4 The relation between slope and inclination

5 Conic sections

6 Parabola

7 Focus, directrix, vertex, and axis of a parabola

8 Ellipse

9 Major and minor axes of an ellipse

10 Foci and vertices of an ellipse

11 Hyperbola

12 Transverse and conjugate axes of a hyperbola

13 Foci and vertices of a hyperbola

14 Asymptotes of a hyperbola

15 Translation of axes

16 Rotation of axes

17 Polar coordinates of a point

18 The relationship between polar and rectangular coordinates

19 Graphs of polar equations

20 Polar equations of conics

21 Plane curve

22 Parametric equations of a curve

Exercises

1 (a) Find the center and radius of the circle which has equation $x^2 + y^2 - 18x + 10y + 6 = 0$.

(b) Find an equation for the circle concentric to the circle of part (a) and passing through the origin.

(c) Find an equation for the circle of radius 6 with center in the fourth quadrant and tangent to both axes.

2 Given the points $A(5, 2)$, $B(-1, 4)$, and $C(-2, -6)$, find each of the following.
(a) An equation for the line through B which is parallel to the line through A and C
(b) An equation for the line through B which is perpendicular to the line through A and C
(c) An equation for the line through C and the midpoint of the line segment AB
(d) An equation for the line through C which is parallel to the y-axis
(e) An equation for the line through B which is perpendicular to the line with equation $5x + 2y - 8 = 0$

In each of Exercises 3–8 find the foci and vertices, and sketch the graph of the conic which has the given equation.

3 $4y^2 = 9x$

4 $y - 5 = 6(x + 7)^2$

5 $y^2 = 144 - 9x^2$

6 $16y^2 = 144 + 9x^2$

7 $y^2 - x^2 - 8 = 0$

8 $x = 8y^2 - 3$

Find equations for the conics in Exercises 9–14.

9 The parabola with focus $(-8, 0)$ and directrix $x = 8$

10 The parabola with vertex at the origin, symmetric to the y-axis and passing through the point $(3, -2)$

11 The ellipse with vertices $V(0, \pm 7)$ and foci $F(0, \pm 3)$

12 The hyperbola with foci $F(\pm 2, 0)$ and vertices $V(\pm 1, 0)$

13 The hyperbola with vertices $V(0, \pm 4)$ and asymptotes $y = \pm 5x$

14 The ellipse with foci $F(\pm 10, 0)$ and passing through the point $(2, \sqrt{2})$

Discuss and sketch the graph of each of the equations in Exercises 15–20 after making a suitable translation of axes.

15 $4x^2 + 9y^2 + 24x - 36y + 36 = 0$

16 $4x^2 - y^2 - 40x - 8y + 88 = 0$

17 $y^2 - 8x + 8y + 32 = 0$

18 $4x^2 + y^2 - 24x + 4y + 36 = 0$

19 $y^2 - 2x^2 + 6y + 8x - 3 = 0$

20 $x^2 - 9y^2 + 8x + 7 = 0$

Sketch the graphs of the equations in Exercises 21–28.

21 $r = -4 \sin \theta$

22 $r = 3 \cos 5\theta$

23 $r = 6 - 3 \cos \theta$

24 $r^2 = 9 \sin 2\theta$

25 $2r = \theta$

26 $r = \dfrac{8}{1 - 3 \sin \theta}$

27 $r = 6 - r \cos \theta$

28 $r = 8 \sec \theta$

Change the equations in Exercises 29–32 to polar equations.

29 $y^2 = 4x$

30 $x^2 + y^2 - 3x + 4y = 0$

31 $2x - 3y = 8$ **32** $x^2 + y^2 = 2xy$

In each of Exercises 33–36 change the equation to an equation in x and y.

33 $r^2 = \tan \theta$ **34** $r = 2\cos \theta + 3\sin \theta$

35 $r^2 = 4\sin 2\theta$ **36** $\theta = \sqrt{3}$

In Exercises 37–39 sketch the graph of the curve and find a rectangular equation of a graph which contains the points on the curve.

37 $x = (1/t) + 1,\ y = (2/t) - t;\ 0 < t \le 4$

38 $x = \cos^2 t - 2,\ y = \sin t + 1;\ 0 \le t \le 2\pi$

39 $x = \sqrt{t},\ y = 2^{-t};\ t \ge 0$

40 Let the curves C_1, C_2, C_3, and C_4 be given parametrically by

$$C_1 : x = t^2,\ y = t$$
$$C_2 : x = t^4,\ y = t^2$$
$$C_3 : x = \sin^2 t,\ y = \sin t$$
$$C_4 : x = 3^{2t},\ y = -3^t$$

where t varies through $\mathbb{R}$. Sketch the graphs of C_1, C_2, C_3, and C_4 and discuss their similarities and differences.

APPLIED PROBLEMS

In this appendix it is assumed that the reader has had experience finding solutions of linear and quadratic equations, that is, equations of the form $ax + b = 0$ or $ax^2 + bx + c = 0$, respectively. We shall consider various applications which make use of such solutions.

Formulas or equations involving variables are used in all fields which deal with numbers. For certain applications it is necessary to solve for a particular variable in terms of the remaining variables which appear in a formula. This is done by treating the equation as if the desired variable were the only one present and transforming the original equation into an equivalent equation in which that variable is isolated on one side, as illustrated in the next three examples.

Example 1 If a sum of money P (the principal) is invested at a simple interest rate of r per cent per year, then the interest I at the end of t years is given by $I = Prt$. Solve for r in terms of the remaining variables.

Solution We begin by writing

$$Prt = I.$$

In order to isolate r we multiply both sides by $1/Pt$, obtaining

$$\frac{1}{Pt} \cdot Prt = \frac{1}{Pt} \cdot I.$$

It follows that
$$r = \frac{I}{Pt}.$$ ∎

Example 2 The relationship between the temperature F on the Fahrenheit scale and the temperature C on the Celsius scale is given by

$$C = \tfrac{5}{9}(F - 32).$$

Solve for F in terms of C.

Solution   We may proceed as follows:

$$C = \tfrac{5}{9}(F - 32)$$
$$\tfrac{9}{5}C = F - 32$$
$$\tfrac{9}{5}C + 32 = F$$
$$F = \tfrac{9}{5}C + 32.$$ ■

Example 3   The formula $R = \dfrac{R_1 R_2}{R_1 + R_2}$ is used in electrical theory, where R_1, R_2, and R are positive. Solve for R_1 in terms of R and R_2.

Solution   The following equations are equivalent to the given equation.

$$(R_1 + R_2)R = (R_1 + R_2)\left(\frac{R_1 R_2}{R_1 + R_2}\right)$$
$$R_1 R + R_2 R = R_1 R_2$$
$$R_1 R - R_1 R_2 = -R_2 R$$
$$R_1(R - R_2) = -R_2 R$$
$$R_1 = \frac{-R_2 R}{R - R_2}$$
$$R_1 = \frac{R_2 R}{R_2 - R}.$$ ■

Problems often occur in everyday life which can be solved by means of equations or other mathematical tools. Some problems are described orally, from one person to another. Others are stated using written words, as is the case in textbooks. For this reason they are often called "word problems" by students and teachers of mathematics. They may also be referred to as "practical problems." We shall use the terminology "applied problem" for any problem which involves an application of mathematics to some field.

Due to the unlimited variety of applied problems it is difficult to state specific rules for finding solutions. However, it is possible to develop a general strategy for attacking such problems. Below are listed some guidelines which may be helpful, provided the problem can be formulated in terms of an equation in one variable.

Guidelines for Solving Applied Problems

---

1. If the problem is stated in written words, read it carefully several times and think about the given facts, together with the unknown quantity that is to be found.

2. Introduce a letter to denote the unknown quantity. This is one of the most crucial steps in the solution! Phrases containing words such as "what," "find," "how much," "how far," or "when" should alert you to the unknown quantity.

3. If possible, draw a picture and label it appropriately.

4. Make a list of known facts together with any relationships involving the unknown quantity. A relationship may often be described by means of an equation in which written statements, instead of letters or numbers, appear on either one or both sides of the equals sign.

5. After analyzing the list in step 4, and perhaps rereading the problem several more times, formulate an equation which describes precisely what is stated in words.

6. Solve the equation formulated in step 5.

7. Check the solutions obtained in step 6 by referring to the original statement of the problem. Carefully note whether the solution agrees with the stated conditions.

8. Don't become discouraged if you are unable to solve a given problem. It takes a great deal of effort and practice to become proficient in solving applied problems. Keep trying!

Example 4 A student has test scores of 64 and 78. What score on a third test will give the student an average of 80?

Solution We shall follow the Guidelines which precede this example. Reading the problem carefully, as suggested in step 1, we note that the unknown quantity is the score on the third test. Accordingly, as in step 2, we introduce a letter as follows:

$$x = \text{score on the third test.}$$

Drawing a picture, as mentioned in step 3, is inappropriate for this problem, so we go on to step 4 and look for relationships involving x. Since the average of the three scores is found by adding them and dividing by 3, we may write

$$\frac{64 + 78 + x}{3} = \text{average of the three scores 64, 78, } x.$$

From the statement of the problem we obtain

$$80 = \text{average desired.}$$

Consequently, x must satisfy the equation

$$\frac{64 + 78 + x}{3} = 80.$$

The last equation is the one referred to in step 5 of the Guidelines. We next solve the equation (see step 6) as follows:

$$64 + 78 + x = 240$$
$$142 + x = 240$$
$$x = 240 - 142$$
$$x = 98.$$

Step 7 tells us to check by referring to the original statement. If three test scores are 64, 78, and 98, then the average is

$$\frac{64 + 78 + 98}{3} = \frac{240}{3} = 80.$$

Hence a score of 98 on the third test will give the student an average of 80. Happily, we can ignore step 8. ∎

In the remaining examples we shall not point out the explicit Guidelines which are used in the solutions. The reader should be able to determine that without being told.

Example 5   A store holding a clearance sale advertises that all prices have been discounted 20%. If a certain article is on sale for $28, what was its price before the sale?

Solution   We begin by noting that the unknown quantity is the presale price. It is convenient to arrange our work as follows, where the quantities are measured in dollars.

$$x = \text{presale price}$$
$$0.20x = \text{discount}$$
$$28 = \text{sale price}$$

The sale price is determined as follows:

$$(\text{presale price}) - (\text{discount}) = (\text{sale price}).$$

This leads to the equation

$$x - 0.20x = 28$$

which we solve as follows:

$$x - \tfrac{1}{5}x = 28$$
$$\tfrac{4}{5}x = 28$$
$$x = (\tfrac{5}{4})28 = 35.$$

Hence the price before the sale was $35.

To check this answer we note that if a $35 article is discounted 20%, then the discount (in dollars) is $(0.20)(35) = 7$, and the selling price is $35 - 7$, or $28. ∎

Example 6   A man has $15,000 to invest. He plans to deposit part of it in a savings account paying 5% simple interest and the remainder in an investment fund yielding 8% simple interest. How much should he invest in each to obtain a 7% return on his money after one year?

Solution   The simple interest formula $I = Prt$ was given in Example 1. In the present example, $t = 1$, and hence the interest is given by $I = Pr$. If we let x denote the amount deposited in the savings account, then the remainder, $15{,}000 - x$, will be put into the investment fund. This leads to the following equalities:

$$x = \text{amount invested at } 5\%$$

$$15{,}000 - x = \text{amount invested at } 8\%$$

$$0.05x = \text{interest on } x \text{ dollars at } 5\%$$

$$0.08(15{,}000 - x) = \text{interest on } 15{,}000 - x \text{ dollars at } 8\%$$

$$0.07(15{,}000) = \text{total interest desired.}$$

Since the total interest must equal the combined interest from the two investments, we see that

$$0.05x + 0.08(15000 - x) = 0.07(15000)$$

$$0.05x + 1200 - 0.08x = 1050$$

$$-0.03x = -150$$

$$x = \frac{-150}{-0.03} = 5000.$$

Consequently, $5,000 should be deposited in the savings account and $10,000 in the investment fund.

Checking, we see that if $5,000 is placed in the savings account, then the interest obtained is $(0.05)(\$5{,}000) = \$250$. If $10,000 is placed in the investment fund then the interest is $(0.08)(\$10{,}000) = \$800$. Hence the total interest is $1,050, which is 7% of $15,000. ∎

An important type of problem which may be solved using linear equations involves mixing two substances to obtain a prescribed mixture. For such problems it is often helpful to draw a picture, as illustrated in the next two examples.

Example 7   A chemist has 10 ml of a solution which contains a 30% concentration of acid. How many ml of pure acid must be added in order to increase the concentration to 50%?

Solution   Since we wish to find the amount of pure acid to add, we let

$$x = \text{ml of acid to be added.}$$

The picture in Figure I.1 is self-explanatory. This leads to the equation

$$3 + x = 0.5(10 + x)$$

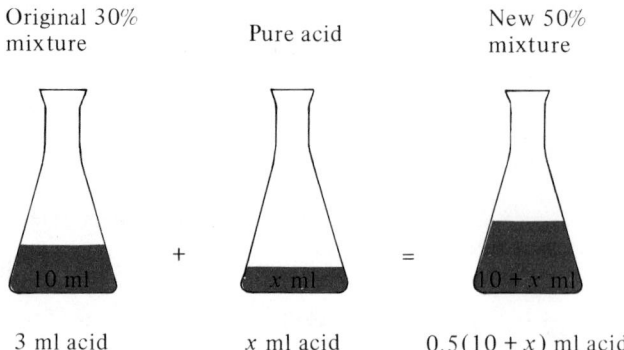

Original 30% mixture   Pure acid   New 50% mixture

+   =

3 ml acid x ml acid $0.5(10 + x)$ ml acid

Figure I.1

or equivalently,

$$3 + x = 5 + 0.5x$$
$$0.5x = 2$$
$$\tfrac{1}{2}x = 2$$
$$x = 4.$$

Hence 4 ml of the acid should be added to the original solution.

To check, we note that if 4 ml of acid is added to the given solution, then the new solution contains 14 ml, 7 of which are acid. This is the desired 50 % concentration.   ■

Example 8   A radiator contains 8 quarts of a mixture of water and antifreeze. If 40 % of the mixture is antifreeze, how much of the mixture should be drained and replaced by pure antifreeze, in order that the resultant mixture will contain 60 % antifreeze?

Solution   Let

$$x = \text{the number of quarts to be drained.}$$

Since there were 8 quarts in the original 40 % mixture, we may picture the problem as shown in Figure I.2. This gives us the equation

$$0.4(8 - x) + x = 0.6(8)$$

or equivalently,

$$3.2 - 0.4x + x = 4.8$$
$$-0.4x + x = 4.8 - 3.2$$
$$0.6x = 1.6$$

$$x = \frac{1.6}{0.6} = \frac{8}{3}.$$

Thus, 8/3 quarts should be drained from the original mixture.

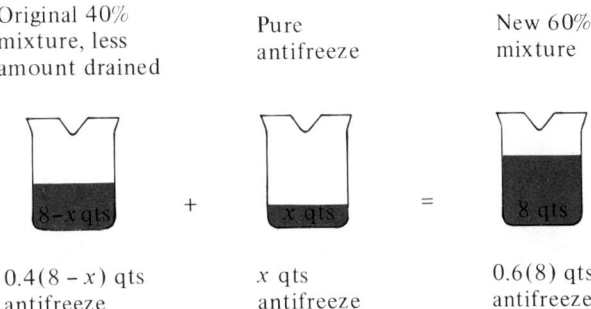

Original 40% mixture, less amount drained

Pure antifreeze

New 60% mixture

8−x qts

+

x qts

=

8 qts

0.4(8 − x) qts antifreeze

x qts antifreeze

0.6(8) qts antifreeze

Figure I.2

To check, let us first note that the amount of antifreeze in the original 8 quart mixture was 0.4(8), or 3.2 quarts. In draining 8/3 quarts of the original 40% mixture, we lose 0.4(8/3) quarts of antifreeze and hence there remain $(3.2) - 0.4(8/3)$ quarts of antifreeze. If we then add 8/3 quarts of pure antifreeze, the amount of antifreeze in the final mixture is $(3.2) - 0.4(8/3) + 8/3$ quarts. This reduces to 4.8, which is 60% of 8. ■

Many applied problems have to do with objects that move at a constant, or uniform, rate. If an object travels at a uniform (or average) rate r, then the distance d traversed in time t is given by $d = rt$. Of course, we assume that the units are properly chosen; that is, if r is in feet per second, then t is in seconds, and so on.

Example 9   Two cities A and B are connected by means of a highway 150 miles long. An automobile leaves A at 1:00 P.M. and travels at a uniform rate of 40 miles per hour toward B. Thirty minutes later, another automobile leaves A and travels toward B at a uniform rate of 55 miles per hour. At what time will the second car overtake the first car?

Solution   Let t denote the time, in hours, *after* 1:30 P.M. At 1:30 P.M. the first automobile has already traveled 20 miles. Hence, at time t *after* 1:30 P.M., the distance it has traveled is $20 + 40t$ miles. Since the second automobile starts the trip at 1:30 P.M., the distance it has traveled at time t is $55t$ miles. We wish to find the time t at which the distances traveled by the two automobiles are equal. This will be true when

$$55t = 20 + 40t.$$

Solving for t, we obtain

$$55t - 40t = 20$$
$$15t = 20$$

$$t = \tfrac{4}{3}.$$

Consequently, $t = 1\tfrac{1}{3}$ hours, or equivalently, 1 hour and 20 minutes. Since this is the amount of time after 1:30 P.M., it follows that the second car overtakes the first at 2:50 P.M.

To check our answer, we note that at 2:50 P.M. the first car has traveled for $1\frac{5}{6}$ hours and its distance from A is $40(11/6) = 220/3$ miles. At 2:50 P.M. the second car has traveled for $1\frac{1}{3}$ hours and is $55(4/3) = 220/3$ miles from A. Hence they are together at 2:50 P.M.   ■

Example 10 A box with a square base and no top is to be made from a square piece of tin by cutting out 3-inch squares from each corner and folding up the sides. If the box is to hold 48 cubic inches, what size piece of tin should be used?

Solution   If we let x denote the length of the side of the piece of tin, then the length of the base of the box is $x - 6$ (see Figure I.3). Since the area of the base is $(x - 6)^2$ and the height is 3, the volume of the box is $3(x - 6)^2$. Moreover, since the box is to hold 48 cubic inches,

$$3(x - 6)^2 = 48.$$

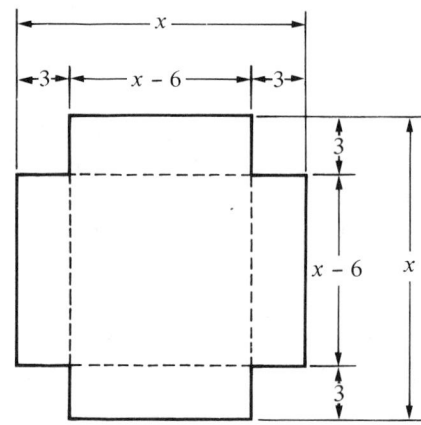
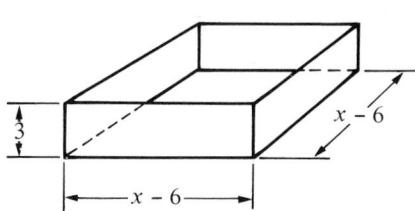

Figure I.3

Solving for x we obtain

$$(x - 6)^2 = 16$$
$$x - 6 = \pm 4$$
$$x = 6 \pm 4.$$

Consequently, either $x = 10$ or $x = 2$.

Let us now check each of these numbers. Referring to Figure I.3 we see that 2 is unacceptable since no box is possible in this case. (Why?) However, if we begin with a 10-inch square of tin, cut out 3-inch corners, and fold, we obtain a box having dimensions 4 inches, 4 inches, and 3 inches. The box has the desired volume of 48 cubic inches. Thus 10 inches is the answer to the problem.   ■

As illustrated in Example 10, even though an equation is formulated correctly, it is possible, owing to the physical nature of a given problem, to arrive at meaningless solutions. These solutions should be discarded. For example, we would not accept the answer -7 years for the age of an individual nor $\sqrt{50}$ for the number of automobiles in a parking lot.

EXERCISES A.I

The formulas in Exercises 1–40 occur in mathematics and its applications. Solve each for the indicated variable in terms of the remaining variables.

1 $A = \frac{1}{2}bh$ for h

2 $C = 2\pi r$ for r

3 $V = \frac{1}{3}\pi r^2 h$ for h

4 $F = g\dfrac{m_1 m_2}{d^2}$ for m_1

5 $\dfrac{1}{R} = \dfrac{1}{R_1} + \dfrac{1}{R_2} + \dfrac{1}{R_3}$ for R_2

6 $\dfrac{x}{a} + \dfrac{y}{b} = 1$ for y

7 $S = P + Prt$ for P

8 $F = \frac{9}{5}C + 32$ for C

9 $V = \frac{1}{3}\pi h^2(3r - h)$ for r

10 $s = \frac{1}{2}gt^2 + v_0 t$ for v_0

11 $S = \dfrac{a - rl}{l - r}$ for r

12 $S = a + (n - 1)d$ for n

13 $Ft = mv_1 - mv_2$ for m

14 $\dfrac{1}{f} = \dfrac{1}{f_1} + \dfrac{1}{f_2}$ for f_1

15 $A = \frac{1}{2}(b_1 + b_2)h$ for b_1

16 $A = 2\pi r(r + h)$ for h

17 $a = \dfrac{v_2 - v_1}{t}$ for v_1

18 $l = l_0(1 + ct)$ for c

19 $R = \dfrac{nE - rI}{nI}$ for n

20 $E = \dfrac{T_1 - T_2}{T_1}$ for T_1

21 $V = \frac{1}{3}\pi r^2 h$ for r

22 $K = \frac{1}{2}mv^2$ for v

23 $F = g\dfrac{m_1 m_2}{d^2}$ for d

24 $V = \frac{4}{3}\pi a^2 b$ for a

25 $s = \frac{1}{2}gt^2 + v_0 t$ for t

26 $A = 2\pi r(r + h)$ for r

27 $\dfrac{x^2}{a^2} - \dfrac{y^2}{b^2} = 1$ for y

28 $s = \sqrt{(r_1 - r_2)^2 + h^2}$ for r_2

29 $S = \pi r \sqrt{r^2 + h^2}$ for h

30 $d = \frac{1}{2}\sqrt{4R^2 - C^2}$ for C

31 $y = \dfrac{b}{a}\sqrt{a^2 - x^2}$ for x

32 $T = 2\pi\sqrt{\dfrac{l}{g}}$ for l

33 $x^{2/3} + y^{2/3} = a^{2/3}$ for y

34 $y = (\sqrt[3]{a} - \sqrt[3]{x})^3$ for x

35 $S = \pi(r + R)s$ for R

36 $S = 2(ab + bc + ac)$ for a

37 $n = \dfrac{\pi PR^4}{8VL}$ for R

38 $V = \frac{4}{3}\pi r^3$ for r

39 $\dfrac{1}{R} = (n - 1)\left(\dfrac{1}{R_1} + \dfrac{1}{R_2}\right)$ for R_2

40 $V = \frac{1}{3}\pi h(R_1^2 + R_2^2 + R_1 R_2)$ for R_1

41 A newspaper boy collects $13.45 in dimes and quarters. If there are 70 coins in all, how many quarters does he have?

42 A girl has 125 coins consisting of nickels and pennies. If the total amount is $4.25, how many coins of each type does she have?

43 Find four consecutive integers whose sum is 550.

44 Find two consecutive integers such that the difference of their squares is 133.

45 The relationship between the temperature F on the Fahrenheit scale and the temperature C on the Celsius scale is given by

$$C = \frac{5}{9}(F - 32).$$

Find the temperature at which the reading is the same on both scales.

46 Refer to Exercise 45. When will the Celsius reading be twice the Fahrenheit reading?

47 A student in an algebra course has test scores of 75, 82, 71, and 84. What score on the next test will raise the student's average to 80?

48 Going into the final exam a student has test scores of 72, 80, 65, 78, and 60. If the final exam counts as 1/3 of the final grade, what score must the student receive in order to end up with an average of 76?

49 A businesswoman wishes to invest $30,000 in two different funds which yield annual profits of 6% and $8\frac{1}{2}\%$, respectively. How much should she invest in each in order to realize a profit of $2,100 after one year?

50 A banker plans to lend part of $24,000 at a simple interest rate of 7% and the remainder at $8\frac{1}{2}\%$. How should he allot the loans in order to obtain a return of $7\frac{1}{2}\%$ after one year?

51 A college student has $3,000 in two different savings accounts, paying interest at the rates of $4\frac{1}{2}\%$ and 5%, respectively. If the total yearly interest is $144.10 how much is deposited in each account?

52 A man has $4,000 more invested at $7\frac{1}{2}\%$ simple interest than he has at 6%. If his total yearly interest is $467.40, how much is invested at each rate?

53 Two boys who are 224 meters apart start walking toward each other at the same time. If they walk at rates of 1.5 and 2 meters per second, respectively, when will they meet? How far will each have walked?

54 A jogger starts from a certain point and runs at a constant rate of 6 mph. Five minutes later a second jogger begins at the same point, running at a rate of 8 mph and following the same course. How long will it take the second jogger to catch up with the first?

55 If the radius of a circle is increased by 2 cm its area increases by 16π cm². What is the original radius of the circle?

56 A rectangle is twice as long as it is wide. If the length and width are decreased by 2 cm and 3 cm, respectively, the area is decreased by 30 cm². Find the original dimensions.

57 After playing 100 games, a major league baseball team has a record of 0.650. If it wins only 50% of its games for the remainder of the season, when will its record be 0.600?

58 A projectile is fired horizontally at a target and the sound of its impact is heard 1.5 seconds later. If the speed of the projectile is 3300 ft/sec and the speed of sound is 1100 ft/sec, how far away is the target?

59 Twenty liters of a solution contains 20% of a certain chemical. How much water should be added so that the resulting solution contains 15% of the chemical?

60 How many grams of an alloy containing 25% silver should be melted with 50 grams of an alloy containing 60% silver in order to obtain an alloy containing 50% silver?

61 How much water should be added to 1 liter of pure acid in order to obtain a solution which is 25% acid?

62 An automobile 20 feet long overtakes a truck that is 40 feet long which is traveling at 50 mph. At what constant speed must the automobile travel in order to pass the truck in 5 seconds?

63 A bus traveled from one city to another at an average rate of 50 mph. On the return trip the average rate was 45 mph and the elapsed time was 15 minutes longer. What was the total distance traveled?

64 At a concert certain tickets sold for $5.50 and others sold for $3.75. If a total of 520 tickets were sold for $2,448.75, how many of each kind were sold?

65 It takes a boy 90 minutes to mow his father's yard, but his sister can do it in 60 minutes. How long would it take them to mow the lawn if they worked together, using two lawnmowers?

66 Using water from one outlet, a swimming pool can be filled in 8 hours. A second, larger, outlet used alone can fill the pool in 5 hours. How long would it take to fill the pool if both outlets are used simultaneously?

67 How much water must be evaporated from 500 grams of a 10% salt solution in order to obtain a solution containing 15% salt?

68 A chemist has two acid solutions, the first containing 20% acid and the second 35% acid. How many ml of each should be mixed to obtain 50 ml of a solution containing 30% acid?

69 Generalize Exercise 67 as follows: Let a, b and d be positive real numbers with $a < b < 100$. Given a solution of d grams of salt water that is a% salt, how much water must be evaporated in order that the resulting mixture will be b% salt? Express the answer in terms of a, b, and d.

70 Let a, b, c, and d be positive real numbers where $a < c < b < 100$. Generalize Exercise 68 to the case where the given solutions contain a% and b% acid respectively, and where d ml of a solution containing c% acid is desired. Express the answer in terms of a, b, c, and d.

71 Find two consecutive odd integers whose product is 255.

72 Find two consecutive even integers, the sum of whose squares is 1060.

73 The diameter of a circle is 10 cm. What change in the radius will decrease the area by 16π cm^2?

74 The hypotenuse of a right triangle is 5 cm long. Find the lengths of the two legs if their sum is 6 cm.

75 A rectangular plot of ground having dimensions 120 ft by 160 ft is surrounded by a walk of uniform width. If the area of the walk is 1425 ft^2, what is its width?

76 It can be shown by mathematical induction (see Chapter 9), that the sum of the first n positive integers $1, 2, 3, \ldots, n$ equals $\frac{1}{2}n(n + 1)$. For what value of n will the sum equal 276?

77 An airplane flying north at 320 mph passed over a point on the ground at 2:00 P.M. Another airplane at the same altitude passed over the point at 2:15 P.M., flying east at 300 mph. At what time were the airplanes 500 miles apart?

78 A box with an open top is to be constructed by cutting out 3-inch squares from a rectangular sheet of tin whose length is twice its width. What size sheet will produce a box having a volume of 60 in^3?

79 A piece of wire 100 inches long is cut into two pieces and then each piece is bent into the shape of a square. If the sum of the enclosed areas is 397 in^2, find the length of each piece of wire.

80 Generalize Exercise 79 to the case where the wire is l inches long and the area is A in^2. Express the lengths of the two pieces in terms of l and A.

81 A projectile is fired straight upward with an initial speed of 800 ft/sec. The number of feet s above the ground after t seconds is given by $s = -16t^2 + 800t$.
(a) When will the projectile be 3200 feet above the ground?
(b) When will it hit the ground?
(c) What is its maximum height?

82 If a projectile is fired upward from a height of s_0 feet above the ground with an initial speed of v_0 feet per second, then its height above the ground at time t is given by $s = -16t^2 + v_0t + s_0$. Solve for t in terms of s, v_0, and s_0. When will the projectile hit the ground? Express your answer in terms of v_0 and s_0.

83 The diagonal of a square is 50 cm long. What change in the length of a side will increase the area by 46 cm^2?

84 The surface area S of a sphere of radius r is given by the formula $S = 4\pi r^2$. If $r = 6$ inches, what change in radius will increase the surface area by 36π square inches?

85 An airplane flew with the wind for 30 minutes and returned the same distance in 45 minutes. If the cruising speed of the airplane is 320 mph, find the speed of the wind.

86 A merchant wishes to mix peanuts costing $1.50 per pound with cashews costing $4.00 per pound, obtaining 50 pounds of a mixture costing $2.40 per pound. How many pounds of each should be used?

87 A chemist has 80 ml of a solution containing 25% acid. How many ml should be removed and replaced by pure acid in order to obtain a solution containing 40% acid?

88 A motorist averaged 45 mph driving outside the city limits and 25 mph within the city limits. If an 80-mile trip took the motorist two hours, how much time was spent driving within the city limits?

89 The width of a page in a book is 2 inches smaller than its length. The printed area is 72 square inches, with 1-inch margins at the top and bottom and $\frac{1}{2}$-inch margins at each side. Find the dimensions of the page.

90 A man puts a fence around a rectangular field and then subdivides the field into three smaller rectangular plots by placing two fences parallel to one of the sides. If the area of the field is 31,250 square yards and 1,000 yards of fencing was used, find the dimensions of the field.

91 A North-South highway intersects an East-West highway at a point P. An automobile crosses P at 10:00 A.M., traveling east at a constant rate of 20 mph. At that same instant another automobile is two miles north of P, traveling south at 50 mph. Find a formula which expresses the distance d between the automobiles at time t (hours) after 10:00 A.M. At what time will the automobiles be 104 miles apart?

92 At 1:00 P.M. two boys with walkie-talkie radios leave from the same point, one walking due west at a rate of 3 mph and the other due south at the rate of 4 mph. If the range of each radio is two miles, at what time will they be unable to communicate with one another?

93 A manufacturer sells a certain article to dealers at a rate of $20 each if less than 50 are ordered. If 50 or more are ordered (up to 600) the price per item is reduced at a rate of 2 cents times the number ordered. How many articles can a dealer purchase for $5,000?

94 A sales representative for a company estimates that gasoline consumption for her automobile averages 28 miles per gallon on the highway and 22 miles per gallon in the city. On a recent trip she covered 627 miles and used 24 gallons of gasoline. How much of the trip was spent driving in the city?

95 The relationship between the Fahrenheit and Celsius temperature scales is given by $C = (5/9)(F - 32)$. If $60 \le F \le 80$, express the corresponding range for C in terms of an inequality.

96 In the study of electricity, Ohm's Law states that $R = E/I$ where E is measured in volts, I in amperes, and R in ohms. If $E = 110$, what values of R correspond to $I \le 10$?

97 According to Hooke's Law, the force F (in pounds) required to stretch a certain spring x inches beyond its natural length is given by $F = (4.5)x$. If $10 \le F \le 18$, what is the corresponding range for x?

98 Boyle's Law for a certain gas states that $pv = 200$, where p denotes the pressure (lbs/in^2) and v denotes the volume (in^3). If $25 \le v \le 50$, what is the corresponding range for p?

99 If a projectile is fired straight upward from level ground with an initial velocity of 72 ft/sec, its altitude s (in feet) after t seconds is given by $s = -16t^2 + 72t$. During what time interval will the projectile be at least 32 feet above the ground?

100 The period T (sec) of a simple pendulum of length l (cm) is given by $T = 2\pi\sqrt{l/g}$, where g is a physical constant. If, under certain conditions, $g = 980$ and $98 \le l \le 100$, what is the corresponding range for T?

Table 2. Common Logarithms

N	0	1	2	3	4	5	6	7	8	9
1.0	.0000	.0043	.0086	.0128	.0170	.0212	.0253	.0294	.0334	.0374
1.1	.0414	.0453	.0492	.0531	.0569	.0607	.0645	.0682	.0719	.0755
1.2	.0792	.0828	.0864	.0899	.0934	.0969	.1004	.1038	.1072	.1106
1.3	.1139	.1173	.1206	.1239	.1271	.1303	.1335	.1367	.1399	.1430
1.4	.1461	.1492	.1523	.1553	.1584	.1614	.1644	.1673	.1703	.1732
1.5	.1761	.1790	.1818	.1847	.1875	.1903	.1931	.1959	.1987	.2014
1.6	.2041	.2068	.2095	.2122	.2148	.2175	.2201	.2227	.2253	.2279
1.7	.2304	.2330	.2355	.2380	.2405	.2430	.2455	.2480	.2504	.2529
1.8	.2553	.2577	.2601	.2625	.2648	.2672	.2695	.2718	.2742	.2765
1.9	.2788	.2810	.2833	.2856	.2878	.2900	.2923	.2945	.2967	.2989
2.0	.3010	.3032	.3054	.3075	.3096	.3118	.3139	.3160	.3181	.3201
2.1	.3222	.3243	.3263	.3284	.3304	.3324	.3345	.3365	.3385	.3404
2.2	.3424	.3444	.3464	.3483	.3502	.3522	.3541	.3560	.3579	.3598
2.3	.3617	.3636	.3655	.3674	.3692	.3711	.3729	.3747	.3766	.3784
2.4	.3802	.3820	.3838	.3856	.3874	.3892	.3909	.3927	.3945	.3962
2.5	.3979	.3997	.4014	.4031	.4048	.4065	.4082	.4099	.4116	.4133
2.6	.4150	.4166	.4183	.4200	.4216	.4232	.4249	.4265	.4281	.4298
2.7	.4314	.4330	.4346	.4362	.4378	.4393	.4409	.4425	.4440	.4456
2.8	.4472	.4487	.4502	.4518	.4533	.4548	.4564	.4579	.4594	.4609
2.9	.4624	.4639	.4654	.4669	.4683	.4698	.4713	.4728	.4742	.4757
3.0	.4771	.4786	.4800	.4814	.4829	.4843	.4857	.4871	.4886	.4900
3.1	.4914	.4928	.4942	.4955	.4969	.4983	.4997	.5011	.5024	.5038
3.2	.5051	.5065	.5079	.5092	.5105	.5119	.5132	.5145	.5159	.5172
3.3	.5185	.5198	.5211	.5224	.5237	.5250	.5263	.5276	.5289	.5302
3.4	.5315	.5328	.5340	.5353	.5366	.5378	.5391	.5403	.5416	.5428
3.5	.5441	.5453	.5465	.5478	.5490	.5502	.5514	.5527	.5539	.5551
3.6	.5563	.5575	.5587	.5599	.5611	.5623	.5635	.5647	.5658	.5670
3.7	.5682	.5694	.5705	.5717	.5729	.5740	.5752	.5763	.5775	.5786
3.8	.5798	.5809	.5821	.5832	.5843	.5855	.5866	.5877	.5888	.5899
3.9	.5911	.5922	.5933	.5944	.5955	.5966	.5977	.5988	.5999	.6010
4.0	.6021	.6031	.6042	.6053	.6064	.6075	.6085	.6096	.6107	.6117
4.1	.6128	.6138	.6149	.6160	.6170	.6180	.6191	.6201	.6212	.6222
4.2	.6232	.6243	.6253	.6263	.6274	.6284	.6294	.6304	.6314	.6325
4.3	.6335	.6345	.6355	.6365	.6375	.6385	.6395	.6405	.6415	.6425
4.4	.6435	.6444	.6454	.6464	.6474	.6484	.6493	.6503	.6513	.6522
4.5	.6532	.6542	.6551	.6561	.6571	.6580	.6590	.6599	.6609	.6618
4.6	.6628	.6637	.6646	.6656	.6665	.6675	.6684	.6693	.6702	.6712
4.7	.6721	.6730	.6739	.6749	.6758	.6767	.6776	.6785	.6794	.6803
4.8	.6812	.6821	.6830	.6839	.6848	.6857	.6866	.6875	.6884	.6893
4.9	.6902	.6911	.6920	.6928	.6937	.6946	.6955	.6964	.6972	.6981
5.0	.6990	.6998	.7007	.7016	.7024	.7033	.7042	.7050	.7059	.7067
5.1	.7076	.7084	.7093	.7101	.7110	.7118	.7126	.7135	.7143	.7152
5.2	.7160	.7168	.7177	.7185	.7193	.7202	.7210	.7218	.7226	.7235
5.3	.7243	.7251	.7259	.7267	.7275	.7284	.7292	.7300	.7308	.7316
5.4	.7324	.7332	.7340	.7348	.7356	.7364	.7372	.7380	.7388	.7396

Table 1. Powers and Roots

n	n^2	$\sqrt{n}$	$\sqrt[3]{n}$	n^3	n	n^2	$\sqrt{n}$	n^3	$\sqrt[3]{n}$
1	1	1.000	1.000	1	51	2,601	7.141	132,651	3.708
2	4	1.414	1.260	8	52	2,704	7.211	140,608	3.733
3	9	1.732	1.442	27	53	2,809	7.280	148,877	3.756
4	16	2.000	1.587	64	54	2,916	7.348	157,464	3.780
5	25	2.236	1.710	125	55	3,025	7.416	166,375	3.803
6	36	2.449	1.817	216	56	3,136	7.483	175,616	3.826
7	49	2.646	1.913	343	57	3,249	7.550	185,193	3.849
8	64	2.828	2.000	512	58	3,364	7.616	195,112	3.871
9	81	3.000	2.080	729	59	3,481	7.681	205,379	3.893
10	100	3.162	2.154	1,000	60	3,600	7.746	216,000	3.915
11	121	3.317	2.224	1,331	61	3,721	7.810	226,981	3.936
12	144	3.464	2.289	1,728	62	3,844	7.874	238,328	3.958
13	169	3.606	2.351	2,197	63	3,969	7.937	250,047	3.979
14	196	3.742	2.410	2,744	64	4,096	8.000	262,144	4.000
15	225	3.873	2.466	3,375	65	4,225	8.062	274,625	4.021
16	256	4.000	2.520	4,096	66	4,356	8.124	287,496	4.041
17	289	4.123	2.571	4,913	67	4,489	8.185	300,763	4.062
18	324	4.243	2.621	5,832	68	4,624	8.246	314,432	4.082
19	361	4.359	2.668	6,859	69	4,761	8.307	328,509	4.102
20	400	4.472	2.714	8,000	70	4,900	8.367	343,000	4.121
21	441	4.583	2.759	9,261	71	5,041	8.426	357,911	4.141
22	484	4.690	2.802	10,648	72	5,184	8.485	373,248	4.160
23	529	4.796	2.844	12,167	73	5,329	8.544	389,017	4.179
24	576	4.899	2.884	13,824	74	5,476	8.602	405,224	4.198
25	625	5.000	2.924	15,625	75	5,625	8.660	421,875	4.217
26	676	5.099	2.962	17,576	76	5,776	8.718	438,976	4.236
27	729	5.196	3.000	19,683	77	5,929	8.775	456,533	4.254
28	784	5.292	3.037	21,952	78	6,084	8.832	474,552	4.273
29	841	5.385	3.072	24,389	79	6,241	8.888	493,039	4.291
30	900	5.477	3.107	27,000	80	6,400	8.944	512,000	4.309
31	961	5.568	3.141	29,791	81	6,561	9.000	531,441	4.327
32	1,024	5.657	3.175	32,768	82	6,724	9.055	551,368	4.344
33	1,089	5.745	3.208	35,937	83	6,889	9.110	571,787	4.362
34	1,156	5.831	3.240	39,304	84	7,056	9.165	592,704	4.380
35	1,225	5.916	3.271	42,875	85	7,225	9.220	614,125	4.397
36	1,296	6.000	3.302	46,656	86	7,396	9.274	636,056	4.414
37	1,369	6.083	3.332	50,653	87	7,569	9.327	658,503	4.431
38	1,444	6.164	3.362	54,872	88	7,744	9.381	681,472	4.448
39	1,521	6.245	3.391	59,319	89	7,921	9.434	704,969	4.465
40	1,600	6.325	3.420	64,000	90	8,100	9.487	729,000	4.481
41	1,681	6.403	3.448	68,921	91	8,281	9.539	753,571	4.498
42	1,764	6.481	3.476	74,088	92	8,464	9.592	778,688	4.514
43	1,849	6.557	3.503	79,507	93	8,649	9.644	804,357	4.531
44	1,936	6.633	3.530	85,184	94	8,836	9.695	830,584	4.547
45	2,025	6.708	3.557	91,125	95	9,025	9.747	857,375	4.563
46	2,116	6.782	3.583	97,336	96	9,216	9.798	884,736	4.579
47	2,209	6.856	3.609	103,823	97	9,409	9.849	912,673	4.595
48	2,304	6.928	3.634	110,592	98	9,604	9.899	941,192	4.610
49	2,401	7.000	3.659	117,649	99	9,801	9.950	970,299	4.626
50	2,500	7.071	3.684	125,000	100	10,000	10.000	1,000,000	4.642

Table 3.   Values of the Trigonometric Functions

t	degrees	sin t	cos t	tan t	cot t	sec t	csc t	degrees	t
.0000	0°00'	.0000	1.0000	.0000	—	1.000	—	90°00'	1.5708
.0029	10	.0029	1.0000	.0029	343.8	1.000	343.8	50	1.5679
.0058	20	.0058	1.0000	.0058	171.9	1.000	171.9	40	1.5650
.0087	30	.0087	1.0000	.0087	114.6	1.000	114.6	30	1.5621
.0116	40	.0116	.9999	.0116	85.94	1.000	85.95	20	1.5592
.0145	50	.0145	.9999	.0145	68.75	1.000	68.76	10	1.5563
.0175	1°00'	.0175	.9998	.0175	57.29	1.000	57.30	89°00'	1.5533
.0204	10	.0204	.9998	.0204	49.10	1.000	49.11	50	1.5504
.0233	20	.0233	.9997	.0233	42.96	1.000	42.98	40	1.5475
.0262	30	.0262	.9997	.0262	38.19	1.000	38.20	30	1.5446
.0291	40	.0291	.9996	.0291	34.37	1.000	34.38	20	1.5417
.0320	50	.0320	.9995	.0320	31.24	1.001	31.26	10	1.5388
.0349	2°00'	.0349	.9994	.0349	28.64	1.001	28.65	88°00'	1.5359
.0378	10	.0378	.9993	.0378	26.43	1.001	26.45	50	1.5330
.0407	20	.0407	.9992	.0407	24.54	1.001	24.56	40	1.5301
.0436	30	.0436	.9990	.0437	22.90	1.001	22.93	30	1.5272
.0465	40	.0465	.9989	.0466	21.47	1.001	21.49	20	1.5243
.0495	50	.0494	.9988	.0495	20.21	1.001	20.23	10	1.5213
.0524	3°00'	.0523	.9986	.0524	19.08	1.001	19.11	87°00'	1.5184
.0553	10	.0552	.9985	.0553	18.07	1.002	18.10	50	1.5155
.0582	20	.0581	.9983	.0582	17.17	1.002	17.20	40	1.5126
.0611	30	.0610	.9981	.0612	16.35	1.002	16.38	30	1.5097
.0640	40	.0640	.9980	.0641	15.60	1.002	15.64	20	1.5068
.0669	50	.0669	.9978	.0670	14.92	1.002	14.96	10	1.5039
.0698	4°00'	.0698	.9976	.0699	14.30	1.002	14.34	86°00'	1.5010
.0727	10	.0727	.9974	.0729	13.73	1.003	13.76	50	1.4981
.0756	20	.0756	.9971	.0758	13.20	1.003	13.23	40	1.4952
.0785	30	.0785	.9969	.0787	12.71	1.003	12.75	30	1.4923
.0814	40	.0814	.9967	.0816	12.25	1.003	12.29	20	1.4893
.0844	50	.0843	.9964	.0846	11.83	1.004	11.87	10	1.4864
.0873	5°00'	.0872	.9962	.0875	11.43	1.004	11.47	85°00'	1.4835
.0902	10	.0901	.9959	.0904	11.06	1.004	11.10	50	1.4806
.0931	20	.0929	.9957	.0934	10.71	1.004	10.76	40	1.4777
.0960	30	.0958	.9954	.0963	10.39	1.005	10.43	30	1.4748
.0989	40	.0987	.9951	.0992	10.08	1.005	10.13	20	1.4719
.1018	50	.1016	.9948	.1022	9.788	1.005	9.839	10	1.4690
.1047	6°00'	.1045	.9945	.1051	9.514	1.006	9.567	84°00'	1.4661
.1076	10	.1074	.9942	.1080	9.255	1.006	9.309	50	1.4632
.1105	20	.1103	.9939	.1110	9.010	1.006	9.065	40	1.4603
.1134	30	.1132	.9936	.1139	8.777	1.006	8.834	30	1.4573
.1164	40	.1161	.9932	.1169	8.556	1.007	8.614	20	1.4544
.1193	50	.1190	.9929	.1198	8.345	1.007	8.405	10	1.4515
		cos t	sin t	cot t	tan t	csc t	sec t	degrees	t

Table 2. (continued)

N	0	1	2	3	4	5	6	7	8	9
5.5	.7404	.7412	.7419	.7427	.7435	.7443	.7451	.7459	.7466	.7474
5.6	.7482	.7490	.7497	.7505	.7513	.7520	.7528	.7536	.7543	.7551
5.7	.7559	.7566	.7574	.7582	.7589	.7597	.7604	.7612	.7619	.7627
5.8	.7634	.7642	.7649	.7657	.7664	.7672	.7679	.7686	.7694	.7701
5.9	.7709	.7716	.7723	.7731	.7738	.7745	.7752	.7760	.7767	.7774
6.0	.7782	.7789	.7796	.7803	.7810	.7818	.7825	.7832	.7839	.7846
6.1	.7853	.7860	.7868	.7875	.7882	.7889	.7896	.7903	.7910	.7917
6.2	.7924	.7931	.7938	.7945	.7952	.7959	.7966	.7973	.7980	.7987
6.3	.7993	.8000	.8007	.8014	.8021	.8028	.8035	.8041	.8048	.8055
6.4	.8062	.8069	.8075	.8082	.8089	.8096	.8102	.8109	.8116	.8122
6.5	.8129	.8136	.8142	.8149	.8156	.8162	.8169	.8176	.8182	.8189
6.6	.8195	.8202	.8209	.8215	.8222	.8228	.8235	.8241	.8248	.8254
6.7	.8261	.8267	.8274	.8280	.8287	.8293	.8299	.8306	.8312	.8319
6.8	.8325	.8331	.8338	.8344	.8351	.8357	.8363	.8370	.8376	.8382
6.9	.8388	.8395	.8401	.8407	.8414	.8420	.8426	.8432	.8439	.8445
7.0	.8451	.8457	.8463	.8470	.8476	.8482	.8488	.8494	.8500	.8506
7.1	.8513	.8519	.8525	.8531	.8537	.8543	.8549	.8555	.8561	.8567
7.2	.8573	.8579	.8585	.8591	.8597	.8603	.8609	.8615	.8621	.8627
7.3	.8633	.8639	.8645	.8651	.8657	.8663	.8669	.8675	.8681	.8686
7.4	.8692	.8698	.8704	.8710	.8716	.8722	.8727	.8733	.8739	.8745
7.5	.8751	.8756	.8762	.8768	.8774	.8779	.8785	.8791	.8797	.8802
7.6	.8808	.8814	.8820	.8825	.8831	.8837	.8842	.8848	.8854	.8859
7.7	.8865	.8871	.8876	.8882	.8887	.8893	.8899	.8904	.8910	.8915
7.8	.8921	.8927	.8932	.8938	.8943	.8949	.8954	.8960	.8965	.8971
7.9	.8976	.8982	.8987	.8993	.8998	.9004	.9009	.9015	.9020	.9025
8.0	.9031	.9036	.9042	.9047	.9053	.9058	.9063	.9069	.9074	.9079
8.1	.9085	.9090	.9096	.9101	.9106	.9112	.9117	.9122	.9128	.9133
8.2	.9138	.9143	.9149	.9154	.9159	.9165	.9170	.9175	.9180	.9186
8.3	.9191	.9196	.9201	.9206	.9212	.9217	.9222	.9227	.9232	.9238
8.4	.9243	.9248	.9253	.9258	.9263	.9269	.9274	.9279	.9284	.9289
8.5	.9294	.9299	.9304	.9309	.9315	.9320	.9325	.9330	.9335	.9340
8.6	.9345	.9350	.9355	.9360	.9365	.9370	.9375	.9380	.9385	.9390
8.7	.9395	.9400	.9405	.9410	.9415	.9420	.9425	.9430	.9435	.9440
8.8	.9445	.9450	.9455	.9460	.9465	.9469	.9474	.9479	.9484	.9489
8.9	.9494	.9499	.9504	.9509	.9513	.9518	.9523	.9528	.9533	.9538
9.0	.9542	.9547	.9552	.9557	.9562	.9566	.9571	.9576	.9581	.9586
9.1	.9590	.9595	.9600	.9605	.9609	.9614	.9619	.9624	.9628	.9633
9.2	.9638	.9643	.9647	.9652	.9657	.9661	.9666	.9671	.9675	.9680
9.3	.9685	.9689	.9694	.9699	.9703	.9708	.9713	.9717	.9722	.9727
9.4	.9731	.9736	.9741	.9745	.9750	.9754	.9759	.9763	.9768	.9773
9.5	.9777	.9782	.9786	.9791	.9795	.9800	.9805	.9809	.9814	.9818
9.6	.9823	.9827	.9832	.9836	.9841	.9845	.9850	.9854	.9859	.9863
9.7	.9868	.9872	.9877	.9881	.9886	.9890	.9894	.9899	.9903	.9908
9.8	.9912	.9917	.9921	.9926	.9930	.9934	.9939	.9943	.9948	.9952
9.9	.9956	.9961	.9965	.9969	.9974	.9978	.9983	.9987	.9991	.9996

TABLE 3. VALUES OF THE TRIGONOMETRIC FUNCTIONS   A15

Table 3. (continued)

Values of the Trigonometric Functions (14° – 20° / 70° – 76°)

t	degrees	sin t	cos t	tan t	cot t	sec t	csc t	degrees	t
.2443	14°00′	.2419	.9703	.2493	4.011	1.031	4.134	76°00′	1.3265
.2473	10	.2447	.9696	.2524	3.962	1.031	4.086	50	1.3235
.2502	20	.2476	.9689	.2555	3.914	1.032	4.039	40	1.3206
.2531	30	.2504	.9681	.2586	3.867	1.033	3.994	30	1.3177
.2560	40	.2532	.9674	.2617	3.821	1.034	3.950	20	1.3148
.2589	50	.2560	.9667	.2648	3.776	1.034	3.906	10	1.3119
.2618	15°00′	.2588	.9659	.2679	3.732	1.035	3.864	75°00′	1.3090
.2647	10	.2616	.9652	.2711	3.689	1.036	3.822	50	1.3061
.2676	20	.2644	.9644	.2742	3.647	1.037	3.782	40	1.3032
.2705	30	.2672	.9636	.2773	3.606	1.038	3.742	30	1.3003
.2734	40	.2700	.9628	.2805	3.566	1.039	3.703	20	1.2974
.2763	50	.2728	.9621	.2836	3.526	1.039	3.665	10	1.2945
.2793	16°00′	.2756	.9613	.2867	3.487	1.040	3.628	74°00′	1.2915
.2822	10	.2784	.9605	.2899	3.450	1.041	3.592	50	1.2886
.2851	20	.2812	.9596	.2931	3.412	1.042	3.556	40	1.2857
.2880	30	.2840	.9588	.2962	3.376	1.043	3.521	30	1.2828
.2909	40	.2868	.9580	.2994	3.340	1.044	3.487	20	1.2799
.2938	50	.2896	.9572	.3026	3.305	1.045	3.453	10	1.2770
.2967	17°00′	.2924	.9563	.3057	3.271	1.046	3.420	73°00′	1.2741
.2996	10	.2952	.9555	.3089	3.237	1.047	3.388	50	1.2712
.3025	20	.2979	.9546	.3121	3.204	1.048	3.356	40	1.2683
.3054	30	.3007	.9537	.3153	3.172	1.049	3.326	30	1.2654
.3083	40	.3035	.9528	.3185	3.140	1.049	3.295	20	1.2625
.3113	50	.3062	.9520	.3217	3.108	1.050	3.265	10	1.2595
.3142	18°00′	.3090	.9511	.3249	3.078	1.051	3.236	72°00′	1.2566
.3171	10	.3118	.9502	.3281	3.047	1.052	3.207	50	1.2537
.3200	20	.3145	.9492	.3314	3.018	1.053	3.179	40	1.2508
.3229	30	.3173	.9483	.3346	2.989	1.054	3.152	30	1.2479
.3258	40	.3201	.9474	.3378	2.960	1.056	3.124	20	1.2450
.3287	50	.3228	.9465	.3411	2.932	1.057	3.098	10	1.2421
.3316	19°00′	.3256	.9455	.3443	2.904	1.058	3.072	71°00′	1.2392
.3345	10	.3283	.9446	.3476	2.877	1.059	3.046	50	1.2363
.3374	20	.3311	.9436	.3508	2.850	1.060	3.021	40	1.2334
.3403	30	.3338	.9426	.3541	2.824	1.061	2.996	30	1.2305
.3432	40	.3365	.9417	.3574	2.798	1.062	2.971	20	1.2275
.3462	50	.3393	.9407	.3607	2.773	1.063	2.947	10	1.2246
.3491	20°00′	.3420	.9397	.3640	2.747	1.064	2.924	70°00′	1.2217
.3520	10	.3448	.9387	.3673	2.723	1.065	2.901	50	1.2188
.3549	20	.3475	.9377	.3706	2.699	1.066	2.878	40	1.2159
.3578	30	.3502	.9367	.3739	2.675	1.068	2.855	30	1.2130
.3607	40	.3529	.9356	.3772	2.651	1.069	2.833	20	1.2101
.3636	50	.3557	.9346	.3805	2.628	1.070	2.812	10	1.2072

(Reverse headers, reading up: t | degrees | csc t | sec t | cot t | tan t | cos t | sin t | degrees | t)

Values of the Trigonometric Functions (7° – 13° / 77° – 83°)

t	degrees	sin t	cos t	tan t	cot t	sec t	csc t	degrees	t
.1222	7°00′	.1219	.9925	.1228	8.144	1.008	8.206	83°00′	1.4486
.1251	10	.1248	.9922	.1257	7.953	1.008	8.016	50	1.4457
.1280	20	.1276	.9918	.1287	7.770	1.008	7.834	40	1.4428
.1309	30	.1305	.9914	.1317	7.596	1.009	7.661	30	1.4399
.1338	40	.1334	.9911	.1346	7.429	1.009	7.496	20	1.4370
.1367	50	.1363	.9907	.1376	7.269	1.009	7.337	10	1.4341
.1396	8°00′	.1392	.9903	.1405	7.115	1.010	7.185	82°00′	1.4312
.1425	10	.1421	.9899	.1435	6.968	1.010	7.040	50	1.4283
.1454	20	.1449	.9894	.1465	6.827	1.011	6.900	40	1.4254
.1484	30	.1478	.9890	.1495	6.691	1.011	6.765	30	1.4224
.1513	40	.1507	.9886	.1524	6.561	1.012	6.636	20	1.4195
.1542	50	.1536	.9881	.1554	6.435	1.012	6.512	10	1.4166
.1571	9°00′	.1564	.9877	.1584	6.314	1.012	6.392	81°00′	1.4137
.1600	10	.1593	.9872	.1614	6.197	1.013	6.277	50	1.4108
.1629	20	.1622	.9868	.1644	6.084	1.013	6.166	40	1.4079
.1658	30	.1650	.9863	.1673	5.976	1.014	6.059	30	1.4050
.1687	40	.1679	.9858	.1703	5.871	1.014	5.955	20	1.4021
.1716	50	.1708	.9853	.1733	5.769	1.015	5.855	10	1.3992
.1745	10°00′	.1736	.9848	.1763	5.671	1.015	5.759	80°00′	1.3963
.1774	10	.1765	.9843	.1793	5.576	1.016	5.665	50	1.3934
.1804	20	.1794	.9838	.1823	5.485	1.016	5.575	40	1.3904
.1833	30	.1822	.9833	.1853	5.396	1.017	5.487	30	1.3875
.1862	40	.1851	.9827	.1883	5.309	1.018	5.403	20	1.3846
.1891	50	.1880	.9822	.1914	5.226	1.018	5.320	10	1.3817
.1920	11°00′	.1908	.9816	.1944	5.145	1.019	5.241	79°00′	1.3788
.1949	10	.1937	.9811	.1974	5.066	1.019	5.164	50	1.3759
.1978	20	.1965	.9805	.2004	4.989	1.020	5.089	40	1.3730
.2007	30	.1994	.9799	.2035	4.915	1.020	5.016	30	1.3701
.2036	40	.2022	.9793	.2065	4.843	1.021	4.945	20	1.3672
.2065	50	.2051	.9787	.2095	4.773	1.022	4.876	10	1.3643
.2094	12°00′	.2079	.9781	.2126	4.705	1.022	4.810	78°00′	1.3614
.2123	10	.2108	.9775	.2156	4.638	1.023	4.745	50	1.3584
.2153	20	.2136	.9769	.2186	4.574	1.024	4.682	40	1.3555
.2182	30	.2164	.9763	.2217	4.511	1.024	4.620	30	1.3526
.2211	40	.2193	.9757	.2247	4.449	1.025	4.560	20	1.3497
.2240	50	.2221	.9750	.2278	4.390	1.026	4.502	10	1.3468
.2269	13°00′	.2250	.9744	.2309	4.331	1.026	4.445	77°00′	1.3439
.2298	10	.2278	.9737	.2339	4.275	1.027	4.390	50	1.3410
.2327	20	.2306	.9730	.2370	4.219	1.028	4.336	40	1.3381
.2356	30	.2334	.9724	.2401	4.165	1.028	4.284	30	1.3352
.2385	40	.2363	.9717	.2432	4.113	1.029	4.232	20	1.3323
.2414	50	.2391	.9710	.2462	4.061	1.030	4.182	10	1.3294

(Reverse headers, reading up: t | degrees | csc t | sec t | cot t | tan t | cos t | sin t | degrees | t)

Table 3. (continued)

t	degrees	sin t	cos t	tan t	cot t	sec t	csc t	degrees	t
.4887	28°00'	.4695	.8829	.5317	1.881	1.133	2.130	62°00'	1.0821
.4916	10	.4720	.8816	.5354	1.868	1.134	2.118	50	1.0792
.4945	20	.4746	.8802	.5392	1.855	1.136	2.107	40	1.0763
.4974	30	.4772	.8788	.5430	1.842	1.138	2.096	30	1.0734
.5003	40	.4797	.8774	.5467	1.829	1.140	2.085	20	1.0705
.5032	50	.4823	.8760	.5505	1.816	1.142	2.074	10	1.0676
.5061	29°00'	.4848	.8746	.5543	1.804	1.143	2.063	61°00'	1.0647
.5091	10	.4874	.8732	.5581	1.792	1.145	2.052	50	1.0617
.5120	20	.4899	.8718	.5619	1.780	1.147	2.041	40	1.0588
.5149	30	.4924	.8704	.5658	1.767	1.149	2.031	30	1.0559
.5178	40	.4950	.8689	.5696	1.756	1.151	2.020	20	1.0530
.5207	50	.4975	.8675	.5735	1.744	1.153	2.010	10	1.0501
.5236	30°00'	.5000	.8660	.5774	1.732	1.155	2.000	60°00'	1.0472
.5265	10	.5025	.8646	.5812	1.720	1.157	1.990	50	1.0443
.5294	20	.5050	.8631	.5851	1.709	1.159	1.980	40	1.0414
.5323	30	.5075	.8616	.5890	1.698	1.161	1.970	30	1.0385
.5352	40	.5100	.8601	.5930	1.686	1.163	1.961	20	1.0356
.5381	50	.5125	.8587	.5969	1.675	1.165	1.951	10	1.0327
.5411	31°00'	.5150	.8572	.6009	1.664	1.167	1.942	59°00'	1.0297
.5440	10	.5175	.8557	.6048	1.653	1.169	1.932	50	1.0268
.5469	20	.5200	.8542	.6088	1.643	1.171	1.923	40	1.0239
.5498	30	.5225	.8526	.6128	1.632	1.173	1.914	30	1.0210
.5527	40	.5250	.8511	.6168	1.621	1.175	1.905	20	1.0181
.5556	50	.5275	.8496	.6208	1.611	1.177	1.896	10	1.0152
.5585	32°00'	.5299	.8480	.6249	1.600	1.179	1.887	58°00'	1.0123
.5614	10	.5324	.8465	.6289	1.590	1.181	1.878	50	1.0094
.5643	20	.5348	.8450	.6330	1.580	1.184	1.870	40	1.0065
.5672	30	.5373	.8434	.6371	1.570	1.186	1.861	30	1.0036
.5701	40	.5398	.8418	.6412	1.560	1.188	1.853	20	1.0007
.5730	50	.5422	.8403	.6453	1.550	1.190	1.844	10	.9977
.5760	33°00'	.5446	.8387	.6494	1.540	1.192	1.836	57°00'	.9948
.5789	10	.5471	.8371	.6536	1.530	1.195	1.828	50	.9919
.5818	20	.5495	.8355	.6577	1.520	1.197	1.820	40	.9890
.5847	30	.5519	.8339	.6619	1.511	1.199	1.812	30	.9861
.5876	40	.5544	.8323	.6661	1.501	1.202	1.804	20	.9832
.5905	50	.5568	.8307	.6703	1.492	1.204	1.796	10	.9803
.5934	34°00'	.5592	.8290	.6745	1.483	1.206	1.788	56°00'	.9774
.5963	10	.5616	.8274	.6787	1.473	1.209	1.781	50	.9745
.5992	20	.5640	.8258	.6830	1.464	1.211	1.773	40	.9716
.6021	30	.5664	.8241	.6873	1.455	1.213	1.766	30	.9687
.6050	40	.5688	.8225	.6916	1.446	1.216	1.758	20	.9657
.6080	50	.5712	.8208	.6959	1.437	1.218	1.751	10	.9628
		cos t	sin t	cot t	tan t	csc t	sec t	degrees	t

t	degrees	sin t	cos t	tan t	cot t	sec t	csc t	degrees	t
.3665	21°00'	.3584	.9336	.3839	2.605	1.071	2.790	69°00'	1.2043
.3694	10	.3611	.9325	.3872	2.583	1.072	2.769	50	1.2014
.3723	20	.3638	.9315	.3906	2.560	1.074	2.749	40	1.1985
.3752	30	.3665	.9304	.3939	2.539	1.075	2.729	30	1.1956
.3782	40	.3692	.9293	.3973	2.517	1.076	2.709	20	1.1926
.3811	50	.3719	.9283	.4006	2.496	1.077	2.689	10	1.1897
.3840	22°00'	.3746	.9272	.4040	2.475	1.079	2.669	68°00'	1.1868
.3869	10	.3773	.9261	.4074	2.455	1.080	2.650	50	1.1839
.3898	20	.3800	.9250	.4108	2.434	1.081	2.632	40	1.1810
.3927	30	.3827	.9239	.4142	2.414	1.082	2.613	30	1.1781
.3956	40	.3854	.9228	.4176	2.394	1.084	2.595	20	1.1752
.3985	50	.3881	.9216	.4210	2.375	1.085	2.577	10	1.1723
.4014	23°00'	.3907	.9205	.4245	2.356	1.086	2.559	67°00'	1.1694
.4043	10	.3934	.9194	.4279	2.337	1.088	2.542	50	1.1665
.4072	20	.3961	.9182	.4314	2.318	1.089	2.525	40	1.1636
.4102	30	.3987	.9171	.4348	2.300	1.090	2.508	30	1.1606
.4131	40	.4014	.9159	.4383	2.282	1.092	2.491	20	1.1577
.4160	50	.4041	.9147	.4417	2.264	1.093	2.475	10	1.1548
.4189	24°00'	.4067	.9135	.4452	2.246	1.095	2.459	66°00'	1.1519
.4218	10	.4094	.9124	.4487	2.229	1.096	2.443	50	1.1490
.4247	20	.4120	.9112	.4522	2.211	1.097	2.427	40	1.1461
.4276	30	.4147	.9100	.4557	2.194	1.099	2.411	30	1.1432
.4305	40	.4173	.9088	.4592	2.177	1.100	2.396	20	1.1403
.4334	50	.4200	.9075	.4628	2.161	1.102	2.381	10	1.1374
.4363	25°00'	.4226	.9063	.4663	2.145	1.103	2.366	65°00'	1.1345
.4392	10	.4253	.9051	.4699	2.128	1.105	2.352	50	1.1316
.4422	20	.4279	.9038	.4734	2.112	1.106	2.337	40	1.1286
.4451	30	.4305	.9026	.4770	2.097	1.108	2.323	30	1.1257
.4480	40	.4331	.9013	.4806	2.081	1.109	2.309	20	1.1228
.4509	50	.4358	.9001	.4841	2.066	1.111	2.295	10	1.1199
.4538	26°00'	.4384	.8988	.4877	2.050	1.113	2.281	64°00'	1.1170
.4567	10	.4410	.8975	.4913	2.035	1.114	2.268	50	1.1141
.4596	20	.4436	.8962	.4950	2.020	1.116	2.254	40	1.1112
.4625	30	.4462	.8949	.4986	2.006	1.117	2.241	30	1.1083
.4654	40	.4488	.8936	.5022	1.991	1.119	2.228	20	1.1054
.4683	50	.4514	.8923	.5059	1.977	1.121	2.215	10	1.1025
.4712	27°00'	.4540	.8910	.5095	1.963	1.122	2.203	63°00'	1.0996
.4741	10	.4566	.8897	.5132	1.949	1.124	2.190	50	1.0966
.4771	20	.4592	.8884	.5169	1.935	1.126	2.178	40	1.0937
.4800	30	.4617	.8870	.5206	1.921	1.127	2.166	30	1.0908
.4829	40	.4643	.8857	.5243	1.907	1.129	2.154	20	1.0879
.4858	50	.4669	.8843	.5280	1.894	1.131	2.142	10	1.0850
		cos t	sin t	cot t	tan t	csc t	sec t	degrees	t

TABLE 3.   VALUES OF THE TRIGONOMETRIC FUNCTIONS   **A17**

Table 3. (continued)

Upper table (42°–48°)

t	degrees	sin t	cos t	tan t	cot t	sec t	csc t	csc t	sec t	cot t	tan t	cos t	sin t	degrees	t
.7330	42° 00'	.6691	.7431	.9004	1.111	1.346	1.494	1.494	1.346	1.111	.9004	.7431	.6691	48° 00'	.8378
.7359	10	.6713	.7412	.9057	1.104	1.349	1.490	1.490	1.349	1.104	.9057	.7412	.6713	50	.8348
.7389	20	.6734	.7392	.9110	1.098	1.353	1.485	1.485	1.353	1.098	.9110	.7392	.6734	40	.8319
.7418	30	.6756	.7373	.9163	1.091	1.356	1.480	1.480	1.356	1.091	.9163	.7373	.6756	30	.8290
.7447	40	.6777	.7353	.9217	1.085	1.360	1.476	1.476	1.360	1.085	.9217	.7353	.6777	20	.8261
.7476	50	.6799	.7333	.9271	1.079	1.364	1.471	1.471	1.364	1.079	.9271	.7333	.6799	10	.8232
.7505	43° 00'	.6820	.7314	.9325	1.072	1.367	1.466	1.466	1.367	1.072	.9325	.7314	.6820	47° 00'	.8203
.7534	10	.6841	.7294	.9380	1.066	1.371	1.462	1.462	1.371	1.066	.9380	.7294	.6841	50	.8174
.7563	20	.6862	.7274	.9435	1.060	1.375	1.457	1.457	1.375	1.060	.9435	.7274	.6862	40	.8145
.7592	30	.6884	.7254	.9490	1.054	1.379	1.453	1.453	1.379	1.054	.9490	.7254	.6884	30	.8116
.7621	40	.6905	.7234	.9545	1.048	1.382	1.448	1.448	1.382	1.048	.9545	.7234	.6905	20	.8087
.7650	50	.6926	.7214	.9601	1.042	1.386	1.444	1.444	1.386	1.042	.9601	.7214	.6926	10	.8058
.7679	44° 00'	.6947	.7193	.9657	1.036	1.390	1.440	1.440	1.390	1.036	.9657	.7193	.6947	46° 00'	.8029
.7709	10	.6967	.7173	.9713	1.030	1.394	1.435	1.435	1.394	1.030	.9713	.7173	.6967	50	.7999
.7738	20	.6988	.7153	.9770	1.024	1.398	1.431	1.431	1.398	1.024	.9770	.7153	.6988	40	.7970
.7767	30	.7009	.7133	.9827	1.018	1.402	1.427	1.427	1.402	1.018	.9827	.7133	.7009	30	.7941
.7796	40	.7030	.7112	.9884	1.012	1.406	1.423	1.423	1.406	1.012	.9884	.7112	.7030	20	.7912
.7825	50	.7050	.7092	.9942	1.006	1.410	1.418	1.418	1.410	1.006	.9942	.7092	.7050	10	.7883
.7854	45° 00'	.7071	.7071	1.0000	1.0000	1.414	1.414	1.414	1.414	1.0000	1.0000	.7071	.7071	45° 00'	.7854
		cos t	sin t	cot t	tan t	csc t	sec t	sec t	csc t	tan t	cot t	sin t	cos t	degrees	t

Lower table (35°–55°)

t	degrees	sin t	cos t	tan t	cot t	sec t	csc t	csc t	sec t	cot t	tan t	cos t	sin t	degrees	t
.6109	35° 00'	.5736	.8192	.7002	1.428	1.221	1.743	1.743	1.221	1.428	.7002	.8192	.5736	55° 00'	.9599
.6138	10	.5760	.8175	.7046	1.419	1.223	1.736	1.736	1.223	1.419	.7046	.8175	.5760	50	.9570
.6167	20	.5783	.8158	.7089	1.411	1.226	1.729	1.729	1.226	1.411	.7089	.8158	.5783	40	.9541
.6196	30	.5807	.8141	.7133	1.402	1.228	1.722	1.722	1.228	1.402	.7133	.8141	.5807	30	.9512
.6225	40	.5831	.8124	.7177	1.393	1.231	1.715	1.715	1.231	1.393	.7177	.8124	.5831	20	.9483
.6254	50	.5854	.8107	.7221	1.385	1.233	1.708	1.708	1.233	1.385	.7221	.8107	.5854	10	.9454
.6283	36° 00'	.5878	.8090	.7265	1.376	1.236	1.701	1.701	1.236	1.376	.7265	.8090	.5878	54° 00'	.9425
.6312	10	.5901	.8073	.7310	1.368	1.239	1.695	1.695	1.239	1.368	.7310	.8073	.5901	50	.9396
.6341	20	.5925	.8056	.7355	1.360	1.241	1.688	1.688	1.241	1.360	.7355	.8056	.5925	40	.9367
.6370	30	.5948	.8039	.7400	1.351	1.244	1.681	1.681	1.244	1.351	.7400	.8039	.5948	30	.9338
.6400	40	.5972	.8021	.7445	1.343	1.247	1.675	1.675	1.247	1.343	.7445	.8021	.5972	20	.9308
.6429	50	.5995	.8004	.7490	1.335	1.249	1.668	1.668	1.249	1.335	.7490	.8004	.5995	10	.9279
.6458	37° 00'	.6018	.7986	.7536	1.327	1.252	1.662	1.662	1.252	1.327	.7536	.7986	.6018	53° 00'	.9250
.6487	10	.6041	.7969	.7581	1.319	1.255	1.655	1.655	1.255	1.319	.7581	.7969	.6041	50	.9221
.6516	20	.6065	.7951	.7627	1.311	1.258	1.649	1.649	1.258	1.311	.7627	.7951	.6065	40	.9192
.6545	30	.6088	.7934	.7673	1.303	1.260	1.643	1.643	1.260	1.303	.7673	.7934	.6088	30	.9163
.6574	40	.6111	.7916	.7720	1.295	1.263	1.636	1.636	1.263	1.295	.7720	.7916	.6111	20	.9134
.6603	50	.6134	.7898	.7766	1.288	1.266	1.630	1.630	1.266	1.288	.7766	.7898	.6134	10	.9105
.6632	38° 00'	.6157	.7880	.7813	1.280	1.269	1.624	1.624	1.269	1.280	.7813	.7880	.6157	52° 00'	.9076
.6661	10	.6180	.7862	.7860	1.272	1.272	1.618	1.618	1.272	1.272	.7860	.7862	.6180	50	.9047
.6690	20	.6202	.7844	.7907	1.265	1.275	1.612	1.612	1.275	1.265	.7907	.7844	.6202	40	.9018
.6720	30	.6225	.7826	.7954	1.257	1.278	1.606	1.606	1.278	1.257	.7954	.7826	.6225	30	.8988
.6749	40	.6248	.7808	.8002	1.250	1.281	1.601	1.601	1.281	1.250	.8002	.7808	.6248	20	.8959
.6778	50	.6271	.7790	.8050	1.242	1.284	1.595	1.595	1.284	1.242	.8050	.7790	.6271	10	.8930
.6807	39° 00'	.6293	.7771	.8098	1.235	1.287	1.589	1.589	1.287	1.235	.8098	.7771	.6293	51° 00'	.8901
.6836	10	.6316	.7753	.8146	1.228	1.290	1.583	1.583	1.290	1.228	.8146	.7753	.6316	50	.8872
.6865	20	.6338	.7735	.8195	1.220	1.293	1.578	1.578	1.293	1.220	.8195	.7735	.6338	40	.8843
.6894	30	.6361	.7716	.8243	1.213	1.296	1.572	1.572	1.296	1.213	.8243	.7716	.6361	30	.8814
.6923	40	.6383	.7698	.8292	1.206	1.299	1.567	1.567	1.299	1.206	.8292	.7698	.6383	20	.8785
.6952	50	.6406	.7679	.8342	1.199	1.302	1.561	1.561	1.302	1.199	.8342	.7679	.6406	10	.8756
.6981	40° 00'	.6428	.7660	.8391	1.192	1.305	1.556	1.556	1.305	1.192	.8391	.7660	.6428	50° 00'	.8727
.7010	10	.6450	.7642	.8441	1.185	1.309	1.550	1.550	1.309	1.185	.8441	.7642	.6450	50	.8698
.7039	20	.6472	.7623	.8491	1.178	1.312	1.545	1.545	1.312	1.178	.8491	.7623	.6472	40	.8668
.7069	30	.6494	.7604	.8541	1.171	1.315	1.540	1.540	1.315	1.171	.8541	.7604	.6494	30	.8639
.7098	40	.6517	.7585	.8591	1.164	1.318	1.535	1.535	1.318	1.164	.8591	.7585	.6517	20	.8610
.7127	50	.6539	.7566	.8642	1.157	1.322	1.529	1.529	1.322	1.157	.8642	.7566	.6539	10	.8581
.7156	41° 00'	.6561	.7547	.8693	1.150	1.325	1.524	1.524	1.325	1.150	.8693	.7547	.6561	49° 00'	.8552
.7185	10	.6583	.7528	.8744	1.144	1.328	1.519	1.519	1.328	1.144	.8744	.7528	.6583	50	.8523
.7214	20	.6604	.7509	.8796	1.137	1.332	1.514	1.514	1.332	1.137	.8796	.7509	.6604	40	.8494
.7243	30	.6626	.7490	.8847	1.130	1.335	1.509	1.509	1.335	1.130	.8847	.7490	.6626	30	.8465
.7272	40	.6648	.7470	.8899	1.124	1.339	1.504	1.504	1.339	1.124	.8899	.7470	.6648	20	.8436
.7301	50	.6670	.7451	.8952	1.117	1.342	1.499	1.499	1.342	1.117	.8952	.7451	.6670	10	.8407
		cos t	sin t	cot t	tan t	csc t	sec t	sec t	csc t	tan t	cot t	sin t	cos t	degrees	t

t	sin t	cos t	tan t	cot t	sec t	csc t
.40	.3894	.9211	.4228	2.365	1.086	2.568
.41	.3986	.9171	.4346	2.301	1.090	2.509
.42	.4078	.9131	.4466	2.239	1.095	2.452
.43	.4169	.9090	.4586	2.180	1.100	2.399
.44	.4259	.9048	.4708	2.124	1.105	2.348
.45	.4350	.9004	.4831	2.070	1.111	2.299
.46	.4439	.8961	.4954	2.018	1.116	2.253
.47	.4529	.8916	.5080	1.969	1.122	2.208
.48	.4618	.8870	.5206	1.921	1.127	2.166
.49	.4706	.8823	.5334	1.875	1.133	2.125
.50	.4794	.8776	.5463	1.830	1.139	2.086
.51	.4882	.8727	.5594	1.788	1.146	2.048
.52	.4969	.8678	.5726	1.747	1.152	2.013
.53	.5055	.8628	.5859	1.707	1.159	1.978
.54	.5141	.8577	.5994	1.668	1.166	1.945
.55	.5227	.8525	.6131	1.631	1.173	1.913
.56	.5312	.8473	.6269	1.595	1.180	1.883
.57	.5396	.8419	.6410	1.560	1.188	1.853
.58	.5480	.8365	.6552	1.526	1.196	1.825
.59	.5564	.8309	.6696	1.494	1.203	1.797
.60	.5646	.8253	.6841	1.462	1.212	1.771
.61	.5729	.8196	.6989	1.431	1.220	1.746
.62	.5810	.8139	.7139	1.401	1.229	1.721
.63	.5891	.8080	.7291	1.372	1.238	1.697
.64	.5972	.8021	.7445	1.343	1.247	1.674
.65	.6052	.7961	.7602	1.315	1.256	1.652
.66	.6131	.7900	.7761	1.288	1.266	1.631
.67	.6210	.7838	.7923	1.262	1.276	1.610
.68	.6288	.7776	.8087	1.237	1.286	1.590
.69	.6365	.7712	.8253	1.212	1.297	1.571
.70	.6442	.7648	.8423	1.187	1.307	1.552
.71	.6518	.7584	.8595	1.163	1.319	1.534
.72	.6594	.7518	.8771	1.140	1.330	1.517
.73	.6669	.7452	.8949	1.117	1.342	1.500
.74	.6743	.7385	.9131	1.095	1.354	1.483
.75	.6816	.7317	.9316	1.073	1.367	1.467
.76	.6889	.7248	.9505	1.052	1.380	1.452
.77	.6961	.7179	.9697	1.031	1.393	1.437
.78	.7033	.7109	.9893	1.011	1.407	1.422
.79	.7104	.7038	1.009	.9908	1.421	1.408

Table 4.   *Trigonometric Functions of Radians and Real Numbers*

t	sin t	cos t	tan t	cot t	sec t	csc t
.00	.0000	1.0000	.0000	—	1.000	—
.01	.0100	1.0000	.0100	99.997	1.000	100.00
.02	.0200	.9998	.0200	49.993	1.000	50.00
.03	.0300	.9996	.0300	33.323	1.000	33.34
.04	.0400	.9992	.0400	24.987	1.001	25.01
.05	.0500	.9988	.0500	19.983	1.001	20.01
.06	.0600	.9982	.0601	16.647	1.002	16.68
.07	.0699	.9976	.0701	14.262	1.002	14.30
.08	.0799	.9968	.0802	12.473	1.003	12.51
.09	.0899	.9960	.0902	11.081	1.004	11.13
.10	.0998	.9950	.1003	9.967	1.005	10.02
.11	.1098	.9940	.1104	9.054	1.006	9.109
.12	.1197	.9928	.1206	8.293	1.007	8.353
.13	.1296	.9916	.1307	7.649	1.009	7.714
.14	.1395	.9902	.1409	7.096	1.010	7.166
.15	.1494	.9888	.1511	6.617	1.011	6.692
.16	.1593	.9872	.1614	6.197	1.013	6.277
.17	.1692	.9856	.1717	5.826	1.015	5.911
.18	.1790	.9838	.1820	5.495	1.016	5.586
.19	.1889	.9820	.1923	5.200	1.018	5.295
.20	.1987	.9801	.2027	4.933	1.020	5.033
.21	.2085	.9780	.2131	4.692	1.022	4.797
.22	.2182	.9759	.2236	4.472	1.025	4.582
.23	.2280	.9737	.2341	4.271	1.027	4.386
.24	.2377	.9713	.2447	4.086	1.030	4.207
.25	.2474	.9689	.2553	3.916	1.032	4.042
.26	.2571	.9664	.2660	3.759	1.035	3.890
.27	.2667	.9638	.2768	3.613	1.038	3.749
.28	.2764	.9611	.2876	3.478	1.041	3.619
.29	.2860	.9582	.2984	3.351	1.044	3.497
.30	.2955	.9553	.3093	3.233	1.047	3.384
.31	.3051	.9523	.3203	3.122	1.050	3.278
.32	.3146	.9492	.3314	3.018	1.053	3.179
.33	.3240	.9460	.3425	2.920	1.057	3.086
.34	.3335	.9428	.3537	2.827	1.061	2.999
.35	.3429	.9394	.3650	2.740	1.065	2.916
.36	.3523	.9359	.3764	2.657	1.068	2.839
.37	.3616	.9323	.3879	2.578	1.073	2.765
.38	.3709	.9287	.3994	2.504	1.077	2.696
.39	.3802	.9249	.4111	2.433	1.081	2.630

TABLE 4. TRIGONOMETRIC FUNCTIONS OF RADIANS AND REAL NUMBERS   A19

t	sin t	cos t	tan t	cot t	sec t	csc t
1.20	.9320	.3624	2.572	.3888	2.760	1.073
1.21	.9356	.3530	2.650	.3773	2.833	1.069
1.22	.9391	.3436	2.733	.3659	2.910	1.065
1.23	.9425	.3342	2.820	.3546	2.992	1.061
1.24	.9458	.3248	2.912	.3434	3.079	1.057
1.25	.9490	.3153	3.010	.3323	3.171	1.054
1.26	.9521	.3058	3.113	.3212	3.270	1.050
1.27	.9551	.2963	3.224	.3102	3.375	1.047
1.28	.9580	.2867	3.341	.2993	3.488	1.044
1.29	.9608	.2771	3.467	.2884	3.609	1.041
1.30	.9636	.2675	3.602	.2776	3.738	1.038
1.31	.9662	.2579	3.747	.2669	3.878	1.035
1.32	.9687	.2482	3.903	.2562	4.029	1.032
1.33	.9711	.2385	4.072	.2456	4.193	1.030
1.34	.9735	.2288	4.256	.2350	4.372	1.027
1.35	.9757	.2190	4.455	.2245	4.566	1.025
1.36	.9779	.2092	4.673	.2140	4.779	1.023
1.37	.9799	.1994	4.913	.2035	5.014	1.021
1.38	.9819	.1896	5.177	.1931	5.273	1.018
1.39	.9837	.1798	5.471	.1828	5.561	1.017
1.40	.9854	.1700	5.798	.1725	5.883	1.015
1.41	.9871	.1601	6.165	.1622	6.246	1.013
1.42	.9887	.1502	6.581	.1519	6.657	1.011
1.43	.9901	.1403	7.055	.1417	7.126	1.011
1.44	.9915	.1304	7.602	.1315	7.667	1.009
1.45	.9927	.1205	8.238	.1214	8.299	1.007
1.46	.9939	.1106	8.989	.1113	9.044	1.006
1.47	.9949	.1006	9.887	.1011	9.938	1.005
1.48	.9959	.0907	10.983	.0910	11.029	1.004
1.49	.9967	.0807	12.350	.0810	12.390	1.003
1.50	.9975	.0707	14.101	.0709	14.137	1.003
1.51	.9982	.0608	16.428	.0609	16.458	1.002
1.52	.9987	.0508	19.670	.0508	19.695	1.001
1.53	.9992	.0408	24.498	.0408	24.519	1.001
1.54	.9995	.0308	32.461	.0308	32.476	1.000
1.55	.9998	.0208	48.078	.0208	48.089	1.000
1.56	.9999	.0108	92.620	.0108	92.626	1.000
1.57	1.0000	.0008	1255.8	.0008	1255.8	1.000

Table 4 (continued)

t	sin t	cos t	tan t	cot t	sec t	csc t
.80	.7174	.6967	1.030	.9712	1.435	1.394
.81	.7243	.6895	1.050	.9520	1.450	1.381
.82	.7311	.6822	1.072	.9331	1.466	1.368
.83	.7379	.6749	1.093	.9146	1.482	1.355
.84	.7446	.6675	1.116	.8964	1.498	1.343
.85	.7513	.6600	1.138	.8785	1.515	1.331
.86	.7578	.6524	1.162	.8609	1.533	1.320
.87	.7643	.6448	1.185	.8437	1.551	1.308
.88	.7707	.6372	1.210	.8267	1.569	1.297
.89	.7771	.6294	1.235	.8100	1.589	1.287
.90	.7833	.6216	1.260	.7936	1.609	1.277
.91	.7895	.6137	1.286	.7774	1.629	1.267
.92	.7956	.6058	1.313	.7615	1.651	1.257
.93	.8016	.5978	1.341	.7458	1.673	1.247
.94	.8076	.5898	1.369	.7303	1.696	1.238
.95	.8134	.5817	1.398	.7151	1.719	1.229
.96	.8192	.5735	1.428	.7001	1.744	1.221
.97	.8249	.5653	1.459	.6853	1.769	1.212
.98	.8305	.5570	1.491	.6707	1.795	1.204
.99	.8360	.5487	1.524	.6563	1.823	1.196
1.00	.8415	.5403	1.557	.6421	1.851	1.188
1.01	.8468	.5319	1.592	.6281	1.880	1.181
1.02	.8521	.5234	1.628	.6142	1.911	1.174
1.03	.8573	.5148	1.665	.6005	1.942	1.166
1.04	.8624	.5062	1.704	.5870	1.975	1.160
1.05	.8674	.4976	1.743	.5736	2.010	1.153
1.06	.8724	.4889	1.784	.5604	2.046	1.146
1.07	.8772	.4801	1.827	.5473	2.083	1.140
1.08	.8820	.4713	1.871	.5344	2.122	1.134
1.09	.8866	.4625	1.917	.5216	2.162	1.128
1.10	.8912	.4536	1.965	.5090	2.205	1.122
1.11	.8957	.4447	2.014	.4964	2.249	1.116
1.12	.9001	.4357	2.066	.4840	2.295	1.111
1.13	.9044	.4267	2.120	.4718	2.344	1.106
1.14	.9086	.4176	2.176	.4596	2.395	1.101
1.15	.9128	.4085	2.234	.4475	2.448	1.096
1.16	.9168	.3993	2.296	.4356	2.504	1.091
1.17	.9208	.3902	2.360	.4237	2.563	1.086
1.18	.9246	.3809	2.427	.4120	2.625	1.082
1.19	.9284	.3717	2.498	.4003	2.691	1.077

ANSWERS TO ODD-NUMBERED EXERCISES

CHAPTER 1

Section 1.1, page 8

1 Commutative **3** Associative **5** Identity **7** Identity **9** Inverse **11** 29/12

13 5/9 **15** 11/12 **17** 13/10 **19** 10/21

21 Two examples are $2 - 3 \neq 3 - 2$ and $2 - (3 - 4) \neq (2 - 3) - 4$.

23 (a) < (b) > (c) = **25** (a) > (b) < (c) > **27** $-8 < -5$ **29** $0 > -1$

31 $x < 0$ **33** $3 < a < 5$ **35** $b \geq 2$ **37** $c \leq 1$ **39** (a) 5 (b) -5 (c) 13

41 (a) 0 (b) $4 - \pi$ (c) -1 **43** (a) 4 (b) 6 (c) 6 (d) 10

45 (a) 12 (b) 3 (c) 3 (d) 9 **47** $x - 5$ **49** $4 + x^2$

51 Not true; for example, $(1/2) + (1/3) \neq 1/(2 + 3)$.

53 If $a = -a$, then $a + a = 0$, or $2a = 0$. Multiplying both sides by 1/2 produces $a = 0$. Conversely, if $a = 0$, then also $(-1)a = (-1)0 = 0$ and hence $-a = 0$. Thus $a = 0 = -a$.

55 If $ab = 0$ and $a \neq 0$, then $(1/a)ab = (1/a)0 = 0$, and hence $b = 0$.

57 $a(b - c) = a[b + (-c)] = ab + a(-c) = ab + [-(ac)] = ab - ac$

59 $\dfrac{-a}{-b} = -\left(\dfrac{a}{-b}\right) = -\left(-\dfrac{a}{b}\right) = \dfrac{a}{b}$

Section 1.2, page 17

1 (a) $-1 < 1$ (b) $-11 < -9$ (c) $-3 < -2$ (d) $3 > 2$

3 $(2, 5)$ **5** $(-1, 3]$ **7** $[1, 4]$

9 $(-1, \infty)$ **11** $(-\infty, 2]$

13 $-1 < x < 7$ **15** $8 < x \le 9$ **17** $5 < x$ **19** $(17/5, \infty)$ **21** $[-2, \infty)$ **23** $(5/2, \infty)$

25 $(-\infty, 44/3]$ **27** $(-3, 1)$ **29** $[1, 5]$ **31** $[2, 8/3)$ **33** $(6/38, \infty)$ **35** $(-2/3, \infty)$

37 $(-7/3, \infty)$ **39** $(1/4, \infty)$ **41** $(-\infty, 1)$ and $(1, \infty)$ **43** $140/9 \le c \le 80/3$

45 $20/9 \le x \le 4$

47 The result is false if $a < 0$ (consider, for example, $a = -2$ and $b = 2$).

49 If $a < b$ and $c < d$, then $b - a$ and $d - c$ are positive and hence $(b - a) + (d - c)$ is positive. Consequently, $(b + d) - (a + c)$ is positive; that is, $a + c < b + d$.

Section 1.3, page 22

 1 $(-2, 2)$ **3** $(6, \infty) \cup (-\infty, -6)$ **5** $[-10, 10]$ **7** $(9.95, 10.05)$ **9** $[-10, 2]$

11 $(-2/3, 4)$ **13** $(-\infty, -6) \cup (2, \infty)$ **15** $(-\infty, 1] \cup [4, \infty)$ **17** $(-2, 3)$

19 $(-\infty, -5/2) \cup (1, \infty)$ **21** $[0, 10]$ **23** $(-\infty, -1/2) \cup (10/3, \infty)$ **25** $(-3, 3)$

27 $(4, \infty) \cup (-4, 0]$ **29** $(-\infty, -3) \cup (2/3, \infty)$ **31** $(-\infty, 1/2) \cup (7/5, \infty)$

33 $(-\infty, -1) \cup (0, 1)$ **35** $[-2, 1] \cup [2, \infty)$ **37** $(-\infty, -1] \cup (1, 2] \cup (3, \infty)$ **39** $[1/2, 4]$

Section 1.4, page 28

1

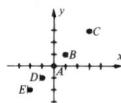

3 The line bisecting quadrants I and III.

5 (a) The line parallel to the y-axis which intersects the x-axis at $(3, 0)$
   (b) The line parallel to the x-axis which intersects the y-axis at $(0, -1)$
   (c) All points to the right of, and on, the y-axis
   (d) All points in quadrants I and III
   (e) All points under the x-axis

7 (a) $\sqrt{29}$ (b) $(5, -1/2)$ **9** (a) $\sqrt{13}$ (b) $(-7/2, -1)$ **11** (a) 4 (b) $(5, -3)$

13 Area $= (1/2)\sqrt{29}\sqrt{116}$ **17** $(13, -28)$ **19** $d(A, P) = d(B, P)$

21 $\sqrt{x^2 + y^2} = 5$. A circle of radius 5 with center at the origin

23 $(0, 3 + \sqrt{11}), (0, 3 - \sqrt{11})$ **25** $(-2, -1)$ **27** $a > 4$ and $a < 2/5$

Section 1.5, page 36

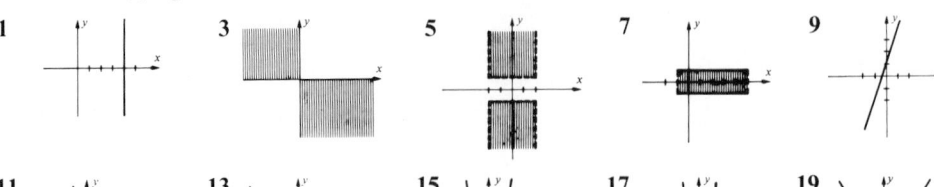

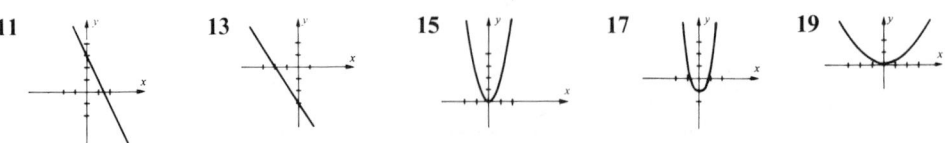

21 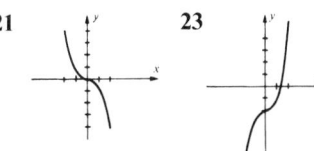 **23** **25** 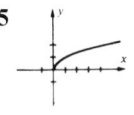 **27**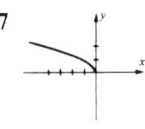

29 Circle of radius 4, center at the origin   **31** Circle of radius 1/3, center at the origin

33 Circle of radius 2, center at $(2, -1)$ **35** Circle of radius 3, center at $(0, 3)$

37 $(x - 3)^2 + (y + 2)^2 = 16$ **39** $(x - 1/2)^2 + (y + 3/2)^2 = 4$ **41** $x^2 + y^2 = 34$

43 $(x + 4)^2 + (y - 2)^2 = 4$ **45** $(x - 1)^2 + (y - 2)^2 = 34$

Section 1.6, Review Exercises, page 37

1 (a) $-5/12$ (b) $39/20$ (c) $-13/56$ (d) $5/8$ **3** (a) $x < 0$ (b) $1/3 < a < 1/2$ (c) $|x| \leq 4$

5 (a) 5 (b) 5 (c) 7   **7** The interval $(-11/4, 9/4)$ **9** $(-\infty, -3/10)$

11 $(-\infty, 11/3] \cup [7, \infty)$ **13** $(-\infty, -3/2) \cup (2/5, \infty)$ **15** $(-\infty, -3/2) \cup (2,9)$ **17** $(1, \infty)$

19 Area $= 10$ **21** The points in quadrants II and IV

23 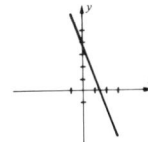 **25** 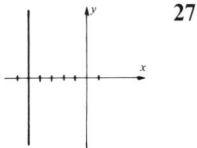 **27** 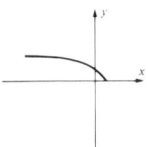 **29**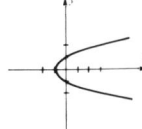

31 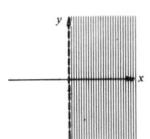 **33**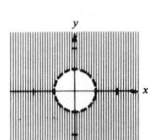

CHAPTER 2

Section 2.1, page 46

1 3, 9, 4, 6   **3** 2, $\sqrt{2} + 6, 12, 23$

5 (a) $5a - 2$ (b) $-5a - 2$ (c) $-5a + 2$ (d) $5a + 5h - 2$ (e) $5a + 5h - 4, 5$

7 (a) $2a^2 - a + 3$ (b) $2a^2 + a + 3$ (c) $-2a^2 + a - 3$ (d) $2a^2 + 4ah + 2h^2 - a - h + 3$
(e) $2a^2 - a + 2h^2 - h + 6$ (f) $4a + 2h - 1$

9 (a) $3/a^2$ (b) $1/3a^2$ (c) $3a^4$ (d) $9a^4$ (e) $3a$ (f) $\sqrt{3a^2}$

11 (a) $2a/(a^2 + 1)$ (b) $(a^2 + 1)/2a$ (c) $2a^2/(a^4 + 1)$ (d) $4a^2/(a^4 + 2a^2 + 1)$
(e) $2\sqrt{a}/(a + 1)$ (f) $\sqrt{2a^3 + 2a}/(a^2 + 1)$

13 $[5/3, \infty)$ **15** $[-2,2]$ **17** All real numbers except 0, 3, and -3

19 All nonnegative real numbers except 4 and 3/2

21 $9/7, (a + 5)/7, \mathbb{R}$ **23** $19, a^2 + 3$, all nonnegative real numbers

25 $\sqrt[3]{4}, \sqrt[3]{a}, \mathbb{R}$ **27** Yes   **29** No   **31** Yes   **33** No   **35** Odd   **37** Even

39 Even   **41** Neither   **43** Neither

45 The equation $y = f(x)$ is not changed if $-x$ is substituted for x.

47 $r = C/2\pi; 6/\pi \approx 1.9$ inches   **49** $V = 4x^3 - 100x^2 + 600x$ **51** $P = 4\sqrt{A}$

53 $d = 2\sqrt{t^2 + 2500}$ **55** $A = 20x$ if $x < 50$ and $A = 20x - 0.02x^2$ if $50 \le x \le 600$

57 Yes   **59** No   **61** Yes   **63** No

Section 2.2, page 55

1 Increasing on $\mathbb{R}$ **3** Decreasing on $\mathbb{R}$ **5** Increasing on $(-\infty, 0]$
Decreasing on $[0, \infty)$

7 Decreasing on $(-\infty, 0]$ **9** Increasing on $[-4, \infty)$ **11** Increasing on $[0, \infty)$
Increasing on $[0, \infty)$

13 Decreasing on $(-\infty, 0)$ **15** Decreasing on $(-\infty, 2]$ **17** Decreasing on $(-\infty, 0]$
and on $(0, \infty)$ Increasing on $[2, \infty)$ Increasing on $[0, \infty)$

19 Neither increasing   **21** 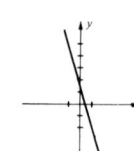   **23** 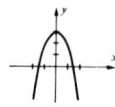   **25**
nor decreasing

27 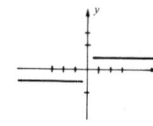   **29**   **31**   **33** 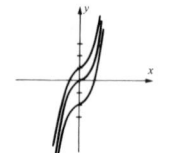   **35**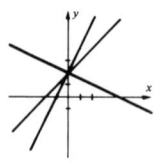

37 $f(x) = -\sqrt{1 - x^2}$

39 If $a < b$, then $f(a) > f(b)$; and if $b < a$, then $f(b) > f(a)$. Hence, if $a \neq b$, then $f(a) \neq f(b)$.

Section 2.3, page 63

1 $m = 8$ **3** $m = 1/11$ **5** $m = -1/4$ **9** $x - 3y - 11 = 0$ **11** $x + 5y + 23 = 0$

13 $4x + y - 7 = 0$ **15** $7x - 3y + 21 = 0$ **17** (a) $y = -3$ (b) $x = 7$ **19** $y = -x$

21 $13x - 8y + 132 = 0$ **23** $8x - 15y - 13 = 0$, $x + 6y + 1 = 0$, $10x - 3y - 11 = 0$

25 $m = 2/5$, $b = 2$ **27** $m = -3/5$, $b = 0$ **29** $m = 0$, $b = -2$ **31** $m = -5/6$, $b = 10/3$

33 $m = 0, b = 0$ **35** $k = -3$

37 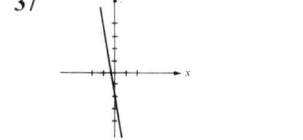   **39**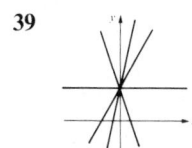

41 $-2/5$ **43** $f(x) = -6x - 2$ **47** $\dfrac{x}{(3/2)} + \dfrac{y}{(-3)} = 1$ **49** $r < 1$ or $r > 2$

Section 2.4, page 70

1 $6x - 1, 6x + 3$ **3** $36x^2 - 5, 12x^2 - 15$ **5** $12x^2 - 8x + 1, 6x^2 + 4x - 1$

7 $x^3 - 1, x^3 - 3x^2 + 3x - 1$ **9** $x + 9 + 9\sqrt{x + 9}, \sqrt{x^2 + 9x + 9}$

11 $x^2/(2 - 5x^2), 4x^2 - 20x + 25$ **13** $5, -5$ **15** $1/x^4, 1/x^4$ **17** x, x

19 and 21 Show $f(g(x)) = x = g(f(x))$ **23** $f^{-1}(x) = (x + 3)/4$

25 $f^{-1}(x) = (1 - 5x)/2x$, $x > 0$ **27** $f^{-1}(x) = \sqrt{9 - x}$, $x \leq 9$ **29** $f^{-1}(x) = \sqrt[3]{(x + 2)/5}$

31 $f^{-1}(x) = (x^2 + 5)/3$, $x \geq 0$ **33** $f^{-1}(x) = (x - 8)^3$ **35** $f^{-1}(x) = x$

37 (a) $f^{-1}(x) = (x - b)/a$ (b) No (not one-to-one) (c) Yes, it is its own inverse.

39 If g and h are both inverse functions of f, then $f(g(x)) = x = f(h(x))$ for all x. Since f is one-to-one this implies that $g(x) = h(x)$ for all x, that is, $g = h$.

Section 2.5, Review Exercises, page 71

1 $-11/19$ **3** $18x + 6y - 7 = 0$ **5** $y + x + 3 = 0$

7 Decreasing on $\mathbb{R}$ **9** Decreasing on $[-1, \infty)$ **11**

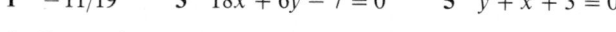
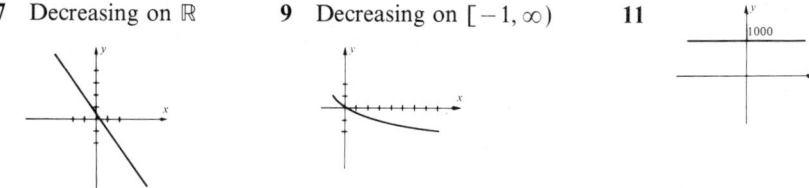

13 (a) $1/2$ (b) $-\sqrt{2}/2$ (c) 0 (d) $-x/\sqrt{3 - x}$ (e) $-x/\sqrt{x + 3}$ (f) $x^2/\sqrt{x^2 + 3}$ (g) $x^2/(x + 3)$

15 $18x^2 + 9x - 1, 6x^2 - 15x + 5$ **17** $f^{-1}(x) = (10 - x)/15$ **19** $V = C^3/8\pi^2$

CHAPTER 3

Section 3.1, page 78

1 $x^4 - 3x^3 + 4x^2 + 4x + 5; 4$ **3** $-8x; 1$ **5** $6x^3 - 25x^2 + 46x - 35; 3$

7 $2x^3 + x^2 + 7; 2x^3 - x^2 - 2x + 3; 2x^5 + 2x^4 + 3x^3 + 4x^2 + 3x + 10$

9 $14x^4 - x^3 + x^2 + 4x - 1; x^3 + x^2 - 4x - 1; 49x^8 - 7x^7 + 7x^6 + 27x^5 - 7x^4 + 5x^3 - 4x$

11 $x + 3; 3 - x; 3x$ **13** Degree of product is $4 + 3$, or 7.

17 Let $f(x) = a_n x^n + \cdots + a_0$ and $g(x) = b_m x^m + \cdots + b_0$, where $a_n \neq 0$ and $n > m$. Then $f(x) + g(x) = a_n x^n + \cdots + a_0 + b_0$ has degree n. If $n = m$, the degree of $f(x) + g(x)$ is not greater than n.

19 $(3x - 2)(x + 4)$ **21** $3x(4x - 5)(x + 3)$ **23** $(5x - 3)(5x + 3)$

25 $(3x - 1)(9x^2 + 3x + 1)$ **27** $(x + 3)(x - 3)(x + 2)(x - 2)$ **29** $(x + 1)^2$ **31** $x^2 + x + 1$

33 $(x^2 + 1)(x + 1)$ **35** $(x^2 + 3)(x^2 - 3)(x^2 + 1)(x + 1)(x - 1)$

37 $(x - 1)(x^2 + x + 1)(x + 1)(x^2 - x + 1)$ **39** $(2x + 3)(2x - 3)(x - 2)$

Section 3.2, page 86

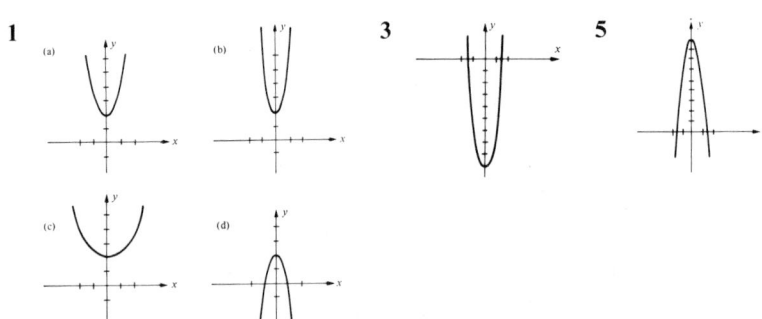

7 $3, -1/4$ **9** $2/3$ **11** $(-1 \pm \sqrt{41})/4$ **13** $(-9 \pm \sqrt{85})/2$ **15** $2(x - 4)^2 - 9$

17 $-5(x + 1)^2 + 8$ **19** $4(x - \frac{1}{2})^2 + 1$

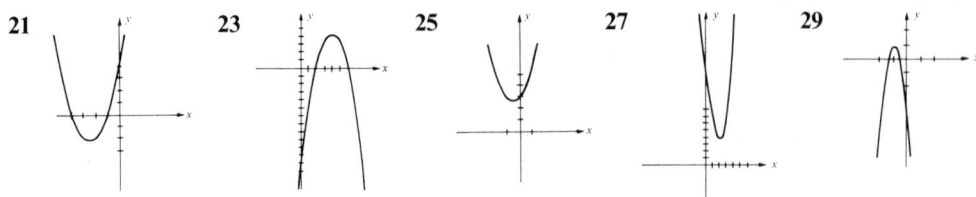

31 $k = 6$ **33** At $t = 20 - 10\sqrt{3}$ and $t = 20 + 10\sqrt{3}$; $t = 40$

35 If $f(x) = ax + b$ and $g(x) = cx + d$, then $f(x)g(x) = (ac)x^2 + (ad + bc)x + bd$.

37 $10, 2$ **39** $-1/2$

Section 3.3, page 94

1

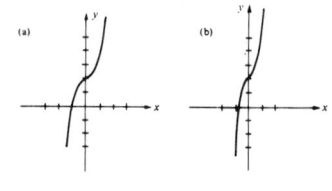

3 $f(x) > 0$ if $x > 2$
$f(x) < 0$ if $x < 2$

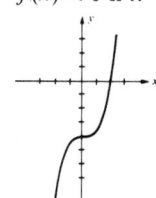

5 $f(x) > 0$ for all x

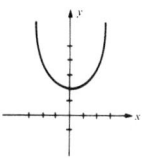

7 $f(x) > 0$ if $-3 < x < 0$ or $x > 3$
$f(x) < 0$ if $x < -3$ or $0 < x < 3$

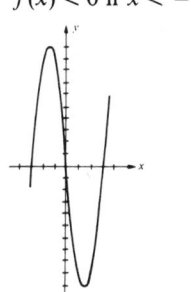

9 $f(x) > 0$ if $x < -2$ or $0 < x < 1$
$f(x) < 0$ if $-2 < x < 0$ or $x > 1$

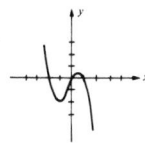

13 $f(x) > 0$ if $x < -4$ or $x > 4$
$f(x) < 0$ if $-4 < x < 4$

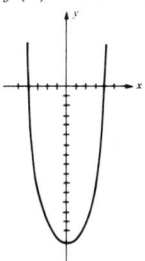

11 $f(x) > 0$ if $-4 < x < 1$ or $x > 5$
$f(x) < 0$ if $x < -4$ or $1 < x < 5$

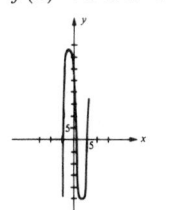

15 $f(x) > 0$ if $-1 < x < 1$
$f(x) < 0$ if $x < -1$ or $x > 1$

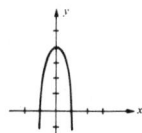

19 $k = -4/3$

17 $f(x) > 0$ if $x < -3$, $-1 < x < 0$, or $x > 2$
$f(x) < 0$ if $-3 < x < -1$ or $0 < x < 2$

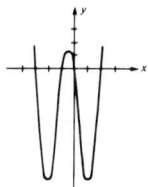

Section 3.4, page 98

1 $x^2 - 5x + 20$, $-84x + 104$ **3** $(5/2)x$, $(-29/2)x$ **5** $0, 9x^3 + 4x - 1$ **7** -37

9 -9 **11** $14/3$ **13** $f(3) = 0$ **15** $f(c) > 0$

17 If $f(x) = x^n - y^n$, then $f(y) = 0$. If n is even, then $f(-y) = 0$.

Section 3.5, page 101

1 $2x^2 + x + 6, 7$ **3** $x^2 - 3x + 1, -8$ **5** $3x^4 - 6x^3 + 12x^2 - 18x + 36, -65$

7 $4x^3 + 2x^2 - 4x - 2, 0$ **9** $x^{n-1} + x^{n-2} + \cdots + x + 1, 0$ **11** $-23, 53$ **13** 3277

15 $8 + 7\sqrt{3}$

Section 3.6, page 107

1 $x^4 - 2x^3 - 23x^2 + 24x + 144$ **3** $x^7 - 3x^6 + 3x^5 - x^4$

5 -5 (multiplicity 2), 1 (multiplicity 1) **7** 0 (multiplicity 3), $(-1 \pm \sqrt{21})/2$ (each of multiplicity 1)

9 3 and -3 (multiplicity 2), 2 and -2 (multiplicity 1)

11 -2 (multiplicity 4), -3, 1 (each of multiplicity 1)

13 $f(x) = (x - 3)^2(x - \sqrt{5})(x + \sqrt{5})$ **15** $f(x) = (x - 1)^3(x + 4); -4$

17 Positive: 3 or 1; Negative: 0 **19** Positive: 0; Negative: 1

21 Positive: 2 or 0; Negative: 2 or 0 **23** Positive: 2 or 0; Negative: 3 or 1

25 Upper 5, lower -2 **27** Upper 2, lower -2

29 Upper 3, lower -3

Section 3.7, page 110

1 $2, -3, 5/2$ **3** $2/3, -2, -1/2$ **5** $4/3$ **7** $-2, 3$ **9** $1/3$ **11** $4, -7, \pm\sqrt{2}$

13 $3, -4, -2, 1/2, -1/2$

Section 3.8, page 122

1 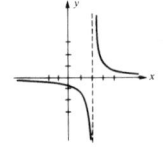 **3** 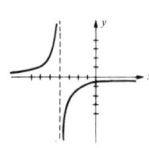 **5** 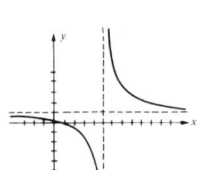 **7**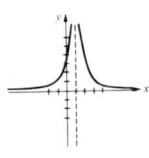

9 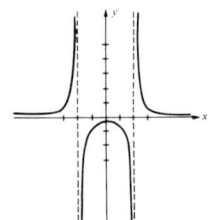 **11** 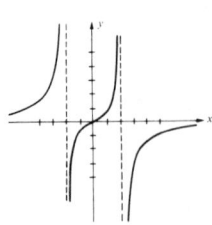 **13**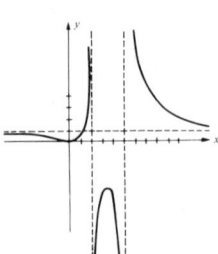

15 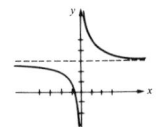 **17** 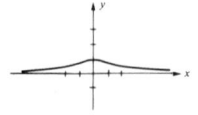 **19**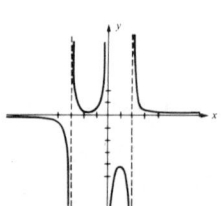

Section 3.9, page 127

1 $3/(x - 2) + 5/(x + 3)$ **3** $5/(x - 6) - 4/(x + 2)$ **5** $2/(x - 1) + 3/(x + 2) - 1/(x - 3)$

7 $(3/x) + 2/(x - 5) - 1/(x + 1)$ **9** $2/(x - 1) + 5/(x - 1)^2$ **11** $(-7/x) + (5/x^2) + 2/(3x - 5)$

13 $(-23/25)/(2x - 1) + (24/25)/(x + 2) + (2/5)/(x + 2)^2$ **15** $(5/x) - 2/(x + 1) + 3/(x + 1)^3$

17 $-2/(x - 1) + (3x + 4)/(x^2 + 1)$ **19** $(4/x) + (5x - 3)/(x^2 + 2)$

21 $(4x - 1)/(x^2 + 1) + 3/(x^2 + 1)^2$ **23** $2x + 3x/(x^2 + 1) + 1/(x - 1)$

Section 3.10, Review Exercises, page 128

1 $-x^3 - 3x^2 + 6x - 9; 3$ **3** $x^3 - 8x^2 + 19x - 20; 3$

5 $-8x^2 + 8x; 2$ **7** $(4x - 1)(2x + 3)$ **9** $x(3x - 2)(x^2 + 4)$

11 $2(x^2 + 4)(x + 2)(x - 2)$ **13** $4x^2 - 10, 12x^2 + 21x - 23$ **15** $3, -14$

17 (a) -65 (b) $f(-3) = 0$ **19** (a) $x^3 - 6x^2 + 11x - 6$ (b) $x^3 - (\sqrt{2} + \sqrt{3})x^2 + 6x$

21 $(x + 2)^3(x - 1)(x + 3)$ **23** (a) Positive: 2 or 0;   Negative: 3 or 1   (b) Upper 2, lower -3

25 $-2/3, -7/2, 3/5$

27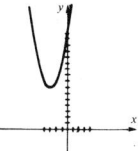

29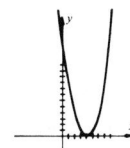

31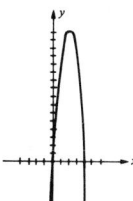

33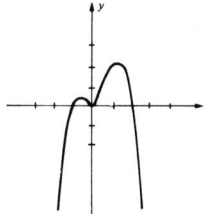

35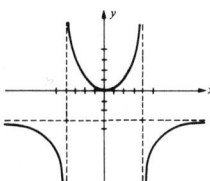

37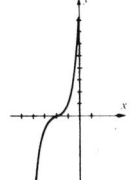

39 $8/(x - 1) - 3/(x + 5) - 1/(x + 3)$

CHAPTER 4

Section 4.1, page 138

1 16/81   **3** 9/8   **5** $-71/9$ **7** 1/8   **9** 0.04   **11** $8x^9$ **13** $6/x$

15 $-2a^{14}$ **17** 9/2   **19** $12u^{11}/v^2$ **21** $4/xy$ **23** $9y^6/x^8$ **25** $81y^6/64$ **27** $s^6/(4r^8)$

29 $20y/x^3$ **31** $9x^{10}y^{14}$ **33** $8a^2$ **35** $24x^{3/2}$ **37** $1/(9a^4)$ **39** $8/x^{1/2}$ **41** $4x^2y^4$

43 $3/x^3y^2$ **45** 1   **47** 9   **49** $-2\sqrt[5]{2}$ **51** $\sqrt[3]{4}/2$ **53** $3y^3/x^2$ **55** $2a^2/b$

57 $(3a^2/b)\sqrt[3]{2ab}$ **59** $\sqrt{3uv}/3u^2v$ **61** $3x^5/y^2$ **63** $(2x/y^2)\sqrt[5]{x^2y^4}$

65 $-3tv^2$ **67** $(\sqrt{a}-\sqrt{b})/(a-b)$ **69** $(y\sqrt{x}+x\sqrt{y})/xy$ **71** $c(1+\sqrt{c})/(1-c)$

73 1.7×10^{-24} **75** (a) 4.27×10^5 (b) 9.8×10^{-8} (c) 8.1×10^8

77 (a) 830,000 (b) 0.0000000000029 (c) 563,000,000   **79** 6.93   **81** 1.125×10^{24}

83 3.6×10^9 **85** 5.87×10^{12} miles

Section 4.2, page 146

1   **3**   **5**   **7**   **9**

11   **13**   **15**   **17**   **19**

21   **23**   **25**

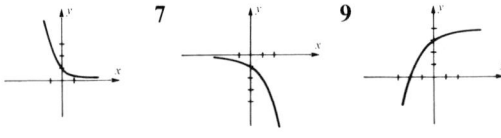

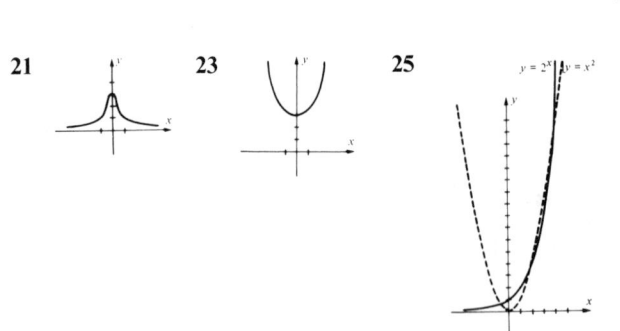

27 $-1/1600$ **29** $1010.00, $1020.10, $1061.52, $1126.83   **31** a^x is not always real if $a<0$.

33 Reflection through the x-axis.

Calculator Exercises

1 (a) $1,061.36 (b) $1,126.49 (c) $1,346.86 (d) $1,814.01

3 (a) 5 years (b) 8 years (c) 12.5 years

5 (a) (b) (c) (d) (e) (f)

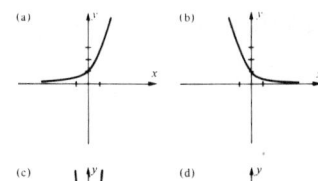 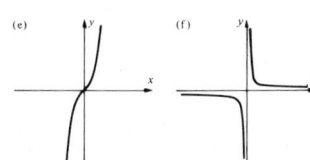

7

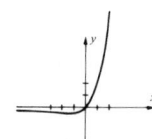

9 2.7048, 2.7169, 2.7181, 2.7183, 2.7183; $f(x)$ gets closer to e as x gets closer to 0   **11** 7.44

Section 4.3, page 153

1 $\log_4 64 = 3$ **3** $\log_2 128 = 7$ **5** $\log_{10}(0.001) = -3$ **7** $\log_t s = r$ **9** $10^3 = 1000$

11 $3^{-5} = 1/243$ **13** $7^0 = 1$ **15** $t^p = r$ **17** -2 **19** 2   **21** 5   **23** 1/3

25 -1 **27** -3 **29** 16   **31** 0.3   **33** 2.4   **35** -0.25 **37** 1.8   **39** 1.25

41 13   **43** 27   **45** $1/5, -1/5$ **47** $2,3$ **49** $\sqrt[3]{7}$ **51** 7/2   **53** No solution

55 1   **57** $2 + \sqrt{8}$ **59** $2\log_a x + \log_a y - 3\log_a z$ **61** $(1/2)\log_a x + 2\log_a z - 4\log_a y$

63 $(2/3)\log_a x - (1/3)\log_a y - (5/3)\log_a z$ **65** $(1/2)\log_a x + (1/4)\log_a y + (3/4)\log_a z$

67 $\log_a(x^2\sqrt[3]{x-2}/(2x+3)^5)$ **69** $\log_a x^4/y^{5/3}$

Section 4.4, page 158

1 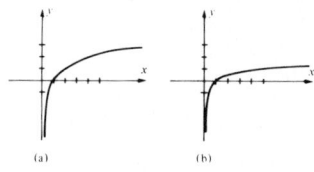   **3**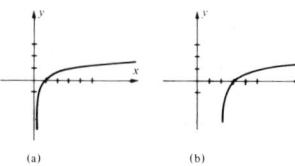
(a)   (b)   (a)   (b)

5 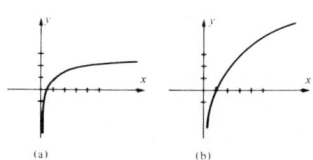   **7**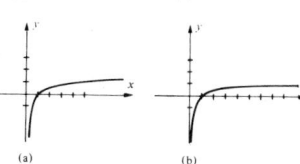
(a)   (b)   (a)   (b)

9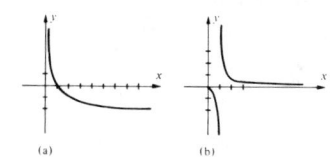
(a)   (b)

11 They are reflections of one another through the line $y = x$. They are inverse functions of one another

13 $t = -1600 \log_2 (q/q_0)$ **15** $t = (1/k)(\ln Q - \ln Q_0)$ **17** (a) 10   (b) 30   (c) 40

Calculator Exercises

1

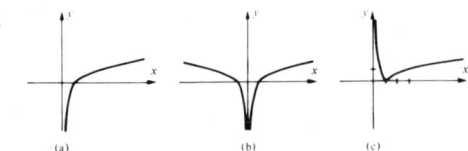

(a)   (b)   (c)

5 (a) 1.2619   (b) 0.3074   (c) 0.1781   (d) 2.5821   (e) -0.1203

Section 4.5, page 164

1 2.5403, 7.5403 $-$ 10, 0.5403   **3** 9.7324 $-$ 10, 2.7324, 5.7324   **5** 1.7796, 5.7796 $-$ 10, 2.7796

7 3.3044, 0.8261, 6.6956 $-$ 10   **9** 9.9235 $-$ 10, 0.0765, 9.9915 $-$ 10   **11** 4.0428   **13** 2.0878

15 3.3759   **17** 4,240   **19** 8.85   **21** 162,000   **23** 0.543   **25** 0.0677   **27** 189

29 0.0237   **31** (a) 2.2   (b) 5   (c) 8.3   **33** Acidic if pH < 7, basic if pH > 7 **35** $\alpha \approx 54.5$

Section 4.6, page 167

1 $\log 7/\log 10$ **3** $\log 3/\log 4$ **5** $4 - (\log 5/\log 3)$ **7** $(\log 2 - \log 81)/\log 24$

9 $-\log 8/\log 2$ **11** 5   **13** $(2/3)\sqrt{1001/111}$ **15** 100, 1   **17** 10^{100} **19** 10,000

21 $x = \log (y \pm \sqrt{y^2 - 1})$ **23** $x = (1/2)\log (1 + y)/(1 - y)$ **25** $x = \ln (y + \sqrt{y^2 + 1})$

27 $x = (1/2)\ln (1 + y)/(1 - y)$ **29** $t = -(L/R)\ln (1 - Ri/E)$ **31** (a) $I = I_0 10^{x/10}$

Calculator Exercises

1 $x \approx 1.2091$ **3** $x \approx \pm 1.347$ **5** 139 months

Section 4.7, page 171

1 1.4062   **3** 3.7294   **5** 7.1000 $-$ 10   **7** 5.0913   **9** 9.8913 $-$ 10   **11** 2.5851

13 9.9760 $-$ 10   **15** 4.8234   **17** 8.6356 $-$ 10   **19** 0.4776   **21** 27.78   **23** 49,330

25 0.1467   **27** 0.006536   **29** 234.7   **31** 1.367   **33** 0.09445   **35** 109,900

37 0.001345   **39** 0.2442

Section 4.8, page 174

1 36,500   **3** 4.26   **5** 1,750,000   **7** 0.876   **9** 6.20   **11** 22.0   **13** 22.1

15 0.000724   **17** 1.37   **19** 1.99   **21** 90.27   **23** 0.9356   **25** 3.068

27 88.6 square units   **29** 1.84 seconds

Section 4.9, Review Exercises, page 175

1 -4 **3** 4   **5** 6

7 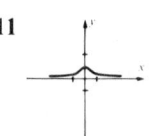 **9** 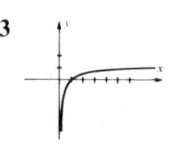 **11** 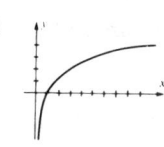 **13** **15**

17 9   **19** 1   **21** $\log(16/3)/\log 2$ **23** $\log(3/8)/\log(32/9)$

25 $4\log x + (2/3)\log y - (1/3)\log z$ **27** $x = (1/2)\log(y+1)/(y-1)$ **29** 1.6796   **31** 5.4780

33 315.9   **35** 0.06266   **37** 25,800   **39** 31.4

CHAPTER 5

Section 5.1, page 181

1 $(-1,0)$ **3** $(0,-1)$ **5** $(1,0)$ **7** $(-1,0)$ **9** $(0,-1)$ **11** $(-\sqrt{2}/2, -\sqrt{2}/2)$

13 $(\sqrt{2}/2, -\sqrt{2}/2)$ **15** $(-\sqrt{2}/2, \sqrt{2}/2)$

17 (a) $(-3/5, -4/5)$ (b) $(-3/5, -4/5)$ (c) $(3/5, -4/5)$ (d) $(-3/5, 4/5)$

19 (a) $(8/17, -15/17)$ (b) $(8/17, -15/17)$ (c) $(-8/17, -15/17)$ (d) $(8/17, 15/17)$

21 (a) $(1,0)$ (b) $(1,0)$ (c) $(-1,0)$ (d) $(1,0)$

23 (a) $(-a/\sqrt{a^2+b^2}, -b/\sqrt{a^2+b^2})$ (b) $(-a/\sqrt{a^2+b^2}, -b/\sqrt{a^2+b^2})$

    (c) $(a/\sqrt{a^2+b^2}, -b/\sqrt{a^2+b^2})$ (d) $(-a/\sqrt{a^2+b^2}, b/\sqrt{a^2+b^2})$

25 $\sqrt{3}/2$ **27** $-\sqrt{5}/3$ **29** $-1/2$ **31** $\sqrt{0.9999}$

Calculator Exercises

1 (a) IV   (b) III   **3** (a) III   (b) III   **5** -0.2419

Section 5.2, page 186

1 Divide both sides of $\sin^2 t + \cos^2 t = 1$ by $\sin^2 t$.

3 $\csc t = -5/3, \sec t = 5/4, \tan t = -3/4, \cot t = -4/3$

5 $\csc t = \sqrt{a^2+b^2}/a, \sec t = \sqrt{a^2+b^2}/b, \tan t = a/b, \cot t = b/a$

Values in Exercises 7–17 are given in the order $\sin t, \cos t, \tan t, \csc t, \sec t, \cot t$; a dash means the value is undefined.

7 $1, 0, -, 1, -, 0$ **9** $\sqrt{2}/2, -\sqrt{2}/2, -1, \sqrt{2}, -\sqrt{2}, -1$ **11** $0, -1, 0, -, -1, -$

13 $-1, 0, -, -1, -, 0$ **15** $0, 1, 0, -, 1, -$ **17** $\sqrt{2}/2, -\sqrt{2}/2, -1, \sqrt{2}, -\sqrt{2}, -1$

19 IV   **21** III   **23** II   **25** III   **27** I

Values in Exercises 29–35 are given in the order $\sin t, \cos t, \tan t, \csc t, \sec t, \cot t$.

29 $3/5, -4/5, -3/4, 5/3, -5/4, -4/3$ **31** $-5/13, 12/13, -5/12, -13/5, 13/12, -12/5$

33 $-2\sqrt{2}/3, -1/3, 2\sqrt{2}, -3\sqrt{2}/4, -3, \sqrt{2}/4$ **35** $\sqrt{15}/4, -1/4, -\sqrt{15}, 4\sqrt{15}/15, -4, -\sqrt{15}/15$

37 No; $|\sin t| \le 1$ **39** (a) $\sqrt{2}/2$ (b) $-1/4$ **41** $(\cos t)(\sec t) = (\cos t)(1/\cos t) = 1$

Calculator Exercises

1 $\tan t = -0.3313, \cot t = -3.018, \sec t = -1.054, \csc t = 3.180$

3 $\sec t = 1.122, \csc t = 2.203, \cot t = 1.963, \sin t = 0.4540$

5 $\sin t = 0.9925, \tan t = -8.142, \sec t = -8.203, \csc t = 1.008, \cot t = -0.1228$

Section 5.3, page 191

1

3 We shall give an indirect proof. Suppose there is a positive number k less than 2π such that $\sin(t + k) = \sin t$ for all t. Letting $t = 0$, we obtain $\sin k = \sin 0 = 0$. Since $0 < k < 2\pi$ it follows that $P(k)$ has coordinates $(-1, 0)$. Consequently, $k = \pi$, and we may write $\sin(t + \pi) = \sin t$ for all t. In particular, if $t = \pi/2$, then $\sin(3\pi/2) = \sin \pi/2$, or $-1 = 1$, an absurdity. This completes the proof.

5 If $y/x = a$, where $x^2 + y^2 = 1$, then $\pm\sqrt{1 - x^2}/x = a$. Hence $(1 - x^2)/x^2 = a^2$, and $1 - x^2 = a^2x^2$ or $1 = a^2x^2 + x^2 = (a^2 + 1)x^2$. Consequently, $x^2 = 1/(a^2 + 1)$ or $x = \pm 1/\sqrt{a^2 + 1}$. If $a > 0$, choose the point $P(x, y)$ on U where $x = 1/\sqrt{a^2 + 1}$. If $a < 0$ choose $P(x, y)$ such that $x = -1/\sqrt{a^2 + 1}$. Thus there is always a point $P(x, y)$ on U such that $y/x = a$.

Section 5.4, page 196

1 $480°, 840°, -240°, -600°$ **3** $495°, 855°, -225°, -585°$ **5** $330°, 690°, -390°, -750°$

7 $260°, 980°, -100°, -460°$ **9** $17\pi/6, 29\pi/6, -7\pi/6, -19\pi/6$ **11** $7\pi/4, 15\pi/4, -9\pi/4, -17\pi/4$

13 $5\pi/6$ **15** $-\pi/3$ **17** $5\pi/4$ **19** $5\pi/2$ **21** $2\pi/5$ **23** $5\pi/9$ **25** $120°$

27 $330°$ **29** $135°$ **31** $-630°$ **33** $1260°$ **35** $20°$ **37** $114° 35' 30''$

39 $286° 28' 44''$ **41** (a) 1.75 (b) $100.27°$ **43** 6.98 meters   **45** 8.59 km

47 Approximations (in miles): (a) 4189 (b) 3142 (c) 2094 (d) 698 (e) 70

49 $1/8$ radian $= (45/2\pi)° \approx 7° 10'$

Calculator Exercises

1 1.2811   **3** 8.4173   **5** 149.7651

7 Approximations (in miles): (a) 4150.26 (b) 3112.69 (c) 2075.13   (d) 691.71 (e) 69.17

9 Approximately 0.126 radians, or $7.219°$

Section 5.5, page 204

Answers to Exercises 1–29 are in the order sin, cos, tan, csc, sec, cot.

1 $-3/5, 4/5, -3/4, -5/3, 5/4, -4/3$ **3** $-5\sqrt{29}/29, -2\sqrt{29}/29, 5/2, -\sqrt{29}/5, -\sqrt{29}/2, 2/5$

5 $4\sqrt{17}/17, -\sqrt{17}/17, -4, \sqrt{17}/4, -\sqrt{17}, -1/4$

7 $-7\sqrt{53}/53, -2\sqrt{53}/53, 7/2, -\sqrt{53}/7, -\sqrt{53}/2, 2/7$

9 $4/5, 3/5, 4/3, 5/4, 5/3, 3/4$ **11** $1, 0, —, 1, —, 0$ **13** $\sqrt{2}/2, -\sqrt{2}/2, -1, \sqrt{2}, -\sqrt{2}, -1$

15 $-\sqrt{2}/2, \sqrt{2}/2, -1, -\sqrt{2}, \sqrt{2}, -1$ **17** $0, 1, 0, —, 1, —$ **21** $4/5, 3/5, 4/3, 5/4, 5/3, 3/4$

23 $2\sqrt{13}/13, 3\sqrt{13}/13, 2/3, \sqrt{13}/2, \sqrt{13}/3, 3/2$ **25** $2/5, \sqrt{21}/5, 2\sqrt{21}/21, 5/2, 5\sqrt{21}/21, \sqrt{21}/2$

27 $a/\sqrt{a^2 + b^2}, b/\sqrt{a^2 + b^2}, a/b, \sqrt{a^2 + b^2}/a, \sqrt{a^2 + b^2}/b, b/a$

29 $b/c, \sqrt{c^2 - b^2}/c, b/\sqrt{c^2 - b^2}, c/b, c/\sqrt{c^2 - b^2}, \sqrt{c^2 - b^2}/b$

Calculator Exercises

1 Table A is exact. Calculator answers are approximations.   **3** 0.743, 0.669, 1.110, 1.346, 1.494, 0.901

Section 5.6, page 214

1 (a) $\pi/4$ (b) $\pi/3$ (c) $\pi/6$ **3** (a) $\pi/4$ (b) $\pi/6$ (c) $\pi/3$ **5** (a) 1.5   (b) $2\pi - 5$

7 (a) $60°$ (b) $20°$ (c) $70°$ **9** (a) $49° 20'$ (b) $45°$ (c) $80° 35'$

11 (a) $\sqrt{3}/2$ (b) $-\sqrt{3}/2$ **13** (a) -1 (b) -1 **15** (a) $-2\sqrt{3}/3$ (b) -2 **17** 0.2644

19 1.540   **21** 1.431   **23** 0.7826   **25** 6.197   **27** -1.019 **29** -0.3035 **31** 0.4440

33 -0.1426 **35** 1.032   **37** 0.7911   **39** 0.5217   **41** 3.436   **43** 0.5315   **45** 1.3521

47 0.7703   **49** $21° 33', 158° 27'$ **51** $26° 45', 206° 45'$ **53** $69° 44', 290° 16'$

55 $43° 48', 316° 12'$

Section 5.7, page 221

1 (c) $[-2\pi, -3\pi/2), (-3\pi/2, -\pi], [0, \pi/2), (\pi/2, \pi]$ (d) $[-\pi, -\pi/2), (-\pi/2, 0], [\pi, 3\pi/2), (3\pi/2, 2\pi]$

3 (a) $\csc(-t) = 1/\sin(-t) = 1/(-\sin t) = -(1/\sin t) = -\csc t$

7 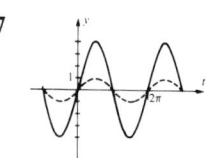   **9** 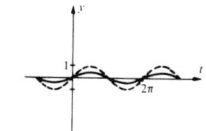   **11**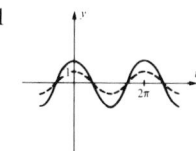

13 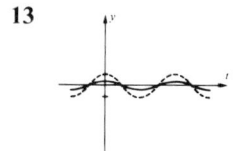   **15**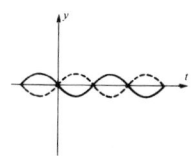

Section 5.8, page 227

1 The amplitudes and periods are: (a) $4, 2\pi$ (b) $1, \pi/2$ (c) $1/4, 2\pi$ (d) $1, 8\pi$ (e) $2, 8\pi$ (f) $1/2, \pi/2$ (g) $4, 2\pi$ (h) $1, \pi/2$

3 The amplitudes and periods are (a) $3, 2\pi$ (b) $1, 2\pi/3$ (c) $1/3, 2\pi$ (d) $1, 6\pi$ (e) $2, 6\pi$ (f) $1/3, \pi$ (g) $3, 2\pi$ (h) $1, 2\pi/3$

5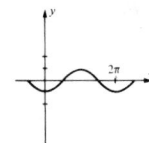

7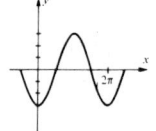

9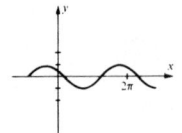

11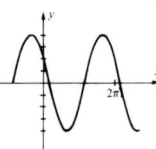

13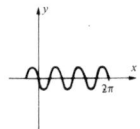

15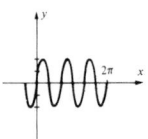

17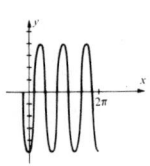

19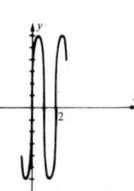

21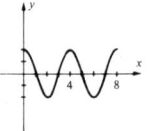

23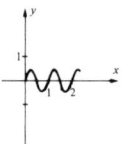

25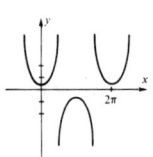

27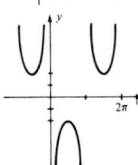

29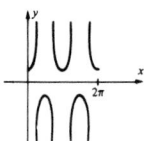

31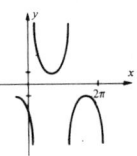

33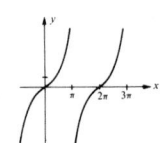

35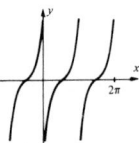

37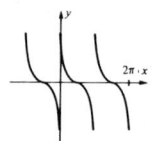

39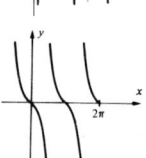

41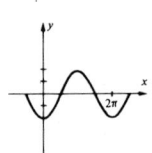

43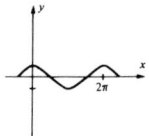

Section 5.9, page 231

1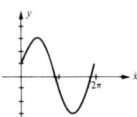

3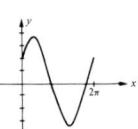

5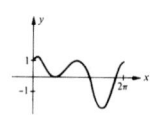

7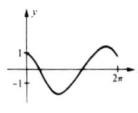

9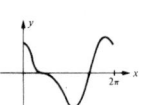

11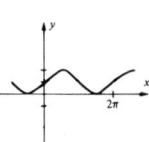

13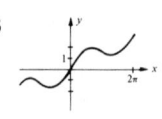

15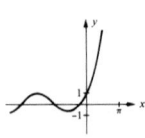

17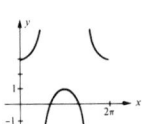

19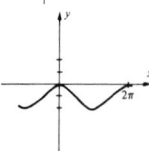

21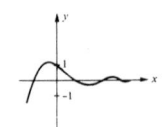

23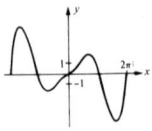

25

27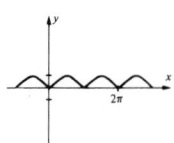

29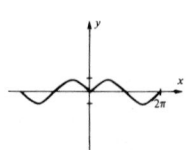

Section 5.10, page 237

1 $\beta = 60°$, $a = 20\sqrt{3}/3 \approx 12$, $c = 40\sqrt{3}/3 \approx 23$ **3** $\alpha = 38°$, $b \approx 19$, $c \approx 24$

5 $\beta = 72°\,20'$, $b \approx 14.1$, $c \approx 14.8$

7 $\alpha = 18°\,9'$, $a \approx 78.7$, $c \approx 252.6$ **9** $\alpha \approx 29°$, $\beta \approx 61°$, $c \approx 51$ **11** $\alpha \approx 69°$, $\beta \approx 21°$, $a \approx 5.4$

13 $\beta = 52°\,14'$, $a \approx 396.7$, $c \approx 647.7$ **15** $\alpha \approx 49°\,40'$, $\beta \approx 40°\,20'$, $b \approx 522$ **17** $51°\,20'$

19 70.6 meters **21** 20.2 meters **23** 29.7 km **25** $d \approx 160$ meters **27** $192(\sin 22°\,30') \approx 73.5$ cm

29 Approximately 9,659 ft **31** Approximately 126.2 mph **33** Approximately 28,800 ft

35 55 miles **37** 325 miles **39** $h = d(\tan\beta - \tan\alpha)$

Calculator Exercises

1 $\beta = 48.73°$, $b \approx 358.5$, $c \approx 476.9$ **3** $\alpha \approx 52.94°$, $b \approx 0.3484$, $c \approx 0.5781$

5 $\alpha \approx 87.29°$, $a \approx 151,000$, $c \approx 151,200$ **7** $c \approx 48.67$, $\alpha \approx 74.36°$, $\beta \approx 15.64°$

9 $\alpha \approx 60.97°$, $\beta \approx 29.03°$, $a \approx 4437$

Section 5.11, Review Exercises, page 239

1 $(-1, 0)$, $(0, -1)$, $(0, 1)$, $(-\sqrt{2}/2, -\sqrt{2}/2)$, $(1, 0)$, $(\sqrt{3}/2, 1/2)$ **3** (a) II (b) III (c) IV

5 The order will be sin, cos, tan, csc, sec, cot:
(a) $1, 0, —, 1, —, 0$ (b) $\sqrt{2}/2, -\sqrt{2}/2, -1, \sqrt{2}, -\sqrt{2}, -1$ (c) $0, 1, 0, —, 1, —$
(d) $-1/2, \sqrt{3}/2, -\sqrt{3}/3, -2, 2\sqrt{3}/3, -\sqrt{3}$

7 $810°$, $-120°$, $315°$, $900°$, $36°$ **9** $65°$, $42°\,50'$, $8°$

11 The order will be sin, cos, tan, csc, sec, cot:
(a) $-4/5, 3/5, -4/3, -5/4, 5/3, -3/4$ (b) $2\sqrt{13}/13, -3\sqrt{13}/13, -2/3, \sqrt{13}/2, -\sqrt{13}/3, -3/2$
(c) $-1, 0, —, -1, —, 0$

13 (a) 0.8500 (b) 2.972 (c) 1.112

15 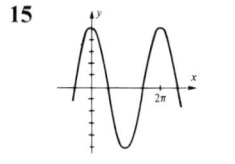 **17** 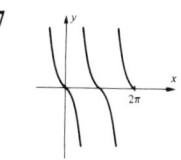 **19** 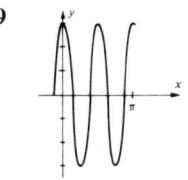 **21**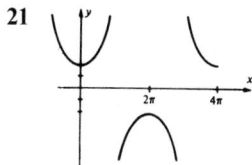

23 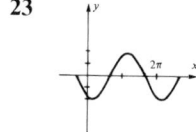 **25** 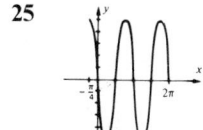 **27** 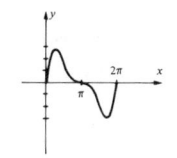 **29**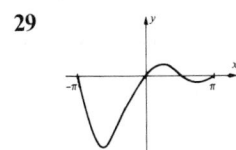

CHAPTER 6

Section 6.1, page 244

The following are arranged in the order $\sin t,\ \cos t,\ \tan t,\ \csc t,\ \sec t,\ \cot t$:

1 $-\sqrt{3}/2,\ 1/2,\ -\sqrt{3},\ -2\sqrt{3}/3,\ 2,\ -\sqrt{3}/3$ **3** $\sqrt{2}/2,\ -\sqrt{2}/2,\ -1,\ \sqrt{2},\ -\sqrt{2},\ -1$

5 $-1/2,\ \sqrt{3}/2,\ -\sqrt{3}/3,\ -2,\ 2\sqrt{3}/3,\ -\sqrt{3}$ **7** $0,\ -1,\ 0,\ -,\ -1,\ -$

9 $-3/5,\ -4/5,\ 3/4,\ -5/3,\ -5/4,\ 4/3$ **11** $12/13,\ -5/13,\ -12/5,\ 13/12,\ -13/5,\ -5/12$

13 $-2/3,\ \sqrt{5}/3,\ -2\sqrt{5}/5,\ -3/2,\ 3\sqrt{5}/5,\ -\sqrt{5}/2$ **15** $1/8,\ -\sqrt{63}/8,\ -\sqrt{63}/63,\ 8,\ -8\sqrt{63}/63,\ -\sqrt{63}$

33 $\cos t = \pm\sqrt{1 - \sin^2 t},\ \tan t = \pm\sin t/\sqrt{1 - \sin^2 t},$

$\csc t = 1/\sin t,\ \sec t = \pm 1/\sqrt{1 - \sin^2 t},\ \cot t = \pm\sqrt{1 - \sin^2 t}/\sin t$

35 $\sin t = \pm\tan t/\sqrt{1 + \tan^2 t}$ **37** $\sec t = \pm\csc t/\sqrt{\csc^2 t - 1}$

Section 6.3, page 254

In the following, n denotes any integer.

1 $(2\pi/3) + 2\pi n,\ (4\pi/3) + 2\pi n$ **3** $(\pi/4) + (\pi/2)n$ **5** $2\pi n,\ (3\pi/2) + 2\pi n$

7 $(\pi/3) + \pi n,\ (2\pi/3) + \pi n$ **9** $(4\pi/3) + 2\pi n,\ (5\pi/3) + 2\pi n$ **11** $(\pi/6) + \pi n,\ (5\pi/6) + \pi n$

13 $(7\pi/6) + 2\pi n,\ (11\pi/6) + 2\pi n$ **15** $\pi n/2,\ (\pi/12) + \pi n,\ (5\pi/12) + \pi n$

17 $\pi/3,\ 2\pi/3,\ 4\pi/3,\ 5\pi/3;\ 60°,\ 120°,\ 240°,\ 300°$ **19** $\pi/6,\ 5\pi/6,\ 3\pi/2;\ 30°,\ 150°,\ 270°$

21 $0,\ \pi,\ \pi/4,\ 3\pi/4,\ 5\pi/4,\ 7\pi/4;\ 0°,\ 180°,\ 45°,\ 135°,\ 225°,\ 315°$

23 $\pi/2,\ 3\pi/2,\ 2\pi/3,\ 4\pi/3;\ 90°,\ 270°,\ 120°,\ 240°$ **25** No solutions

27 $7\pi/6,\ 11\pi/6,\ \pi/2;\ 210°,\ 330°,\ 90°$ **29** $0,\ \pi/2,\ \pi,\ 3\pi/2;\ 0°,\ 90°,\ 180°,\ 270°$ **31** $0;\ 0°$

33 $7\pi/6,\ 11\pi/6,\ \pi/2,\ 3\pi/2;\ 210°,\ 330°,\ 90°,\ 270°$ **35** $3\pi/4,\ 7\pi/4;\ 135°,\ 315°$ **37** $15°\ 30',\ 164°\ 30'$

39 $41°\ 50',\ 138°\ 10',\ 194°\ 30',\ 345°\ 30'$

Section 6.4, page 261

1 (a) $\cos 43°\ 23'$ (b) $\sin 16°\ 48'$ (c) $\cot \pi/3$ **3** (a) $\sin 3\pi/20$ (b) $\cos (2\pi - 1)/4$
(c) $\cot (\pi - 2)/2$

5 (a) $(\sqrt{2} + 1)/2$ (b) $(\sqrt{2} - \sqrt{6})/4$ **7** (a) $\sqrt{3} + 1$ (b) $-2 - \sqrt{3}$

9 (a) $(\sqrt{2} - 1)/2$ (b) $(\sqrt{6} + \sqrt{2})/4$ **11** $\cos 25°$ **13** $\sin (-5°)$ **15** $-\sin 5$

17 $36/85,\ 77/85;\ \text{I}$ **19** $3/5,\ 4/5,\ 3/4,\ -117/125,\ 44/125,\ -117/44$

41 $\sin u \cos v \cos w + \cos u \sin v \cos w + \cos u \cos v \sin w - \sin u \sin v \sin w$

49 $0,\ \pi/3,\ 2\pi/3,\ \pi,\ 4\pi/3,\ 5\pi/3;\ 0°,\ 60°,\ 120°,\ 180°,\ 240°,\ 300°$

51 $f(x) = 2\cos (2x - \pi/6);\ 2,\ \pi,\ \pi/12$ **53** $f(x) = 2\sqrt{2}\cos (3x + \pi/4);\ 2\sqrt{2},\ 2\pi/3,\ -\pi/12$

Section 6.5, page 268

1 $24/25$, $-7/25$, $-24/7$ **3** $-4\sqrt{2}/9$, $-7/9$, $4\sqrt{2}/7$ **5** $\sqrt{10}/10$, $3\sqrt{10}/10$, $1/3$

7 $-\sqrt{2+\sqrt{2}}/2$, $\sqrt{2-\sqrt{2}}/2$, $-\sqrt{2}-1$ **9** (a) $\sqrt{2-\sqrt{2}}/2$ (b) $\sqrt{2-\sqrt{3}}/2$ (c) $\sqrt{2}+1$

29 $\frac{3}{8}+\frac{1}{2}\cos\theta+\frac{1}{8}\cos 2\theta$ **31** 0, $2\pi/3$, π, $4\pi/3$; $0°$, $120°$, $180°$, $240°$ **33** $\pi/3$, π, $5\pi/3$; $60°$, $180°$, $300°$

35 0, π; $0°$, $180°$ **37** 0, $\pi/3$, $5\pi/3$; $0°$, $60°$, $300°$

Calculator Exercises

1 -0.8217

Section 6.6, page 273

1 $\sin 12\theta + \sin 6\theta$ **3** $(1/2)\cos 4t - (1/2)\cos 10t$ **5** $(1/2)\cos 10u + (1/2)\cos 2u$

7 $(3/2)\sin 3x + (3/2)\sin x$ **9** $2\cos 2\theta \sin 4\theta$ **11** $-2\sin 4x \sin x$ **13** $-2\cos 5t \sin 2t$

15 $2\cos(3x/2)\cos(x/2)$ **25** $(1/2)\sin(a+b)x + (1/2)\sin(a-b)x$ **27** $n\pi/4$ where n is any integer

29 $n\pi/2$ where n is any integer.

Section 6.7, page 281

1 (a) $\pi/6$ (b) $\pi/3$ **3** (a) 0 (b) $\pi/2$ **5** (a) $\pi/2$ (b) 0 **7** (a) $\pi/3$ (b) $-\pi/3$

9 (a) $-\pi/4$ (b) $-\pi/6$ **11** -0.7069 **13** 1.1403 **15** $\sqrt{3}/2$ **17** $3/5$ **19** $-\pi/4$

21 0 **23** $-77/36$ **25** $-24/25$ **27** $u\sqrt{1+u^2}/(1.+u^2)$ **29** $\sqrt{2+2u}/2$

39 $\cot^{-1} u = v$ if and only if $\cot v = u$.

43   **45**   **47**   **49**   **51**

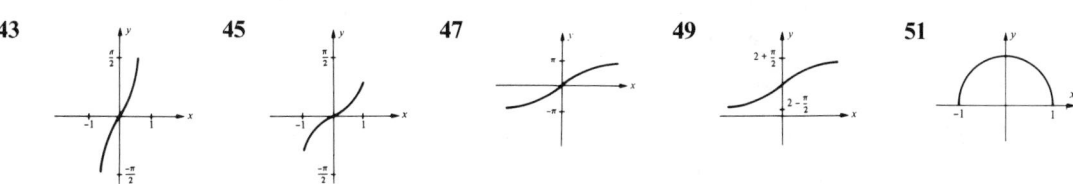

55 $\arctan(-9 \pm \sqrt{57})/4$ **57** $\arccos(\pm\sqrt{15}/5)$, $\arccos(\pm\sqrt{3}/3)$ **59** $\arcsin(\pm\sqrt{30}/6)$

Calculator Exercises

1 -0.3478, -1.3336 **3** 0.6847, 2.4569, 0.9553, 2.1863 **5** ± 1.1503

Section 6.8, page 288

1 $\beta \approx 62°$, $b \approx 14.1$, $c \approx 15.6$ **3** $\gamma \approx 100°\,10'$, $b \approx 55.1$, $c \approx 68.7$

5 $\alpha \approx 58°\,40'$, $a \approx 487$, $b \approx 442$ **7** $\beta \approx 53°\,40'$, $\gamma \approx 61°\,10'$, $c \approx 20.6$

9 $\alpha \approx 77°\,30'$, $\beta \approx 49°\,10'$, $b \approx 108$; $\alpha \approx 102°\,30'$, $\beta \approx 24°\,10'$, $b \approx 58.7$

11 $\alpha \approx 20°\,30'$, $\gamma \approx 46°\,20'$, $a \approx 94.5$ **13** 219 **15** 50 feet   **17** 2.7 miles

19 Approximately 3.7 miles from A and 5.4 miles from B **21** Approximately 627 meters

Calculator Exercises

1 $\alpha = 119.7°$, $\alpha \approx 371.5$, $c \approx 243.5$ **3** $\beta = 49.36°$, $a \approx 49.78$, $c \approx 23.39$

5 $\alpha \approx 25.993°$, $\gamma \approx 32.383°$, $a \approx 0.146$

Section 6.9, page 291

1 $\alpha \approx 26$, $\beta \approx 41°$, $\gamma \approx 79°$ **3** $b \approx 177$, $\alpha \approx 25°\,10'$, $\gamma \approx 4°\,50'$

5 $c \approx 2.8$, $\alpha \approx 21°\,10'$, $\beta \approx 43°\,40'$ **7** $\alpha \approx 29°$, $\beta \approx 46°\,30'$, $\gamma \approx 104°\,30'$

9 $\alpha \approx 12°\,30'$, $\beta \approx 136°\,30'$, $\gamma \approx 31°$ **11** 63, 87 inches   **13** 92 feet   **15** 24 miles   **17** 39 miles

19 Approximately 2.3 miles   **21** Approximately N55° 31′ E

Calculator Exercises

1 $\alpha \approx 40.8$, $\beta \approx 26.9°$, $\gamma \approx 104.8°$ **3** $c \approx 0.487$, $\beta \approx 5.12°$, $\alpha \approx 168.03°$

5 $\alpha \approx 157°\,16'$, $\beta \approx 7°\,49'$, $\gamma \approx 14°\,55'$

Section 6.10, page 300

1 $\langle 3,1 \rangle$, $\langle 1, -7 \rangle$, $\langle 13, 8 \rangle$, $\langle 3, -32 \rangle$ **3** $\langle -15, 6 \rangle$, $\langle 1, -2 \rangle$, $\langle -68, 28 \rangle$, $\langle 12, -12 \rangle$

5 $4\mathbf{i} - 3\mathbf{j}$, $-2\mathbf{i} + 7\mathbf{j}$, $19\mathbf{i} - 17\mathbf{j}$, $-11\mathbf{i} + 33\mathbf{j}$ **7** $-2\mathbf{i} - 5\mathbf{j}$, $-6\mathbf{i} + 7\mathbf{j}$, $-6\mathbf{i} - 26\mathbf{j}$, $-26\mathbf{i} + 34\mathbf{j}$

9 $-3\mathbf{i} + 2\mathbf{j}$, $3\mathbf{i} + 2\mathbf{j}$, $-15\mathbf{i} + 8\mathbf{j}$, $15\mathbf{i} + 8\mathbf{j}$ **31** $3\sqrt{2}, 7\pi/4$ **33** $5, \pi$ **35** $\sqrt{41}, \arccos(-4\sqrt{41}/41)$

37 $18, 3\pi/2$ **39** Approximately 12.3, S38° 26′ W

Section 6.11, Review Exercises, page 301

17 $\pi/2$, $3\pi/2$, $\pi/4$, $3\pi/4$, $5\pi/4$, $7\pi/4$; 90°, 270°, 45°, 135°, 225°, 315°   **19** $0, \pi$; 0°, 180°

21 $0, \pi, 2\pi/3, 4\pi/3$; 0°, 180°, 120°, 240°   **23** $\pi/2, 7\pi/6, 11\pi/6$; 90°, 210°, 330°

25 $\pi/6, 5\pi/6, \pi/3, 5\pi/3$; 30°, 150°, 60°, 300°   **27** $\pi/3, 5\pi/3$; 60°, 300°

29 $\sqrt{2-\sqrt{3}}/2$ or $(\sqrt{6} - \sqrt{2})/4$ **31** $-\sqrt{2-\sqrt{3}}/2$ or $(\sqrt{2} - \sqrt{6})/4$

33 84/85   **35** $-36/77$ **37** 240/289   **39** 24/7   **41** 1/3

43 (a) $(1/2)\cos 3t - (1/2)\cos 11t$ (b) $(1/2)\cos(u/12) + (1/2)\cos(5u/12)$ (c) $3\sin 8x - 3\sin 2x$

45 $5\pi/6$ **47** π **49** 1/2   **51** $-7/25$ **53** $\pi/2$

55

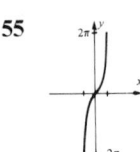

57 $a = 50\sqrt{6}$, $c = 50(1 + \sqrt{3})$, $\gamma = 75°$

59 $a = \sqrt{43}$, $\beta = \arccos(4\sqrt{43}/43) \approx 52°\,25'$, $\gamma = \arccos(5\sqrt{43}/86) \approx 67°\,35'$

CHAPTER 7

Section 7.1, page 307

1 $-2 + 6i$ **3** $-2 + 4i$ **5** $7 - 5i$ **7** $-4 - i$ **9** $4 + 7i$ **11** $-6 + 3i$

13 $-10 + 5i$ **15** $20 - 10i$ **17** 25 **19** $-72 - 36i$ **21** $32 - 44i$ **23** 15 **25** 10

27 $5 + 12i$ **29** $7 + 17i$ **31** 1 **33** -1 **35** i **37** $x = -8/5, y = 3$

39 $x = -1/2, y = -1/7$

Section 7.2, page 311

1 $(3/13) - (2/13)i$ **3** $(35/61) + (42/61)i$ **5** $(-1/5) - (11/10)i$ **7** $(-7/13) + (17/13)i$

9 $-7 - 21i$ **11** $(12/17) - (54/17)i$ **13** $(42/5) - (9/5)i$ **15** $(-1/4) - (1/4)i$

17 $-2 - i$ **19** $(1/125)i$ **21** 5 **23** $\sqrt{85}$ **25** 8 **27** 1

The geometric representations in Exercises 29–37 are the following points:

29 $P(4, 2)$ **31** $P(3, -5)$ **33** $P(-3, 6)$ **35** $P(-6, 4)$ **37** $P(0, 2)$

39 $(26/29) - (7/29)i$ **41** $(-1/2)i$

Section 7.3, page 318

1 -5 **3** $-3 + 7i$ **5** 54 **7** $-2 + 2i$ **9** $-10\sqrt{10}i$ **11** $(11/10) - (13/10)i$

13 $(3 \pm \sqrt{31}i)/2$ **15** $-1 \pm 2i$ **17** $(-1 \pm \sqrt{47}i)/8$ **19** $5, (-5 \pm 5\sqrt{3}i)/2$

21 $2, -2, -1 \pm \sqrt{3}i, 1 \pm \sqrt{3}i$ **23** $\pm 2i, \pm(3/2)i$ **25** $0, (-3 \pm \sqrt{7}i)/2$ **27** $x^2 - 8x + 17 = 0$

29 $x^2 + 6x + 13 = 0$ **31** $x^2 + 25 = 0$ **33** $x^2 - (10 + i)x + 10i = 0$ **35** $x^2 + 1 = 0$

37 $x^2 - 5ix - 6 = 0$ **39** $f(x) = (x - 3)^2(x - 2i)(x + 2i)$ **41** $x^2 - 4x + 13$

43 $x^4 - 6x^3 + 18x^2 - 30x + 25$

Section 7.4, page 322

1 $\sqrt{2}(\cos 7\pi/4 + i \sin 7\pi/4)$ **3** $8(\cos 5\pi/6 + i \sin 5\pi/6)$ **5** $20(\cos 3\pi/2 + i \sin 3\pi/2)$

7 $10(\cos 4\pi/3 + i \sin 4\pi/3)$ **9** $7(\cos 0 + i \sin 0)$ **11** $\sqrt{5}(\cos \theta + i \sin \theta)$, where $\theta = \arctan(1/2)$

13 $4\sqrt{2}(\cos 5\pi/4 + i \sin 5\pi/4)$ **15** $6(\cos \pi/2 + i \sin \pi/2)$ **17** $8(\cos 5\pi/6 + i \sin 5\pi/6)$

19 $12(\cos 0 + i \sin 0)$ **21** $-2, i$ **23** $10\sqrt{3} - 10i, (-2\sqrt{3}/5) + (2/5)i$ **25** $40, 5/2$

27 $8 - 4i, (8/5) + (4/5)i$

29 $z_1 z_2 z_3 = r_1 r_2 r_3 [\cos(\theta_1 + \theta_2 + \theta_3) + i \sin(\theta_1 + \theta_2 + \theta_3)]$,
$z_1 z_2 \cdots z_n = (r_1 r_2 \cdots r_n)[\cos(\theta_1 + \theta_2 + \cdots + \theta_n) + i \sin(\theta_1 + \theta_2 + \cdots + \theta_n)]$

Section 7.5, page 327

1 $-972 - 972i$ **3** $-32i$ **5** -8 **7** $-(\sqrt{2}/2) - (\sqrt{2}/2)i$ **9** $(-1/2) - (\sqrt{3}/2)i$

11 $64 + 64\sqrt{3}i$ **13** $(\sqrt{6}/2) + (\sqrt{2}/2)i, -(\sqrt{6}/2) - (\sqrt{2}/2)i$

15 $(\sqrt[4]{2}/2) + (\sqrt[4]{18}/2)i, -(\sqrt[4]{18}/2) + (\sqrt[4]{4}/2)i, (\sqrt[4]{18}/2) - (\sqrt[4]{2}/2)i, -(\sqrt[4]{2}/2) - (\sqrt[4]{18}/2)i$

17 $3i, -(3\sqrt{3}/2) - (3/2)i, (3\sqrt{3}/2) - (3/2)i$ **19** $\pm 1, (1/2) \pm (\sqrt{3}/2)i, (-1/2) \pm (\sqrt{3}/2)i$

21 $\sqrt[5]{2}\,[\cos\theta + i\sin\theta]$, where $\theta = 9°, 81°, 153°, 225°, 297°$ **23** $\pm 2, \pm 2i$

25 $\pm 2i, \pm\sqrt{3} + i, \pm\sqrt{3} - i$ **27** $2i, -\sqrt{3} - i, \sqrt{3} - i$

29 $3\cos\theta + 3\sin\theta i$, where $\theta = 0, 2\pi/5, 4\pi/5, 6\pi/5, 8\pi/5$

Section 7.6, Review Exercises, page 327

1 $-1 + 8i$ **3** $-28 + 6i$ **5** $40 + 56i$ **7** 97 **9** $16 + 15i$

11 $(7/13) + (17/13)i$ **13** $(5/82) + (2/41)i$ **15** $2\sqrt{34}$ **17** $-19 - 7i$ **19** $-i$

21 $(1 \pm \sqrt{14}i)/5$ **23** $\pm(\sqrt{14}/2)i, \pm(2\sqrt{3}/3)i$ **25** $x^2 - 6x + 34 = 0$ **27** $x^2 + 8x + 17 = 0$

29 $(3 + 2x)(3 - 2x)(9 + 4x^2), (3 + 2x)(3 - 2x)(2x + 3i)(2x - 3i)$

31 $x(x^2 + 4)(3x - 2), x(3x - 2)(x + 2i)(x - 2i)$ **33** $x^4 + 7x^2 + 12$

35 $10\sqrt{2}\,[\cos 3\pi/4 + i\sin 3\pi/4]$ **37** $17[\cos\pi + i\sin\pi]$ **39** $10[\cos 7\pi/6 + i\sin 7\pi/6]$

41 $-512i$ **43** $-972 + 972i$ **45** $-3, (3/2) \pm (3\sqrt{3}/2)i$

CHAPTER 8

Section 8.1, page 334

1 $(3,5), (-1,-3)$ **3** $(1,0), (-3,2)$ **5** $(0,0), (1/8, 1/128)$ **7** $(3,-2)$

9 No solutions   **11** $(-4,3), (5,0)$ **13** $(-2,2)$

15 $((-6 - \sqrt{86})/10, (2 - 3\sqrt{86})/10), ((-6 + \sqrt{86})/10, (2 + 3\sqrt{86})/10)$ **17** $(-4,0), (12/5, 16/5)$

19 $(0,1), (4,-3)$ **21** $(\pm 2, 5), (\pm\sqrt{5}, 4)$ **23** $(\sqrt{2}, \pm 2\sqrt{3}), (-\sqrt{2}, \pm 2\sqrt{3})$ **25** $(2\sqrt{2}, \pm 2),$

$(-2\sqrt{2}, \pm 2)$ **27** $x = 3, y = -1, z = 2$ **29** $(\log 3/\log 5, 7)$ **31** $r = 3, s = -4$

33 $x = 35/22, y = 43/44$ **35** $13, 9$ **37** 8 inches, 12 inches   **39** $5 - \sqrt{19}, 5 + \sqrt{19}, 10$

Section 8.2, page 339

1 $(4,-2)$ **3** $(8,0)$ **5** $(-1, 3/2)$ **7** $(76/53, 28/53)$ **9** $(51/13, 96/13)$

11 $(8/7, -3\sqrt{6}/7)$

13 No solution   **15** All ordered pairs (m, n) such that $3m - 4n = 2$

17 $(0,0)$ **19** $(-22/7, -11/5)$

21 $a = 2, b = 4$ **23** 137 adults, 313 children   **25** 68 **27** 25 dimes, 18 quarters

29 $\$6,650$ at 6.5% and $\$3,325$ at 8%

Section 8.3, page 346

1 $(2, 3, -1)$ **3** $(-2, 4, 5)$ **5** No solution **7** $(2/3, 31/21, 1/21)$

There are other forms for the answers in Exercises 9–15.

9 $(2c, -c, c)$ where c is any real number **11** $(0, -c, c)$ where c is any real number

13 $((12 - 9c)/7, (8c - 13)/14, c)$ where c is any real number

15 $((7c + 5)/10), (19c - 15)/10, c)$ where c is any real number

17 $(1, 3, -1, 2)$ **19** $(1/11, 31/11, 3/11)$ **21** $(-2, -3)$ **23** No solution **25** $(2, 1, 2)$

27 24 **29** 17 1 of 10%, 11 1 of 30%, 22 1 of 50% **31** $x^2 + y^2 - x + 3y - 6 = 0$

Section 8.4, page 352

29 $x = 2, y = -1, z = 3, s = 4, t = 1$

Section 8.5, page 356

1 $\begin{bmatrix} 9 & -1 \\ -2 & 5 \end{bmatrix}, \begin{bmatrix} 1 & -3 \\ 4 & 1 \end{bmatrix}, \begin{bmatrix} 10 & -4 \\ 2 & 6 \end{bmatrix}, \begin{bmatrix} -12 & -3 \\ 9 & -6 \end{bmatrix}$

3 $\begin{bmatrix} 9 & 0 \\ 1 & 5 \\ 3 & 4 \end{bmatrix}, \begin{bmatrix} 3 & -2 \\ 3 & -5 \\ -9 & 4 \end{bmatrix}, \begin{bmatrix} 12 & -2 \\ 4 & 0 \\ -6 & 8 \end{bmatrix}, \begin{bmatrix} -9 & -3 \\ 3 & -15 \\ -18 & 0 \end{bmatrix}$

5 $[11 \quad -3 \quad -3], [-3 \quad -3 \quad 7], [8 \quad -6 \quad 4], [-21 \quad 0 \quad 15]$

7 $\begin{bmatrix} -3 & 4 & 1 & 6 \\ 3 & 2 & 7 & -7 \end{bmatrix}, \begin{bmatrix} 3 & 4 & -1 & 0 \\ -1 & 2 & -7 & -3 \end{bmatrix}, \begin{bmatrix} 0 & 8 & 0 & 6 \\ 2 & 4 & 0 & -10 \end{bmatrix}, \begin{bmatrix} 9 & 0 & -3 & -9 \\ -6 & 0 & -21 & 6 \end{bmatrix}$

9 $\begin{bmatrix} 16 & 38 \\ 11 & -34 \end{bmatrix}, \begin{bmatrix} 4 & 38 \\ 23 & -22 \end{bmatrix}$ **11** $\begin{bmatrix} 3 & -14 & -3 \\ 16 & 2 & -2 \\ -7 & -29 & 9 \end{bmatrix}, \begin{bmatrix} 3 & -20 & -11 \\ 2 & 10 & -4 \\ 15 & -13 & 1 \end{bmatrix}$

13 $\begin{bmatrix} 4 & 8 \\ -18 & 11 \end{bmatrix}, \begin{bmatrix} 3 & -4 & 4 \\ -5 & 2 & 2 \\ -51 & 26 & 10 \end{bmatrix}$ **15** $\begin{bmatrix} 1 & 2 & 3 \\ 4 & 5 & 6 \\ 7 & 8 & 9 \end{bmatrix}, \begin{bmatrix} 1 & 2 & 3 \\ 4 & 5 & 6 \\ 7 & 8 & 9 \end{bmatrix}$

17 $[15], \begin{bmatrix} -3 & 7 & 2 \\ -12 & 28 & 8 \\ 15 & -35 & -10 \end{bmatrix}$ **19** $\begin{bmatrix} 4 \\ 12 \\ -1 \end{bmatrix}$ **21** $\begin{bmatrix} 18 & 0 & -2 \\ -40 & 10 & -10 \end{bmatrix}$

Section 8.6, page 361

1 $\frac{1}{10}\begin{bmatrix} 3 & 4 \\ -1 & 2 \end{bmatrix}$ **3** Does not exist **5** $\frac{1}{8}\begin{bmatrix} 2 & 1 & 0 \\ -2 & 3 & 0 \\ 0 & 0 & 2 \end{bmatrix}$ **7** $\frac{1}{3}\begin{bmatrix} -4 & -5 & 3 \\ -4 & -8 & 3 \\ 1 & 2 & 0 \end{bmatrix}$

9 $\begin{bmatrix} \frac{1}{2} & 0 & 0 \\ 0 & \frac{1}{4} & 0 \\ 0 & 0 & \frac{1}{6} \end{bmatrix}$ **11** $\frac{1}{6}\begin{bmatrix} -8 & -7 & 4 & -9 \\ -8 & -4 & 4 & -6 \\ -4 & -5 & 2 & -3 \\ 6 & 3 & 0 & 3 \end{bmatrix}$ **13** $ab \neq 0; \begin{bmatrix} 1/a & 0 \\ 0 & 1/b \end{bmatrix}$

17 $(13/10, -1/10)$ **19** $(-25/3, -34/3, 7/3)$

Section 8.7, page 367

1 $M_{11} = -14$, $M_{21} = 7$, $M_{31} = 11$, $M_{12} = 10$, $M_{22} = -5$, $M_{32} = 4$, $M_{13} = 15$, $M_{23} = 34$, $M_{33} = 6$;
$A_{11} = -14$, $A_{21} = -7$, $A_{31} = 11$, $A_{12} = -10$, $A_{22} = -5$, $A_{32} = -4$, $A_{13} = 15$, $A_{23} = -34$, $A_{33} = 6$

3 $M_{11} = 0$, $M_{12} = 5$, $M_{21} = -1$, $M_{22} = 7$; $A_{11} = 0$, $A_{12} = -5$, $A_{21} = 1$, $A_{22} = 7$ **5** -83 **7** 5

9 2   **11** 0   **13** -125 **15** 48   **17** -216 **19** $abcd$

Section 8.8, page 371

Answers to Exercises 1, 3, 5, 9, and 13 refer to the Theorem on Row and Column Transformations of a Determinant.

1 (i)   **3** (iii)   **5** (ii)   **7** Two rows are identical.   **9** (ii)

11 Every number in column 2 is 0; also two rows are identical.   **13** (iii)   **15** -10

17 -142 **19** -183 **21** 44   **23** 359

Section 8.10, page 382

1 The region above the graph of $2y = 3x - 6$

3 The region above and on the graph of $y = 1 - 2x$

5 The region under the graph of $y = x^2 - 2$

7 The region above and on the graph of $y = x^2 + 1$

9 The region above and on the graph of $y = 1/x^2$

11 The region under the graph of $3x + y = 3$ and also above the graph of $y = 4 - 2x$

13 The region under the graph of $y = x$ and also under the graph of $2x + 5y = 10$

15 The quadrilateral bounded by the graphs of $3x + y = 6$, $y - 2x = 1$, $x = -2$, and $y = 4$

17 The region above and on the line $x + y = 1$ which is also inside or on the circle $x^2 + y^2 = 4$

19 The region under or on the graph of $y = 1 - x^2$ and also under or on the line $x = 1 + y$

21 The region in the first quadrant between the graphs of $y = 2^x$ and $y = 3^x$

23 The region bounded by the graphs of $y = \log x$, $y + x = 1$, and $x = 10$

25 Let y denote the number of bats and x the number of balls. Then $x \geq 2$, $y \geq 3$, and $5x + 7y \leq 80$.

27 If x and y denote the number of brand A and brand B, respectively, then $x \geq 20$, $y \geq 10$, $x \geq 2y$, and $x + y \leq 100$.

29 If x and y denote the amounts in the first and second accounts, respectively, then $x \geq 2000$, $y \geq 2000$, $y \geq 3x$, and $x + y \leq 15{,}000$.

Section 8.11, page 388

1 50 Double Fault and 30 Set Point   **3** 3.5 pounds of S and 1 pound of T

5 Send 25 from W_1 to X and 0 from W_1 to Y. Send 10 from W_2 to X and 60 from W_2 to Y.

7 45 acres of crop A and 50 acres of crop B

9 Minimum cost: 16 ounces X, 4 ounces Y, 0 ounces Z; maximum cost: 0 ounces X, 8 ounces Y, 12 ounces Z

Section 8.12. Review Exercises, page 389

1 $(19/23, -18/23)$ **3** $(-3, 5), (1, -3)$ **5** $(2\sqrt{3}, \pm\sqrt{2}), (-2\sqrt{3}, \pm\sqrt{2})$ **7** $(14/17, 14/27)$

9 $(6/11, -7/11, 1)$ **11** $(-2c, -3c, c)$ where c is any real number

13 $(5c - 1, (-19c + 5)/2, c)$ where c is any real number   **15** $(-1, 1/2, 1/3)$

17 The region inside the circle $x^2 + y^2 = 16$ which is also above the graph of $y = x^2$

19 The triangular region bounded by the graphs of $x - 2y = 2$, $y - 3x = 4$, and $2x + y = 4$

21 -6 **23** 48 **25** -84 **27** 120 **29** 0 **31** $a_{11}a_{22}a_{33}\cdots a_{nn}$

33 $\left(-\dfrac{1}{2}\right)\begin{bmatrix} 2 & 4 \\ 3 & 5 \end{bmatrix}$ **35** $\dfrac{1}{7}\begin{bmatrix} 2 & 1 & 0 & 0 \\ -1 & 3 & 0 & 0 \\ 0 & 0 & 3 & 2 \\ 0 & 0 & -5 & -1 \end{bmatrix}$ **37** $\begin{bmatrix} 4 & -5 & 6 \\ 4 & -11 & 5 \end{bmatrix}$

39 $\begin{bmatrix} 0 & 4 & -6 \\ 16 & 22 & 1 \\ 12 & 11 & 9 \end{bmatrix}$ **41** $\begin{bmatrix} -12 & 4 & -11 \\ 6 & -11 & 5 \end{bmatrix}$ **43** $\begin{bmatrix} a & 3a \\ 2b & 4b \end{bmatrix}$ **45** $\begin{bmatrix} 5 & 9 \\ 13 & 19 \end{bmatrix}$

CHAPTER 9

Section 9.2, page 406

1 $a^6 + 6a^5b + 15a^4b^2 + 20a^3b^3 + 15a^2b^4 + 6ab^5 + b^6$

3 $a^8 - 8a^7b + 28a^6b^2 - 56a^5b^3 + 70a^4b^4 - 56a^3b^5 + 28a^2b^6 - 8ab^7 + b^8$

5 $81x^4 - 540x^3y + 1350x^2y^2 - 1500xy^3 + 625y^4$

7 $u^{10} + 20u^8v + 160u^6v^2 + 640u^4v^3 + 1280u^2v^4 + 1024v^5$

9 $r^{-12} - 12r^{-9} + 60r^{-6} - 160r^{-3} + 240 - 192r^3 + 64r^6$

11 $1 + 10x + 45x^2 + 120x^3 + 210x^4 + 252x^5 + 210x^6 + 120x^7 + 45x^8 + 10x^9 + x^{10}$

13 $(3^{25})c^{10} + 25(3^{24})c^{52/5} + 300(3^{23})c^{54/5} + 2300(3^{22})c^{56/5}$

15 $60(3^{14})b^{13} - (3^{15})b^{15}$ **17** $30618a^{10}b^2$

19 $840u^4v^6$ **21** $924x^2y^2$ **23** -61236 **25** $24x^2y^6$ **27** $210c^3d^2$

29 $1 + 0.2 + 0.018 + 0.00096 = 1.21896$. If $n = (1.02)^{10}$, then $\log n = 10\log(1.02) \approx 0.0860$ and $n \approx 1.2190$.

Section 9.3, page 412

1 $9, 6, 3, 0, -3; -12$ **3** $1/2, 4/5, 7/10, 10/17, 13/26; 22/65$

5 $9, 9, 9, 9, 9; 9$ **7** $1.9, 2.01, 1.999, 2.0001, 1.99999; 2.00000001$

9 $4, -9/4, 5/3, -11/8, 6/5; -15/16$ **11** $2, 0, 2, 0, 2; 0$ **13** $2/3, 2/3, 8/11, 8/9, 32/37; 128/33$

15 $4, 9, 25, 49, 121; 289$ **17** $7, 19, 29, 37, 47; 79$ **19** $1, 2, 3, 4, 5; 8$ **21** $2, 1, -2, -11, -38$

23 $-3, 3^2, 3^4, 3^8, 3^{16}$ **25** $5, 5, 10, 30, 120$ **27** $2, 2, 4, 4^3, 4^{12}$

29 $a_n = 2n + (1/24)(n - 1)(n - 2)(n - 3)(n - 4)(a - 10)$. There are many other answers.

31 -5 **33** 10   **35** 25   **37** $-17/15$ **39** 61   **41** 10,000

43 $(2n^3 + 12n^2 + 40n)/6$ **45** $(4n^3 - 12n^2 + 11n)/3$ **47** $\sum_{k=1}^{5} (4k - 3)$

49 $\sum_{k=1}^{4} k/(3k - 1)$ **51** $1 + \sum_{k=1}^{n} (-1)^k \dfrac{x^{2k}}{2k}$ **53** $\sum_{k=1}^{7} (-1)^{k-1}/k$ **55** $\sum_{n=1}^{99} 1/n(n + 1)$

Section 9.4, page 417

1 $18, 38, 4n - 2$ **3** $1.8, 0.3, 3.3 - 0.3n$ **5** $5.4, 20.9, (3.1)n - (10.1)$

7 $\log 3^5, \log 3^{10}, \log 3^n$ **9** -8.5 **11** -9.8 **13** -105 **15** 30   **17** 530

19 $423/2$ **21** $60; 12{,}780$ **23** 24

25 $10/3, 14/3, 6, 22/3, 26/3$ **27** $2, 2, 2^2, 2^6, 2^{24}$

29 $23.25, 22.5, 21.75, 21, 20.25, 19.5, 18.75$

Section 9.5, page 424

1 $1/2, 1/6; 8(1/2)^{n-1} = 2^{4-n}$ **3** $0.03, -0.00003; 300(-0.1)^{n-1}$ **5** $3125, 5^8; 5^n$

7 $81/4, -3^7/2^5; 4(-3/2)^{n-1}$ **9** $x^8, -x^{14}; (-1)^{n-1}x^{2n-2}$ **11** $2^{4x+1}, 2^{7x+1}; 2^{nx-x+1}$

13 $243/8$ **15** $a_1 = 1/81, S_5 = 211/1296$ **17** $(-3/2)(1 - 3^{10}) = 88{,}572$

19 $(-1/3)(1 - 2^{-10})$ **21** $25/256\%$ **23** $10{,}000(6/5)^t, 10{,}000(6/5)^{10}$ **25** $2/3$ **27** $50/33$

29 $2 - \sqrt{2}$ **31** $23/99$ **33** $2393/990$ **35** $5141/999$ **37** $16123/9999$ **39** 30 meters

Section 9.6, Review Exercises, page 425

7 $x^{12} - 18x^{10}y + 135x^8y^2 - 540x^6y^3 + 1215x^4y^4 - 1458x^2y^5 + 729y^6$ **9** $-(63/16)b^{12}c^{10}$

11 $5, -2, -1, -20/29, -7/19$ **13** $1, 1/2, 5/4, 7/8, 65/64$ **15** $10, 11/10, 21/11, 32/21, 53/32$

17 $9, 3, \sqrt{3}, \sqrt[4]{3}, \sqrt[8]{3}$ **19** 75   **21** 1000   **23** $\sum_{k=1}^{5} 3k$ **25** $\sum_{k=0}^{4} (-1)^k 5(20 - k)$

27 $13 + 10\sqrt{3}, 85 + 55\sqrt{3}$ **29** $-31, 50$ **31** 64

33 $4\sqrt{2}$ **35** 530   **37** 2041   **39** $5/7$

CHAPTER 10

Section 10.1, page 429

1 $x^2 + y^2 - 6x + 8y + 21 = 0$ **3** $x^2 + y^2 - 12x + 4y + 31 = 0$

5 $x^2 + y^2 - 6x - 2y - 7 = 0$ **7** $(-1, 5); 4$ **9** $(0, 3); 2$ **11** $(5, -4); \sqrt{41}$

13 $(-1, 1); 1/2$ **15** $(1/2, -1/3); 1/6$ **17** $x^2 + y^2 - 3x + 9y - 20 = 0$

Section 10.2, page 434

1 (a) 135° (b) 30° (c) $\tan^{-1} 2$ (d) 20°   **3** (a) 0 (b) $-\sqrt{3}$ (c) 1 (d) 4/3

7 $5x + 2y - 29 = 0$ **9** $5x - 3y + 41 = 0$ **11** $5x - 7y + 15 = 0$

13 $x + 6y - 9 = 0; 4x + y - 4 = 0; 3x - 5y + 5 = 0; (15/23, 32/23)$

15 (a) 10/11 (b) 1/8   **17** $83°, 23°, 74°$ **19** $m = (6 \pm 5\sqrt{3})/3$

Section 10.4, page 441

1 $F(3,0); x = -3$ **3** $F(0, -2); y = 2$ **5** $F(-5/16, 0); x = 5/16$

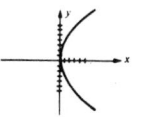   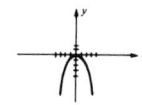   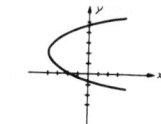

7 $F(0, 1/36), y = -1/36$ **9** $F(1, -3); x = 0$ **11** $F(-7/2, 2); x = -9/2$

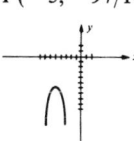   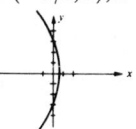

13 $F(-5, -97/16), y = -95/16$ **15** $F(-9/2, 0); x = 11/2$

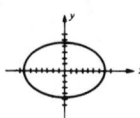   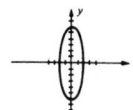

17 $y^2 = -24x$ **19** $(x - 2)^2 = 12(y - 2)^2$ **21** $2x^2 = -9y$ **23** 9/16 feet from the vertex

Section 10.5, page 446

1 $V(\pm 7, 0); F(\pm\sqrt{24}, 0)$ **3** $V(0, \pm 6); F(0, \pm 4\sqrt{2})$ **5** $V(\pm 2, 0); F(\pm 1, 0)$

   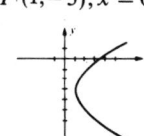

7 $V(\pm 1/2, 0); F(\pm\sqrt{5}/6, 0)$ **9** $x^2/39 + y^2/64 = 1$ **11** $x^2/25 + y^2/9 = 1$

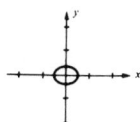

13 $91x^2 + 16y^2 = 1600$ **15** $\{(2, 2), (4, 1)\}$ **17** $2\sqrt{21}$ feet

19 If the eccentricity is close to 1, the ellipse is very flat. If the eccentricity is close to zero, the ellipse is almost circular.

Section 10.6, page 452

1 $V(\pm 7, 0), F(\pm\sqrt{74}, 0), y = \pm 5x/7$ **3** $V(0, \pm 7), F(0, \pm\sqrt{74}), y = \pm 7x/5$

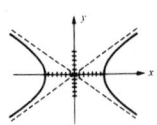

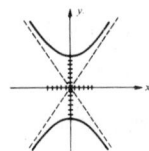

5 $V(0, \pm 6), F(0, \pm 2\sqrt{10}), y = \pm 3x$

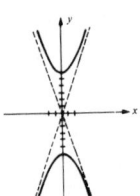

7 $V(\pm 1, 0), F(\pm\sqrt{2}, 0), y = \pm x$

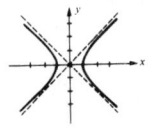

9 $V(\pm 1/4, 0), F(\pm\sqrt{13}/12, 0), y = \pm 2x/3$

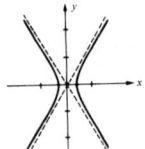

11 $x^2/25 - y^2/39 = 1$ **13** $4y^2 - 5x^2 = 20$ **15** $x^2 - 4y^2 = 100$ **17** $\{(0, 4), (8/3, 20/3)\}$

19   Conjugate hyperbolas have the same asymptotes.

Section 10.7, page 458

1   Parabola; $V(1, -3), F(4, -3)$

3   Ellipse; center $(-5, 4)$, vertices $(-9, 4)$ and $(-1, 4)$, endpoints of minor axis $(-5, 7)$ and $(-5, 1)$

5   Hyperbola; center $(2, -7)$, vertices $(-4, -7)$ and $(8, -7)$, endpoints of conjugate axis $(2, 0)$ and $(2, -14)$

7   Ellipse; center $(-2, 2)$, vertices $(-2, 3)$ and $(-2, 1)$, endpoints of minor axis $(-5/2, 3)$ and $(-3/2, 3)$

1   **3**   **5**   **7**

9   Hyperbola; center $(3, 0)$, vertices $(3, \pm 10)$, endpoints of conjugate axis $(-1, 0)$ and $(7, 0)$

11   Ellipse; center $(-3, 5)$, endpoints of major axis $(-3, 9)$ and $(-3, 1)$, endpoints of minor axis $(-5, 5)$ and $(-1, 5)$

13   Ellipse; center $(6, 2)$, vertices $(6, 5)$ and $(6, -1)$, endpoints of minor axis $(5, 2)$ and $(7, 2)$

15   Parabola; $V(-2, 10), F(-2, 85/8)$

9 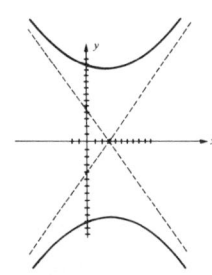 **11** 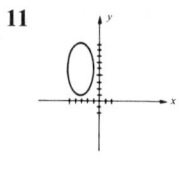 **13** 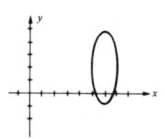 **15**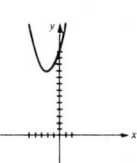

17 Hyperbola; center $(-5, -3)$, vertices $(-5, 1)$ and $(-5, -7)$, endpoints of conjugate axis $(-3, -3)$ and $(-7, -3)$

19 Hyperbola; center $(2, 6)$, vertices $(0, 6)$ and $(4, 6)$, endpoints of conjugate axis $(2, 0)$ and $(2, 12)$

17 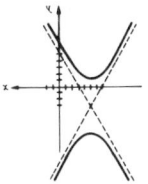 **19**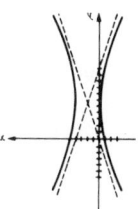

21 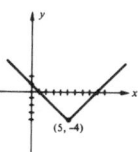 **23** 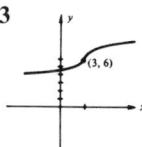 **25**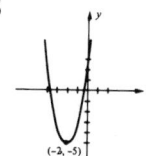

27 $(x - h)^2/a^2 - (y - k)^2/b^2 = 1$

Section 10.8, page 463

Answers 1–7 contain equations in x' and y' resulting from a rotation of axes.

1 Ellipse, $(x')^2 + 16(y')^2 = 16$ **3** Hyperbola, $4(x')^2 - (y')^2 = 1$ **5** Ellipse, $(x')^2 + 9(y')^2 = 9$

7 Parabola, $(y')^2 = 4(x' - 1)$

9 Sketch of proof: It can be shown that $B^2 - 4AC = B'^2 - 4A'C'$. For a suitable rotation of axes we obtain $B' = 0$ and the general quadratic equation has the form $A'x'^2 + C'y'^2 + D'x' + E'y' + F' = 0$. Except for degenerate cases, the graph of the latter equation is an ellipse, hyperbola, or parabola if $A'C' > 0$, $A'C' < 0$, or $A'C' = 0$, respectively. However, if $B' = 0$, then $B^2 - 4AC = -4A'C'$ and hence the graph is an ellipse, hyperbola, or parabola if $B^2 - 4AC < 0$, $B^2 - 4AC > 0$, or $B^2 - 4AC = 0$, respectively.

Section 10.9, page 468

1 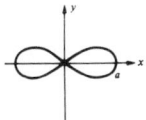 **3** 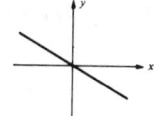 **5** 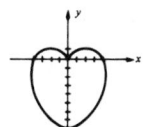 **7** 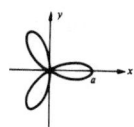 **9**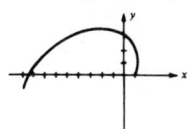

11  **13** 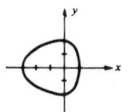 **15** **17**

19 $r = -3\sec\theta$ **21** $r = 4$ **23** $r = 6\csc\theta$ **25** $r^2 = 6\sec 2\theta$ **27** $x = 5$

29 $x^2 + y^2 - 6y = 0$ **31** $x^2 + y^2 = a^2$ **33** $x^2(x^2 + y^2) = y^2$

Section 10.10, page 472

1 Ellipse; vertices $(3/2, \pi/2)$ and $(3, 3\pi/2)$, foci $(0,0)$ and $(3/2, 3\pi/2)$
3 Hyperbola; vertices $(-3,0)$ and $(3/2, \pi)$, foci $(0,0)$ and $(-9/2, 0)$
5 Parabola; $V(3/4, 0)$, $F(0,0)$ **7** Ellipse; vertices $(-4,0)$ and $(-4/3, \pi)$, foci $(0,0)$ and $(-8/3, 0)$

9 Hyperbola (except for the points $(\pm 3, 0)$); vertices $(6/5, \pi/2)$ and $(-6, 3\pi/2)$, foci $(0,0)$ and $(-36/5, 3\pi/2)$

11 $9x^2 + 8y^2 + 12y - 36 = 0$ **13** $y^2 - 8x^2 - 36x - 36 = 0$ **15** $4y^2 = 9 - 12x$

17 $3x^2 + 4y^2 + 8x - 16 = 0$ **19** $4x^2 - 5y^2 + 36y - 36 = 0$

21 $r = 2/(1 + \cos\theta), e = 1, r = 2\sec\theta$ **23** $r = 4/(1 + 2\sin\theta), e = 2, r = 2\csc\theta$

25 $r = 2/(3 + \cos\theta), e = 1/3, r = 2\sec\theta$ **27** $r = 2/(3 + \cos\theta)$ **29** $r = 12/(1 - 4\sin\theta)$

31 $r = 5/(1 + \cos\theta)$ **33** $r = 8/(1 + \sin\theta)$

Section 10.11, page 477

1 $y = 2x + 7$ **3** $y = x - 2$ **5** $x = y^2 - 6y + 4$ **7** $y = 1/x^2$

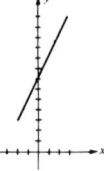 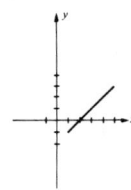 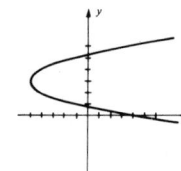 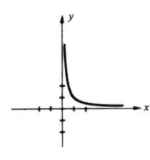

9 $x^2/4 + y^2/9 = 1$ **11** $x^2 - y^2 = 1$ **13** $y = \log x$ **15** $y = 1/x$

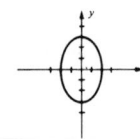 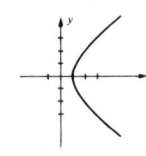 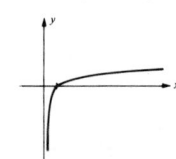 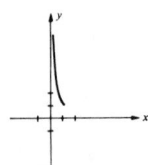

17 $x = (x_2 - x_1)t^n + x_1, y = (y_2 - y_1)t^n + y_1$, are parametric equations for l if n is any odd positive integer.

Section 10.12, Review Exercises, page 478

1 (a) $(9, -5)$; 10   (b) $x^2 + y^2 - 18x + 10y = 0$ (c) $x^2 + y^2 - 12x + 12y = 0$

3 $V(0,0); F(9/16, 0)$ **5** $V(0, \pm 12), F(0, \pm 8\sqrt{2})$ **7** $V(0, \pm\sqrt{8}), F(0, \pm 4)$

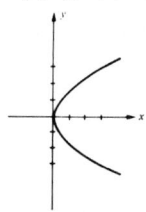 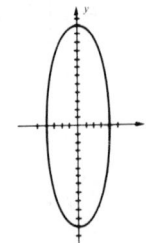 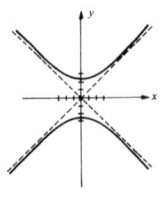

9 $y^2 = -32x$ **11** $y^2/49 + x^2/40 = 1$ **13** $y^2 - 25x^2 = 16$

15 Ellipse; center $(-3, 2)$, vertices $(-6, 2)$ and $(0, 2)$, endpoints of minor axis $(-3, 0)$ and $(-3, 4)$

17 Parabola; $V(2, -4)$, $F(4, -4)$

19 Hyperbola; center $(2, -3)$, vertices $(2, -1)$ and $(2, -5)$, endpoints of conjugate axis $(2 \pm \sqrt{2}, -3)$

15   **17**   **19**

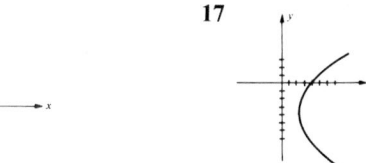   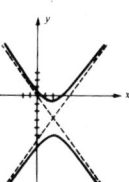

21   **23**   **25**   **27**

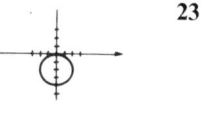   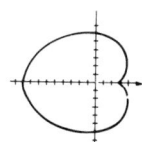   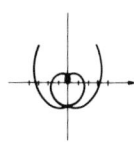   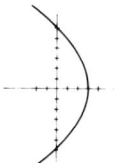

29 $r \sin^2 \theta = 4 \cos \theta$ **31** $r(2 \cos \theta - 3 \sin \theta) = 8$ **33** $x^3 + xy^2 = y$ **35** $(x^2 + y^2)^2 = 8xy$

37 $y = 2(x - 1) - 1/(x - 1)$ **39** $y = 2^{-x^2}$

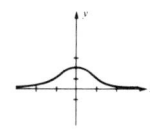

APPENDIX I, page A9

1 $h = 2A/b$ **3** $h = 3V/\pi r^2$ **5** $R_2 = RR_1 R_3/(R_1 R_3 - RR_3 - RR_1)$ **7** $P = S/(1 + rt)$

9 $r = (3V + \pi h^3)/3\pi h^2$ **11** $r = (a - Sl)/(l - S)$ **13** $m = Ft(v_1 - v_2)$

15 $b_1 = (2A - hb_2)/h$ **17** $v_1 = v_2 - at$ **19** $n = -rI/(IR - E)$ **21** $r = \sqrt{3\pi h V}/\pi h$

23 $d = \sqrt{gm_1 m_2 F}/F$ **25** $t = (-v_0 + \sqrt{v_0^2 + 2gs})/g$ **27** $y = \pm(b/a)\sqrt{x^2 - a^2}$

29 $h = \sqrt{S^2 - \pi^2 r^4}/\pi r$ **31** $x = \pm(a/b)\sqrt{b^2 - y^2}$ **33** $y = (a^{2/3} - x^{2/3})^{3/2}$

35 $R = (S/\pi s) - r$ **37** $R = \sqrt[4]{8nVL/\pi P}$ **39** $R_2 = (n - 1)R_1 R/(R_1 - R(n - 1))$

41 43 quarters, 27 dimes   **43** 136, 137, 138, 139   **45** -40 **47** 88

49 \$12000 at 8.5%; \$18000 at 6%   **51** \$1820 at 5%; \$1180 at 4.5%

53 64 seconds; 96 meters and 128 meters, respectively   **55** 3 cm

57 After an additional 50 games   **59** 20/3 liters   **61** 3 liters   **63** 225 miles

65 36 minutes   **67** 500/3 gm   **69** $d(b - a)/b$ gm   **71** 15,17   **73** 2 cm decrease

75 2.5 ft   **77** 3:15 P.M.   **79** 24 in., 76 in.

81 (a) After $25 - 5\sqrt{17}$ seconds and also after $25 + 5\sqrt{17}$ seconds   (b) After 50 seconds   (c) 10,000 feet

83 $36 - 25\sqrt{2}$ cm ≈ 0.645 cm   **85** 64 mph   **87** 16 ml   **89** 9 in., 11 in.

91 $d = \sqrt{4 - 200t + 2900t^2}$; $t = (1 + \sqrt{30})/29 \approx 0.22$ hr after 10:00 A.M.

93 500   **95** $140/9 \le C \le 80/3$ **97** $20/9 \le x \le 4$ **99** $[1/2, 4]$

INDEX

The Trigonometric Functions

I. *Of Real Numbers*

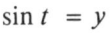

$$\sin t = y \qquad\qquad \csc t = \frac{1}{y}$$

$$\cos t = x \qquad\qquad \sec t = \frac{1}{x}$$

$$\tan t = \frac{y}{x} \qquad\qquad \cot t = \frac{x}{y}$$

II. *Of Arbitrary Angles*

$$\sin \theta = \frac{b}{r} \qquad\qquad \csc \theta = \frac{r}{b}$$

$$\cos \theta = \frac{a}{r} \qquad\qquad \sec \theta = \frac{r}{a}$$

$$\tan \theta = \frac{b}{a} \qquad\qquad \cot \theta = \frac{a}{b}$$

III. *Of Acute Angles*

$$\sin \theta = \frac{\text{opp}}{\text{hyp}} \qquad\qquad \csc \theta = \frac{\text{hyp}}{\text{opp}}$$

$$\cos \theta = \frac{\text{adj}}{\text{hyp}} \qquad\qquad \sec \theta = \frac{\text{hyp}}{\text{adj}}$$

$$\tan \theta = \frac{\text{opp}}{\text{adj}} \qquad\qquad \cot \theta = \frac{\text{adj}}{\text{opp}}$$